Cloud Radio Access Networks

Principles, Technologies, and Applications

Understand the fundamental theory, current techniques, and potential applications of cloud radio access networks (C-RANs) with this unique text. Leading experts from academia and industry guide you through all of the key elements of C-RANs, including system architecture, performance analysis, technologies in both the physical and medium access control layers, self-organizing and green networking, standards development, and standardization perspectives. Recent developments in the field are covered, as well as open research challenges and possible future directions.

The first book to focus exclusively on cloud radio access networks, this is essential reading for engineers in academia and industry working on future wireless networks.

Tony Q. S. Quek is an Associate Professor in the Information Systems Technology and Design Pillar at the Singapore University of Technology and Design. He is a Senior Member of the IEEE and a co-editor of the book *Small Cell Networks* (Cambridge University Press, 2013).

Mugen Peng is a Professor in the School of Information and Communication Engineering at the Beijing University of Posts and Telecommunications. He is a Fellow of the IET and a recipient of the First Grade Award of Technological Invention from both the Ministry of Education of China and the Ministry of Industry and Information Technology of China.

Osvaldo Simeone is a Professor affiliated with the Center for Information Processing Research at the New Jersey Institute of Technology. He is a Fellow of the IEEE.

Wei Yu is a Professor and Canada Research Chair in Information Theory and Wireless Communications at the University of Toronto. He is a Fellow of the IEEE.

"This is the first book that covers the fundamental theory, current techniques, and applications of cloud radio access networks. The coverage is balanced and the topics are very timely... This book should be an essential reading for graduate students, engineers, and researchers who are interested in exploring this important field of wireless communication research."

Zhi-Quan Luo, The Chinese University of Hong Kong, Shenzhen

Cloud Radio Access Networks

Principles, Technologies, and Applications

TONY Q. S. QUEK
Singapore University of Technology and Design

MUGEN PENG
Beijing University of Posts and Telecommunications

OSVALDO SIMEONE
New Jersey Institute of Technology

WEI YU
University of Toronto

CAMBRIDGE
UNIVERSITY PRESS

Shaftesbury Road, Cambridge CB2 8EA, United Kingdom

One Liberty Plaza, 20th Floor, New York, NY 10006, USA

477 Williamstown Road, Port Melbourne, VIC 3207, Australia

314–321, 3rd Floor, Plot 3, Splendor Forum, Jasola District Centre, New Delhi – 110025, India

103 Penang Road, #05–06/07, Visioncrest Commercial, Singapore 238467

Cambridge University Press is part of Cambridge University Press & Assessment,
a department of the University of Cambridge.

We share the University's mission to contribute to society through the pursuit of
education, learning and research at the highest international levels of excellence.

www.cambridge.org
Information on this title: www.cambridge.org/9781107142664

DOI: 10.1017/9781316529669

First published 2017

A catalogue record for this publication is available from the British Library

ISBN   978-1-107-14266-4   Hardback

Contents

Acknowledgments

First, we would like to express our sincere gratitude to all our contributors, without whom this book would never have been produced. Indeed, it was a great pleasure for us to have such high-quality contributions from prominent researchers in the field of wireless networks. Our contributors, as they appear in the book, are as follows: Tony Q. S. Quek, Mugen Peng, Osvaldo Simeone, Wei Yu (editors), Chih-Lin I, Jinri Huang, Ran Duan, Sadayuki Abeta, Wuri A Hapsari, Kazuaki Takeda, Peter Rost, Matthew C. Valenti, Salvatore Talarico, Andreas Mäder, Pratik Patil, Binbin Dai, Yuhan Zhou, Zhendong Mao, Yourrong Ban, Di Chen, Shi Jin, Jun Zhang, Kai-Kit Wong, Hongbo Zhu, Yuanming Shi, Jun Zhang, Khaled B. Letaief, Bo Bai, and Wei Chen, Seok-Hwan Park, Onur Sahin, Shlomo Shamai Shitz, Thang X. Vu, Hieu D. Nguyen, Shao-Yu Lien, Shao-Chou Hung, Chih-Hsiu Zeng, Hsiang Hsu, Qimei Cui, Kwang-Cheng Chen, Ahmed Douik, Hayssam Dahrouj, Oussama Dhifallah, Tareq Y. Al-Naffouri, Mohamed-Slim Alouini, Jian Li, Hongyu Xiang, Yuling Yu, Jian Zhao, Zhongding Lei, Sheng Zhou, Jingchu Liu, Tao Zhao, Zhisheng Niu, Yuanzhang Xiao, Mihaela van der Schaar, Haijun Zhang, Julian Cheng, Victor C. M. Leung, Kenza Hamidouche, Walid Saad, Merouane Debbah, Sau-Hsuan Wu, Hsi-Lu Chao, Hsin-Li Chiu, Chun-Hsien Ko, Yun-Ting Li, Ting-Wei Chang, Tong-Lun Tsai, Che Chen, Min Yan, Xiaogen Jiang. We would like to thank Cambridge University Press staff and in particular Julie Lancashire and Heather Brolly for their continuous encouragement and support during the course of this project.

Tony Q. S. Quek would like to thank his family and colleagues at the Singapore University of Technology and Design (SUTD) for their encouragement and support. He would also like to acknowledgment funding support from the MOE ARF Tier 2 under grant MOE2014-T2-2-0 the MOE ARF Tier 2 under grant MOE2015-T2-2-10 and the SUTD-ZJ4 Research Collaboration under grant SUTD-ZJ4/RES/01/2014.

Mugen Peng would like to thank his family and to acknowledge fund support from the National Natural Science Foundation of China (Grant No. 61222103), the National High Technology Research and Development Program of China (Grant No. 2014AA01A701), and the National Basic Research Program of China (973 Program) (Grant No. 2013CB336600).

Osvaldo Simeone would like to acknowledge support from the US NSF under grant CCF-1525629.

Wei Yu wishes to acknowledge the support of the Natural Sciences and Engineering Research Council (NSERC) of Canada through a Collaborative Research and Development grant, an E.W.R. Steacie Memorial Fellowship, and the Canada Research Chairs program. Wei Yu also wishes to acknowledge the support of Huawei Technologies Canada Co. Ltd.

Last, but not least, we would also like to sincerely thank Professor Vincent Poor for writing the foreword to the book.

Foreword

Wireless networking is one of the most advanced and rapidly advancing technologies of our time. The modern wireless era has produced an array of technologies of tremendous economic and social value and almost ubiquitous market penetration. A major contemporary focus of the community of researchers and engineers working on the development of new wireless technologies is the specification and design of the fifth generation (5G) of mobile communications. Among the envisioned features of 5G are an extremely large and heterogeneous population of connected devices communicating with humans or machines (or both) and having highly varied quality of service requirements in terms of latency, data rates, etc., leading to the so-called Internet of Things. To deal with this scale, density, and variety of use, new network architectures are being proposed for 5G. One of the most promising of these is the Cloud Radio Access Network (C-RAN), in which radio connectivity to end-users is provided via densely deployed low-complexity radio heads, and most signal processing tasks are performed in the cloud. This architecture enables the provision of much greater capacity, by allowing both the densification of radio resources and the implementation of sophisticated signal processing algorithms at scale, and doing so at substantially lower capital and energy costs than conventional base-station-centric cellular architectures would require. Edited by four leaders in the field, *Cloud Radio Access Networks: Principles, Technologies, and Applications* provides a comprehensive treatment of C-RANs, describing in the depth the overall C-RAN architecture, and the many physical layer, resource allocation, and networking challenges that arise in this important and innovative concept, together with potential solutions to these challenges. These advances are described in chapters written by leading contributors to their development, thus providing a clear and up-to-date exposition of the state of the art in C-RANs. As such, this volume should be of considerable interest to researchers and engineers looking to develop the next generation of mobile networking technologies.

H. Vincent Poor
Princeton, New Jersey

Preface

Cloud radio access networks (C-RANs) refer to a wireless cellular architecture in which all network functionalities of conventional base stations, apart from radio frequency operations and possibly analog–digital conversions, are carried out at a central cloud processor. The idea was relegated for many years to the realm of information- and communication-theoretic studies, which promised gains in terms of spectral efficiency thanks to the possibility of implementing joint baseband processing at the central processor. The main obstacles to the deployment of C-RAN-type systems were thought to be the high complexity of the necessary cloud processor as well as the limited availability of high-speed backhaul links connecting edge and cloud.

In recent years, advances in cloud computing and a more pervasive deployment of fiber optic cables and high-frequency wireless backhaul links towards the network edge have spurred the reconsideration, and eventually the implementation, of cloud-based radio access systems. In fact, as argued in the seminal white paper by China Mobile, not only can the C-RAN architecture reap the spectral efficiency gains promised by academic studies, it can also crucially reduce capital and operating expenses. This is a consequence of the centralization of network resources in the cloud: the complexity and cost of edge nodes can be drastically reduced with respect to conventional base stations, and updates and maintenance can be performed solely at the cloud.

As C-RAN moves from paper to the real world, industry and academia are working towards the definition of protocols and algorithms at all layers of the communication protocol stack, so as to enable cost-effective and high-performance cloud-based systems to be widely adopted as a leading solution for 5G networks.

This book is intended to provide a broad overview of the current research activity in the industry and academia on the subject of C-RANs. While this is an active field of study, involving theoreticians and practitioners, the editors believe that the current state of the art is sufficiently mature to warrant a monographic treatment. The book covers the architecture, physical-layer design, resource allocation, and networking of C-RAN systems, in separate parts each consisting of various chapters authored by leading researchers in both industry and academia.

It is our hope that this book will serve as a useful reference for engineers and students and that it will motivate more researchers to undertake the numerous open problems highlighted in the following pages.

Tony Quek, Mugen Peng, Osvaldo Simeone, and Wei Yu

List of Contributors

Sadayuki Abeta
NTT DoCoMo, Kanagawa, Japan

Tareq Y. Al-Naffouri
King Abdullah University of Science and Technology, Thuwal, Saudi Arabia, and also King Fahd University of Petroleum and Minerals, Dhahran, Saudi Arabia

Mohamed-Slim Alouini
Division of Computer, Electrical and Mathematical Sciences and Engineering, King Abdullah University of Science and Technology, Thuwal, Saudi Arabia

Bo Bai
Department of Electronic Engineering, Tsinghua University, Beijing, China

Yourrong Ban
Beijing University of Posts and Telecommunications, Beijing, China

Hsi-Lu Chao
Institute of Network Engineering, National Chiao Tung University, Taiwan

Ting-Wei Chang
Institute of Network Engineering, National Chiao Tung University, Taiwan

Che Chen
Institute of Network Engineering, National Chiao Tung University, Taiwan

Di Chen
Beijing University of Posts and Telecommunications, Beijing, China

Kwang-Cheng Chen
Graduate Institute of Communication Engineering, National Taiwan University, Taiwan

Julian Cheng
Department of Electrical and Computer Engineering, The University of British Columbia, Vancouver, Canada

Wei Chen
Department of Electronic Engineering, Tsinghua University, Beijing, China

Hsin-Li Chiu
Institute of Communications Engineering, National Chiao Tung University, Taiwan

Qimei Cui
Beijing University of Post and Telecommunications, Beijing, China

Hayssam Dahrouj
Department of Electrical and Computer Engineering, Effat University, Jeddah, Saudi Arabia

Binbin Dai
Electrical and Computer Engineering Department, University of Toronto, Toronto, Canada

Merouane Debbah
Large Networks and Systems Group (LANEAS), CentraleSupélec, Gif-sur-Yvette, France

Oussama Dhifallah
Division of Computer, Electrical and Mathematical Sciences and Engineering, King Abdullah University of Science and Technology, Thuwal, Saudi Arabia

Ahmed Douik
Department of Electrical Engineering, California Institute of Technology, Pasadena, CA, USA

Ran Duan
Green Communication Technology Research Center, China Mobile Research Institute, Beijing, China

Kenza Hamidouche
Large Networks and Systems Group (LANEAS), CentraleSupélec, Gif-sur-Yvette, France

Wuri A. Hapsari
NTT DoCoMo, Kanagawa, Japan

Jinri Huang
Green Communication Technology Research Center, China Mobile Research Institute, Beijing, China

Ran Huang
Green Communication Technology Research Center, China Mobile Research Institute, Beijing, China.

Hsiang Hsu
Graduate Institute of Communication Engineering, National Taiwan University, Taiwan

Shao-Chou Hung
Graduate Institute of Communication Engineering, National Taiwan University, Taiwan

Chih-Lin I
Green Communication Technology Research Center, China Mobile Research Institute, Beijing, China

Xiaogen Jiang
Alcatel-Lucent Shanghai Bell Co., China

Shi Jin
National Mobile Communications Research Laboratory, Southeast University, Nanjing, China

Chun-Hsien Ko
Institute of Communications Engineering, National Chiao Tung University, Taiwan

Zhongding Lei
Huawei Technologies, Singapore

Khaled B. Letaief
Department of Electronic and Computer Engineering, Hong Kong University of Science and Technology, Hong Kong

Victor C. M. Leung
Department of Electrical and Computer Engineering, The University of British Columbia, Vancouver, Canada

Jian Li
Beijing University of Posts and Telecommunications, Beijing, China

Yun-Ting Li
Institute of Network Engineering, National Chiao Tung University, Taiwan

Juigchu Liu
Tsinghua University, Beijing

Shao-Yu Lien
National Formosa University, Taiwan

Andreas Mäeder
Nokia Networks, Technology and Innovation, Radio Systems Research, Munich, Germany

Zhendong Mao
Beijing University of Posts and Telecommunications, Beijing, China

Hieu D. Nguyen
Swiss Re, Singapore

Zhisheng Niu
Tsinghua University, Beijing

Seok-Hwan Park
Chonbuk National University, Jeonju, Korea

Pratik Patil
Electrical and Computer Engineering Department, University of Toronto, Toronto, Canada

Mugen Peng
Beijing University of Posts and Telecommunications, Beijing, China

Tony Q. S. Quek
Singapore University of Technology and Design, Information Systems Technology and Design Pillar, Singapore

Peter Rost
Nokia Networks, Technology and Innovation, Radio Systems Research, Munich, Germany

Walid Saad
Wireless@VT, Bradley Department of Electrical and Computer Engineering, Virginia Tech, USA

Onur Sahin
Interdigital, NY, USA

Mihaela van der Schaar
Department of Electrical Engineering, University of California, Los Angeles, CA, USA

Shlomo Shamai (Shitz)
Technion, Haifa, Israel

Yuanming Shi
School of Inoformation Science and Technology, ShanghaiTech University, Shanghai, China

Osvaldo Simeone
New Jersey Institute of Technology, NJ, USA

Salvatore Talarico
Huawei Technologies, CA, USA

Kazuaki Takeda
NTT DoCoMo, Kanagawa, Japan

Tong-Lun Tsai
Institute of Communications Engineering, National Chiao Tung University, Taiwan

Matthew C. Valenti
West Virginia University, Morgantown, WV, USA

Thang X. Vu
Interdisciplinary Centre for Security, Reliability and Trust (SnT), University of Luxembourg, Luxembourg

Kai-Kit Wong
Department of Electronic and Electrical Engineering, University College London, London, United Kingdom

Sau-Hsuan Wu
Institute of Communications Engineering, National Chiao Tung University, Taiwan

Hongyu Xiang
Beijing University of Posts and Telecommunications, Beijing, China

Yuanzhang Xiao
Department of Electrical Engineering, University of California, Los Angeles, CA, USA

Min Yan
Green Communication Technology Research Center, China Mobile Research Institute, Beijing, China

Wei Yu
Electrical and Computer Engineering Department, University of Toronto, Toronto, Canada

Yuling Yu
Beijing University of Posts and Telecommunications, Beijing, China

Chih-Hsiu Zeng
Graduate Institute of Communication Engineering, National Taiwan University, Taiwan

Haijun Zhang
Engineering and Technology Research Center for Convergence Networks, University of Science and Technology, Beijing, China

Jun Zhang
Department of Electronic and Computer Engineering, Hong Kong University of Science and Technology, Hong Kong

Jun Zhang
Jiangsu Key Laboratory of Wireless Communications, Nanjing University of Posts and Telecommunications, Nanjing, China

Jian Zhao
Department of Electronic Science and Engineering, Nanjing University, Nanjing, China

Tao Zhao
Tsinghua University, Beijing

Sheng Zhou
Tsinghua University, Beijing, China

Yuhan Zhou
Electrical and Computer Engineering Department, University of Toronto, Toronto, Canada

Hongbo Zhu
Jiangsu Key Laboratory of Wireless Communications, Nanjing University of Posts and Telecommunications, Nanjing, China

Part I

Architecture of C-RANs

1 Overview of C-RAN

Chih-Lin I, Jinri Huang, and Ran Duan

1.1 Introduction

In 2008, as the specification for long-term evolution (LTE) Release 8 was frozen in the Third Generation Partner Project (3GPP), operators began to shift the network deployment focus to 4G. In 2009, the world's first commercial LTE network was launched by TeliaSonera in Norway and Sweden. As of today, there are several hundred LTE networks in operation, providing unprecedented user experiences to customers. Consequently, we are witnessing the recent mobile traffic explosion in the telecom industry. It is expected that by 2020 consumer Internet traffic will increase by a factor of over one thousand [1].

As operators roll out and expand 4G networks, more and more challenges arise.

First, network deployment is becoming more and more difficult simply due to an insufficient number of equipment rooms. Traditional base stations (BSs) comprise either a co-located baseband unit (BBU) with a radio unit or a distributed BBU with a remote radio unit (RRU) connected via fiber. For either case, a separate equipment room with supporting facilities such as air conditioning is required in order for BS deployment. However, since the operating frequency of LTE is usually higher than that of 2G and 3G, the coverage of an LTE cell is smaller than that of a 2G or 3G cell. As a result, more LTE cells are needed to cover the same area, meaning that more equipment rooms are required. Unfortunately, this is increasingly difficult since available real estate is becoming scarcer and more expensive. Traditional deployment puts a lot of pressure on capital expenditure (CAPEX).

Second, in a society where people are promoting energy conservation and environment protection, power consumption has become a sensitive word and a major concern for operators. It is estimated that the carbon footprint of the ICT industry accounts for 2% of the global total, which is the same as that of the aviation industry. For the telecom industry, further analysis has shown that a large percentage of power consumption in mobile networks comes from radio access networks (RANs) [1, 2]. Take China Mobile's networks, for example. The largest mobile network in the world consumed over 14 billion kWh of energy in 2012 in its network of 1.1 million base stations. It can be seen that saving energy in RANs could directly lower the operating expense (OPEX) of the network.

Last, but not least, a concern comes from interference issues. Long-term evolution is expected to have much more interference owing to the increased number of cells, i.e., a shortened intercellular distance than with 2G or 3G. In addition, the interference issue will become increasingly urgent when heterogeneous networks with high densities of small cells are introduced. In order to mitigate interference, various cooperative algorithms such as coordinated multi-point (CoMP) [2] have been proposed. However, efficient CoMP algorithms such as Joint Transmission (JT) cannot achieve their maximum performance gain using traditional LTE architecture with X2 interfaces, owing to their high latency and low bandwidth [3, 4]. There is a need to facilitate information exchange in an efficient way to enable and maximize the effect of CoMP from an architecture perspective.

In order to address the aforementioned challenges, both industry and academia are proactively investigating and researching potential technologies, one of which is Centralized, Collaborative, Cloud, and Clean RAN or, in brief, Cloud RAN (C-RAN). In this chapter, we will provide an overview of C-RAN, including its basic concept, benefits, and challenges as well as the evolving C-RAN architecture based on a new fronthaul interface.

1.2   C-RAN Basic

In 2009, China Mobile (CMCC) proposed the concept of C-RAN for the first time. The "C" here has four meanings, "centralized, collaborative, cloud, and clean". The basic idea of C-RAN starts from centralization, which is to aggregate different BBUs, which in traditional deployment are geographically separated, into the same location. It is clear to see that the base stations that C-RAN supports should be of the distributed type, i.e., the BBU and the RRU are separate and are connected via fiber.

The advantages of centralization are very straightforward. First, the number of equipment rooms for BS placement is greatly reduced, leading to CAPEX reduction. Furthermore, the facilities, especially the air conditioning, could be shared by BBUs in the same central office. Given that power consumption by air conditioning usually accounts for over half the total for operators, extensive facility sharing helps to save energy. According to the report in [2], such saving could be as high as 70% compared with the traditional deployment method in a 3G trial network. Therefore, the OPEX could be greatly reduced.

Further, centralization leads to C-RAN's second namesake, "collaborative". The idea behind it is that once the BBUs are aggregated in the same place, communication between them will become much easier, faster, and more effective. In fact, like in a data center, it is convenient to connect different BBUs together with switches of high bandwidth and low latency. In this way, the information exchange among BBUs can be completed in a timely manner, which facilitates the implementation of joint processing technologies. As a result, system performance is expected to be improved.

The ultimate goal of C-RAN is to realize the "cloud"feature, which is similar to the cloudification concept in data centers. The essence of cloudification is to soften baseband processing resources, which are of a "hard" nature in traditional BSs in the sense that they are developed on proprietary platforms. In these platforms the processing resources present hardware such as digital signal processing (DSP) and application specific integrated circuits (ASICs), making it difficult to achieve resource sharing. In C-RAN with cloudification, the processing resources are supposed to be "soft" and flexible enough that they can be dynamically managed with different operations on BBU such as instantiation, scale-in, and scale-out. In this sense, the BS in C-RAN could be called "soft" to distinguish it from the "hard" BS in traditional systems.

Figure 1.1 illustrates a C-RAN architecture based on a commercial off-the-shelf (COTS) platform. A C-RAN system centralizes different processing resources together to form a pool so that the resources can be managed and dynamically allocated on demand. The key enabler towards C-RAN is the virtualization technology widely used in modern data centers. With virtualization, standard IT servers are used as the general platform with computation and storage as the common resources. As shown in

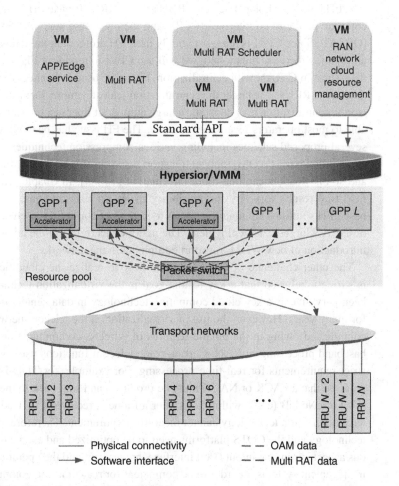

Figure 1.1 Illustration of the C-RAN architecture.

Fig. 1.1, on top of the servers different applications run in the form of virtual machines (VMs). The indispensible applications in C-RAN are those that realize different radio access technologies (RATs) including 2G, 3G, 4G, and future 5G. Additional user applications such as content delivery network (CDN) and Web Cache can also be deployed on the open virtualized platform. In addition, the C-RAN platform could provide a set of standard application program interfaces (APIs), which opens an opportunity for new service provision and deployment. In this way, C-RAN is no longer a single RAT processing entity but rather a platform for the coexistence of diverse services.

1.3   Challenges

Towards the realization of C-RAN lie two major challenges: the fronthaul (FH) issue and virtualization for cloudification.

A FH link is the link between the BBU and the RRU. Typical FH interfaces include the common public radio interface (CPRI) and the open base station architecture initiative (OBSAI). Since CPRI is the most widely used FH protocol in the industry, in this chapter, unless otherwise mentioned, we will use CPRI to represent FH in order to describe the issues. In C-RAN with centralization, fibers are used to connect the BBU pool with the remote RRUs. The larger the centralization scale, the more fibers are needed. In other words, centralization may consume a large number of fiber resources, which is unaffordable to most operators given fiber scarcity. The FH issue has been widely studied, with several proposed schemes including various compression techniques, wavelength division multiplexing (WDM), optical transport networks (OTNs), microwave transmission, and so on. Readers can find more information in [2, 5]. In Chapter 18 we will present field trial results to verify the feasibility of WDM FH solutions. In general, WDM-based FH solutions are mature enough to save fiber consumption effectively in support of C-RAN large-scale deployment. The major concern lies in the additional cost of the introduction of new WDM transport equipment in the networks.

The other challenge of C-RAN lies in how to realize the cloudification feature. It is strongly believed that a key to this goal is the virtualization technology which has been pervasive as a key cloud computing technology in data centers in the IT industry for many years. However, the use of virtualization in the telecom networks is far more complicated owing to the unique features of wireless communications, especially the baseband processing in RAN. Carrier-grade telecom functions usually have extremely strict requirements for real-time processing. For example, for TDD-LTE systems it is required that an ACK or NACK must be produced and sent back to the user equipment (UE) or eNodeB (eNB) within 3 ms after a frame is received [5]. Traditional data center virtualization technology cannot meet this requirement. Therefore, the virtualization technology and the COTS platforms need to be optimized and even customized in various aspects, from the in/out (I/O) interface, hypervisor, and the operating systems to the management systems, in order to be competent for real-time and computation-intensive baseband processing.

1.4   Evolved C-RAN with NGFI

As mentioned in the previous sections, the FH issue has been one of the key challenges for C-RAN. Several solutions including CPRI compression and WDM transport technologies have been proposed. In essence, the idea of all the solutions is to "accommodate" the FH without changing the FH interfaces themselves. It should be realized that the root cause for the FH challenge lies exactly at the FH interfaces themselves. Take the CPRI as an example: CPRI specification has defined several classes of line rates. For a TD-LTE carrier with 20 MHz and eight antennas, the CPRI rate could be as high as 9.8 Gb/s [6]. Moreover, the rate is constant regardless of the dynamically changing mobile traffic, which leads to low transmission efficiency. In addition, existing CPRI interfaces have other shortcomings, such as low scalability and flexibility, which impede C-RAN large-scale deployment. Therefore in [7–9], the authors proposed to redefine the CPRI and brought forward a new concept called next generation fronthaul interface (NGFI). This concept possesses the following desirable features [7, 8].

- Its data rate should be traffic-dependent and therefore support statistical multiplexing.
- The mapping between BBU and RRH should be one-to-many and flexible.
- It should be independent of the number of antennas.
- It should be packet-based, i.e. the FH data could be packetized and transported via packet-switched networks.

The key way to achieve NGFI is to repartition the function layout between the BBUs and the RRUs (see Fig. 1.2). Traditionally, all the baseband functions, including the physical layer (PHY), media access control (MAC), and the packet data convergence protocol (PDCP) are processed on the BBU side while the RRU mainly deals with the

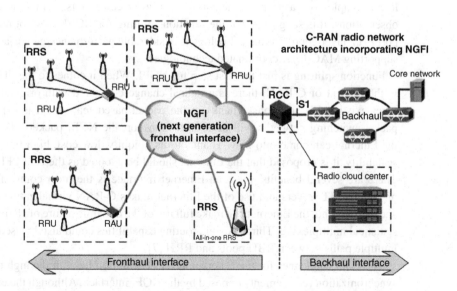

Figure 1.2 NGFI-based C-RAN architecture.

radio-related functions. The signal transmitted by the CPRI is the high-bandwidth I/Q sampling signal. From the effective-information perspective, any data between the base-band protocol stacks (e.g., between MAC and PHY) could be transported. The basic idea of function splitting is to move partial baseband functions to the RRU to reduce the bandwidth without losing any information.

There have been some related studies on this topic in the literature. To achieve NGFI in general, the function splitting should decouple the bandwidth from the antennas, which can be achieved by moving antenna-related functions (downlink antenna mapping, FFT, channel estimation, equalization, etc.) to the RRU. It is shown that the FH bandwidth of an LTE carrier may decrease to the order of 100 Mb/s no matter how many antennas are used [10]. In addition, it is suggested that the user equipment (UE) processing functions should be decoupled from cell-processing functions. In this way, the FH bandwidth will be lowered and, more importantly, load-dependent. The load-dependent feature gives an opportunity to exploit the statistical multiplexing gain when it comes to FH transport network design for C-RAN deployment. Thanks to statistical multiplexing, the bandwidth needed for the transport of a number of FH links in C-RAN can be reduced greatly, subsequently decreasing the cost.

The support of collaborative technologies is another key factor for the design of function splitting. The coordinated multi-point algorithm has been viewed as one of the key technologies in 4G and 5G to mitigate interference. It can be divided into two classes: MAC-layer coordination and physical-layer coordination. For example, collaborative schedule (CS) is an MAC-layer coordinated mechanism. Joint reception (JR) and joint transmission (JT) are physical-layer coordinated technologies. In [11] it was found that the performance gain of JR and JT decreases significantly as the number of antennas increases. Moreover, in [7] the authors found through field trial data that MAC-level collaborative technologies can bring comparable performance gains with lower complexity, easier implementation, and fewer constraints. On the basis of these observations, it is suggested that the function splitting for NGFI does not have to support PHY-layer coordination technology. Considerable performance gain is achieved by supporting MAC-layer coordinated technologies.

Function splitting is just the first step for NGFI. When it comes to the FH networks in the context of C-RAN, there is a radical change compared with original WDM or other existing FH solutions. Thanks to the packet-based features, it is expected that packet switching networks will be used to transport the NGFI packets. This is when the Ethernet can come into play. Thanks to its ubiquity, low cost, high flexibility, and scalability, it is proposed that the Ethernet should be adopted as the NGFI FH solution. There are several benefits. First, an Ethernet interface is the most common interface on standard IT servers and use of the Ethernet makes C-RAN virtualization easier and cheaper. Second, the Ethernet can make full use of the dynamic nature of NGFI to realize statistical multiplexing. Third, flexible routing capabilities could also be used to realize multiple paths between BBU pools and RRH [7].

The main challenges for the Ethernet as an FH solution lie with the high timing and synchronization requirements imposed by the NGFI interface. Although the exact NGFI has so far not been specified, it is possible that NGFI may keep some requirements of the CPRI, such as the synchronization requirements. The allowable radio frequency error for a CPRI link is 2 ppb and the timing alignment error must not exceed 65 ns in order to support multiple-input multiple-output (MIMO) and transmission diversity [12]. In order to meet the timing requirements, both the BBU and the RRUs should be perfectly synchronized, which therefore requires a very accurate clock distribution mechanism. Potential solutions may include any combination of the Global Positioning System (GPS), IEEE 1588, and synchronous Ethernet (Sync-E). Finally, the transport protocols on top of the Ethernet such as Multi-Protocol Label Switching (MPLS) and Packet Transport Network (PTN) that establish transport paths for FH traffic need to be defined.

As proposed by the authors in [7], the C-RAN architecture is also evolving as traditional FH interfaces change to NGFI. As shown in Fig. 1.2, the evolved C-RAN consists of three parts [7]:

- a radio aggregation unit (RAU) With function split, the moved partial BB functions form a new entity which is called the radio aggregation unit. This is a logical concept and its realization depends on implementation solutions. For example, the RAU could be integrated into the RRU to form a new type of RRU. Alternatively, it could be an independent hardware entity.
- remote radio systems (RRS) An RRS consists of an RAU and multiple RRUs. It is expected that collaboration could happen among different RRHs via the RAU within the same area coverage of an RRS. There could be multiple RRSs in a C-RAN network.
- a radio cloud center (RCC) The remaining BB functions together with high-layer functionalities constitute an RCC. The RCC is the place where all the processing resources are pooled into a cloud with virtualization technology.

1.5 Deployment Cases and Standardization Activities

Since its proposal in 2009, C-RAN has gradually become a hotter and hotter topic in both industry and academia. Centralization has been tested and deployed by many major operators. China Mobile, for example, as the originator of C-RAN, has been actively conducting C-RAN centralization field trials in more than ten cities across 2G, 3G, and 4G since 2010. The two biggest carriers in South Korea, SK Telecom and Korea Telecom, have adopted the C-RAN centralization method to deploy commercial LTE networks since 2011. In Japan, DoCoMo has successfully completed an outdoor-commercial-environment verification of its Advanced C-RAN, achieving a 240 Mbps downlink using 35 MHz bandwidth in February 2015. There are many other operators experimenting with C-RAN including Orange, China Telecom, and China Unicom.

At the same time, several C-RAN projects have been initiated in many organizations including Next Generation Mobile Networks (NGMN) and the European Commission's Seventh Framework Program (EU 7FP). In NGMN, a dedicated C-RAN project named

P-CRAN was founded in 2010 [13]. Led by China Mobile and receiving extensive support from both operators and vendors, including KT, SKT, Orange, Intel, ZTE, Huawei, and Alcatel-Lucent, this project aimed at promoting the concept of C-RAN, collecting requirements from operators, and helping build the ecosystem. The project was closed at the end of 2012, releasing four deliverables into the industry. Through the deliverables, the advantages of C-RAN in saving the total cost of ownership (TCO) and speeding up site construction are well understood. In 2013, NGMN extended the study on C-RAN into a C-RAN work stream under the project RAN Evolution. On the basis of previous C-RAN projects, this work stream aimed at a more detailed study on the key technologies critical to C-RAN implementation, including BBU pooling, RAN sharing, function splitting between the BBU and the RRU, and C-RAN virtualization. In addition, there are several C-RAN related projects under EU FP7. For example, the iJOIN project deals with the interworking and joint design of an open access and backhaul network architecture for small cells on cloud networks [14]. Another project, Mobile Cloud Networking (MCN), aims at exploiting cloud computing as the infrastructure for future mobile network deployment and operation and innovative value-added services [15].

There are also many efforts in the fronthaul area, especially for NGFI. In NGMN, schemes of BBU-RRH function splitting are being analyzed, aiming at reducing the FH bandwidth to facilitate C-RAN deployment [13]. Open Radio Interface (ORI) is studying the compression technologies with the aim of reducing the CPRI data rate. The CPRI forum has begun a discussion of "Radio over Ethernet", whose idea is to use the Ethernet to transport CPRI streams. In the IEEE a task force called IEEE 1904.3 was founded, targeting the design of CPRI encapsulation on Ethernet packets [16]. There has also been heated discussion regarding FH in the IEEE 802.1 time-sensitive networking task group and the IEEE 1588 working group. In addition, IEEE is considering founding a dedicated NGFI working group to promote and study NGFI comprehensively. There are some EU-funded research projects including convergence of fixed and mobile broadband access/aggregation networks (COMBO) [17], intelligent converged network consolidating radio and optical access around user equipment (iCIRRUS) [18], and X-Fronthaul.

Among the projects, it is worth mentioning the IEEE 1904.3 task force that deals with the FH data encapsulation in the form of Ethernet packets. The IEEE 1904.3 Radio over Ethernet (RoE) project is an ongoing effort to standardize a versatile encapsulation solution for transporting radio samples with the associated control traffic over a switched Ethernet network. The RoE project concerns only transport-level encapsulation with a flow-level multiplexing capability and the required enablers for the time synchronization of transported radio and control data flows. This project is by design agnostic to the transport technologies and the functional splitting between the BBU and RRU, which implicitly allows its use for existing and future 5G radio technologies. For incremental deployments, RoE also offers mechanisms for transporting and mapping existing fronthaul solutions such as CPRI into its native transport service. Furthermore, RoE can be transported over any networking technology that carries Ethernet packets, assuming that the NGFI timing requirements can be met.

References

[1] C. I, C. Rowell, S. Han, Z. Xu, G. Li, and Z. Pan, "Toward green and soft: a 5G perspective." *IEEE Commun. Mag.*, vol. 52, no. 2, pp. 66–73, 2014.

[2] C. M. R. Institute, "C-RAN: The road towards green RAN." Online, 2014.

[3] CMCC, "Simulation results for CoMP phase i, evaluation in homogeneous network." R1-111301, 3GPP TSG-RAN WG1 #65, May 2011.

[4] Q. X. Wang, D. J. Jiang, G. Y. Liu, and Z. G. Yan, "Coordinated multiple points transmission for lte-advanced systems," in *Proc. 5th Int. Conf. on Wireless Communications, Networking and Mobile Computing*, pp. 1–5, 2009.

[5] C. I, J. Huang, R. Duan, C. Cui, J. Jiang, and L. Li, "Recent progress on C-RAN centralization and cloudification," *IEEE Access*, vol. 2, pp. 1030–1039, 2014.

[6] CPRI, "Common public radio interface (CPRI) specification (v6.0)." *Technical Report*, August 2013.

[7] C. M. R. Institute, "White paper of next generation fronthaul interface." Online, 2015.

[8] C. I, Y. Yuan, J. Huang, S. Ma, R. Duan, and C. Cui, "Rethink fronthaul for soft RAN," *IEEE Commun., Mag.*, vol. 53, no. 9, pp. 82 –88, September 2015.

[9] C. I, Y. Yuan, J. Huang, S. Ma, and R. Duan, "NGFI, the xhaul," *Proc. IEEE GLOBECOM Workshop*, pp. 1–6, 2015.

[10] D. Wubben, P. Rost, J. Bartelt, M. Lalam, and V. Savin, "Benefits and impact of cloud computing on 5G signal processing: flexible centralization through cloud-RAN," *IEEE Signal Process. Mag.*, pp. 35–44, November 2014.

[11] A. Davydov, G. Morozov, I. Bolotin, and A. Papathanassiou, "Evaluation of joint transmission comp in C-RAN based LTE-A hetnets with large coordination areas," in *Proc. IEEE Globecom Workshops*, December 2013.

[12] G. 36.104, "Base station (BS) radio transmission and reception (release 11)," March 2013.

[13] Online. Available at www.ngmn.org.

[14] Online. Available at www.ict-ijoin.eu.

[15] Online. Available at www.mobile-cloud-networking.eu.

[16] Online. Available at www.ieee1904.org.

[17] Online. Available at www.ict-combo.eu.

[18] Online. Available at www.icirrus-5gnet.eu.

2 Advanced C-RAN for Heterogeneous Networks

Sadayuki Abeta, Wuri A. Hapsari, and Kazuaki Takeda

2.1 Introduction

Motivated by the increase in user demand for high data rates and new service applications due to the fast market penetration of smartphones, a large number of mobile operators in the world are introducing long-term evolution (LTE) into their networks [1]. In accordance with the further growth of mobile data traffic, these operators are deploying, or plan to deploy, their LTE networks with multiple-frequency-band operation in order to provide satisfactory user experience to their customers. Therefore, from the viewpoint of mobile operators, technologies that achieve high capacity LTE networks deployed with multiple-frequency-band operation are essential.

In order to achieve high capacity by utilizing multiple LTE frequency bands, carrier aggregation (CA) was specified as one of the new features for LTE in 3GPP Release 10 (i.e., LTE-advanced) [2]. The CA feature will enable operators to provide improved user throughput in their LTE networks by simultaneously using multiple LTE carriers. It can support large bandwidths (up to 100 MHz) and the flexible use of a fragmented spectrum in different frequency bands, where multiple LTE carriers do not have to be contiguous in a frequency band and can even be located in different frequency bands. The increase in user throughput with CA is achieved by assigning available radio resources over multiple LTE carriers to a single user. However, in a high-load network condition due to a large number of connected users, the increase in user throughput would be limited as the radio resources that could be assigned to a single user would not be changed irrespectively of whether CA is employed. Therefore, the utilization of CA only will not contribute to an increase in network capacity.

One conventional way to increase network capacity is to increase the number of cell sites in a certain area (i.e., to employ a densification of cells). However, the densification way of using macro cell deployment is becoming less efficient especially in dense urban areas since it has become difficult to find sites (a building or tower) in which new macro base stations can be installed. To cope with this problem, the deployment of heterogeneous networks, in which multiple small cells are deployed over a macro-cell area, is considered to be a promising option. In this deployment, the frequency band of the small cells is the same as that of the macro cell. A small cell will support smaller coverage areas served by smaller-size equipment with reduced transmission power, e.g., 1 W, compared with that of conventional macro-cell base stations. Mobile operators can easily improve network capacity even in dense urban areas by using multiple small-cell base stations with small-size equipment and by installing it on, e.g., walls of buildings, telephone poles, billboards, etc., which are relatively easy to find compared with the installation sites for macro-cell base stations.

The concept of using a heterogeneous network deployment has already been investigated in other literature, e.g., [3–7]. These studies have assumed usage of the same frequency band in both macro and small cells together with a new functionality called enhanced inter-cell interference coordination (eICIC) specified in 3GPP in order to improve the spectrum efficiency. However, such single-frequency heterogeneous network deployment has the following practical difficulties and disadvantages from a mobile operator's viewpoint:

- The coverage area supported by small cells will be limited due to severe co-channel inter-cell interference from the macro cell. In particular, a degraded performance will be observed for legacy LTE user equipment (UE) without a new functionality to alleviate inter-cell interference (e.g., eICIC) for heterogeneous networks.
- Frequent handovers are foreseen when a UE passes areas supported by small cells and, without fine-tuned handover control parameters, this may increase the number of radio link failures and signaling overhead [8].

This chapter presents the deployment concept, performance evaluation, and application of an unconventional network deployment that would allow mobile operators to achieve high network capacity while avoiding the above-mentioned difficulties and that would be suitable for the multiple-LTE-band operation observed in real LTE-advanced networks. The concept employs a heterogeneous network deployment with combination of CA between a macro cell and small cells, which is realized by an "advanced C-RAN architecture" and small-cell deployment called add-on cells. The rest of the chapter is organized as follows. Section 2.2 describes in detail the concepts of the advanced C-RAN architecture and add-on cells. Section 2.3 explains the results of performance evaluation of our proposed concept. Section 2.4 describes a new cell structure called smart-cell adaptation using add-on cells as an application of the "advanced C-RAN architecture" concept. Section 2.5 explains the results of performance evaluation with the smart-cell structure.

2.2 Advanced C-RAN Architecture and Add-On Cells

2.2.1 Advanced C-RAN Architecture and Add-On Cells

To overcome the difficulties described in Section 2.1, we proposed an add-on cell concept based on an "advanced C-RAN architecture", which is suitable for heterogeneous networks employing multiple-LTE frequency-band operation. As shown in Fig. 2.1, in the advanced C-RAN architecture, high-capacity base station equipment provides a deployment of macro cells covering relatively broad areas and of multiple numbers of small add-on cells covering small but high-traffic areas.

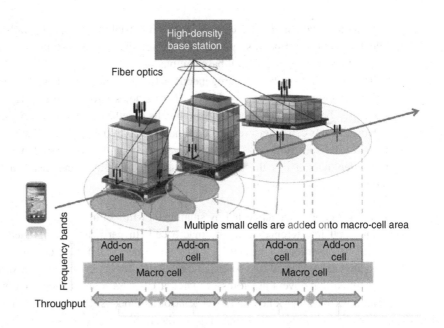

Figure 2.1 Advanced C-RAN architecture and add-on cell.

In the advanced C-RAN architecture, a macro cell is served by one LTE component carrier (CC), whereas a small cell is served by other CCs. To enable CA operation between these CCs, the small cells are deployed by using a remote radio head (RRH), which is connected via optical fiber cables to the same baseband processing unit (eNB) that is serving the macro cell. The optical fiber cable connections to the RRH from the eNB provide a small signaling delay, which enables a joint packet scheduling process (details of which are given in Section 2.2.3) and effective traffic load balancing between the macro and small cells. Furthermore, for a given mobility coverage area served by a macro cell, a number of small cells that are deployed within that mobility coverage area are connected to the eNB that serves the corresponding macro cell. This is to guarantee the mobility performance of UE, i.e., to ensure that frequent handover does not occur when the UE moves from one small cell to another. Instead of the handover operation, when a UE enters or leaves a small cell, the radio resource connection (RRC) management of multiple CCs is performed within the same eNB. More specifically, when a UE enters a small cell area, a CC of the small cell is added and the CA operation between the macro cell and small cell is started; when the UE moves out of the small-cell area, the CC of the small cell is removed. Further details of the CA operation will be described below, in Section 2.2.2.

With the adoption of the advanced C-RAN architecture, with the utilization of CA between small cells and a macro cell, the drawback foreseen in the single-frequency heterogeneous network deployment can be avoided, and the following benefits are foreseen.

- No co-channel inter-cell interference between a macro and small cells would be observed owing to the usage of different LTE carriers between the macro and small cells.
- No degradation of the mobility performance of a UE would be experienced. In the advanced C-RAN architecture, a UE is always connected to a macro cell and only the addition or removal of a CC is performed when the UE enters or leaves a small cell area.
- The flexible installation of small cells in a spot area added onto an existing macro cell is possible (i.e., the add-on cell concept). This provides increased network capacity in a congested area through the installation of multiple small cells.
- Traffic load balancing between a macro cell and add-on cells is achieved through joint packet scheduling between these cells, which results in improved cell throughput and user throughput.

2.2.2 Carrier Aggregation Operation in Advanced C-RAN Architecture

In the CA mode of operation, two or more CCs, which are compatible with the LTE carrier, are aggregated. When a UE first establishes or re-establishes an RRC connection, only one serving cell is configured. This cell is called the primary serving cell (PCell). Then the UE is configured with additional serving cells, called secondary serving cells (SCells), through RRC signaling. The main restriction for the CA operation is that the UE can only aggregate the CCs served by the same base station. Specifically, in the C-RAN architecture with add-on cells, multiple numbers of SCells may exist within the macro-cell coverage. The relative radio quality of a given add-on cell will change depending on the UE position in relation to that cell; e.g., due to UE mobility, an add-on cell set used as an SCell some time ago may not be a suitable SCell any more. If an SCell no longer offers sufficient radio quality for significantly good data transmission, such an SCell should be deleted for the sake of UE battery saving. In order to select the most optimum add-on cell and to set or remove it as an SCell, a series of control procedures are needed [9]. These control procedures, which utilize UE measurements and reporting mechanisms specified in 3GPP [10], are used to ensure the measurement quality of the cell and to achieve relevant configurations of the SCell, i.e., SCell addition, SCell modification, and SCell removal. These control procedures are sometimes referred to as CC management. Figure 2.2 shows the overall procedures for SCell addition, modification, and removal in CA.

SCell Addition

Secondary serving cell addition is performed when the UE is connected only to a macro cell and not configured with CA. As shown in Fig. 2.2 (left-hand lower figure), the eNB configures the UE to perform neighbor-cell measurements in order to be aware of the radio quality of the neighboring cells. If a neighbor cell exists, the UE will return a measurement report to the eNB including the radio quality of the surrounding add-on cells. If the reported radio quality of a neighboring add-on cell satisfies certain conditions, the eNB configures the UE to add the relevant add-on cell as a SCell. Upon completion of

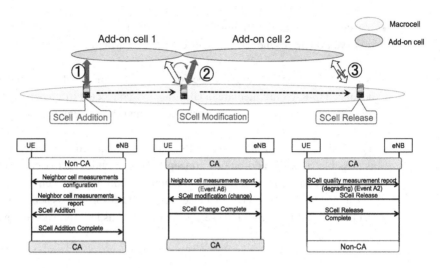

Figure 2.2 SCell (CC) management in advanced C-RAN.

this procedure, the UE is configured with CA, which enables it to be simultaneously connected to the macro cell and the add-on cell, which act as the PCell and SCell, respectively.

SCell Modification

SCell modification is performed when a UE configured with CA, i.e., connected to both macro cell and add-on cell, experiences a change of SCell radio quality compared with its neighboring add-on cells. Changes in SCell radio quality happen in cases where the UE moves from one add-on cell area (SCell) area to another add-on cell area (e.g., one of the neighboring cells of the SCell). The control procedure for this case is described in Fig. 2.2 (middle panel in lower figure). During the SCell addition procedure the eNB also configures the UE with a measurement event such that UE will send a measurement report back to the eNB if a neighboring cell of the same frequency as the SCell becomes better in radio quality than the present SCell (e.g. an event A6 measurement report). This usually occurs when the UE moves into the boundary area between add-on cell 1 and add-on cell 2, so that the radio quality of the SCell (i.e., add-on cell 1) deteriorates while the radio quality of add-on cell 2 improves. Since the UE is configured with the event A6 measurement configuration described above, this condition will trigger the UE to report the radio quality of add-on cell 2 to the eNB. On the basis of this report, the eNB sends a configuration to change the present SCell (add-on cell 1) to add-on cell 2. This control procedure makes it possible to keep up the UE mobility and at the same time to ensure that the UE is always configured with the most optimum add-on cell as an SCell.

SCell Release

The SCell release procedure is performed when a UE configured with CA moves out of an SCell area. The control procedure for this case is shown in Fig. 2.2 (right-hand panel in lower figure). During the SCell addition procedure, the eNB also configures the UE with a measurement event such that the UE will send a measurement report back to the eNB if the currently configured SCell radio quality deteriorates below a certain threshold (e.g., an event A2 measurement). Event A2 measurements are continued even there is a change in the SCell. This usually occurs when the UE moves out from an SCell area. As shown in the figure, if the UE currently connected to the macro cell and add-on cell 2 moves out of the add-on cell 2 area, the radio quality of add-on cell 2 will deteriorate. Since the UE is configured with measurement event A2, this condition will trigger the UE to report the radio quality of add-on cell 2 to the eNB. On the basis of this report, the eNB understands that the radio quality of add-on cell 2 has deteriorated and in this case there is no other add-on cell that can be a candidate for the SCell. Therefore, the eNB sends a configuration to release the current SCell, which results in a condition where the UE is configured only with the PCell (i.e., a conventional non-CA LTE configuration). Applying this kind of SCell release procedure based on the add-on-cell radio quality can reduce the use of eNB resources and save battery consumption on the UE.

In the advanced C-RAN architecture, a UE in a heterogeneous network can maintain a connection with the macro cell (PCell) even if it loses the connection to the small cell (SCell). In other words, there is some flexibility in how to choose the threshold values for CC management, i.e., SCell addition and removal, based on the mobile operator's network-operation policy. For example, if a higher threshold value based on a higher signal-to-interference-and-noise ratio (SINR) criterion is used for SCell addition, only the UEs experiencing good SINR conditions will employ CA between the macro and small cells. Thus, the measured cell throughput in small cells would be improved; however, the cell-edge user throughput and fairness of the user throughput among UEs may be degraded.

2.2.3 Joint Packet Scheduling for Traffic Load Balancing between Macro and Small Cells

In the advanced C-RAN architecture, employing joint packet scheduling across macro and small cells (i.e., across all configured CCs) is effective in achieving traffic load balancing between these cells. In joint packet scheduling [11], the average data rate (ADR) to be used for a proportional fairness (PF) criterion is calculated across all the configured CCs, as shown in (2.1):

$$j = \arg\max \left\{ \frac{R_{n,m}(i,k)}{ADR} \right\} = \arg\max \left\{ \frac{R_{n,m}(i,k)}{\sum_{m}^{\text{all configured CCs}} T_{n,m}(i-1)} \right\}. \qquad (2.1)$$

In this equation, $R_{n,m}(i,k)$ is the expected throughput of the nth UE in the kth resource block of the mth CC in the ith scheduling period; $T_{n,m}(i-1)$ is the ADR achieved using the radio resources assigned to the nth UE in the mth CC in the $(i-1)$th scheduling period. The function arg feeds back the UE number which has the maximum value of $R_{n,m}(i,k)/T_{n,m}(i-1)$ and j is the UE selected to be scheduled according to (2.1). After every scheduling period, the ADR value is updated on the basis of the results of radio resource allocations. It should be noted that the numerator in (2.1) does not need to be changed in order to derive the expected throughput for each resource block in the respective configured CCs based on the principle of the PF scheduler.

By using joint scheduling across CCs where the scheduling results of a specific UE in other CCs are reflected in calculating the ADR value, UEs are liable to be scheduled in cells experiencing good SINR conditions across all the configured CCs and the expected throughput for each resource block in the respective configured CCs as expressed in Eq. (2.1), which will eventually improve the overall throughput performance. The simulation performed in Section 2.3 applies such a joint packet scheduling algorithm.

2.3   Performance Evaluation of Advanced C-RAN Architecture and Add-On Cells

2.3.1   Simulation Conditions

In order to evaluate the system-level downlink throughput performance of the add-on cells using the advanced C-RAN architecture, a system level simulation based on Fig. 2.3 was performed.

We employed a 19-hexagonal macro-cell layout with three macro cells per site and compared the performance of the deployment scenarios shown in Figs. 2.4(a) (CA between macro cells) and 2.4(b) (CA between macro cells and add-on cells). Each macro cell is partitioned into three sectors. For the deployment scenario of Fig. 2.4(b), the following three cases were further defined for comparison: (1) one add-on cell per sector of a macro cell, (2) two add-on cells per sector of a macro cell, and (3) four add-on cells per sector of a macro cell.

In these cases, the add-on cells are randomly placed within each sector of the macro cell with uniform distribution. The cell radius of the macro cell is set to 289 m. For the UE distribution in Fig. 2.4(b), among the total of 30 UEs, 20 UEs are randomly located within a distance of 40 m from the center of each add-on cell in uniform distribution, while the remaining 10 UEs are randomly located in uniform distribution over each sector of a macro-cell area. This distribution is prepared under the assumption that small cells are deployed in congested areas with many UEs.

In the propagation model, line-of-sight (LOS) and non-line-of-sight (NLOS) conditions are statistically selected as a function of the distance between the UE and an add-on cell, as in [12]. This assumption is applicable when the add-on cells are placed in dense urban outdoor congested areas, where such environments include both LOS and NLOS conditions owing to large buildings. We also take into account distance-dependent path loss, lognormal shadowing, and instantaneous multipath fading, assuming the six-ray typical urban (TU) channel model [13]. The maximum Doppler frequency, f_D, is set to 5.55 Hz for one CC and 9.71 Hz for the other CC, which corresponds to 3 km/h at the carrier frequencies of 2 GHz and 3.5 GHz, respectively. It is assumed that the distance-dependent path loss is constant, while the time-varying shadowing and instantaneous fading variations are added to each sample in the system simulation. The MIMO transmission mode employed is an open loop spatial multiplexing (TM3) scheme for two antennas, specified in [14], that selects single- or dual-layer transmission depending on the channel conditions. The threshold used for SCell addition or removal is at the SCells

Number of CCs for CA			2 CCs
Carrier bandwidth per CC			10 MHz
Macro-cell layout			Hexagonal grid 19 sites, 3 sectors per site
Macro-cell radius			289 m
Macro-cell antenna			Horizontal: 70-degree beam width Vertical: 10-degree beam width/ 15-degree down-tilt 14 dBi
Small-cell antenna			Omni, 5dBi
Minimum site distance			75 m (between a macro and a small cell), 40 m (between small cells), 35 m (between UE and macro cell), 10 m (between UE and small cell)
Number of UEs			30 / sector
Traffic model			3GPP FTP traffic model
Transmission power of base station			46 dBm (macro) 30 dBm (small)
Penetration loss			20 dB
LOS / NLOS conditions			Based on Rep. ITU-R M.2135 [12]
Distance-dependent path loss decay factor			3.76 (macro cell) 2.2 (small cell, LOS), 3.67 (small cell, NLOS)
Shadowing	Deviation		8 dB (macro cell), 3 dB(small cell, LOS), 4 dB(small cell, NLoS)
	Correlation		Between macro cells: 0.5 Between small cells: 0.5 Between sectors in the same site: 1.0 Between CCs in the same site: 1.0
	Auto-correlation distance (d_{corr})		50 m (macro cell), 10 m (small cell, LOS), 13 m (small cell, NLOS)
Multipath delay profile			6-ray typical urban
Antenna configuration			Tx: 2 Rx: 2 (uncorrelated)
MIMO transmission mode			Transmission mode 3
MIMO signal detection algorithm			Minimum mean square error (MMSE)
Modulation and coding sets			QPSK ($R = 1/8 - 5/6$) 16QAM ($R = 1/2 - 5/6$) 64QAM ($R = 3/5 - 4/5$)
HARQ retransmission limit			4
Threshold for adding/removing an SCell			0 dB in SINR

Figure 2.3 Major system-level simulation parameters.

SINR value 0 dB, where the SCells SINR value is assumed to be measured ideally in the simulation, which is averaged over the carrier bandwidth (10 MHz) and every frame (1 ms). In other words, the SINR values are measured ideally taking into account channel variation due to the distant-dependent path loss and shadowing; however, these values do not take into account the instantaneous fading variation. The SINR value 0 dB is assumed to be the case where the UE is located near the cell edge. We also employed a joint packet scheduling algorithm between CCs that jointly calculates the average data rate of the UE, as explained in Section 2.4.

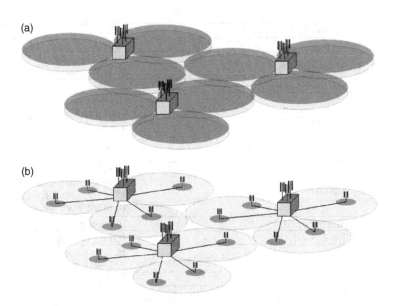

Figure 2.4 Carrier aggregation (CA) deployment scenarios for evaluation: (a) CA between macro cells; (b) CA between macro cells and add-on cells.

2.3.2   Simulation Results

Figure 2.5 shows the improvement in user throughput performance achieved by installing the add-on cells using the advanced C-RAN architecture. Figures 2.5(a) and 2.5(b) show a "snapshot" of the achievable user throughput in the downlink based on the deployment of Figs. 2.4(a) and 2.4(b), respectively. As shown in these figures, the achievable user throughput performance in Fig. 2.5(b) based on the proposed deployment concept is significantly improved through the installation of multiple add-on cells (i.e., the densification of cells), in particular in the areas close to the add-on cells. Furthermore, the achievable user throughput is improved even in those areas mainly served by the macro cell, which are relatively far from each add-on cell. This is due to the effect of traffic off-loading to the add-on cells associated with the joint packet scheduling; consequently the macro cell can effectively assign more resources to the traffic of users who are connected only to it.

Figure 2.6 shows the cumulative distribution function (CDF) for the user throughput performance in the downlink obtained through the system-level simulation. As shown in the figure, the five percentile user throughput for the case (1) (i.e., one small cell per sector) in Fig. 2.4(b) is lower than that for the scenario in Fig. 2.4(a). The reason is that the add-on cell effect using the SCells for case (1) in Fig. 2.4(b) cannot cover the entire coverage area, and consequently the throughput for cell-edge users in the macro cell that cannot be connected to an add-on cell (the SCell) becomes low. However, the 50 percentile and 95 percentile user throughput for this case in Fig. 2.4(b) is higher than that for the scenario in Fig. 2.4(a), because the UEs connected to an add-on cell through the SCell can achieve high throughput performance through CA. Meanwhile,

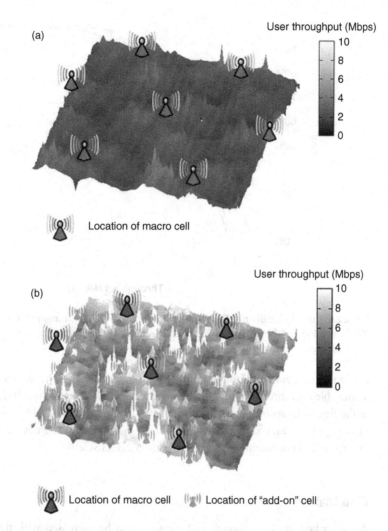

Figure 2.5 Snapshot of achievable user throughput performance: (a) CA between macro cells; (b) CA between macro cell and add-on cell.

the UE throughput for case (2) or case (3), i.e., the two or four small cells per sector scenario in Fig. 2.4(b) is always higher than that for the scenario in Fig. 2.4(a). For example, the 50 percentile UE throughput is improved by 78% for the two small cells per sector scenario and by 243% for the four small cells per sector scenario, and the 5%-tile UE throughput is also improved by 8% for the two small cells per sector scenario and by 28% for the four small cells per sector scenario. In these two scenarios, although approximately 20% of UEs still cannot connect to an add-on cell, these UEs can be assigned more PCell resources and achieve higher user throughput owing to the traffic-load balancing effect obtained through joint packet scheduling.

In order to further demonstrate the improved user throughput in the macro cell associated with the traffic-load balancing effect (i.e., the effect of traffic off-loading to the

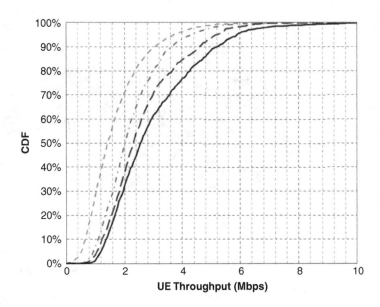

Figure 2.6 Cumulative distribution function for achievable user throughput. Left to right: Fig. 2.4(a), and Fig. 2.4(b) with one, two, and four small cells per sector.

add-on cells) obtained through joint packet scheduling, Fig. 2.7 shows the CDF for the achievable user throughput for the UEs connected only to a macro cell (PCell). As shown in the figure, in accordance with the increase in the number of add-on cells, higher user throughput is achieved for these UEs because more PCell resources can be used through the traffic-load balancing effect of the add-on cells (SCells).

2.3.3   Field Trial

As described above, although cell capacity can be increased with the use of add-on cells, installing multiple add-on cells within a macro cell area also means creating a higher number of cell edges. In such an environment, a moving user will frequently move between two add-on cells coverages. If the UE is connected to the add-on cell only at a given time (i.e., the CA between macro cell and add-on cell is not configured for the UE), the UE will experience a drop in communication quality every time it moves between the areas of the two add-on cells owing to interference between those cells and handover. In contrast, advanced C-RAN architecture enables a UE to be simultaneously connected to an add-on cell and macro cell through CA, which means that stable communications can be ensured and the noticeable degradation in quality due to mobility can be suppressed. Figure 2.8 illustrates the details of the above environment. At point 1 in the left-hand figure, an LTE UE connects to add-on cell A and achieves the throughput provided by LTE. In the right-hand figure, an LTE-Advanced UE, in contrast, can be configured with CA in such a way that the UE connects to both a macro cell and add-on cell A, which means that it can achieve higher throughput than the LTE UE. Looking at

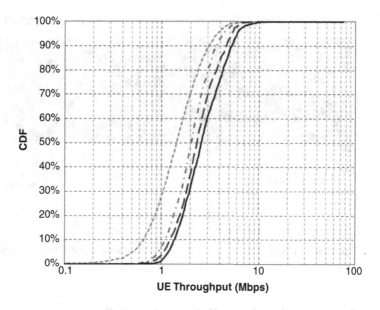

Figure 2.7 Cumulative distribution function for achievable user throughput for UEs connected only to a macro cell. Left to right: Fig. 2.4(a), and Fig. 2.4(b) with one, two, and four cells per sector.

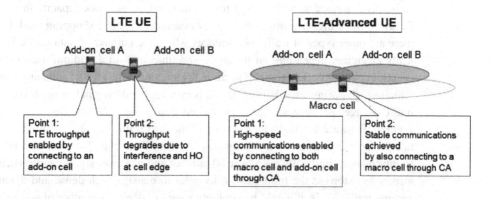

Figure 2.8 UE mobility in Advanced C-RAN architecture.

point 2 in the left-hand figure, at the edge of the add-on cells A and B the LTE UE experiences a drop in throughput owing to interference between the add-on cells and handover (HO). However, as shown in the right-hand figure, the LTE-Advanced UE is always connected to the macro cell although it also switches between add-on cells A and B. This prevents a dramatic deterioration in throughput and enables a certain level of communication quality to be maintained. In short, the advanced C-RAN architecture enables flexible switching between add-on cells while keeping an LTE-Advanced UE connected to a macro cell, thereby achieving high-speed and stable communications. This gain has been verified by field tests performed in an outdoor commercial environment, as shown in Fig. 2.9.

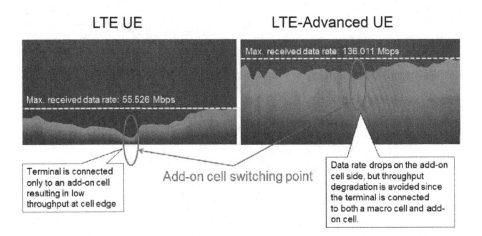

Figure 2.9 LTE and LTE-Advanced terminals switching between add-on cells in a commercial environment.

2.4   Smart-Cell Adaptation Using Add-On Cells

In Section 2.3 the proposed concept of the Advanced C-RAN architecture and add-on cells was described as a promising tool to increase the network capacity in dense traffic areas where higher traffic is constantly observed, e.g., in a shopping mall. However, there are other types of traffic occurrences where the amount of data traffic is dynamically changing, depending on the place and time, e.g., in a stadium used for various events. For example, in a football stadium the amount of data traffic suddenly surges and remains high during a football game and is even higher during half time. More dynamic traffic changes are seen around train stations during the rush hours in the morning and evening. Figure 2.10 shows an example of the data traffic variation around a train station in Japan, observed over 30 minutes. As seen from Fig. 2.10, the amount of data traffic changes drastically even over 30 minutes and increases significantly when a train arrives and stops at the train station. In order to manage such dense and dynamically varying traffic loads, it would be straightforward to deploy a number of add-on cells in dense traffic areas. However, from the operator's viewpoint, how and where to install those add-on cells should be carefully determined, considering the needs for capacity enhancement, as well as OPEX and CAPEX, to deploy and operate the add-on cells. For instance, if the add-on cells are deployed to satisfy peak traffic for a limited time, it would not be cost efficient since there is less utilization of the add-on cells most of the time. Namely, it would cost much more than needed, although the requirement for the capacity would be satisfied. Therefore, from the CAPEX and OPEX perspective, it would be desirable to deploy the add-on cells by targeting the average network capacity and to cope with severe traffic variations by means of technical solutions.

A promising solution is to dynamically change serving areas of cells, e.g., by adaptive antenna array, so that dense traffic loads can be handled by multiple cells. However, changing the shape of the serving areas of these cells would create coverage holes in

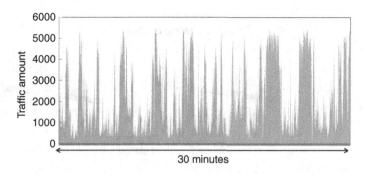

Figure 2.10 Snapshot of traffic distribution over 30 minutes in a train station.

some places and would impair the mobility of some UEs; these are undesirable situations for mobile operators. Therefore, in this section we further propose a smart cell adaptation using the advanced C-RAN architecture and add-on cells using multiple frequencies to cope with variable traffic conditions. In the advanced C-RAN and add-on cell concept, the shapes of areas served by add-on cells can be dynamically changed without sacrificing coverage, since coverage as well as mobility is maintained in the macro-cell layer. In our proposed smart cell adaptation using add-on cells, the add-on cells adaptively create transmission antenna beamforming toward the area where the traffic volume is suddenly significantly increased. The key technology to realize this smart cell adaptation includes adaptive antenna array, control of antenna tilting, and full dimension (FD)-MIMO.

2.4.1 Concept of Smart-Cell Adaptation Using Advanced C-RAN and Add-On Cells

With smart-cell adaptation using add-on cells, transmit-antenna beamforming of the add-on cells is adaptively achieved and directed to a dense traffic area where the traffic has abruptly increased, e.g., owing to the arrival of trains during rush hours, as shown in Fig. 2.11. The proposed smart-cell adaptation is enabled by utilizing different frequency bands for the macro cell and add-on cell layers. More specifically, the cell adaptation is applied to the add-on cell layer only in order to boost the capacity and to cope with dynamic traffic variation, while coverage and mobility are ensured by the macro-cell layer. Cell adaptation, i.e., transmit antenna beamforming toward the dense traffic area, is actualized by means of an adaptive antenna array, antenna tilting, and full dimension (FD)-MIMO. An application of these transmit antenna beamforming techniques would be feasible and beneficial for add-on cells installed on the walls of buildings and telegraph poles. With antenna tilting, transmit antenna beamforming would be mechanically or electronically controlled in a semi-static manner such that the transmit antenna beamforming can track to the semi-static change in the traffic conditions, e.g., for a few minutes as a train approaches and stops at the station. With adaptive antenna arrays, more dynamic antenna beamforming would be possible by multiplying the transmission signal with a transmission weight obtained using some sort of channel-state information. Furthermore, for the FD-MIMO, more accurate and sharpened antenna beamforming

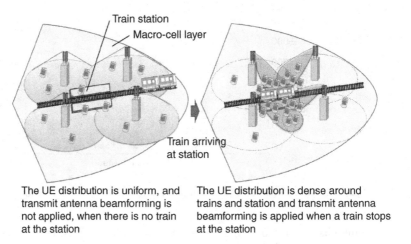

The UE distribution is uniform, and transmit antenna beamforming is not applied, when there is no train at the station

The UE distribution is dense around trains and station and transmit antenna beamforming is applied when a train stops at the station

Figure 2.11 Smart-cell adaptation applied in a train station area.

is realized by installing a larger number of transmit antenna elements at the eNodeB, which would be a valid deployment scenario for a higher frequency band to be allocated in the near future. For example, the study item for LTE-Advanced in 3GPP Release 13 considers the FD-MIMO targeting up to 64 transmit antennas [15].

Since it is a target scenario for smart-cell adaptation, we have focused here on a train station. When there are no trains at the station, no specific transmit antenna beamforming of the add-on cells is created and the antenna pattern is simply determined by the transmit antenna configurations. However, the traffic is significantly increased when the train arrives and stops for a few minutes. In this case, by beamforming, the transmit antennas of the add-on cells are directed to the platform at the train station. Figure 2.11 illustrates the behavior of smart-cell adaption in a train station such that the coverage of add-on cells within the same macro-cell area changes to accommodate traffic before train arrival and during/after train arrival.

2.4.2   Control for Smart-Cell Adaptation

In smart-cell adaptation, knowledge regarding the traffic load on the add-on cells is required at the eNodeB and needs to be updated regularly. There are several methods to obtain a knowledge of the traffic load. Below, we describe how this knowledge, needed to control the transmission beamforming of the add-on cells, is obtained.

Using Measurement Reports from UEs

User equipment that is connected to the macro cell can be instructed by the eNodeB to perform measurements for the add-on cells on the different frequency bands and report the corresponding measurement results, i.e., the reference signal received power (RSRP) and the reference signal received quality (RSRQ), to the eNodeB. On the basis of those measurement results, through the macro cell, the eNodeB configures the UE with an add-on cell as the secondary cell (SCell) for the CA. Since the UEs report the set of measurement results for candidate SCells which satisfy a certain threshold level, the eNodeB knows roughly which add-on cell is congested. In addition, from the average traffic load, and/or the number of connected UEs observed in each add-on cell, the level of congestion would be known. With these information, one possible method is to direct the transmit antenna beamforming of the surrounding add-on cells toward the congested cell area. For example, predetermined patterns or parameters of transmit antenna beamforming are stored in each add-on cell and selected according to the level of congestion.

For this method, a moderate number of surrounding add-on cells need to be measured by the UE. In order for the UE to perform efficient measurement of the surrounding add-on cells, the small-cell discovery signal specified in Release 12 LTE-Advanced would be used directly. For this small-cell discovery signal in Release 12, which is an extension of the channel state information RS (CSI-RS) and zero-power CSI-RS (resource muting) supported in the Release 10 LTE-Advanced specification, a larger number of orthogonal reference signals (RSs) can be used, and thus accurate measurement reports can be expected.

Using Geographical Information from the UE

If a GPS of the UE is assumed to be always available, the geographical locations of the UEs can be obtained. Another approach is to utilize the positioning RS in the LTE specification. Using this location information reported from the UE, transmit antenna beamforming of the add-on cells is created and directed to the relevant area such as a recently arrived train.

Using Extrinsic Information

If it is likely that the traffic load is increased when trains stop at the station during rush hours, a train timetable can be utilized. Then, transmit-antenna beamforming is applied when an increase in the traffic is expected from the timetable. However, this information is not always reliable since, e.g., trains may be delayed. In this case, further information may be needed to improve the accuracy of this method.

Although we have described separately the three different approaches to obtaining knowledge of the traffic load when performing smart cell adaptation, these approaches can be applied jointly.

2.5 Simulation Results

2.5.1 Simulation Conditions

We conducted a system-level simulation to assess the effectiveness of the smart-cell adaptation from the viewpoint of the UE throughput performance. We employed a seven-hexagonal-cell layout with three sectors per macro-cell site. As for the add-on cell deployment, we assumed that three add-on cells are placed within each sector of a macro cell, as shown in Fig. 2.12. The cell radius of the macro cells is set to 289 m.

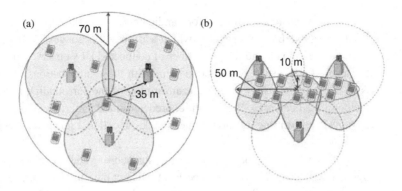

Figure 2.12 The UE distribution for the system-level simulation. (a) Uniform UE distribution; (b) non-uniform UE distribution.

The system bandwidth is set to 10 MHz for both the macro and add-on cell layers. The total base station transmission power is set to 46 dBm and 30 dBm for the macro and add-on cells, respectively.

In the propagation model, line-of-sight (LOS) and non-line-of-sight (NLOS) conditions are statistically selected as a function of the distance between a UE and an add-on cell. We also took into account distance-dependent path loss, lognormal shadowing, and instantaneous multipath fading, assuming the ITU UMi channel model with a three-dimensional distance between a base station on the small-cell layer and a UE, as shown in [16]. The maximum Doppler frequency, f_D, was set to 5.55 Hz for the macro-cell layer and 9.71 Hz for the add-on cell layer, which corresponds to 3 km/h at the carrier frequencies 2 GHz and 3.5 GHz, respectively. It was assumed that the distance-dependent path loss is constant, while the time-varying shadowing and instantaneous fading variations are added to each sample in the system simulation. The MIMO transmission mode employed was the closed-loop spatial multiplexing (transmission mode 9) scheme for two antennas, which selects single- or dual-layer transmission depending on the channel conditions. We assumed an interference-rejection combining (IRC) receiver based on the minimum-mean-square-error (MMSE) criterion that works to mitigate the dominant inter-cell interference and to maximize the signal-to-noise power ratio (SINR). The FTP traffic model 1 with packet arrival rate $\lambda = 8$ was also assumed. For the UE distribution, we used the cases shown in Fig. 2.12.

Case (a): Uniform traffic distribution assuming there is no train at the station.

Case (b): Non-uniform traffic distribution assuming a train has stopped at the station.

In case 1 the UEs are uniformly and randomly dropped within a radius of 70 m, assuming a UE distribution when there is no train on the platform. On the other hand, in case 2 the UEs are uniformly and randomly dropped within an ellipse to model the non-uniform UE distribution due to the train arriving and stopping at the station. We also considered a cell association between the macro and add-on cell layers. For the purpose of this cell association, the RSRP and RSRQ of each add-on cell, which are defined in [14], were used. The largest RSRQ value of the add-on cells is compared with a predetermined threshold value. If this largest RSRQ value exceeds the threshold, the corresponding add-on cell is selected. Otherwise, the macro cell with the largest RSRP value is selected. In this way, UEs are offloaded to the add-on cells as far as possible, and the macro cell is used only when the RSRQ of the add-on cell layer is very low, i.e., the SINR of the add-on cell layer is very low.

2.5.2 Throughput Performance

Figures 2.13(a) and 2.13(b) show the UE throughput performances of the add-on cell layer for the uniform and non-uniform UE distributions, respectively. The macro-cell layer is taken into account for cell-association purposes only, and the UE throughput performance of the macro-cell layer is not shown in the figures. Instead, the percentage of UEs associated with add-on cells is provided below. The two types of antenna patterns, i.e., with transmit antenna beamforming and without transmit antenna beamforming, are compared for the two UE distributions. Transmit antenna beamforming simply applies a fixed antenna pattern with a 3 dB beam-width of 25 degrees in both the vertical and horizontal directions. When the UE distribution is uniform, we observe that the performance without transmit antenna beamforming is better than that with transmit-antenna beamforming. This is so because in this case each transmit antenna beam from the add-on cells has a certain direction, and an application of beamforming causes coverage holes, and thus the received SINR becomes worse for most UEs. However, when the UE distribution is not uniform, the observation is different. The performance without beamforming is significantly degraded since dense traffic in a certain cell reduces the amount of available radio resources per UE. In such a case, transmit antenna beamforming toward the dense traffic area is very beneficial in improving the UE throughput performance. The gains from transmit antenna beamforming to achieve

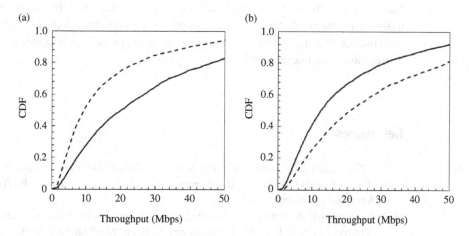

Figure 2.13 CDFs for UE throughput performance. (a) Uniform UE distribution: upper curve, without beamforming; lower curve, with beamforming. (b) Non-uniform UE distribution; upper curve, with beamforming; lower curve, without beamforming.

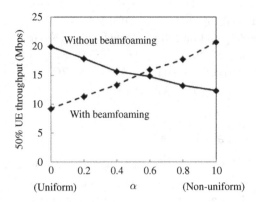

Figure 2.14 The UE throughput performance for mixed UE distributions as a function of α.

a 5% (50%) UE throughput performance is about 41% (68%). Another aspect to consider is the impact on the percentage of UEs associated with add-on cells. We observe that 95% of the UEs are connected to the add-on cell layer when beamforming is not applied, while 88% of the UEs are connected to the add-on cell layer when beamforming is applied. Although the percentage of UEs associated with add-on cells layer is slightly reduced after applying smart-cell adaptation, the impact of this would be negligible considering that there are still many UEs connected to the add-on cells and a greater throughput gain is achieved in a dense traffic area.

We have so far shown throughput performances for the extreme cases, of uniform and non-uniform UE distributions. However, both distributions could coexist in a practical scenario. Hence we also considered another UE distribution, where the uniform and non-uniform UE distributions are mixed. The probability of non-uniform traffic occurrence is given by α. Figure 2.14 shows the 50% UE throughput performances with α as the parameter. It is found to be beneficial to change the shape of cells depending on the traffic distribution. More specifically, beamforming provides a higher UE throughput performance than that without beamforming when the probability of non-uniform traffic distribution starts to exceed 0.5.

References

[1] "GSA – The Global Mobile Suppliers Association." Available at www.gsacom.com.
[2] 3GPP TR 36.913 (V12.0.0), "Requirements for further advancements for E-UTRA (LTE-Advanced)," December 2014.
[3] A. Khandekar, N. Bhushan, J. Tingfang, and V. Vanghi, "LTE-Advanced: heterogeneous networks," in *Proc. European Wireless Conf. 2010*, pp. 978–982, April 2010.
[4] N. Miki, Y. Saito, M. Shirakabe, A. Morimoto, and T. Abe, "Investigation on interference coordination employing almost blank subframes in heterogeneous networks for LTE-Advanced downlink," *IEICE Trans. Commun.*, vol. E95-B, no. 4, April 2012.

[5] G. K. Tran, S. Tajima, R. Ramamonjison, K. Sakaguchi, K. Araki, S. Kaneko *et al.*, "Study on resource optimization for heterogeneous networks," *IEICE Trans. Commun.*, vol. E95-B, no. 4, April 2012.

[6] S. Brueck, "Heterogeneous networks in LTE-Advanced," in *Proc. ISWCS 2011*.

[7] S. Xu, J. Han, and T. Chen, "Enhanced inter-cell interference coordination in heterogeneous networks for LTE-Advanced," in *Proc. IEEE VTC 2012*.

[8] 3GPP TR 36.839 (V11.1.0), "Mobility enhancements in heterogeneous networks," 3GPP Standard, January 2013.

[9] K. Kiyoshima, T. Takiguchi, Y. Kawabe and Y. Sasaki, "Commercial development of LTE-Advanced applying advanced C-RAN architecture expanded capacity by add-on cells and stable communications by advanced inter-cell coordination," *NTT DOCOMO Tech. J.*, vol. no. 17/2, October, 2015.

[10] 3GPP TS 36.331 V10.15.0, "Evolved universal terrestrial radio access (E-UTRA); radio resource control (RRC); protocol specification," December 2014.

[11] T. Takiguchi, K. Kiyoshima, Y. Sagae, K. Yagyu, H. Atarashi, and S. Abeta, "Performance evaluation of LTE-Advanced heterogeneous network deployment using carrier aggregation between macro and small cells," *IEICE Trans. Commun.*, vol. E96-B, no. 6, June 2013.

[12] Report ITU-R M.2135, "Guidelines for evaluation of radio interface technologies for IMT-Advanced," December 2009.

[13] 3GPP TS 45.005 (V8.3.0), "Radio transmission and reception." 3GPP Standard, June 2008.

[14] 3GPP TR 36.814 (V9.0.0), "Further advancements for E-UTRA physical layer aspects." 3GPP Standard, March 2010.

[15] 3GPP RP-141644, "Study on elevation beamforming/full-dimension (FD) MIMO for LTE." 3GPP Standard, September 2014.

[16] 3GPP TR 25.892, "Feasibility study for OFDM for UTRAN enhancement." 3GPP Standard, September 2004.

References

Part II

Physical-Layer Design in C-RANs

3   The Tradeoff of Computational Complexity and Achievable Rates in C-RANs

Peter Rost, Matthew C. Valenti, Salvatore Talarico, and Andreas Mäder

3.1   Introduction

The trend of increased centralization holds the potential to transform mobile networks in two ways. First, centralization enables the exploitation of common channel knowledge, which in turn allows for significant improvements in the performance of a communication channel by, for instance, performing the joint transmission and reception of signals or allocating resources jointly amongst adjacent cells [1]. Second, centralized processing leverages the trend towards deploying mobile networks on low-cost commodity hardware that is running commodity or open-source software solutions. Deploying software-based implementations increases implementation flexibility, reduces service-creation time, and enables the flexible usage of processing resources through virtualization. In this chapter we use the term *Cloud-RAN* (C-RAN) to refer to a flexible use of commodity solutions that combines gains in both the telecommunication and information technology domains.

Before implementing the protocol stack of a RAN on a cloud-computing platform, we must also take the required effort into account, e.g., commodity hardware is considered to be less performant and energy efficient than dedicated hardware such as ASIC, DSP, or FPGA. Furthermore, resource virtualization implies an overbooking of resources while satisfying joint resource requirements of all processed base stations (BSs), which is in contrast with fulfilling individual processing constraints at each BS. Centralized signal processing may further impose stringent requirements on the fronthaul network between a radio access point (RAP) and the data center.

So far, research in the area of Cloud-RAN has focused on the telecommunication domain, e.g., the applicability of joint processing approaches, gains from centralization, and optimal degrees of centralization under different side constraints. In this chapter the focus is on the impact of limiting and virtualizing the data processing resources on the communication rate, i.e., the quantitative coupling of the required computational resources and communication rates [2]. After introducing basic notation and definitions, we consider metrics and an analytical framework that allows one to determine the data processing demand. Interestingly, the data processing requirements depend not only on the number of information bits but also to a large extent on the quality of a user's communication channel. In this chapter we discuss and quantify multi-user gains, which lower the requirements on the data processing resources to be provided. Using this framework, we discuss joint resource allocation strategies, considering the computational constraints and their impact on the communication rate. The allocation of resources in both domains follows similar optimal strategies, which are evaluated numerically.

3.2   Basics

3.2.1   Cloud-RAN System Model

The RAN functionality can be categorized in different layers, i.e., analog processing, physical layer, medium access control, radio link control, and higher layers such as the radio resource control and packet data convergence protocol layer [3]. One central aspect of Cloud-RAN is the split of RAN functionality into one part that is executed at the RAP and one part that is performed at the central data center. In this chapter we consider the split illustrated in Fig. 3.1 [4], where anything below forward error correction (FEC) is executed at the RAP, while FEC and anything above is centralized. Forward error correction decoding consumes a major portion of the resources necessary to process the RAN protocol stack and is therefore the focus of this chapter [5]. Centralizing FEC and the functionality above offers a number of benefits, i.e., joint decoding allows for improving the performance on each link, a major part of the computational resources is centralized, and the required fronthaul capacity and computational resources scale with the number of connected user terminals and their data rates, which is important for exploiting statistical multiplexing gains [6].

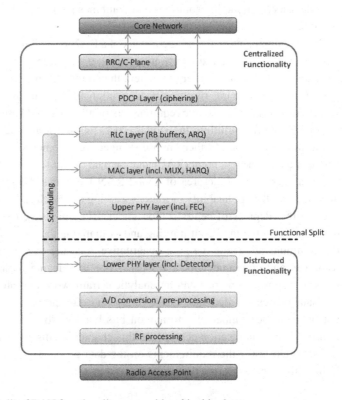

Figure 3.1 Split of RAN functionality as considered in this chapter.

For the following discussion we assume an underlying processing platform that allows for an elastic and dynamic assignment of computational resources to FEC decoding processes. The currently available processing platforms do not satisfy this requirement, owing to the high interface latencies in multi-processor systems, processing jitter caused by operating system interrupts, or jitter in accessing shared memory. However, research in the information technology domain is progressing quickly and is likely to address these drawbacks, e.g., through advanced multi-core and co-processor architectures as well as virtualization concepts such as containers. Furthermore, new paradigms for designing RAN functions are being developed [7] which take the underlying processing platforms and their restrictions into account. Hence, "cloud-native" implementations may be developed which exploit the possibilities of cloud computing efficiently while maintaining the required quality of service in a mobile network.

The following discussion focuses on a setup where the FEC decoding associated with N_c RAPs is performed centrally at a data center. To simplify the notation, we assume that each RAP serves exactly one user terminal and assigns all available resources to it. Throughout this chapter, the instantaneous SNR for the link between the ith user terminal and its RAP is given by γ_i, with an average SNR $\overline{\gamma}_i = \mathbb{E}\{\gamma_i\}$. The channel gain remains fixed for the duration of one transport block (TB) but varies independently from TB to TB, which corresponds to a *block-fading model*. For each TB the channel is conditionally subject to additive white Gaussian noise (AWGN). The discussion will focus on the impact of the limitation of processing resources on communication performance; hence, we do not consider imperfect backhaul conditions or gains from the joint processing of signals.

3.2.2   Link Adaptation

In the following sections we discuss the tradeoff of communication complexity and communication rate. A pivotal roll in this tradeoff is taken by the link adaptation in a wireless network. In an ideal system, the transmitter could adapt perfectly to the channel state and transmit, theoretically, at channel capacity [8]. However, in a practical system, the transmitter can choose from only N_R different modulation and coding schemes (MCSs). Each MCS is characterized by a modulation scheme, i.e. QPSK, 16-QAM or 64-QAM, and a coding rate. Details on both can be found in [9] for a 3GPP LTE system. The combination of modulation and coding results in a unique rate R_k of the kth MCS, $k \in \{1, \ldots, N_R\}$. In line with 3GPP LTE, we assume that a TB using the kth MCS is segmented into C_k code blocks (CBs), each of which conveys D_k information bits. We further assume that the transmitter has perfect knowledge of the SNR in order that it can adapt the MCS perfectly.

At a given SNR γ_i, the highest MCS is selected such that a given average outage constraint is guaranteed for a given receiver structure, e.g. a turbo-decoder with $L_{\max}$ turbo-decoding iterations. The minimum SNR such that this constraint is fulfilled for MCS k is denoted by γ_k^R. The ordered set of tuples $\left[\left(\gamma_1^R, R_1 \right), \ldots, \left(\gamma_{N_R}^R, R_{N_R} \right) \right]$ is referred to as a link adaptation table. This table can be obtained as follows. For each possible MCS k the code block error rate (CBLER) curves are determined as functions of the SNR for a relatively large value of $L_{\max}$ (usually $L_{\max} = 8$). Then γ_k^R is the highest SNR for which the channel outage constraint $\hat{\epsilon}_{\text{channel}}$ is satisfied (a TB error occurs when any CB in a TB fails).

As shown later, the link adaptation table may be used to control the required decoding complexity by shifting the individual SNRs γ_k^R by a constant factor $\Delta\gamma$. Hence, the SNR required to select the kth MCS is then $\Delta\gamma$ above γ_k^R. With this margin, the rate selection for channel SNR γ can be represented by the function

$$r(\gamma, \Delta\gamma) = \begin{cases} R_1 & \text{if } \gamma/\Delta\gamma \le \gamma_1^R, \\ R_k & \text{if } \gamma_k^R < \gamma/\Delta\gamma \le \gamma_{k+1}^R, \\ R_{N_R} & \text{if } \gamma_{N_R}^R < \gamma/\Delta\gamma. \end{cases} \tag{3.1}$$

3.2.3 Resource Allocation

While *link adaptation* determines the chosen MCS, i.e. the actual communication rate, *resource allocation* determines the number of assigned resources and therefore the amount of information per TB. Hence, the two processes are tightly coupled and so we must consider constraints that are usually expressed by quality of service classes in terms of latency, data rate, and reliability. In our case, the resource allocation problem is multi-dimensional, i.e., we need to consider assigning wireless channel resources, as well as data processing resources, jointly.

In this chapter we focus on the problem of assigning data processing resources under an outage probability side-constraint. In particular, we guarantee a maximum computational outage probability; a computational outage implies that insufficient data processing resources are available to process all user terminal signals. From a network point of view, there is no difference between channel and computational outage. However, because of the previously mentioned correspondence between the assigned MCS and the required data processing resources, the rate allocation problem is equivalent to allocating data processing resources.

For the sake of simplicity we assume that each user receives all the resources of its BS (time-division multiple access) and that opportunistic scheduling is not used here [10]. However, the main approaches and conclusions are equally applicable to a system using an opportunistic scheduling, although the interaction with data processing resources needs to be considered. In the following, let $\mathcal{R}_i = \{r_1, \ldots, r_{N_c}\}$ be the set of MCS rates allocated to each uplink for the ith resource allocation strategy. Then, the set $\mathcal{R}^* = \{\mathcal{R}_1, \ldots, \mathcal{R}_{N_A}\}$ with $N_A \le N_R^{N_c}$ denotes the set of all feasible rate allocations. The goal of our resource allocation strategy is to find an $\mathcal{R}_k \in \mathcal{R}^*$ which maximizes the system throughput while satisfying the given outage constraint.

3.3 Complexity Model and Metrics

3.3.1 Complexity Model

In modern wireless systems, including cellular networks, capacity-approaching codes (i.e., turbo or low density parity check (LDPC)) are commonly used and the decoders for such codes are iterative. Iterative decoders have a natural tradeoff between complexity and performance: the complexity is directly proportional to the number of iterations while executing more iterations can potentially correct more errors, allowing the system to operate at a lower SNR. In the context of Cloud-RAN, the iterative decoder dominates the computational complexity of the uplink, requiring far more operations than any other single process. It follows that the computational effort required by a receiver that includes an iterative decoder is (approximately) linear in the number of iterations and in the number of information bits. Thus, a reasonable metric for computational effort is the *bit-iteration*; i.e., the amount of computation required per information bit to execute one decoder iteration.

Let $L_r(\gamma, \Delta\gamma)$ be the number of iterations required to decode a particular CB for a given SNR γ at a fixed SNR margin $\Delta\gamma$. Even when the SNR and SNR margins are fixed, this value will fluctuate from one CB to the next, and it follows that $\mathbb{E}\{L_r(\gamma, \Delta\gamma)\}$ is the average number of decoding iterations, which can be found using the same set of simulation curves as those used to determine the MCS thresholds γ_k^R. In particular, for a given rate $r(\gamma, \Delta\gamma)$, the simulation results will show the CBLER for each iteration up to the L_{max}th iteration. These CBLER curves can be interpreted as the probability mass function (PMF) of the number of required iterations at SNR γ and, from the PMF coefficients, the average number of iterations is easily computed.

Having found the average number of iterations, the expected decoding complexity (averaged over the number of iterations required per CB) expressed in bit-iterations per channel use (pcu) can be found using

$$C(\gamma, \Delta\gamma) = \frac{D_k C_k \mathbb{E}\{L_r(\gamma, \Delta\gamma)\}}{S_{re}} \tag{3.2}$$

where S_{re} is the number of channel uses required to convey the TB, and D_k and C_k are the number of information bits per code block and number of code blocks, respectively, associated with the MCS selected by (3.1) for the given instantaneous SNR γ and SNR margin $\Delta\gamma$.

While (3.2) can be used to evaluate the computational requirements of a Cloud-RAN system by using simulations [11], a more analytical approach can be pursued by using the groundbreaking work provided by [12]. In this work, the authors provided a model that accurately predicts the average power consumption of an iterative decoder for a given code rate, coding scheme, channel, and decoder type. Utilizing [12, Eqs. (4) and (9)], the expected decoding complexity can be approximated by

$$C(\gamma, \Delta\gamma) \approx \frac{r(\gamma, \Delta\gamma)}{\log_2(\zeta - 1)} \left[\log_2\left(\frac{2 - \zeta}{K'} \log_{10}(\hat{\epsilon}_{channel})\right) - 2\log_2 l_k(\gamma, \Delta\gamma) \right] \tag{3.3}$$

where ζ is a parameter of the model that is related to the connectivity of the decoder when represented as a graph, K' is another free parameter of the model, $\hat{\epsilon}_{channel}$ is the channel outage constraint, and

$$l_k(\gamma, \Delta\gamma) = \log_2(1 + \gamma) - r(\gamma, \Delta\gamma). \tag{3.4}$$

When evaluating (3.3), the function $r(\gamma, \Delta\gamma)$ is defined by (3.1). The relationship between each threshold γ_k^R and the corresponding rate R_k in (3.1) can be modeled as

$$R_k = \log_2\left(1 + \frac{\gamma_k^R}{\nu}\right) \tag{3.5}$$

where ν is a parameter that models the gap between the capacity at γ_k^R and the SNR needed for the actual code to meet the performance objective at rate R_k. For a given iterative decoder, the values of (ζ, K', ν) can be found for each MCS by statistically fitting the empirically observed complexity to the complexity predicted by the model. Alternatively, a common set of (ζ, K', ν) can be found for all MCSs by finding the fit that minimizes the error averaged across all MCS schemes.

3.3.2   Performance Metrics

In a Cloud-RAN network, the amount of computational resources available at the centralized unit that are used to centrally process the uplink signals of a pool of BSs affects the communications performance. If the total computational effort required by the iterative decoder to complete its decoding within a deadline exceeds the available computational resources, the centralized processor is unable to finish the decoding of at least one uplink user. This event is called a *computational outage* [11] and from the operator's perspective its effect is no different from a channel outage, because in neither case is the information made available to the end user.

As discussed earlier, the amount of computational resources required by the centralized processor is a random variable. It fluctuates over time, depending on the changing channel conditions and the corresponding changes in the selected MCS. Let C_{max} represent the amount of computational resources available per RAP. More specifically, it is the maximum number of bit-iterations that can be processed within the decoding deadline. When N_c RAPs are jointly processed, the total available computational budget is $N_c C_{max}$. If we let $C(\gamma_i, k)$ denote the computational load required by the ith RAP, then it follows that the computational outage probability is

$$\epsilon_{comp}(N_c) = P\left[\sum_{i=1}^{N_c} C(\gamma_i, k) > N_c C_{max}\right]. \tag{3.6}$$

Conventionally, wireless systems and their resource-allocation policies are designed to guarantee that a channel-outage constraint is satisfied, with little regard paid to computational constraints. However, in a Cloud-RAN environment, constraints on the available computational resources affect the occurrence of computational outages. In a poorly dimensioned system, computational outage could become more prevalent than channel outages. Hence, the system needs to be also dimensioned based on a constraint on the per-cell computational outage probability, which we denote $\hat{\epsilon}_{comp}$.

We define *outage complexity* as the minimum amount of computational resources per RAP required to guarantee that the computational outage probability satisfies a given target computational constraint. The target outage constraint $\hat{\epsilon}_{\text{comp}}$ must be satisfied for each RAP. However, (3.6) is the computational outage probability for a group of N_c RAPs. Since the cloud group will experience a computational outage if any of its constituent RAPs experiences an outage, it follows that the constraint on the computational outage of the cluster is $1 - (1 - \hat{\epsilon}_{\text{comp}})^{N_c}$, i.e., this is the probability that *any* of the RAPs suffers a computational outage. Thus, the outage complexity $\mathcal{C}_{\text{outage}}(\hat{\epsilon}_{\text{comp}}, N_c)$ may be mathematically defined as

$$\mathcal{C}_{\text{outage}}(\hat{\epsilon}, N_c) = \arg \min_{\mathcal{C}_{\max}} \left[\epsilon_{\text{comp}}(N_c) \leq 1 - \left(1 - \hat{\epsilon}_{\text{comp}}\right)^{N_c} \right] \tag{3.7}$$

where $\hat{\epsilon}_{\text{comp}}$ is the per-cell computational outage constraint.

Using the definitions of the computational outage (3.6) and the outage complexity (3.7), the impact on the performance from centralizing the communication-processing resources can be measured in two ways: (1) by considering the reduction of the required computational resources made possible by the centralization of processing, which we refer here to as the *computational gain*; and (2) by examining the rate of computational outage improvement as more BSs are jointly processed, which we refer to as the *computational diversity*.

When N_c BSs are jointly processed, the computational gain is formally defined by

$$g_{\text{comp}}(N_c) = \frac{N_c \mathcal{C}_{\text{outage}}(\hat{\epsilon}_{\text{comp}}, 1)}{\mathcal{C}_{\text{outage}}(\hat{\epsilon}_{\text{comp}}, N_c)} \tag{3.8}$$

and its asymptotic behavior is found by letting the number of BSs go to infinity,

$$g_{\text{comp}}^{\infty} = \lim_{N_c \to \infty} g_{\text{comp}}(N_c). \tag{3.9}$$

The computational diversity is found by determining the rate at which the computational outage probability is improved as a function of N_c and is formally defined by

$$d_{\text{comp}}(N_c) = -\frac{\partial \log_{10} \epsilon_{\text{comp}}(N_c)}{\partial N_c}. \tag{3.10}$$

The computational diversity is the slope of the curve that relates the computational outage probability to the number of jointly processed BSs. The larger this negative slope is, the faster the maximum computational gain (3.9) is achieved. The computational diversity provides a good indication of the required number of centralized BSs per data center. When more BSs must be centralized, the fronthaul requirements will increase accordingly. Since the slope of the computational diversity has its maximum magnitude at $N_c = 1$, in the following the derivative of (3.10) is evaluated at $N_c = 1$ and it is indicated by d_{comp}, dropping the dependence on N_c.

While the previously defined metrics are able to capture and highlight the benefit of a centralized architecture, the interplay between the computational complexity and the throughput has thus far been ignored. As the margin $\Delta \gamma$ increases, the computational complexity decreases, but so does the rate since the system is operating farther away from capacity. The metric that captures this relationship is the complexity–rate tradeoff (CRT), which measures how much additional complexity is required in order to further improve the rate. The CRT is formally defined by

$$t_{\text{comp}}(N_c) = \lim_{\Delta\gamma \to 0} \frac{\partial r(\gamma, \Delta\gamma)}{\partial \Delta\gamma} \left[\frac{\partial \mathcal{C}_{\text{outage}}(\hat{\epsilon}_{\text{cloud}}, N_c)}{\partial \Delta\gamma} \right]^{-1}. \tag{3.11}$$

3.4   Complexity Analysis Framework

In order to evaluate the previously introduced performance metrics, this section describes an analytical framework to quantify the expected outage complexity in a Cloud-RAN environment. In the following, the analysis is first built assuming that processing is done locally for just a single BS, interference is negligible, and the average SNR $\overline{\gamma}$ is fixed, which is true when the uplink is fully power controlled. Subsequently, the analysis is expanded to take into account the path loss and the fractional-power-control policy when users are distributed uniformly within a circle centered at their serving BS. Finally, on the basis of these derivations, the analytical framework is further expanded to quantify the expected outage complexity for a Cloud-RAN composed of N_c BSs.

3.4.1   Local Processing

Consider an uplink with a fixed average SNR $\overline{\gamma}$, and assume that the corresponding signal is processed locally (i.e., $N_c = 1$). Let Y_i denote both the ith BS and its location. Each BS Y_i serves only one user, which is located at X_i and is a distance $|Y_i - X_i|$ away. For a Rayleigh-fading channel, the received uplink signal y_i from the user X_i to its serving BS Y_i can be expressed as follows:

$$y_i = \sqrt{P_i |Y_i - X_i|^{-\alpha}} h_i x_i + \sum_{j \neq i} \sqrt{P_j |Y_i - X_j|^{-\alpha}} h_j x_j + w_i, \tag{3.12}$$

where h_i is the channel attenuation with complex normal distribution $h_i \sim \mathcal{CN}(0, 1)$, x_i is the unit-power signal transmitted from user X_i modeled as $x_i \sim \mathcal{CN}(0, 1)$, P_i is its transmit power, α denotes the path-loss exponent, and w_i is the AWGN with variance σ_n^2 and is modeled as $w_i \sim \mathcal{CN}(0, \sigma_n^2)$.

The value of P_i is selected according to a fractional-power-control policy:

$$P_i = P_0 |Y_i - X_i|^{s\alpha} \tag{3.13}$$

where P_0 is a reference power (typically taken to be the power received at unit distance from the transmitter) and $s, 0 \leq s \leq 1$, is the *compensation factor* for fractional power control. The instantaneous SINR is then given by

$$\gamma_i = \frac{|h_i|^2 |Y_i - X_i|^{-\alpha(1-s)}}{\overline{\gamma}_{\text{ud}}^{-1} + \sum_{j \neq i} |h_j|^2 |Y_i - X_j|^{-\alpha} |Y_j - X_j|^{\alpha s}} \tag{3.14}$$

where $\overline{\gamma}_{\text{ud}} = P_0/\sigma_n^2$ is the average SNR assuming unit-distance signal transmission. In the following, as mentioned above, it is assumed that the channel gain remains fixed for the duration of one TB, but varies independently from TB to TB which corresponds to a block-Rayleigh-fading model.

Single User with Full Power Control

Initially assume that full power control is used, i.e., $s = 1$, and that inter-cell interference is neglected. It follows that γ is an exponential random variable with probability density function (PDF) given by:

$$f_\gamma(\gamma) = \frac{1}{\overline{\gamma}_{ud}} \exp\left(-\frac{\gamma}{\overline{\gamma}_{ud}}\right), \quad \gamma \geq 0. \tag{3.15}$$

For this $f_\gamma(\gamma)$, the expected value and variance of the required computational complexity can be found using Theorems 1 and 2, respectively, from [2].

When each user is processed locally, the outage complexity is the smallest value C_{outage} that satisfies the following expression:

$$\mathbb{P}\{C_i \geq C_{\text{outage}}(\hat{\epsilon}_{\text{comp}}, 1)\} \leq \hat{\epsilon}_{\text{comp}}, \tag{3.16}$$

which can be evaluated given the cumulative distribution function (CDF) of the complexity, defined as

$$F_C(x) = \mathbb{P}\{C_i \leq x\}. \tag{3.17}$$

A detailed expression for (3.17) is given in [2, Eq. (20)].

Partial Power Control

Let us now account for the effect of partial power control. Consider a user located randomly within the cell. For the sake of tractability, assume a circular cell of unit radius and that the user is placed according to a uniform distribution. Again assuming that inter-cell interference is negligible, the average SNR received by Y_i for the ith user is given by

$$\overline{\gamma}_i = \overline{\gamma}_{ud} |Y_i - X_i|^{-\alpha(1-s)}. \tag{3.18}$$

In this specific case, it is not possible to determine a closed-form expression or a suitable approximation for both the expected value and the variance of the required complexity. However, knowing that the PDF of $r_i = |X_i - Y_i|$ is

$$f_{r_i}(\omega) = \begin{cases} 2\omega, & 0 \leq \omega \leq 1, \\ 0, & \text{otherwise,} \end{cases} \tag{3.19}$$

the expected complexity $\mathbb{E}\{C_i\}$ and variance $\text{Var}\{C_i\}$ can be obtained by performing a numerical integration over r as detailed in [2, Theorems 1 and 2]. Under this scenario, the CDF of γ averaged over the spatial distribution can be evaluated in closed form using [2]:

$$F_\gamma(\gamma) = 1 - 2\left(\frac{\gamma}{\overline{\gamma}_{ud}}\right)^{-2/[\alpha(1-s)]} \left[\Gamma\left(\frac{2}{\eta(1-s)}, \frac{\gamma}{\overline{\gamma}_{ud}}\right) \right. \\ \left. - \Gamma\left(\frac{2}{\alpha(1-s)}, 0\right)\right][\alpha(1-s)]^{-1}, \tag{3.20}$$

where $\Gamma(z, x)$ is the incomplete gamma function [13]. Since the CDF of γ is equal to (3.20), the outage complexity can be determined analytically. Furthermore, the average achievable rate can be determined by substituting (3.20) into

$$\mathbb{E}\{R\} = \sum_{k=1}^{N_R} \left[F_\gamma(\gamma_{k+1}^R) - F_\gamma(\gamma_k^R) \right] r \left(\gamma_k^R, \Delta\gamma \right). \tag{3.21}$$

3.4.2   Centralized Processing

When the signals of N_c RAPs are processed centrally, the required complexity is the sum of the individual complexities. The required complexity to process the signal from the ith RAP is a random variable with mean $\mathbb{E}\{C_i\}$ and variance $\text{Var}\{C_i\}$. Assuming that N_c is sufficiently large, the central limit theorem implies that the required complexity to process the signals centrally is Gaussian. Using this Gaussian approximation, it follows that the outage complexity is

$$C_{\text{outage}}(\hat{\epsilon}_{\text{comp}}, N_c) = \sqrt{\frac{\text{Var}\{C_i\}}{N_c}} \underbrace{\sqrt{2} Q^{-1} \left(2(1 - \hat{\epsilon}_{\text{comp}})^{N_c} - 1 \right)}_{\text{inv. norm. CDF}} + \mathbb{E}\{C_i\} \tag{3.22}$$

where $Q^{-1}(\cdot)$ is the inverse error function [13].

3.4.3   Numerical Examples

In this section, the performance metrics introduced in Section 3.3.2 and the analytical framework described above are used to perform an evaluation of a centralized system compliant with the 3GPP LTE standard. Consider an LTE system that uses a normal cyclic prefix and a 10 MHz bandwidth, corresponding to 50 resource blocks (RBs). Assuming that five RBs are reserved for the physical uplink control channel [9], the transmitted signal occupies 45 RBs. Since there are 12 information-bearing single-carrier frequency-division multiple access (SC-FDMA) symbols per subframe and 12 subcarriers per RB, it follows that $S_{\text{re}} = 45 \times 12 \times 12 = 6480$. The MCS selection is implemented according to (3.1) with the threshold found for $L_{\text{max}} = 8$ iterations; a typical outage constraint for an LTE network of $\hat{\epsilon}_{\text{channel}} = 0.1$ is used and no excess margin ($\Delta\gamma = 0$ dB) is applied. For this scenario, the aforementioned complexity model was calibrated by choosing the set of values for (ζ, K', ν), which are common to all MCSs, that gives the best fit with the empirical complexity obtained from a simulator that is compliant with the 3GPP LTE standard. Good agreement [2] with the empirical data is obtained when $\zeta = 6$, $K' = 0.2$, and $\nu = 0.2$ dB.

Figure 3.2(a) shows the computational gain as function of the number of centralized BSs, while Fig. 3.2(b) shows the computational gain as function of the target computational outage probability. The results were calculated for a path loss exponent $\alpha = 2$, compensation factor $s = 0.1$, and average SNR at unit distance $\gamma_{\text{ud}} = 0$ dB. Figure 3.2(a) shows that the computational gain relative to local processing increases sharply for small pools of centralized BSs but, as the dimension of the pool increases, the computational gain becomes flatter. The notches at the right-hand side of Fig. 3.2(a) show

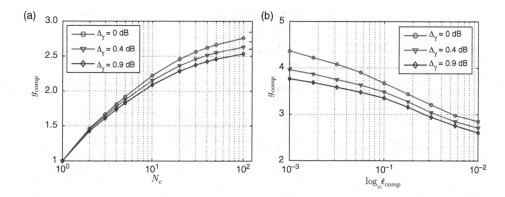

Figure 3.2 Complexity gain. The three notches at the right-hand side of (a) show the behavior as $N_c \to \infty$. ©IEEE 2015, reprinted with permission.

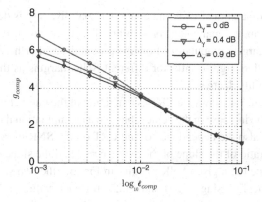

Figure 3.3 Computational diversity as function of the target computational outage probability. ©IEEE 2015, reprinted with permission.

the asymptotic behavior ($N_c \to \infty$). Figure 3.2(b) shows that the computational gain increases as the target computational outage probability decreases, the reason being that there is more variability in the load offered from one RAP to the next when the outage complexity constraint is tight. This is a consequence of the fact that the complexity scales with $- \log_{10}(\epsilon_{comp})$ in (3.3). In this case, centralization provides more benefits as diversity effects can be exploited to balance the load among the individual BSs. Furthermore, the computational gain differs more significantly for different SNR margins as the target computational outage probability decreases. This effect is mainly due to the fact that at a lower target outage probability it is more challenging to operate close to capacity, because the decoder is then required to have a steeper error rate curve.

Figure 3.3 shows the computational diversity as a function of the target computational outage probability. For each SNR margin $\Delta\gamma$ and target computational outage probability, the computational diversity is computed by numerically evaluating the derivative in (3.10), once the computational outage probability is obtained from (3.22). As with

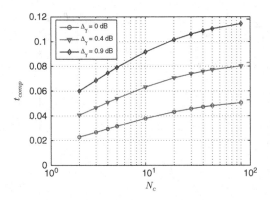

Figure 3.4 The CRT as a function of N_c. The notches at the right-hand side of the figure show the behavior as $N_c \to \infty$. ©IEEE 2015, reprinted with permission.

the computational gain, the computational diversity is decreased by increasing the target computational outage probability. This is due to the overprovisioning of resources at lower target computational outage probabilities. Furthermore, the computational diversity differs significantly for different SNR margins as the target computational outage probability decreases.

Figure 3.4 shows the CRT as a function of the number of centralized BSs, which quantifies the increase in the achievable rate as computational resources are added. Figure 3.4 shows also the dependence of the CRT on the SNR margin. The CRT defined in (3.11) is the ratio of the slope of the average rate and the slope of the outage complexity, both computed at a given SNR margin. In Fig. 3.4 the two slopes are individually evaluated numerically using (3.21) and (3.22) for each value of N_c and SNR margin $\Delta\gamma$. The notches at the right-hand side of the figure show its asymptotic behavior ($N_c \to \infty$). The figure shows that the CRT increases as the SNR margin is increased, and this is due to the fact that the decoder is operating farther away from capacity and therefore does not cause a high peak complexity. Hence, additional complexity is beneficial for gaining additional achievable rate. In other words, the closer the decoder operates to capacity, the more resources must be invested to increase the achievable rate.

3.5　　Joint RAN and Cloud Scheduling

3.5.1　　Practical Aspects

The computational demand at the data center changes randomly over time, with a granularity of one transmission time interval (TTI), i.e., 1 ms in the case of LTE. A Cloud-RAN system with dynamic resource provisioning could, in theory, assign the required computational resources on the same time scale. However, owing to practical limitations, a cloud-computing platform may assign resources only on a much more coarse time scale, i.e., every few seconds. Nonetheless a dynamic resource allocation is still important: on the one hand, over-provisioning of resources should be avoided

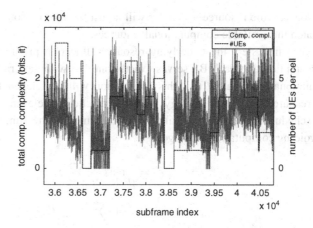

Figure 3.5 Computational complexity versus time in scenario with mobility.

as much as possible; on the other hand, the performance of the system should not be negatively affected. This can be achieved by adjusting the resource provisioning to the large-scale behavior of the complexity process while avoiding computational outage.

It is therefore important to know the characteristics of the computational complexity process, taking into account the effects of mobility, link adaptation, the number of UEs, the network deployment, and hand-over. Figure 3.5 shows an extract from a trace of the computational complexity demands per TTI for a single cell in a heterogeneous network scenario with 104 cells (50% macro cells and 50% small cells) and 10 UEs on average per macro cell, when each UE moves with a speed of 30 km/h. The trace shows that a dependence exists between the number of active UEs (dashed line) and computational complexity (solid line). This is trivially the case if there are no UEs in the cell, but also e.g., at time index 3.73×10^4, where three UEs arrive in a short time at the cell. However, one cannot necessarily conclude that there is strong correlation between the number of UEs and the complexity, since the main impact factor is the mobility of the UEs and the corresponding variations of the channel.

In the following, a mechanism is introduced which allows for adapting the scheduling process in the data center in such a way that the total computational complexity demand does not exceed a certain target threshold. Note that the target threshold should be set according to the expected user traffic demand and can vary over time as well.

3.5.2 Scheduling under Complexity Constraints

Using the framework introduced in Section 3.4, the required computational resources are estimated for a given computational outage probability and SNR per UE. As discussed earlier, the computational complexity depends not only on extrinsic factors such as mobility and number of UEs but also on the link adaptation and scheduling process in the BS. For this reason, the joint scheduling of RAN and data processing resources is applied: if the required computational resources may exceed the available computational resources, the resource scheduler will adjust the radio resource assignment in order to match the available computational resources.

In a first step, the underlying resource allocation problem is formulated with the assumption that each BS serves exactly one UE with data rate r_k and associated computational complexity C_k. This assumption will be relaxed later in order to develop a resource allocation strategy which is practically relevant. The objective is to find a scheduling metric which allows for maximizing the sum-rate of the system while avoiding computational outage:

$$r^* = \max_{\forall \mathcal{R}: \mathcal{R} \in \mathcal{R}^*} \sum_{r_k \in \mathcal{R}} r_k \tag{3.23}$$

$$\text{s.t. } \sum_{r_k \in \mathcal{R}} C_k \leq C_{\max}. \tag{3.24}$$

In the following, two computational complexity-aware resource allocation schemes are compared: scheduling with complexity cut-off (SCC) and scheduling with water filling (SWF). The SCC scheme is an heuristic approach where the data rates of the UEs with the highest computational complexities are consecutively reduced. The SWF scheme is based on the insight that the optimization problem resembles the well-known water-filling problem, used to solve the equivalent power-allocation problem in [14]. In the following, both methods are described in more detail.

Scheduling with Complexity Cut-off

Scheduling with complexity cut-off is based on the insight that UEs operating close to capacity require significantly more computational resources than those operating farther away from capacity. However, the gain in terms of data rates is only small, as expressed by the CRT. The idea is therefore to reduce the MCS of those UEs which occupy most computational resources in order to satisfy the computational constraints. The method is formally described in Algorithm 1.

Algorithm 1 Scheduling with complexity cut-off

Set $\mathcal{R}$ such that each user k receives the maximum possible rate r_k.
while $\sum_{r_k \in \mathcal{R}} C_k > C_{\max}$ **do**
    Select the user k^* with the highest complexity: $k^* = \text{argmax}_i\, C_k$
    Reduce the MCS of k^* by one such that $r'_{k^*} < r_{k^*}$
end while

As a result of this procedure, we obtain a rate allocation which satisfies the computational complexity constraint and always reduces the rate for users that incur the highest complexity. Although this metric is efficient in the sense that it requires only low implementation and also computational effort, it is not necessarily sum-rate optimal. In the following, we refer to this empirical metric as SCC.

Scheduling with Water Filling

In order to obtain a sum-rate optimal solution, the corresponding resource allocation optimization needs to be tackled. The solution to this problem requires the derivative $\partial C_k / \partial r_k$, which is rather complex. Therefore, the following linearization of $l(\gamma_k, r_k)$ is applied:

$$l(\gamma_k, r_k) \approx a_k r_k + b_k, \tag{3.25}$$

with parameters

$$a_k = \frac{\partial l(\gamma_k, r_k)}{\partial r_k} = \frac{-1}{\log(2) \left[\log_2 (1 + \gamma_k) - r_k \right]},$$
$$b_k = \log_2 \left[\log_2 (1 + \gamma_k) - r_k \right] - a_k r_k.$$

Using this linearization, we can rephrase the required computational complexity C_k as

$$C_k = \alpha_k r_k^2 + \beta_k r_k, \tag{3.26}$$

where

$$\alpha_k = -\frac{2a_k}{\log_2 (\zeta - 1)},$$

and

$$\beta_k = \frac{1}{\log_2 (\zeta - 1)} \left[\log_2 \left(\frac{\zeta - 2}{K(\hat{\epsilon}_{\text{channel}})\zeta} \right) - 2b_k \right].$$

THEOREM 3.1 *The solution to the rate allocation problem in (3.24) can be well approximated by*

$$r_k = \frac{1}{2\alpha_k} \left(\frac{1}{\eta} - \beta_k \right)^+, \tag{3.27}$$

where $0 \leq \eta \leq 1/\beta_k$ and $\sum_{r_k \in \mathcal{R}} C_k \leq N_c C_{\max}$.

The details of the derivation are provided in [15]. In (3.27), the parameter $1/\eta$ determines the "water-level" which is used as the decision criterion for serving a UE in a given TTI or not. The corresponding cost in terms of complexity for decoding the data from a UE is denoted by β_k. If the difference between the selected rate r_k and the channel capacity $\log(1 + \gamma_k)$ is small, then the term β_k becomes very large (owing to the steep slope of a_k). In this case, the UE is unlikely to be served if the aggregated complexity is close to the maximum complexity, $N_c C_{\max}$. Correspondingly, the variable α_k scales the rate, i.e. if the rate r_k is close to capacity then α_k becomes very large and therefore scales down the assigned data rate in order to reduce the required computational complexity.

The solution in (3.27) is now extended to the case of multiple UEs per cell and discrete data rates as selected by the link-adaptation function in the BS. From (3.27) and (3.26) follows that

$$\frac{1}{2\alpha_k} \left(\frac{1}{\eta - \beta_k} \right) \geq \sqrt{\frac{C_k}{\alpha_k} + \left(\frac{\beta_k}{2\alpha_k} \right)^2} - \frac{\beta_k}{2\alpha_k} \tag{3.28}$$

must be fulfilled, which can be simplified to

$$\frac{1}{\eta} \geq \sqrt{4\alpha_k C_k + \beta_k^2}. \tag{3.29}$$

This equation provides the required "water-level" for each user in order to satisfy the constraints in Theorem 3.1. Algorithm 2 describes how the resource scheduling is performed using the previously described complexity metric.

Algorithm 2 Scheduling with water-filling

1: Set $\mathcal{R}$ such that each user k receives the maximum possible rate r_k
2: **while** $\sum_{r_k \in \mathcal{R}} C_k > C_{\max}$ **do**
3:    Compute $k^* = \mathrm{argmax}_k \left(4\alpha_k C_k + \beta_k^2\right)$
4:    Reduce the MCS of k^* by one such that $r'_{k^*} < r_{k^*}$
5: **end while**

3.5.3 Numerical Examples

The previously introduced scheduling metrics were numerically evaluated. The considered performance metrics were the sum-rate and the corresponding computational complexity requirement. The evaluation was conducted with a 3GPP LTE compliant system-level simulator in a wide-area scenario. The details of the evaluation parameters are shown in Table 3.1.

Figure 3.6(a) shows the CDF of the computational complexity and Figure 3.6(b) shows the CDF of the achieved system throughput. The figures give the curves for both SWF and SCC, when the system is parameterized in such a way that a computational outage $\epsilon_{\mathrm{out}} \in \{10, 1, 0.1\}\%$ holds (the notches in the inset show the corresponding value of $C_{\max}$). The figures also show as a benchmark the curve for the unconstrained case, which selects the maximum possible rate that can be used.

Table 3.1 Simulation parameters

Parameter	Value
Spatial distribution of users	Poisson point process
Density of UEs per unit area	$\lambda = 1$ UEs/km^2
Path loss exponent	$\alpha = 3.7$
Number of centrally processed base stations	$N_c = 10$
Target computational outage	$\hat{\epsilon}_{\mathrm{comp}} \in \{10, 1, 0.1\}\%$
Channel outage constraint	$\epsilon_{\mathrm{channel}} = 10\%$
Fading model	Rayleigh
Fractional-power-control factor	$s = 0.1$
Transmit power	$P_0 = 10\,\mathrm{W}$
Noise power	$W = 100\,\mathrm{mW}$
Simulation trials	$N_{\mathrm{trials}} = 10^7$

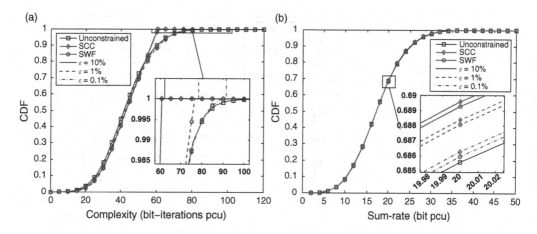

Figure 3.6 CDFs for the computational complexity and achieved system throughput. ©IEEE 2015, reprinted with permission.

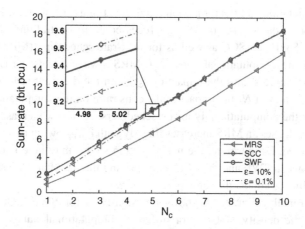

Figure 3.7 Mean sum-rate over the number of centralized base stations. ©IEEE 2015, reprinted with permission.

First, it can be observed that a stronger constraint on the computational outage implies the need for more computational resources. However, the figures highlight that when the computational complexity is reduced (for higher $\hat{\epsilon}_{comp}$), the sum-rate only decreases slightly for both SWF and SCC, i.e. the average sum-rate only decreases by 0.28 % for $\hat{\epsilon}_{comp} = 10\%$ and by 0.07 % for $\hat{\epsilon}_{comp} = 0.1\%$. Furthermore, Figure 3.6 shows that both schedulers are able to avoid computational outage, which would lead to reduced service performance otherwise.

So far, the systems have been dimensioned for a very low target computational outage, i.e. $\hat{\epsilon}_{comp} = 10^{-6}$. This has the drawback that such a system will be significantly over-provisioned and most of the time the allocated resources are under-utilized. By contrast, complexity-aware schedulers are allowed to avoid computational outage while

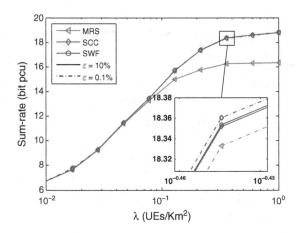

Figure 3.8 Mean sum-rate versus user density. ©IEEE 2015, reprinted by permission.

maintaining a high utilization efficiency. Figure 3.7 shows the mean sum-rate as a function of the number of BSs N_c processed at the central unit. The performance is shown for SWF and SCC as well as for max-rate scheduling (MRS), which does not account for the computational constraint; MRS always selects the UE with the highest MCS. The figure shows the impact of $\hat{\epsilon}_{comp}$ on the different schedulers. As can be seen, for all values of N_c the impact of the constraint on the computational resources is marginal for the computationally aware schedulers. As $\hat{\epsilon}_{comp}$ increases, the impact on a system which uses an MRS increases linearly with $\hat{\epsilon}_{comp}$ owing to the increasing computational outage. Furthermore, the magnification shows that there is only a marginal difference between SWF and the SCC, emphasizing the fact that the less complex SCC algorithm achieves almost the same performance as SWF.

Finally, Figure 3.8 shows the sum-rates for SWF, SCC, and MRS as functions of the density of UEs, for different computational outage values (fulfilled for $\lambda = 10^{-1}$ UEs/km²). The computationally aware schedulers are able to serve all the UEs while marginally sacrificing system throughput as the user density increases. Furthermore, the figure shows that the proposed scheduling algorithms are able to accommodate the increasing traffic demand, while the MRS suffers gradually from increasing computational outage at higher user densities.

3.6   Summary

This chapter has focused on the interaction of the data rates and the computational resources required for the uplink of a centralized RAN. It shows that there are significant fluctuations of complexity depending on the current channel quality experienced by a UE. In order to evaluate this dependency and to control the required computational resources, an analytical framework has been described that predicts of the occupied computational resources. In addition, this framework can be applied to a computationally aware scheduler that avoids computational outage at the cost of a marginal system throughput reduction. In a practical system, this scheduler allows for scaling the occupied computational resources of a centralized RAN in order to adapt to changing parameters within the network. Since such a resource scaling actually happens within a few seconds, this scheduler allows the avoidance of any serious service impact, such as call drops, during the transition period.

References

[1] R. Irmer, H. Droste, P. Marsch, M. Grieger, G. Fettweis, S. Brueck et al., "Coordinated multipoint: concepts, performance, and field trial results," *IEEE Commun. Mag.*, vol. 49, no. 2, pp. 102–111, February 2011.

[2] P. Rost, S. Talarico, and M. C. Valenti, "The complexity–rate tradeoff of centralized radio access networks," *IEEE Trans. Wireless Commun.*, vol. 14, no. 11, pp. 6164–6176, November 2015.

[3] 3GPP, "Evolved universal terrestrial radio access (e-utra) and evolved universal terrestrial radio access network (E-UTRAN); overall description; stage 2 (release 12)." 3GPP, Technical Report, September 2014.

[4] A. Domenico, M. D. Girolamo, M. Consonni, U. Salim, J. Ortin, A. Banchs et al., "D5.2: Final definition of iJOIN requirements and scenarios." EU FP7 project iJOIN, Technical Report, November 2014.

[5] S. Bhaumik, S. P. Chandrabose, M. K. Jataprolu, G. Kumar, A. Muralidhar, P. Polakos et al., "CloudIQ: a framework for processing base stations in a data center," in *Proc. 18th Int. Conf. on Mobile Computing and Networking*, Istanbul, August 2012.

[6] F. Kelly, *Notes on Effective Bandwidths*. Oxford University Press, 1996, pp. 141–168.

[7] D. Wuebben, P. Rost, J. Bartelt, M. Lalam, V. Savin, M. Gorgoglione et al., "Benefits and impact of cloud computing on 5G signal processing," *IEEE Signal Process. Mag.*, vol. 31, no. 6, pp. 35–44, November 2014.

[8] C. Shannon, "Mathematical theory of communication," *Bell System Tech. J.*, vol. 27, pp. 379–423 and 623–656, July and October 1948.

[9] 3GPP, "Radio access network; evolved universal terrestrial radio access; (e-utra); physical channels and modulation (release 10)." 3GPP, Technical Report, December 2012.

[10] J. Choi and S. Bahk, "Cell-throughput analysis of the proportional fair scheduler in the single-cell environment," *IEEE Trans. Vehi. Technology*, vol. 56, no. 2, pp. 766–778, March 2007.

[11] M. C. Valenti, S. Talarico, and P. Rost, "The role of computational outage in dense cloud-based centralized radio access networks," in *Proc. IEEE Global Conf. on Communications*, Austin, TX, December 2014.

[12] P. Grover, K. A. Woyach, and A. Sahai, "Towards a communication-theoretic understanding of system-level power consumption," *IEEE J. Sel. Areas Commun.*, September 2011.

[13] M. Abramowitz and I. A. Stegun, *Handbook of Mathematical Functions: With Formulas, Graphs, and Mathematical Tables*. Dover Publications, 1965.

[14] W. Yu and J. M. Cioffi, "On constant power water-filling," in *Proc. IEEE Int. Conf. on Communications*, Helsinki, June 2001.

[15] P. Rost, A. Maeder, M. Valenti, and S. Talarico, "Computationally aware sumrate optimal scheduling for centralized radio access networks," in *Proc. IEEE Global Conf. on Communication,* San Diego, CA, November 2015.

4 Cooperative Beamforming and Resource Optimization in C-RANs

Wei Yu, Pratik Patil, Binbin Dai, and Yuhan Zhou

Cloud radio access network (C-RAN) architecture offers two key advantages as compared with traditional radio access networks (RANs) from the physical-layer transmission point of view. First, the centralization and virtualization of RANs allow the coordination of base stations (BSs) across a large geographic area, thereby enabling coordinated physical-layer resource allocation across the BSs. The physical-layer resources here refer to the frequency, time, and spatial dimensions that can be utilized by radio transmission. Second, and more importantly, the C-RAN architecture also opens up the possibility of the joint transmission and joint reception of user signals across multiple BSs, thereby fundamentally addressing the issue of inter-cell interference. As interference is the main bottleneck in modern densely deployed wireless networks, the C-RAN architecture offers significant advantages in that it provides the possibility of interference mitigation leading to performance enhancement without the need for additional site and bandwidth acquisition.

This chapter provides an optimization framework for cooperative beamforming and resource allocation in C-RANs. We begin by identifying frequency, time, and spatial resources in wireless cellular networks and defining the overall spectrum allocation, scheduling, and beamforming problem in a cooperative network. The chapter then provides a network model for the C-RAN architecture and illustrates typical network objective functions and constraints for network utility maximization. A key characteristic of the C-RAN architecture is that the fronthaul connections between the cloud and the BSs may have limited capacities. One of the main goals of this chapter is to illustrate the impact of limited fronthaul capacity on the cooperative beamforming and resource allocation in C-RANs.

The chapter explores the optimization of the design variables associated with C-RANs, depending on the transmission strategies at the cooperative BSs. For the uplink C-RAN, we illustrate compress-forward as the main strategy at the BSs and focus on the impact of the choice of quantization noise levels at the BSs and possible joint transmit optimization strategies. For the downlink C-RAN, we compare a compression-based strategy and a data-sharing strategy and illustrate the problem formulation and solution strategy in both cases. Throughout the chapter, key optimization techniques for solving resource-allocation problems in C-RANs are presented.

4.1   C-RAN Model

In the C-RAN architecture, the baseband processing, traditionally performed locally at each BS, is aggregated and performed centrally at a cloud computing center. This is enabled by high-speed connections, referred to as *fronthaul* links, between the BSs and the cloud. Such centralized signal processing allows for the possibility of interference cancellation and interference pre-compensation across all the users in the uplink and downlink, respectively. The C-RAN architecture thus facilitates the implementation of network multiple-input multiple-output (network MIMO) [1], also known as coordinated multi-point (CoMP) or multi-cell processing (MCP) in the literature [2, 3]. The main focus of this chapter is on the interference mitigation capability enabled by C-RAN architecture. Toward this end we abstract a physical-layer channel model in order to allow an information-theoretic understanding of the capacity limits of the C-RAN model as compared with traditional RAN.

4.1.1   System Model

To highlight key benefits of the C-RAN architecture, we focus on the network topology of one central processor (CP) coordinating a cluster of BSs serving users over a certain geographic area, as illustrated in Fig. 4.1. The BSs in the C-RAN architecture are

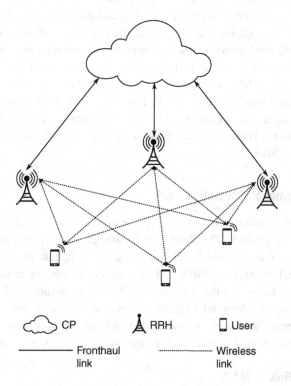

Figure 4.1 C-RAN system model.

also termed remote radio heads (RRHs) as their functionality is often restricted to the transmission and reception of radio signals. These RRHs are managed by the cloud-computing-based CP, which communicates with the RRHs via fronthaul links. The fronthaul connections can be dedicated fiber optic cables, or they can be wireless links. Although analog transport is a possible option, this chapter models the fronthaul links as finite-capacity noiseless digital links. Our aim is to understand the impact of limited fronthaul capacity on the overall system performance, and subsequently to design efficient transmission and relaying strategies that account for the limited available fronthaul capacity.

As a concrete setup, in this chapter we consider a C-RAN model consisting of a CP coordinating a total of L RRHs, each equipped with M antennas, serving K users each equipped with N antennas. The analysis developed in this chapter can be extended easily to the case with unequal numbers of antennas at different terminals. The main resources in the system are the fronthaul link capacities and the power budgets at the RRHs and at the users. We denote by C_l the capacity of the fronthaul link connecting RRH l to the CP. The power spectrum density constraint at the user k in the uplink is denoted as P_k^{ul} and that at RRH l in the downlink as P_l^{dl}. The precise uplink and downlink channel models are abstracted out in the next section for an information-theoretic study of the C-RAN architecture.

To enable signal-level cooperation for joint signal processing, it is crucial to be able to precisely synchronize the signals of different users. In this chapter, perfect synchronization among the RRHs in the downlink and among the users in the uplink is assumed. The impact of synchronization error in the context of uplink C-RANs is considered in [4]. In addition, instantaneous and perfect channel state information (CSI) is assumed to be available to all the RRHs and the users and also at the CP. In practice, the amount of CSI available is limited by the coherence time of the channel and the overhead of communicating CSI to the CP. The effect of partial CSI and channel estimation errors are taken into account in [5]. The main focus in this chapter is to illustrate the different fundamental transmission strategies in C-RAN and their interference mitigation capabilities.

4.1.2   Information-Theoretic Model

From an information-theoretic point of view, the C-RAN model is best understood as a relay network. The RRHs can be thought of as relays that facilitate communication between the CP and the mobile users. In the uplink, multiple users communicate with the CP through the RRHs and thus can be modeled as an instance of a multiple-access relay channel. In the downlink, the CP communicates with multiple users through the RRHs. The downlink C-RAN can thus be modeled as an instance of a broadcast relay channel. We assume frequency-flat channels for now. The difference from the case of frequency-selective channels is discussed later in the section.

Uplink C-RAN
Let $\mathbf{x}_k^{ul} \in \mathbb{C}^{N \times 1}$ be the signal transmitted from user k and $\mathbf{y}_l^{ul} \in \mathbb{C}^{M \times 1}$ be the received

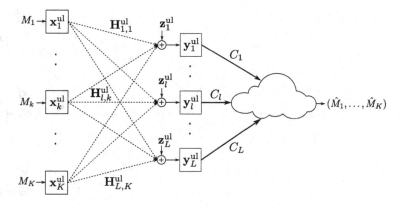

Figure 4.2 Information-theoretic uplink C-RAN model.

radio signal at RRH l. Assuming additive Gaussian noise at the RRH receivers, the channel response at RRH l can be modeled as

$$\mathbf{y}_l^{ul} = \sum_k \mathbf{H}_{l,k}^{ul} \mathbf{x}_k^{ul} + \mathbf{z}_l^{ul}, \tag{4.1}$$

where $\mathbf{H}_{l,k}^{ul} \in \mathbb{C}^{M \times N}$ is the channel from user k to RRH l and $\mathbf{z}_l^{ul} \in \mathbb{C}^{M \times 1} \sim \mathcal{CN}(\mathbf{0}, \sigma_{ul}^2 \mathbf{I})$ is the additive Gaussian noise. Figure 4.2 illustrates the uplink system model.

In traditional RAN, after receiving the radio signals each BS independently decodes the messages of its scheduled users, treating the combined signal from all other users as interference. In the C-RAN architecture, however, the RRHs have the flexibility to relay some information about their observed signals to the CP, which can then jointly process the information from all the RRHs for decoding. Joint processing has the advantage that the effect of inter-user interference can be mitigated. There are various different possible relaying strategies, depending on the information that the RRHs relay to the CP and the eventual decoding strategy. These strategies are discussed in detail in Section 4.2.

Downlink C-RAN

Let $\mathbf{x}_l^{dl} \in \mathbb{C}^{M \times 1}$ be the signal transmitted from RRH l. Assuming additive Gaussian noise, the received signal at user k, $\mathbf{y}_k^{dl} \in \mathbb{C}^{N \times 1}$, is represented as

$$\mathbf{y}_k^{dl} = \sum_l \mathbf{H}_{k,l}^{dl} \mathbf{x}_l^{dl} + \mathbf{z}_k^{dl}, \tag{4.2}$$

where $\mathbf{H}_{k,l}^{dl} \in \mathbb{C}^{N \times M}$ is the channel from RRH l to user k and $\mathbf{z}_k^{dl} \in \mathbb{C}^{N \times 1} \sim \mathcal{CN}(\mathbf{0}, \sigma_{dl}^2 \mathbf{I})$ is the additive Gaussian noise. Figure 4.3 illustrates the downlink system model.

In traditional RAN, the user messages from the core network are sent directly to the BSs, which independently encode the messages for users within the cells. As consequence, the transmit signals from the neighboring BSs interfere with each other. In contrast, in the C-RAN architecture, the fact that the CP has access to the messages of all the users enables joint encoding across the cooperating cluster, thereby allowing inter-cell interference to be mitigated. Depending on the specific ways that the CP utilizes

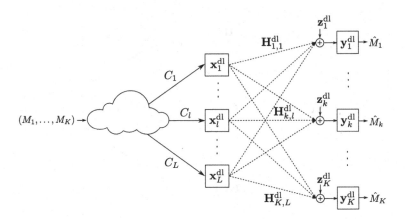

Figure 4.3 Information-theoretic downlink C-RAN model.

the capacity-limited fronthaul to enable joint encoding, different downlink strategies are possible. These strategies are discussed in more detail in Section 4.3.

4.1.3   Achievable Rate Region

The different transmission, relaying, and decoding strategies for both uplink and downlink result in different achievable rate tuples for the users. As multiple users share radio resources, an increase in user rate for one user usually comes at the cost of the rates of other users. The concept of a rate region captures this tradeoff. The rate region is defined as the set of all the achievable user rates, $\mathcal{R} = \{R_1, \ldots, R_K\}$. Given a transmission and relaying strategy in C-RAN, the rate region $\mathcal{R}$ is a function of the underlying channels $\mathbf{H}_{l,k}^{\mathrm{ul}}$, the fronthaul capacities C_l, and the power constraints P_k^{ul} in the uplink and similarly in the downlink. In allocating these resources to different users, a desirable operating point is to be chosen depending on the overall system objective. With that in mind, the overall goal of this chapter is to provide an optimization framework to maximize certain system objectives under achievable rate regions for different strategies and subsequently to point out the overall design insight obtained from such resource allocation perspective. Towards this end, we describe below the widely used system objective based on network utility considered in this chapter.

4.1.4   Network Utility Maximization

Network utility maximization is an optimization framework that takes into account the physical-layer tradeoffs in terms of the rate region as well as the application layer tradeoffs in terms of the varying usefulness of the rates of different users (e.g., the value of the additional rate increase for video application might be very different from that for file transfer). In the network utility maximization framework, each user has an utility $U_k(\bar{R}_k)$, as a function of its average user rate $\bar{R}_k$, that captures the value of having such a rate for user k. Most common utility functions are concave-increasing functions. Overall network utility maximization is the problem of maximizing the sum of the utilities over all the users in the system, with respect to operating parameters such as scheduling, beamforming, and quantization.

More specifically, the network utility maximization problem considered in this chapter aims to solve the following problem:

$$\text{maximize} \quad \sum_{k=1}^{K} U_k(\bar{R}_k) \tag{4.3a}$$

$$\text{s.t.} \quad (R_1, \ldots, R_K) \in \mathcal{R}, \tag{4.3b}$$

where the objective function depends on the average user rates $\bar{R}_k$, while the optimization parameters affect the instantaneous rate R_k. The average rate is usually computed in a windowed fashion. For example, with exponential weighting the average rate is obtained as

$$\bar{R}_k^{\text{updated}} = (1 - \alpha)\bar{R}_k^{\text{prior}} + \alpha R_k, \tag{4.4}$$

where $\bar{R}_k^{\text{prior}}$ is the average rate prior to the present time slot and R_k is the instantaneous rate of the current time slot. The above optimization problem is solved repeatedly for each time slot under the rate-region constraint on the instantaneous rates, until the average user rates eventually converge.

A common user utility function U_k is the logarithm function, i.e., $U_k(\bar{R}_k) = \log(\bar{R}_k)$. Under such a logarithmic utility function and exponentially weighted rate averaging, and assuming that the new contribution to the instantaneous rate αR_k is small, optimization of the network utility objective function can be achieved approximately by a maximization of the instantaneous *weighted* sum-rate, with the weights updated as the inverses of the average rates, as follows:

$$\text{maximize} \quad \sum_{k=1}^{K} w_k R_k \tag{4.5a}$$

$$\text{s.t.} \quad (R_1, \ldots, R_K) \in \mathcal{R}, \tag{4.5b}$$

where $w_k = 1/\bar{R}_k^{\text{prior}}$. The above weighted sum-rate maximization problem is solved for each time slot over the transmission strategies, with weights updated after each iteration. This transmit optimization problem under a logarithmic utility function is known as the proportionally fair resource-allocation problem. The rest of this chapter focuses on this weighted sum-rate maximization problem for C-RAN.

We remark that the log-utility is not the only possible choice of utility function. For delay-sensitive applications, it is often desirable to maximize the minimum rate or to guarantee a minimum rate while maximizing the sum-rate. Different choices of utility functions would lead to different optimization formulations.

4.1.5   Resource Allocation Problem

The resource-allocation problem for C-RAN consists of solving the above optimization problem over the operating parameters and under the system constraints. The operating parameters to be optimized can include not only cellular transmission parameters such as scheduling (i.e., to which users a non-zero rate should be assigned), beamforming, bandwidth, and power allocation but also relay strategies such as quantization noise levels in the context of C-RAN. The system constraints are the fronthaul link capacities and the transmit power spectral density constraints at the users for the uplink and at the RRHs for the downlink.

4.1.6   Disjoint versus User-Centric Clustering

While defining the system model for C-RAN, we have implicitly assumed that the RRHs are clustered into disjoint clusters and that the RRHs within each cluster serve the users in the cluster cooperatively. Such a model has explicit cluster boundaries, and the users near the cluster boundaries still suffer from inter-cluster interference. One way to further reduce inter-cluster interference is to let each user form a user-centric cluster of RRHs. Different clusters for different users may overlap in this case, and there are no explicit cluster boundaries. Such user-centric clustering typically improves the fairness in the system.

4.1.7   Frequency-Selective Channels

The chapter mainly considers a frequency-flat channel model, but wireless channels are often frequency selective. In this case, one can employ an orthogonal frequency division multiplex (OFDM) to divide the total bandwidth into a number of flat subchannels. Then each subchannel can independently employ the relay strategies for the frequency-flat channel model considered in this chapter.

The OFDMA-based C-RAN presents an additional dimension for resource allocation, namely among the frequency subchannels. This includes the assignment of the subchannels to the different users and the allocation of fronthaul capacities as well as transmit power among the different subchannels. Some initial work on resource allocation for C-RANs employing OFDM has been carried out in [6] under certain simplifying modeling assumptions.

4.2   Uplink C-RAN

The ability to manage interference is one of the main benefits of the C-RAN architecture. In the uplink, different users in the cluster communicate their messages to the CP through RRHs. The RRHs, instead of decoding the messages locally, can relay information about their observations to the CP for centralized processing. This enables an interference mitigation capability for the uplink C-RAN.

In the ideal case, where the fronthaul links have infinite capacities, the RRHs can convey their exact observations to the CP. The resulting channel model reduces to a MIMO multiple-access channel. Practical systems, however, have capacity-constrained fronthaul links. This limits the amount of information that the RRHs can relay. A key question is then to decide what information about the observed signals is most useful at the CP, so as to enable as much interference cancellation as possible.

This section discusses different strategies for relaying and centralized processing in the uplink C-RAN and then formulates their respective resource optimization problems and indicates methods to solve these problems. We provide key insights obtained from such optimization throughout the chapter.

4.2.1 Compress, Decode, versus Compute-Forward

From the perspective of maximizing the overall capacity of the network, the aim of an RRH should be to preserve as much information as possible in relaying its observation to the CP under the finite fronthaul capacity constraint. A natural strategy is for the RRHs to describe the observed signals by compressing the received analog signals and relaying their digital representations to the CP [7–9]. The resolution of the compression determines the amount of fronthaul capacity needed. Higher fronthaul capacity leads to lower quantization noise, which in turn leads to higher achievable user rates. At the CP, the user messages can be jointly decoded on the basis of the compression indices received from all the RRHs in the cluster. Such joint processing at the CP enables effective interference cancellation. This relaying strategy is known as *compress-forward* in the literature. Note that the compress-forward strategy also inevitably forwards some part of the receiver noise at the RRHs to the CP.

There are different ways of performing compression and decompression depending on whether some side information is utilized in the compression process, leading to either independent or Wyner–Ziv compression strategies. There are also different ways of performing decoding at the CP, depending on how the user messages and the compression codewords are decoded successively. These possibilities are discussed in detail in the next section. We mention here that, in theory, there is also the possibility of performing decompression and message decoding at the CP jointly [3]. Doing so is in fact information-theoretically better justified, but it also has very high complexity and is impractical to implement. For this reason, in this chapter we restrict consideration to the successive-decoding type of strategy.

As an alternative to compress-forward, some RRHs can attempt to decode the messages of the users closest to them and relay the messages themselves (rather than the compressed version of their observations) to the CP. The users being decoded at the RRHs cannot benefit from the joint processing capabilities of C-RAN, but these decoded messages can help the decoding of other users at the CP. This type of relaying strategy can be broadly referred to as a version of *decode-forward*.

Finally, the RRHs may opt to decode some linear combination of the user messages, or more generally some function of the user messages, and forward the result to the CP. This is called the *compute-forward* strategy [10]. In compute-forward, the users choose the transmit codewords from a structured lattice codebook. The benefit of using structured codebook is that linear integer combinations of different codewords are still codewords. After receiving the signals, the RRHs compute functions of the user codewords from the received signal. Typically, functions that closely mimic the channel output at the RRHs are those that give the best computation rate. The indices corresponding to the function values are sent over the fronthaul links. After receiving all such function values, the CP inverts the functions to recover the original user messages.

The main advantage of decode-forward and compute-forward is that they eliminate noise at the RRHs. In practice, however, there are only a limited number of scenarios in which they outperform compress-forward. Further, compute-forward is quite sensitive to channel estimation error [11]. With this in mind, this chapter mostly focuses on the compress-forward strategy. We refer the reader to [12] for details regarding the achievable rate region and network optimization for the compute-forward strategy in the context of uplink C-RAN.

The use of compress-forward for C-RAN can also be justified from information-theoretic considerations. For a Gaussian multi-message multicast network, it can be shown that compress-forward (and its variations, called quantize-map-forward [13] and noisy network coding [14]) can achieve the information-theoretic capacity of the network to within a constant gap which depends only on the network topology and is independent of other channel parameters.

The rest of this section focuses on compress-forward as the main relaying strategy for uplink C-RAN and presents variants and their corresponding achievable rates and resource optimization.

4.2.2 Compress-Forward Strategy

In the compress-forward strategy, the received signals $\mathbf{y}_l^{ul}$ are compressed at the RRHs, and the compression indices are sent to the CP. The CP then decodes the original user signals $\mathbf{x}_k^{ul}$ from these indices.

There are different ways of performing compression at the RRHs and different ways of decoding the user messages at the CP, leading to variations of the compress-forward strategy. The two main compression methods are *independent compression* and *Wyner–Ziv compression*. In independent compression, the observations at the RRHs are compressed and decompressed independently. In Wyner–Ziv compression it is possible to take advantage of the fact that the observed signals at the RRHs are correlated, in order to reduce the amount of fronthaul capacity needed.

The processing at the CP can also take different forms. For example, after decoding the compression codewords, the CP may perform linear beamforming across the RRH signals for the independent decoding of user messages, or the CP may perform successive interference cancellation (SIC). Alternatively, the CP may even interleave the decoding of user messages and compression codewords, using the decoded user messages as side information in subsequent processing.

To characterize the achievable rates and the required fronthaul capacities for the different compress-forward strategies, we model the user transmission and the compression process below. These models are based on information-theoretic considerations; they provide accurate yet simplified rate expressions for variants of the compress-forward strategy.

We assume that the input signals $\mathbf{x}_k^{ul}$ at the users are chosen according to a Gaussian codebook. While the choice of a Gaussian-like input is not necessarily optimal for the compress-forward strategy [7], it makes the evaluation of the rate region tractable. Let $\mathbf{U}_k \in \mathbb{C}^{N \times d_k}$ denote the transmit beamformer that user k utilizes to transmit the message signal $\mathbf{s}_k^{ul} \in \mathbb{C}^{d_k \times 1} \sim \mathcal{CN}(\mathbf{0}, \mathbf{I})$ to the CP. Here d_k denotes the number of data streams per user k. The transmit signal at user k is then given by $\mathbf{x}_k^{ul} = \mathbf{U}_k \mathbf{s}_k^{ul}$ with covariance matrix $\mathbb{E}[\mathbf{x}_k^{ul}(\mathbf{x}_k^{ul})^H] = \mathbf{U}_k \mathbf{U}_k^H$. The total transmission power consumed at user k is expressed as $\mathrm{Tr}(\mathbf{U}_k \mathbf{U}_k^H)$. With the linear Gaussian channel model as described earlier in the chapter, the received signal at RRH l in the uplink can thus be expressed as

$$\mathbf{y}_l^{ul} = \sum_k \mathbf{H}_{l,k}^{ul} \mathbf{U}_k \mathbf{s}_k^{ul} + \mathbf{z}_l^{ul}. \tag{4.6}$$

For the compression process, we again assume a Gaussian codebook. Let $\hat{\mathbf{y}}_l^{ul}$ denote the compressed signal for RRH l. Then, the quantization process at RRH l is modeled as the addition of independent Gaussian quantization noises, as follows:

$$\hat{\mathbf{y}}_l^{ul} = \mathbf{y}_l^{ul} + \mathbf{q}_l^{ul}, \tag{4.7}$$

where $\mathbf{q}_l^{ul} \in \mathbb{C}^{M \times 1} \sim \mathcal{CN}(\mathbf{0}, \mathbf{Q}_l^{ul})$ and $\mathbf{Q}_l^{ul}$ is the covariance matrix of the quantization noise in the compressed signal corresponding to the RRH l. We point out that, even though it may seem that a more general linear additive model for compression is to first process the received signal $\mathbf{y}_l^{ul}$ using a transformation matrix $\mathbf{A}_l$ and then compress the resulting transformed output $\mathbf{A}_l \mathbf{y}_l^{ul}$ (perhaps of even lower dimension than $\mathbf{y}_l^{ul}$), with an appropriate choice of $\mathbf{Q}_l^{ul}$ the model in (4.7) can be shown to be equivalent to such a linear model and is therefore without loss of generality.

4.2.3   Compression Strategies

The full benefit of joint processing, in terms of its interference cancellation capability, would be achieved if each RRH were able to convey the exact $\mathbf{y}_l^{ul}$ to the CP. In practice, the more accurately the compressed signal $\hat{\mathbf{y}}_l^{ul}$ resembles the actual received signals $\mathbf{y}_l^{ul}$ at the RRHs, higher the achievable rate would be for the overall network. There is, however, a cost for transmitting high-fidelity versions of the $\mathbf{y}_l^{ul}$ through the digital fronthaul link. This cost can be modeled using information-theoretic rate-distortion theory.

The rate–distortion tradeoff can be most easily understood in terms of the quantization noise $\mathbf{q}_l^{ul}$ introduced in the compression process. On one hand, the quantization noise level directly provides an indication of the accuracy of $\hat{\mathbf{y}}_l^{ul}$; it enters the achievable rate expression as additional noise introduced by the quantization process. On the other hand, the level of the quantization noise indicates the amount of fronthaul capacity needed for compression. Higher fronthaul capacity leads to better compression resolution and smaller quantization noise. The precise relationship between fronthaul capacity and quantization noise can be understood via rate–distortion theory, as follows. Consider a single RRH l. In order to keep the statistical variance of the quantization noise to a certain level $\mathbf{Q}_l^{\mathrm{ul}}$, the amount of fronthaul capacity needed must satisfy

$$C_l^{\mathrm{indep,\,ul}} \geq I\left(\mathbf{y}_l^{\mathrm{ul}}; \hat{\mathbf{y}}_l^{\mathrm{ul}}\right) \tag{4.8}$$

$$= \log \frac{\left|\sum_{k=1}^{K} \mathbf{H}_{l,k}^{\mathrm{ul}} \mathbf{U}_k \mathbf{U}_k^H (\mathbf{H}_{l,k}^{\mathrm{ul}})^H + \sigma_{\mathrm{ul}}^2 \mathbf{I} + \mathbf{Q}_l^{\mathrm{ul}}\right|}{\left|\mathbf{Q}_l^{\mathrm{ul}}\right|}. \tag{4.9}$$

As expected, the above is a decreasing function of $\mathbf{Q}_l^{\mathrm{ul}}$. The superscript 'indep' refers to the fact that the quantization process is done independently for each RRH without utilizing any potential side information at the CP.

The above fronthaul rate can be improved using a more sophisticated compression technique that utilizes the fact that signals received at different RRHs are often highly correlated, as they come from the same set of user messages. Thus, once some of the quantization codewords are decoded, they can serve as side information in the subsequent decoding of other quantization codewords. As a result, the fronthaul capacity needed for compression can be reduced. This compression technique is referred to as Wyner–Ziv compression. Assuming that the compressed signals from RRHs are recovered in the order 1 to L, the fronthaul capacity required for the Wyner–Ziv compression of the received signal at RRH l is given as follows:

$$C_l^{\mathrm{WZ,ul}} \geq I\left(\mathbf{y}_l^{\mathrm{ul}}; \hat{\mathbf{y}}_l^{\mathrm{ul}} | \hat{\mathbf{y}}_1^{\mathrm{ul}}, \dots, \hat{\mathbf{y}}_{l-1}^{\mathrm{ul}}\right) \tag{4.10}$$

$$= \log \frac{\left|\sum_{k=1}^{K} \mathbf{H}_{1:l,k}^{\mathrm{ul}} \mathbf{U}_k \mathbf{U}_k^H (\mathbf{H}_{1:l,k}^{\mathrm{ul}})^H + \sigma_{\mathrm{ul}}^2 \mathbf{I}_{1:l} + \mathbf{Q}_{1:l}^{\mathrm{ul}}\right|}{\left|\sum_{k=1}^{K} \mathbf{H}_{1:l-1,k}^{\mathrm{ul}} \mathbf{U}_k \mathbf{U}_k^H (\mathbf{H}_{1:l-1,k}^{\mathrm{ul}})^H + \sigma_{\mathrm{ul}}^2 \mathbf{I}_{1:l-1} + \mathbf{Q}_{1:l-1}^{\mathrm{ul}}\right|} - \log |\mathbf{Q}_l^{\mathrm{ul}}|. \tag{4.11}$$

Here and throughout the rest of this section, the notation $\mathbf{H}_{\mathcal{S},\mathcal{T}}^{\mathrm{ul}}$ denotes the channel matrix from the users in the set $\mathcal{T}$ to the RRHs in the set $\mathcal{S}$, $\mathbf{Q}_{\mathcal{S}}^{\mathrm{ul}}$ denotes the block-diagonal matrix formed from the quantization covariance matrices of the RRHs belonging to the set $\mathcal{S}$, and $1:l$ denotes the set $\{1, \dots, l\}$. In the mutual information expression above, because the signals already recovered at the CP $\hat{\mathbf{y}}_1^{\mathrm{ul}}, \dots, \hat{\mathbf{y}}_{l-1}^{\mathrm{ul}}$ can serve as side information they can be included in the conditioning in order to reduce the fronthaul rate for the compression at RRH l. We remark that the above compression rates are the information-theoretic limit for compression with side information. The practical implementation of Wyner–Ziv compression is non-trivial.

We further remark that, in the case where some user messages are decoded before the compressed signals for some other subset of RHH signals are recovered, we can include the decoded user messages as side information in the decompression process in order to further lower the fronthaul capacity requirements for these RRHs.

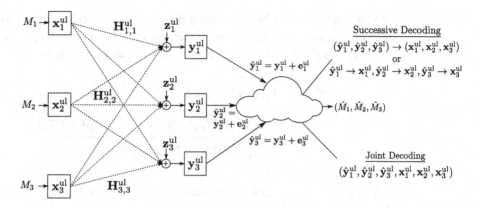

Figure 4.4 Illustration of compress-forward strategies for uplink C-RAN.

4.2.4 Decoding Strategies

The goal of the CP in compress-forward in the uplink C-RAN is to decode the user messages on the basis of the compression indices sent from the RRHs. The CP has various options for decoding user messages. The CP can choose to first recover all the compressed signals at the RRHs and then subsequently decode the user messages from on the compressed versions of the received signals. Alternatively, the CP can arbitrarily interleave the decoding of the user messages and the compression codewords. Doing so can benefit the users decoded later in the process at the expense of earlier users. The achievable rates of these various options are discussed in detail in this section.

In the first option, the CP first recovers the compressed signals $\hat{\mathbf{y}}_l^{\mathrm{ul}}$ from all the RRHs and then uses these compressed signals to decode the user messages, which are encoded in $\mathbf{x}_k^{\mathrm{ul}}$. Such a successive decoding strategy essentially converts the uplink C-RAN setup into a virtual multiple-access model, with the CP receiving codewords $(\hat{\mathbf{y}}_1^{\mathrm{ul}}, \dots, \hat{\mathbf{y}}_L^{\mathrm{ul}})$ for decoding the user messages. The achievable rate region of this successive-decoding strategy thus resembles the rate region of a multiple-access channel with additional quantization noise.

For example, with all the compression codewords $\hat{\mathbf{y}}_l^{\mathrm{ul}}$ already decoded, the decoding of the $\hat{\mathbf{x}}_k^{\mathrm{ul}}$ can be done independently for each user, resulting in the following achievable rate region:

$$R_k^{\mathrm{linear,ul}} \le I\left(\mathbf{x}_k^{\mathrm{ul}}; \hat{\mathbf{y}}_1^{\mathrm{ul}}, \dots, \hat{\mathbf{y}}_L^{\mathrm{ul}}\right) \tag{4.12}$$

$$= \log \frac{\left|\sum_{j=1}^K \mathbf{H}_{1:L,j}^{\mathrm{ul}} \mathbf{U}_j \mathbf{U}_j^H (\mathbf{H}_{1:L,j}^{\mathrm{ul}})^H + \sigma_{\mathrm{ul}}^2 \mathbf{I} + \mathbf{Q}_{1:L}^{\mathrm{ul}}\right|}{\left|\sum_{j \ne k} \mathbf{H}_{1:L,j}^{\mathrm{ul}} \mathbf{U}_j \mathbf{U}_j^H (\mathbf{H}_{1:L,j}^{\mathrm{ul}})^H + \sigma_{\mathrm{ul}}^2 \mathbf{I} + \mathbf{Q}_{1:L}^{\mathrm{ul}}\right|}. \tag{4.13}$$

In writing down the above achievable rate region, we have implicitly assumed that a linear minimum-mean-squared-error (MMSE) network-wide beamforming is performed across the signals received from the RRHs. The above rate region therefore already includes the capability of inter-RRH interference cancellation to a certain extent.

The rate region (4.13) can be improved if successive interference cancellation (SIC) is implemented across the users. In particular, assuming that user messages are decoded in the order 1 to K, the SIC achievable rate for user k can be written as

$$R_k^{\text{SIC,ul}} \le I\left(\mathbf{x}_k^{\text{ul}}; \hat{\mathbf{y}}_1^{\text{ul}}, \dots, \hat{\mathbf{y}}_L^{\text{ul}} | \mathbf{x}_1^{\text{ul}}, \dots, \mathbf{x}_{k-1}^{\text{ul}}\right) \tag{4.14}$$

$$= \log \frac{\left| \sum_{j=k}^{K} \mathbf{H}_{1:L,j}^{\text{ul}} \mathbf{U}_j \mathbf{U}_j^H (\mathbf{H}_{1:L,j}^{\text{ul}})^H + \sigma_{\text{ul}}^2 \mathbf{I} + \mathbf{Q}_{1:L}^{\text{ul}} \right|}{\left| \sum_{j>k} \mathbf{H}_{1:L,j}^{\text{ul}} \mathbf{U}_j \mathbf{U}_j^H (\mathbf{H}_{1:L,j}^{\text{ul}})^H + \sigma_{\text{ul}}^2 \mathbf{I} + \mathbf{Q}_{1:L}^{\text{ul}} \right|}. \tag{4.15}$$

Note that the above achievable rate reduces to the successive decoding rate region of a multiple-access channel if the quantization noise is ignored.

Alternatively, instead of recovering all the compressed signals before decoding any user messages, the decoding can also be done on a per-RRH basis [15]. More specifically, once the compressed signals from RRH l, $\hat{\mathbf{y}}_l^{\text{ul}}$, are recovered, the messages of the users associated with that RRH can be decoded immediately. Such decoding resembles the traditional per-BS decoding except that, since the decoding of all users is done centrally at the CP, previously decoded user messages can serve as side information in subsequent decoding, so their interference can be subtracted. Assuming that $K = L$ and that user k is associated with RRH k, the achievable rate for user k in this case can be written as

$$R_k^{\text{perRRH,ul}} \le I\left(\mathbf{x}_k^{\text{ul}}; \hat{\mathbf{y}}_k^{\text{ul}} | \mathbf{x}_1^{\text{ul}}, \dots, \mathbf{x}_{k-1}^{\text{ul}}\right) \tag{4.16}$$

$$= \log \frac{\left| \sum_{j=k}^{K} \mathbf{H}_{k,j}^{\text{ul}} \mathbf{U}_j \mathbf{U}_j^H (\mathbf{H}_{k,j}^{\text{ul}})^H + \sigma_{\text{ul}}^2 \mathbf{I} + \mathbf{Q}_k^{\text{ul}} \right|}{\left| \sum_{j>k} \mathbf{H}_{k,j}^{\text{ul}} \mathbf{U}_j \mathbf{U}_j^H (\mathbf{H}_{k,j}^{\text{ul}})^H + \sigma_{\text{ul}}^2 \mathbf{I} + \mathbf{Q}_k^{\text{ul}} \right|}. \tag{4.17}$$

Note that the above rate expression for per-RRH decoding can be further improved by including the compressed signals of the RRHs recovered before the RRH k in the conditioning in the mutual information expression. Moreover, the Wyner–Ziv compression rate (4.11) can also benefit from the conditioning of the user signals already decoded before user k, in per-RRH decoding.

We remark that the main benefit of C-RAN, namely inter-RRH interference mitigation, is achieved in the uplink through beamforming, i.e., the decoding of user messages on the basis of the received signals across multiple RRHs, or through SIC, i.e., the previously decoded user messages serve as side information for subsequent decoding, or through both. In general, the benefit of network-wide beamforming is more important than that of successive decoding alone as in per-RRH SIC. This is so because per-RRH SIC necessarily requires some users to be decoded first; these users therefore cannot benefit from centralized processing. The largest achievable rates are obtained if both beamforming and SIC are implemented. With this in mind, the rest of this section focuses on the achievable rates involving network-wide beamforming, i.e., either $R_k^{\text{linear,ul}}$ in (4.13) or $R_k^{\text{SIC,ul}}$ in (4.15).

4.2.5   Optimization Framework for Compress-Forward

Within the framework of network utility maximization, the optimization of the compress-forward strategy for uplink C-RAN is essentially involves solving a weighted-sum-rate maximization problem (4.5a) over the transmission and relaying strategies. The underlying optimization variables are the user scheduling, user transmit powers, and beamformers, and the quantization codebook – constrained by the input power and fronthaul capacity constraints.

User scheduling is usually determined by network-layer protocols as a function of user priorities, traffic delay constraints, and also physical-layer channel conditions. While in theory user scheduling should be included in the weighted sum-rate maximization, doing so rigorously is often difficult especially when there are a large number of potential users in the system. In practice, it is often desirable to use an heuristic approach that combines user traffic demand with channel strength considerations to schedule users. For example, users with longer queues of data to transmit should be scheduled first; users with stronger channels should be given priority; grouping users with near-orthogonal channels to the cluster of RRHs is a sensible strategy.

When successive decoding of the user signals and the compressed signals (in the cases of SIC and Wyner–Ziv compression, respectively) is implemented, the decoding orders are additional variables to be optimized. Exhaustive searches for a C-RAN cluster of K users and L RRHs would involve $K!$ user orderings and $L!$ RRH orderings, respectively, which is clearly impractical, but sensible heuristic strategies often exist. For example, for SIC, users with strong channels should usually be decoded first in order to help the weak users and to improve fairness. For maximizing the weighted sum-rate, normally the SIC user decoding order should be chosen to be in ascending order of the user priority weights. Likewise, good heuristic ordering for Wyner–Ziv compression is also possible. For example, in [16] it was proposed that the signals from those RRHs with either a higher value of the fronthaul capacity or a lower value of the average received signal power should be decompressed first. The rationale here is that the already decompressed signals can serve as side information for subsequent decompression, so this ordering helps balance the effective quantization noise levels across the RRHs.

To simplify the problem, we now fix the set of users to be scheduled, and fix the order in which the users decoding is performed. Without loss of generality, assume that the user signals are decoded in order from 1 to K. Similarly, in the case of Wyner–Ziv compression, assume that the signals from RRHs are decompressed in order from 1 to L. In this case, the joint transmitter and quantization noise covariance optimization problem can be formulated as follows:

$$\underset{\mathbf{U}_k, \mathbf{Q}_l^{\text{ul}}}{\text{maximize}} \sum_{k=1}^{K} w_k R_k^{\text{ul}} \tag{4.18a}$$

$$\text{s.t.} \quad R_k^{\text{ul}} = (4.13) \quad \text{or} \quad R_k^{\text{ul}} = (4.15), \quad \forall k, \tag{4.18b}$$

$$(4.9) \quad \text{or} \quad (4.11) \quad \leq C_l, \quad \forall l, \tag{4.18c}$$

$$\text{Tr}\left(\mathbf{U}_k \mathbf{U}_k^H\right) \leq P_k^{\text{ul}}, \quad \forall k, \tag{4.18d}$$

$$\mathbf{Q}_l^{\text{ul}} \succeq \mathbf{0}, \quad \forall l. \tag{4.18e}$$

Here the w_k are the priority weights in the weighted sum-rate maximization framework. The optimization has two sets of design variables, the transmit beamformer for user k, $\mathbf{U}_k$, which is constrained by the power budget, and the quantization noise covariance matrix for RRH l, $\mathbf{Q}_l^{\text{ul}}$, which is constrained by the fronthaul capacity. This optimization problem is non-convex; it is in general challenging to find its global optimum solution.

In formulating the above optimization problem, we have implicitly assumed that both the transmit strategy at the user side and also the compression process at the RRHs can be done adaptively, in the sense that the users can adaptively choose their transmit power level, beamformers, and rate and the RRHs can adaptively choose different quantization codebooks, according to the network condition. While transmit optimization is invariably included in modern cellular networks, adaptive quantization may not be included. In the analysis below we discuss adaptive quantization noise optimization first and then transmit beamforming and power optimization.

4.2.6   Optimization of Quantization at RRHs

In this section, we analyze the quantization-noise-optimization component of problem (4.18). To illustrate the key ideas, we first consider one instance of the optimization problem (4.18) with independent compression and the successive decoding of user messages ordered according to the user priority weights (i.e., we assume that $w_1 \leq \cdots \leq w_K$). Similar analysis can be obtained for Wyner–Ziv coding with linear MMSE beamforming. Denote by $\mathbf{\Sigma}_k = \mathbf{U}_k \mathbf{U}_k^H$ the transmit signal covariance matrix for user k. The weighted sum-rate maximization problem thus becomes

$$\underset{\mathbf{\Sigma}_k, \mathbf{Q}_l^{\text{ul}}}{\text{maximize}} \quad \sum_{k=1}^{K} w_k \log \frac{\left| \sum_{k=1}^{K} \mathbf{H}_{1:L,k}^{\text{ul}} \mathbf{\Sigma}_k (\mathbf{H}_{1:L,k}^{\text{ul}})^H + \sigma_{\text{ul}}^2 \mathbf{I} + \mathbf{Q}_{1:L}^{\text{ul}} \right|}{\left| \sum_{j>k} \mathbf{H}_{1:L,j}^{\text{ul}} \mathbf{\Sigma}_j (\mathbf{H}_{1:L,j}^{\text{ul}})^H + \sigma_{\text{ul}}^2 \mathbf{I} + \mathbf{Q}_{1:L}^{\text{ul}} \right|} \tag{4.19a}$$

$$\text{s.t.} \quad \log \frac{\left| \sum_{k=1}^{K} \mathbf{H}_{l,k}^{\text{ul}} \mathbf{\Sigma}_k (\mathbf{H}_{l,k}^{\text{ul}})^H + \sigma_{\text{ul}}^2 \mathbf{I} + \mathbf{Q}_l^{\text{ul}} \right|}{\left| \mathbf{Q}_l^{\text{ul}} \right|} \leq C_l, \quad \forall l; \tag{4.19b}$$

$$\text{Tr}\,(\mathbf{\Sigma}_k) \leq P_k^{\text{ul}}, \quad \forall k; \tag{4.19c}$$

$$\mathbf{Q}_l^{\text{ul}} \succeq \mathbf{0}, \quad \forall l. \tag{4.19d}$$

First we focus on the optimization over $\mathbf{Q}_l^{\text{ul}}$ with fixed $\mathbf{\Sigma}_k$. The main difficulty in solving the above optimization problem stems from the fact that the objective function is not concave and the fronthaul capacity constraints are not convex functions of $\mathbf{Q}_l^{\text{ul}}$. A method based on successive convex approximation (SCA) is proposed in [16] to solve this problem. The basic idea behind SCA is first to approximate the original problem into a convex program by linearizing the non-convex parts in the objective function and the constraints at a suitable starting point. Then, after solving the convex program, a new convex approximation is made around the updated solution from the previous iteration. This procedure is iterated until convergence and can be shown to reach the local optimum of the original optimization problem.

The optimal solution $\mathbf{Q}_l^{\text{ul}}$ obtained from the procedure above comprises a set of positive semidefinite matrices. These optimized quantization-noise covariance matrices can be implemented using an architecture where the received vector signal at the RRH is first beamformed; this is followed by compression across the components of the resulting signal. Assuming the eigenvalue decomposition $\mathbf{Q}_l^{ul} = \mathbf{A}_l^H \mathbf{\Lambda}_l \mathbf{A}_l$, where $\mathbf{A}_l$ is a unitary matrix and $\mathbf{\Lambda}_l$ is a diagonal matrix, the quantization process with $\mathbf{Q}_l^{ul}$ is equivalent to first beamforming $\mathbf{y}_l^{ul}$ with $\mathbf{A}_l$ and then performing compression across each element of the newly beamformed vector $\mathbf{A}_l \mathbf{y}_l^{ul}$. The diagonal entries in $\mathbf{\Lambda}_l$ represent the quantization noise levels in each resulting component. If some of these noise levels in the optimal $\mathbf{\Lambda}_l$ are nearly infinite, this implies that those corresponding components are not useful for decoding at the CP, in which case the effective beamforming matrix essentially projects the received signal at the RRH into a lower-dimensional space.

We remark that the optimized beamformers $\mathbf{A}_l$ and the quantization noise levels $\mathbf{\Lambda}_l$ depend on the channels $\mathbf{H}_{l,k}^{ul}$ and the transmit beamformers $\mathbf{U}_k$, which often change as the user scheduling, user locations, etc. change. To implement jointly optimized transmission and quantization therefore requires an adaptive compression architecture at RRHs that dynamically adapts to the changing transmission and channel parameters. There is, however, a special case where such adaptive design is not necessary. For a high signal-to-quantization-noise ratio and assuming that as many users as the total number of RRH antennas are scheduled, then adopting uniform quantization noise levels across the antennas, i.e., setting $\mathbf{Q}_l^{ul} = \gamma_l \mathbf{I}$, can be shown to be a reasonable strategy for maximizing the sum-rate [16]; the proportionality constant γ_l is chosen to satisfy the fronthaul capacity constraint at RRH l. Thus, under this special condition, adaptive quantization at the RRHs is not needed; independent quantization on a per-antenna basis is already an approximately optimal design.

4.2.7   Fronthaul-Aware Transmit Beamforming

We now address the optimization of transmit beamforming in fronthaul capacity-limited uplink C-RANs. Consider again the optimization problem (4.19), but over the transmit covariance matrices $\mathbf{\Sigma}_k$. If we assume that the quantization-noise covariance matrices $\mathbf{Q}_l^{ul}$ are fixed then the maximization of the weighted sum-rate subject to the input power constraints resembles a conventional MIMO multiple-access channel input optimization problem, but with additional quantization noise $\mathbf{Q}_l^{ul}$.

For the optimization problem (4.19) it is assumed that SIC is implemented. The objective function in this case is concave in the transmit covariance matrices $\mathbf{\Sigma}_k$, and the problem can be solved using efficient convex optimization methods. When linear MMSE receive beamforming is implemented, the optimization problem is non-convex but a class of algorithms known as weighted minimum-mean-square error (WMMSE) algorithms [17] are well suited for this scenario. The WMMSE algorithm is capable of reaching a locally optimal transmit beamforming solution for the problem.

The above discussion assumes that the quantization-noise covariance matrices $\mathbf{Q}_l^{ul}$ are fixed. In the general case, where the transmit covariance matrices $\mathbf{\Sigma}_k$ and the quantization-noise covariance matrices $\mathbf{Q}_l^{ul}$ are optimized jointly, a method called

WMMSE-SCA, which incorporates SCA into the WMMSE algorithm, can be used to arrive at a stationary point of the weighted sum-rate maximization problem [16].

We conclude this section by pointing out the importance of being fronthaul aware when designing transmit beamformers, particularly for the heterogeneous C-RAN architecture, where the fronthaul capacities of different RRHs can be quite different. Transmit beamforming serves to steer the radio transmission in certain spatial directions. Intuitively, if certain RRHs have more limited fronthaul capacities than others then the beamformers should steer away from them and instead point toward RRHs with higher fronthaul capacities.

In fact, as the joint optimization framework of the transmit and quantization noise covariance matrices for the uplink C-RAN model shows, for optimized performance the transmit beamformers should adapt to the quantization noise levels, and conversely the quantization noise levels should also adapt to the transmit beamforming.

4.3   Downlink C-RAN

In the downlink C-RAN, messages intended for different users in the cluster originate from the CP. Since the CP has access to all the user data, it can send useful information about the user messages to multiple RRHs in order to facilitate cooperation among different RRHs in such a way as to minimize unwanted interference seen by the users.

In the ideal case with infinite fronthaul capacities, the data of all the users in the entire cluster can be provided to all the RRHs. This reduces the downlink model to a MIMO broadcast channel with distributed antennas. However, the practical case with finite fronthaul capacities allows for only limited information transfer. As in the uplink, a key question is to decide the most useful information about the user messages to be sent to the RRHs in order to enable as much interference pre-subtraction as possible.

In this section we discuss various relaying strategies that utilize the limited fronthaul capacities in different manners for the downlink C-RAN, along with their corresponding optimization frameworks and methods for finding solutions. We conclude by providing design insights learned from such optimization.

4.3.1   Data-Sharing, Compression, Versus Compute-Forward

In the downlink, the benefit of the C-RAN architecture arises from the ability to cooperatively transmit signals from RRHs to minimize the effect of unwanted interference at the users. Cooperative transmission from multiple RRHs takes the form of network-wide beamforming. A straightforward way for the CP to enable cooperation is simply to share each user message with multiple RRHs, which can then form a cooperative cluster to serve the users. Ideally, to enable full cooperation, the each user's message needs to be shared with all the RRHs in the entire network. However, such full cooperation may not be feasible owing to the corresponding fronthaul capacity constraints. One way to reduce the fronthaul consumption is to share each user's message with only a subset of

RRHs, which then locally form beamformed signals to serve the users. This strategy is termed *data sharing*.

Another way to achieve cooperation is to compute centrally the beamformed signals to be transmitted by the RRHs at the CP. These signals are then compressed and sent to the individual RRHs for transmission to the users. Since the CP has the messages of all the users, the signals computed at the CP can mimic full cooperation. However, since these signals are analog, they need to be compressed before they can be sent to the RRHs. This introduces quantization noises that limit the system performance. Such a strategy is termed a *compression-based stategy* in this chapter.

Instead of sharing direct user messages or beamformed signals, there is also the possibility of sharing some function of the user messages to the RRHs. In the *reverse compute-forward* strategy [18], linear functions of user signals, chosen from a structured lattice codebook, are sent to the RRHs. These functions are computed in such a way that, after the messages have passed through the channels, each user can effectively retrieve its own message.

In the data-sharing strategy, the finite fronthaul capacity limits the size of the cooperation cluster while in the compression-based strategy the limited fronthaul capacity adds additional quantization noise. Further, as with compute-forward in the uplink, the performance of the reverse compute-forward strategy in the downlink is quite sensitive to the channel gain. With this in mind, in this chapter we focus the data-sharing and compression-based strategies. Readers are referred to [12] for details about optimization in the reverse compute-forward strategy.

From an information-theoretic perspective, the downlink C-RAN is an instance of the broadcast-relay channel. While it reduces to a broadcast channel if the fronthaul links have infinite capacities, the capacity characterization for the practical case with finite fronthaul capacities is very challenging. Approximate capacity and approximately optimal relaying strategies for the general broadcast-relay network are studied in [19, 20], but the exact characterization of capacity for the downlink C-RAN remains an open problem.

4.3.2 Data-Sharing Strategy

In traditional RANs, each BS receives raw data for users in its cell and computes the transmit signal on the basis of that data, independently of the other BSs. From a user's perspective, it receives useful signal from its serving BS and overhears interference from other nearby BSs. In C-RANs, the fronthaul connections from the CP to RRHs open up the possibility of signal-level cooperative transmission. Since the CP has access to the data of all the users in its cluster, a straightforward way to enable such cooperative transmission is to share the data of each user with all the RRHs. This essentially coverts the overall C-RAN downlink setup into a large antenna array, with the antennas distributed over the network or equivalently as a MIMO broadcast channel. However, sharing the data of each user with all the RRHs requires very high fronthaul capacity. In the more practical case where the fronthaul links have limited capacities, each RRH can receive data for only a subset of users or equivalently each user is served by only a subset of

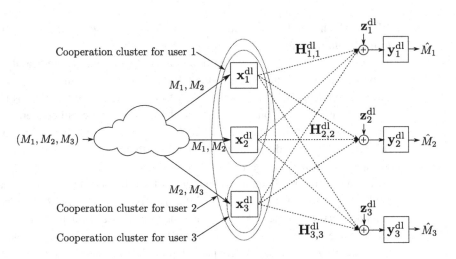

Figure 4.5 The data-sharing strategy for downlink C-RAN.

RRHs, as illustrated in Fig. 4.5. The effect of such limited cooperation is characterized below.

To illustrate the key ideas, we assume Gaussian signaling and use linear beamforming. Let $\mathbf{V}_{k,l} \in \mathbb{C}^{M \times d_k}$ denote the transmit beamformer matrix that conveys d_k data streams from RRH l to user k. The transmit signal from RRH l is given by $\mathbf{x}_l^{\mathrm{dl}} = \sum_k \mathbf{V}_{k,l} \mathbf{s}_k^{\mathrm{dl}}$, where $\mathbf{s}_k^{\mathrm{dl}} \in \mathbb{C}^{d_k \times 1} \sim \mathcal{CN}(\mathbf{0}, \mathbf{I})$ is the message for user k. The covariance matrix of the signal transmitted by RRH l is given by $\mathbb{E}\left\{ \mathbf{x}_l^{\mathrm{dl}} (\mathbf{x}_l^{\mathrm{dl}})^H \right\} = \sum_k \mathbf{V}_{k,l} \mathbf{V}_{k,l}^H$, with total transmit power $\sum_k \mathrm{Tr}\left(\mathbf{V}_{k,l} \mathbf{V}_{k,l}^H \right)$. Note that if user k's message $\mathbf{s}_k$ is not available at RRH l then the corresponding beamformer $\mathbf{V}_{k,l}$ is zero. Finally, with the linear Gaussian channel model described earlier in the chapter, the received signal at user k can be written as

$$\mathbf{y}_k^{\mathrm{dl}} = \sum_l \mathbf{H}_{k,l}^{\mathrm{dl}} \mathbf{V}_{k,l} \mathbf{s}_k^{\mathrm{dl}} + \sum_l \sum_{j \neq k} \mathbf{H}_{k,l}^{\mathrm{dl}} \mathbf{V}_{j,l} \mathbf{s}_j^{\mathrm{dl}} + \mathbf{z}_k^{\mathrm{dl}}. \tag{4.20}$$

Given (4.20), the achievable rate for user k under the data-sharing strategy, treating interference as noise, can be expressed as

$$R_k^{\mathrm{data, \, dl}} = I(\mathbf{s}_k^{\mathrm{dl}}; \mathbf{y}_k^{\mathrm{dl}}) \tag{4.21}$$

$$= \log \frac{\left| \sum_j \mathbf{H}_k^{\mathrm{dl}} \mathbf{V}_j \mathbf{V}_j^H \left(\mathbf{H}_k^{\mathrm{dl}} \right)^H + \sigma_{\mathrm{dl}}^2 \mathbf{I} \right|}{\left| \sum_{j \neq k} \mathbf{H}_k^{\mathrm{dl}} \mathbf{V}_j \mathbf{V}_j^H \left(\mathbf{H}_k^{\mathrm{dl}} \right)^H + \sigma_{\mathrm{dl}}^2 \mathbf{I} \right|}, \tag{4.22}$$

where $\mathbf{H}_k^{\mathrm{dl}} \in \mathbb{C}^{N \times LM} = \left[\mathbf{H}_{k,1}^{\mathrm{dl}}, \dots, \mathbf{H}_{k,L}^{\mathrm{dl}} \right]$ and $\mathbf{V}_k \in \mathbb{C}^{LM \times d_k} = \left[\mathbf{V}_{k,1}^T, \dots, \mathbf{V}_{k,L}^T \right]^T$ are the combined channel gains and transmit beamformers from all the RRHs to user k.

To support these user rates, the fronthaul capacity must support the aggregate data of the users to which each RRH beamforms. The fronthaul capacity required to send data to RRH l is thus simply the sum of the rates of the users served by RRH l. To write this mathematically, we make use of the fact that the transmit beamformer from

RRH l to user k is zero, i.e. $\mathbf{V}_{k,l} = \mathbf{0}$, if RRH does not serve user k or, equivalently, $\text{Tr}\left(\mathbf{V}_{k,l}\mathbf{V}_{k,l}^H\right) = 0$. Writing it in this way is useful for the optimization of the data-sharing strategy considered later on. The total fronthaul required for RRH l can now be written as $\sum_k \mathbb{1}\left\{\text{Tr}\left(\mathbf{V}_{k,l}\mathbf{V}_{k,l}^H\right)\right\} R_k^{\text{data, dl}}$, where $\mathbb{1}\left\{\text{Tr}\left(\mathbf{V}_{k,l}\mathbf{V}_{k,l}^H\right)\right\}$ is the indicating function defined as

$$\mathbb{1}\left\{\text{Tr}\left(\mathbf{V}_{k,l}\mathbf{V}_{k,l}^H\right)\right\} = \begin{cases} 0 & \text{if } \text{Tr}\left(\mathbf{V}_{k,l}\mathbf{V}_{k,l}^H\right) = 0, \\ 1 & \text{otherwise.} \end{cases} \tag{4.23}$$

This function determines whether user k's message is revealed to RRH l.

Note that, to participate in the beamforming to user k, there is also the overhead of transmitting the beamformer coefficients of the user to all the RRHs involved in order for them to be combined with the user data. In practice, sending the beamforming coefficients usually requires much less fronthaul capacity than sending user messages, especially in a slowly varying environment, as the beamforming coefficients typically need to be updated only as the user channels vary.

We remark further that the fronthaul consumption model (4.20) assumes that all the data streams of user k are either completely available or are not available at all, at RRH l, and ignores the possibility that only part of the data stream is revealed to an RRH. If such a possibility is considered then a user may receive different data streams from different serving RRHs, and the fronthaul consumption model (4.23) needs to be adjusted by using the indicator function and the rate expression for each individual data stream instead.

Finally, we point out that, instead of linear beamforming, a non-linear precoding technique (e.g. dirty paper coding) can also be utilized to improve the achievable user rates. The optimization framework developed in the next section can be extended easily to such a case.

4.3.3   Optimization Framework for Data-Sharing

Given (4.22) and (4.23), the weighted sum-rate maximization problem for the data-sharing strategy can be formulated as

$$\underset{\mathbf{V}_{k,l}}{\text{maximize}} \quad \sum_k w_k R_k^{\text{data, dl}} \tag{4.24a}$$

$$\text{s.t.} \quad \sum_k \mathbb{1}\left\{\text{Tr}\left(\mathbf{V}_{k,l}\mathbf{V}_{k,l}^H\right)\right\} R_k^{\text{data, dl}} \le C_l, \quad \forall l, \tag{4.24b}$$

$$\sum_k \text{Tr}\left(\mathbf{V}_{k,l}\mathbf{V}_{k,l}^H\right) \le P_l^{\text{dl}}, \quad \forall l. \tag{4.24c}$$

Here w_k in (4.24a) is the priority weight associated with user k.

The above optimization problem is non-convex, so finding its globally optimal solution is challenging. One source of non-convexity arises from the indicator function in (4.24b). A way to tackle this issue is to recast the indicator function into an expression involving an ℓ_0-norm, which can be further approximated as a convex weighted ℓ_1-norm using the compressive sensing concept [21]. Another source of non-convexity is the rate $R_k^{\text{data, dl}}$ expressed in (4.22). To resolve this difficulty, $R_k^{\text{data, dl}}$ in (4.24b) can be fixed and then updated iteratively to new constant values. This then turns the fronthaul constraint into a convex constraint for a given iteration. Then the WMMSE algorithm [17, 22] can be applied to reach a stationary-point solution of the beamforming problem. The details of such an approach can be found in [23]. Although this algorithm does not have a theoretical convergence proof, it is numerically observed to converge and performs as well as other algorithms with theoretical convergence guarantees [24].

In the problem formulation (4.24), it is assumed that the RRH cluster for each user can be updated dynamically in each time slot. In the case where the RRH clustering is static and is updated only when the user locations change, the compressive sensing idea can still be applied to address the fronthaul constraints [23]. But, in this case, the optimal static RRH clustering design problem needs to be formulated on the basis of loading considerations and is non-trivial to solve.

One way to form such static RRH clusters is simply to partition the entire set of RRHs geographically into different groups. The RRHs within the same group form a cooperative array of antennas and jointly serve the users that fall into that geographic area [25]. In such a user–RRH association, however, users near the boundary of the partitions still suffer from considerable interference.

In an alternative way, each individual user can decide on static and fixed sets of serving RRHs. The criteria to select the best RRHs need to be based on both the channel strengths as well as the loading at the RRHs. We refer to [23–27] for details on possible ways to form such static user-centric RRH associations.

4.3.4 Compression-Based Strategy

In the data-sharing strategy, the limited fronthaul capacities restrict the cooperation size of the RRH cluster in serving a user. However, since all the user data are available at the CP, the CP can centrally compute the beamformed signals that the RRHs should transmit. Such signals computed at the CP can in principle mimic the effect of full cooperation. The downside to such an approach is that the beamformed signals are no longer discrete (unlike the raw data in the data-sharing strategy) but instead are analog in nature. So these signals need to be compressed before they can be sent over digital fronthaul links of finite capacity. The process of compression introduces compression noise. The amount of such noise is determined by the available fronthaul capacities. Higher fronthaul capacity leads to finer compression and less quantization noise. Figure 4.6 illustrates the compression-based strategy. In the following, we characterize the effect of such quantization noise on the performance of the downlink C-RAN system.

We make transmission assumptions similar to those in the case of the data-sharing strategy. With $\mathbf{V}_{k,l}$ as the matrix of beamforming vectors for user k from RRH l, we can write the precoded signal computed at the CP and intended for transmission by RRH l as

$$\hat{\mathbf{x}}_l^{\text{dl}} = \sum_k \mathbf{V}_{k,l} \mathbf{s}_k^{\text{dl}}. \tag{4.25}$$

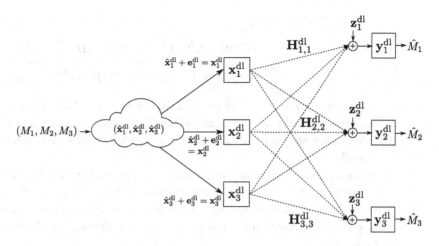

Figure 4.6 Compression-based strategy for downlink C-RAN.

These signals are then compressed and sent to the RRHs. As with the compress-forward strategy in the uplink, we model the compression process mathematically as an additive process:

$$\mathbf{x}_l^{\mathrm{dl}} = \hat{\mathbf{x}}_l^{\mathrm{dl}} + \mathbf{e}_l^{\mathrm{dl}}, \tag{4.26}$$

where $\mathbf{x}_l^{\mathrm{dl}}$ is the reconstructed signal that RRH l actually transmits to the users and the additional noise $\mathbf{e}_l^{\mathrm{dl}} \in \mathbb{C}^{M \times 1}$ (assumed to be independent of the signals to be compressed) captures the effect of quantization. We assume a Gaussian quantization model with $\mathbf{e}_l^{\mathrm{dl}} \sim \mathcal{CN}(\mathbf{0}, \mathbf{Q}_{l,l}^{\mathrm{dl}})$. We remark that, as for the uplink, the additive model for the compression process above is without loss of generality and includes the possibility of processing $\hat{\mathbf{x}}_l^{\mathrm{dl}}$ with a beamformer $\mathbf{B}_l$ (possibly to reduce the rank) prior to quantization. Note that the transmit power at RRH l can be represented as $\sum_k \mathrm{Tr}\left(\mathbf{V}_{k,l}\mathbf{V}_{k,l}^H\right) + \mathrm{Tr}\left(\mathbf{Q}_{l,l}^{\mathrm{dl}}\right)$; it accounts for the contribution due to quantization noise. It is also worth noting that the quantization noises of different RRHs are not necessarily independent, as the signals for all the RRHs are compressed jointly at the CP.

Let $\mathbf{Q}^{\mathrm{dl}} \in \mathbb{C}^{LM \times LM}$ denote the covariance matrix of the jointly Gaussian quantization noise of all the RRH signals, with $\mathbf{Q}_{l,l}^{\mathrm{dl}}$ as the lth diagonal block submatrix in $\mathbf{Q}^{\mathrm{dl}}$. The received signal at user k under the compression strategy can now be expressed as

$$y_k^{\mathrm{dl}} = \sum_l \mathbf{H}_{k,l}^{\mathrm{dl}}\mathbf{V}_{k,l}s_k^{\mathrm{dl}} + \sum_l \sum_{j \neq k} \mathbf{H}_{k,l}^{\mathrm{dl}}\mathbf{V}_{j,l}s_j^{\mathrm{dl}} + \mathbf{H}_k^{\mathrm{dl}}\mathbf{Q}^{\mathrm{dl}}\left(\mathbf{H}_k^{\mathrm{dl}}\right)^H + e^{\mathrm{dl}}, \tag{4.27}$$

where $\mathbf{e}^{\mathrm{dl}} = [\mathbf{e}^{\mathrm{dl}} \cdots \mathbf{e}_L^{\mathrm{dl}}]$. As can be seen from (4.27), the received signal in the compression strategy has an additional noise term due to the quantization noise in the signals transmitted to the RRHs.

Given (4.27), the achievable rate for user k under the compression strategy, again treating interference as noise, can be expressed as

$$R_k^{\text{comp, dl}} = I(s_k^{\text{dl}}; y_k^{\text{dl}}) \tag{4.28}$$

$$= \log \frac{\left| \sum_j \mathbf{H}_k^{\text{dl}} \mathbf{V}_j \mathbf{V}_j^H \left(\mathbf{H}_k^{\text{dl}} \right)^H + \mathbf{H}_k^{\text{dl}} \mathbf{Q}^{\text{dl}} \left(\mathbf{H}_k^{\text{dl}} \right)^H + \sigma_{\text{dl}}^2 \mathbf{I} \right|}{\left| \sum_{j \neq k} \mathbf{H}_k^{\text{dl}} \mathbf{V}_j \mathbf{V}_j^H \left(\mathbf{H}_k^{\text{dl}} \right)^H + \mathbf{H}_k^{\text{dl}} \mathbf{Q}^{\text{dl}} \left(\mathbf{H}_k^{\text{dl}} \right)^H + \sigma_{\text{dl}}^2 \mathbf{I} \right|}. \tag{4.29}$$

As compared with the rate in the data-sharing strategy (4.22), the rate (4.29) in the compression-based strategy has an additional term that represents the combined quantization noise after it passes through the channel. This quantization noise lowers the achievable rate.

On the plus side, since the beamformers are computed at the CP there are no specific constraints on $\mathbf{V}_{k,l}$ that limit the participation of RRHs in serving the users. As long as the CSI from the serving RRHs to the users is available at the CP, the CP can precompute all the beamformers and describe the beamformed signals to the RRHs in an efficient way.

We now look at the relationship between the quantization noise levels and the fronthaul capacities. The precise relationship depends on the compression technique used at the CP. We start with the case where the signals of different RRHs are compressed independently. In such a scenario the quantization noise at different RRHs is uncorrelated, and the quantization-noise covariance matrix $\mathbf{Q}^{\text{dl}}$ is a block-diagonal matrix with $\mathbf{Q}_{l,l}^{\text{dl}}$ on the diagonal blocks. Using results from rate-distortion theory, as in the case of independent compression in the uplink, the fronthaul capacity required for independent compression at RRH l is given by

$$C_l^{\text{indep,ul}} \geq I(\mathbf{x}_l^{\text{dl}}; \hat{\mathbf{x}}_l^{\text{dl}}) \tag{4.30}$$

$$= \log \left| \sum_k \mathbf{V}_{k,l} \mathbf{V}_{k,l}^H + \mathbf{Q}_{l,l}^{\text{dl}} \right| - \log \left| \mathbf{Q}_{l,l}^{\text{dl}} \right|. \tag{4.31}$$

Note that when independent compression is performed across signals of different RRHs, i.e., with a block-diagonal $\mathbf{Q}^{\text{dl}}$, the aggregated effect of the quantization noise at the users, $\mathbf{H}_k^{\text{dl}} \mathbf{Q}^{\text{dl}} \left(\mathbf{H}_k^{\text{dl}} \right)^H$, is just the sum of the contributions $\mathbf{H}_{k,l}^{\text{dl}} \mathbf{Q}_{l,l}^{\text{dl}} (\mathbf{H}_{k,l}^{\text{dl}})^H$ from each RRH. However, it is possible to improve the achievable rates by considering a more general compression scheme that allows for arbitrary correlation between quantization noise in the signals of different RRHs. Such a correlation allows the possibility of non-zero off-diagonal block matrices in $\mathbf{Q}^{\text{dl}}$, which can potentially lead to terms that eventually cancel each other at the user side. This type of compression is termed multivariate compression, as first proposed in [28], and is discussed below.

Assuming a compression order from RRH 1 to RRH L, the fronthaul required to compress the signals for RRH l for multivariate compression can be expressed as

$$C_l^{\text{mult,dl}} \geq I(\mathbf{x}_l^{\text{dl}}; \hat{\mathbf{x}}_l^{\text{dl}}) + I(\mathbf{e}_l^{\text{dl}}; \mathbf{e}_1^{\text{dl}}, \dots, \mathbf{e}_{l-1}^{\text{dl}}) \tag{4.32}$$

$$= \log \left| \sum_k \mathbf{V}_{k,l} \mathbf{V}_{k,l}^H + \mathbf{Q}_{l,l}^{\text{dl}} \right|$$

$$- \log \left| \mathbf{Q}_{l,l}^{\text{dl}} - \mathbf{Q}_{l,1:l-1}^{\text{dl}} \left(\mathbf{Q}_{1:l-1,1:l-1}^{\text{dl}} \right)^{-1} (\mathbf{Q}_{l,1:l-1}^{\text{dl}})^H \right|. \tag{4.33}$$

Here, $\mathbf{Q}^{dl}_{\mathcal{A},\mathcal{B}}$ denotes the covariance submatrix of $\mathbf{Q}^{dl}$ indexed by the RRHs in the sets $\mathcal{A}$, and $\mathcal{B}$ and $1:l$ as before denotes the set $\{1,\ldots,l\}$. As can be seen from the expression above, introducing correlations between the quantization noise of different RRHs actually costs more fronthaul capacity as compared with the case of independent compression. The benefit of such correlations is that, since the quantization noises pass through the channel and add up at the end users, we can potentially design the noise correlations in such a way as to make the noise cancel at the user side, thereby improving the overall system performance.

As with the Wyner–Ziv compression in the uplink, different ordering of the RRHs results in different fronthaul requirements and quantization noise covariance matrices. For a fixed order, a practical implementation of the multivariate compression is proposed in [28].

4.3.5   Optimization Framework for Compression

Under the different compression strategies described above, the weighted sum-rate maximization problem for the compression-based strategy in the downlink C-RAN can be formulated differently as follows:

$$\underset{\mathbf{V}_{k,l},\mathbf{Q}}{\text{maximize}} \quad \sum_k w_k R_k^{\text{comp, dl}} \tag{4.34a}$$

$$\text{s.t.} \quad (4.31) \quad \text{or} \quad (4.33) \quad \leq C_l, \quad \forall l, \tag{4.34b}$$

$$\sum_k \text{Tr}\left(\mathbf{V}_{k,l}\mathbf{V}_{k,l}^H\right) + \text{Tr}\left(\mathbf{Q}^{dl}_{l,l}\right) \leq P_l^{dl}, \quad \forall l. \tag{4.34c}$$

Here $R_k^{\text{comp, dl}}$ in (4.34a) is defined in (4.29). Note that additional constraints on the format of the covariance matrix $\mathbf{Q}^{dl}$ are to be imposed depending on the compression strategy. For example, in (4.31) $\mathbf{Q}^{dl}$ needs to be a block-diagonal matrix with diagonal matrices $\mathbf{Q}^{dl}_{l,l}$ that are positive semidefinite, i.e. $\mathbf{Q}^{dl}_{l,l} \succeq \mathbf{0}, \forall l$; in (4.33) $\mathbf{Q}^{dl}$ needs to be a positive semidefinite matrix, i.e. $\mathbf{Q}^{dl} \succeq \mathbf{0}$.

Unfortunately, none of the above optimization problems is a convex optimization program. In [28] the optimization problems (4.34) under (4.31) and (4.33) are solved through the majorize-minimization (MM) method. The main observation that allows such a method is that both the non-convex objective and the fronthaul relation can be represented as a difference of convex functions. To implement the MM-based method proposed in [28], first the transmit beamforming variables are converted into transmit covariance matrices and the rank constraints on the covariance matrices are relaxed in the subsequent optimization. Then a sequence of convex programs is solved over the covariance matrices by repeatedly linearizing the convex parts in the objective function and the concave parts in the fronthaul constraints until some convergence criterion is met. Such a method can be shown to reach a local optimum of the rank-relaxed problem. In the end, to get back the appropriate beamformers the eigenvectors corresponding to the largest eigenvalues of the final transmit covariance matrices are selected.

4.3.6   Hybrid Strategy

The data-sharing and compression-based strategies utilize the fronthaul capacity in two distinct ways. In data sharing the fronthaul links carry raw user messages for RRHs to compute the beamformed signals, while in the compression-based strategy the fronthaul links carry compressed bits from the previously computed beamformed signals. The advantage of the data-sharing approach is that the RRHs receive clean messages to be used for joint transmission. However, the fronthaul capacity constraint limits the cooperation cluster size. The main advantage of the compression-based approach is that the fronthaul capacity is more efficiently utilized when beamformed signals of multiple user messages are transmitted through the fronthaul. However, it pays a price in the extra quantization noise levels term in the resulting rate expression.

On the basis of the above comparison, a hybrid compression and data-sharing strategy is proposed in [29] to obtain the benefit of both strategies. In the hybrid strategy, a part of the fronthaul capacity is used to carry direct messages for some users and the remainder is used to carry the compressed beamformed signal of the rest of the users.

The rationale behind such an approach is the following. The desired precoded signal typically consists of both strong and weak signals and both high-rate and low-rate data streams. It would be beneficial to directly carry clean messages for the relatively strong signal with relatively low rate, because in this case it is typically more efficient to send the information bits themselves than to compress such signals. With these strong signals separated out, the amplitude of the rest of the signal is now lower. It would therefore require fewer bits to compress.

From the RRH's perspective, each RRH receives the direct messages for the strong users and the compressed precoded signals for the rest of the weak users in the network. It can compute a beamformed signal from the direct messages and the decompressed signal and transmit the result on its antennas. An optimization framework to design such a hybrid strategy is discussed in [29]. The key design parameters in such a hybrid approach are the selection of users that are suitable for direct data sharing, in addition to the beamforming and quantization noise levels.

4.3.7   Data-Sharing versus Compression

Two fundamentally distinct strategies, of data sharing and compression, are presented in this chapter for the downlink C-RAN. A natural question to ask is, which performs better in a realistic wireless network? The answer to this question depends on the amount of fronthaul capacity available.

In theory, to achieve full cooperation across the cluster managed by the CP, the amount of fronthaul capacity required for the data-sharing strategy at each RRH is simply the sum of the achievable rates of all the users across the cluster, which is finite. However, for the compression-based strategy to achieve full cooperation, infinite fronthaul capacity would be needed in order to bring the quantization noise to zero. Thus, at extremely high fronthaul capacities, data sharing has an advantage compared with compression.

At extremely low fronthaul capacities, data sharing also has an advantage. This is so because this case reduces to traditional single-cell processing, where each user's data is sent to one RRH only. Since the user data is discrete, it is more efficient to send messages rather than the compressed version of the analog signal.

However, for most realistic network settings, where the fronthaul capacity is moderately high, the compression-based strategy almost always outperforms the data-sharing strategy. The reason is that the effect of quantization noise is usually quite small. Further, compression is a more efficient utilization of the fronthaul capacity than data sharing, because the latter essentially replicates the same user message across multiple fronthaul links, which is inefficient. Numerical comparison of the two strategies is investigated in [30] under a realistic network topology, for different fronthaul capacities. When the fronthaul capacity is moderate and the two strategies are comparable, a hybrid of the two can bring additional gains [29].

In the downlink C-RAN, the gains due to cooperation depends crucially on the ability of the CP to obtain the CSI for the users in its cluster. The discussion so far assumes that the CSI for all users in the cluster is available at the CP. But, in practice, CSI acquisition and sharing consume significant fronthaul capacity and are expected to be major factors in limiting the size of cooperation cluster in the C-RAN architecture. Note that, at a given cluster size, the data-sharing strategy achieves higher rates than the compression strategy, owing to the additional quantization noise in compression. So, to achieve the same rate, the compression strategy requires a larger cluster size and hence more CSI. In a typical deployment, the cooperation cluster size under the compression strategy is mostly limited by CSI availability, while for data sharing it is mostly limited by the fronthaul capacity.

As a concluding remark, we note that the implementations of the data-sharing and compression strategies have key differences, in that the RRHs need to have knowledge of the modulation and coding format for implementing data sharing but such codebook knowledge is not needed for compression. Thus, the RRHs for implementing the compression strategy can be made much simpler.

4.4 Summary

This chapter illustrates cooperative beamforming and relaying strategies and the associated resource allocation for both uplink and downlink C-RANs. In the uplink, we show compress-forward as the fundamental strategy and provided an optimization framework for transmit beamforming at the users and quantization at the RRHs. In the downlink, we demonstrate data sharing and compression as two competing and fundamentally different strategies. The data-sharing optimization framework for RRH clustering and transmit beamforming and the compression optimization framework for cooperative beamforming and quantization at the CP are discussed. In all cases, the finite fronthaul capacity has a major impact on the analysis and design of different transmission and relaying strategies in the C-RAN architecture.

The achievable user rate and the fronthaul rate expressions used throughout the chapter are based on information-theoretic analysis and assume the use of capacity-achieving and rate-distortion-achieving codes. The codes used in practice usually operate below the information-theoretic limit. However, to a good approximation, the performance due to such practical codes can be captured by incorporating gap factors in the respective user-rate and fronthaul-rate expressions. The optimization algorithms developed in this chapter can be easily extended with such factors taken into account.

References

[1] M. K. Karakayali, G. J. Foschini, and R. A. Valenzuela, "Network coordination for spectrally efficient communications in cellular systems," *IEEE Wireless Commun. Mag.*, vol. 13, no. 4, pp. 56–61, August 2006.

[2] S. Shamai and B. M. Zaidel, "Enhancing the cellular downlink capacity via co-processing at the transmitting end," in *Proc. IEEE Vehicular Technology Conf.*, vol. 3, May 2001, pp. 1745–1749.

[3] D. Gesbert, S. Hanly, H. Huang, S. Shamai (Shitz), O. Simeone, and W. Yu, "Multi-cell MIMO cooperative networks: a new look at interference," *IEEE J. Sel. Areas Commun.*, vol. 28, no. 9, pp. 1380–1408, December 2010.

[4] E. Heo, O. Simeone, and H. Park, "Optimal fronthaul compression for synchronization in the uplink of cloud radio access networks," *CoRR*, vol. abs/1510.01545, 2015. Online. Available at http://arxiv.org/abs/1510.01545.

[5] S.-H. Park, O. Simeone, O. Sahin, and S. Shamai, "Robust layered transmission and compression for distributed uplink reception in cloud radio access networks," *IEEE Trans. Veh. Technol.*, vol. 63, no. 1, pp. 204–216, January 2014.

[6] L. Liu, S. Bi, and R. Zhang, "Joint power control and fronthaul rate allocation for throughput maximization in OFDMA-based cloud radio access network," *IEEE Trans. Commun.*, vol. 63, no. 11, pp. 4097–4110, November 2015.

[7] A. Sanderovich, S. Shamai, Y. Steinberg, and G. Kramer, "Communication via decentralized processing," *IEEE Trans. Inf. Theory*, vol. 54, no. 7, pp. 3008–3023, July 2008.

[8] S.-H. Park, O. Simeone, O. Sahin, and S. Shamai, "Robust and efficient distributed compression for cloud radio access networks," *IEEE Trans. Veh. Technol.*, vol. 62, no. 2, pp. 692–703, February 2013.

[9] Y. Zhou and W. Yu, "Optimized backhaul compression for uplink cloud radio access network," *IEEE J. Sel. Areas Commun.*, vol. 32, no. 6, pp. 1295–1307, June 2014.

[10] B. Nazer and M. Gastpar, "Compute-and-forward: harnessing interference through structured codes," *IEEE Trans. Inf. Theory*, vol. 57, no. 10, pp. 6463–6486, October 2011.

[11] B. Nazer, A. Sanderovich, M. Gastpar, and S. Shamai, "Structured superposition for backhaul constrained cellular uplink," in *Proc. IEEE Int. Symp. on Inf. Theory*, June 2009, pp. 1530–1534.

[12] S.-N. Hong and G. Caire, "Compute-and-forward strategies for cooperative distributed antenna systems," *IEEE Trans. Inf. Theory*, vol. 59, no. 9, pp. 5227–5243, September 2013.

[13] A. Avestimehr, S. Diggavi, and D. Tse, "Wireless network information flow: a deterministic approach," *IEEE Trans. Inf. Theory*, vol. 57, no. 4, pp. 1872–1905, April 2011.

[14] S. H. Lim, Y.-H. Kim, A. El Gamal, and S.-Y. Chung, "Noisy network coding," *IEEE Trans. Inf. Theory*, vol. 57, no. 5, pp. 3132–3152, May 2011.

[15] L. Zhou and W. Yu, "Uplink multicell processing with limited backhaul via per-base-station successive interference cancellation," *IEEE J. Sel. Areas Commun.*, vol. 31, no. 10, pp. 1981–1993, October 2013.

[16] L. Zhou and W. Yu, "Fronthaul compression and transmit beamforming optimization for multi-antenna uplink C-RAN," *IEEE Trans. Signal Process.*, vol. 64, no. 16, pp. 4138–4151, August 2016.

[17] Q. Shi, M. Razaviyayn, Z.-Q. Luo, and C. He, "An iteratively weighted MMSE approach to distributed sum-utility maximization for a MIMO interfering broadcast channel," *IEEE Trans. Signal Process.*, vol. 59, no. 9, pp. 4331–4340, September 2011.

[18] S.-N. Hong and G. Caire, "Reverse compute and forward: a low-complexity architecture for downlink distributed antenna systems," in *Proc. IEEE Int. Symp. Information Theory*, 2012, pp. 1147–1151.

[19] S. Kannan, A. Raja, and P. Viswanath, "Approximately optimal wireless broadcasting," *IEEE Trans. Inf. Theory*, vol. 58, no. 12, pp. 7154–7167, December 2012.

[20] S. H. Lim, K. T. Kim, and Y. Kim, "Distributed decode-forward for relay networks," *CoRR*, vol. abs/1510.00832, 2015 Online. Available at http://arxiv.org/abs/1510.00832.

[21] E. Candès, M. Wakin, and S. Boyd, "Enhancing sparsity by reweighted ℓ_1 minimization," *J. Fourier Anal. Appl.*, vol. 14, no. 5, pp. 877–905, 2008.

[22] S. Christensen, R. Agarwal, E. de Carvalho, and J. Cioffi, "Weighted sum-rate maximization using weighted MMSE for MIMO-BC beamforming design," *IEEE Trans. Wireless Commun.*, vol. 7, no. 12, pp. 4792–4799, December 2008.

[23] B. Dai and W. Yu, "Sparse beamforming and user-centric clustering for downlink cloud radio access network," *IEEE Access*, vol. 2, pp. 1326–1339, 2014.

[24] B. Dai and W. Yu, "Backhaul-aware multicell beamforming for downlink cloud radio access network," in *Proc. IEEE Int. Commun. Conf. Workshop*, June 2015, pp. 2689–2694.

[25] H. Huang, M. Trivellato, A. Hottinen, M. Shafi, P. Smith, and R. Valenzuela, "Increasing downlink cellular throughput with limited network MIMO coordination," *IEEE Trans. Wireless Commun.*, vol. 8, no. 6, pp. 2983–2989, June 2009.

[26] S. Kaviani, O. Simeone, W. Krzymien, and S. Shamai, "Linear precoding and equalization for network MIMO with partial cooperation," *IEEE Trans. Veh. Technol.*, vol. 61, no. 5, pp. 2083–2096, June 2012.

[27] C. T. K. Ng and H. Huang, "Linear precoding in cooperative MIMO cellular networks with limited coordination clusters," *IEEE J. Sel. Areas Commun.*, vol. 28, no. 9, pp. 1446–1454, December 2010.

[28] S.-H. Park, O. Simeone, O. Sahin, and S. Shamai, "Joint precoding and multivariate backhaul compression for the downlink of cloud radio access networks," *IEEE Trans. Signal Process.*, vol. 61, no. 22, pp. 5646–5658, November 2013.

[29] P. Patil and W. Yu, "Hybrid compression and message-sharing strategy for the downlink cloud radio-access network," in *Proc. Inf. Theory Applic. Workshop*, pp. 1–6, February 2014.

[30] P. Patil, B. Dai, and W. Yu, "Performance comparison of data-sharing and compression strategies for cloud radio access networks," in *Proc. European Signal Processing Conf.*, July 2015, pp. 2456–2460.

5 Training Design and Channel Estimation in C-RANs

Mugen Peng, Zhendong Mao, Yourrong Ban, and Di Chen

5.1 Background Overview

In the era of mobile Internet, the explosive growth of data traffic imposes a heavy strain on mobile operators to manage networks efficiently while maintaining the quality of service (QoS). It is envisioned that more diverse intelligent mobile devices will be incorporated in the future fifth generation (5G) networks [1].

Cloud-Ran (C-RAN) is a novel RAN architecture, proposed by China Mobile with the aim of achieving the goal of green communication, which has demonstrated its superiority over the conventional cellular networks in terms of energy and spectral efficiencies [2]. Cloud C-RANs are still faced with challenges for successful implementation and commercial operation. One of the major key technical issues awaiting timely resolution in C-RANs is the lack of instantaneous channel state information (CSI), which is essential for beamforming and for coherent reception. As a matter of fact, nearly all the advanced technologies operated in C-RANs demand an accurate knowledge of CSI. The assumption that perfect CSI is known at the baseband unit (BBU) pool is also not practical owing to the time-varying nature of the radio channel. Consequently, developing accurate and reliable channel estimation techniques is vital to support the feasible design of C-RANs.

5.1.1 The Introduction of Channel Estimation

Channel estimation as a fundamental problem in wireless communication systems has been extensively investigated in the past few decades [3]. In general, channel estimation can be basically categorized into two kinds of strategies, i.e., training-based and non-training-based channel estimation. Training-based channel estimation is the most commonly used method owing to its flexibility and low complexity. By periodically inserting the training sequences into data frames in the time or frequency domain, the receiver can easily obtain a reliable estimate of the radio channel impulse responses by means of prior knowledge of the training sequences [4]. However, the use of training sequences will degrade the system performance in terms of the spectral and energy efficiencies, as additional spectrum resource is consumed by the training transmission. Therefore, non-training-based channel estimation is proposed as an alternative way to obtain CSI, by adequately using the intrinsic characteristics of the data signal itself instead of using training [5]. Despite the high spectral efficiency, there still exist many restrictions of non-training-based channel estimation, e.g., a slow convergence speed, high computation complexity and sensitivity to noise, which to some extent limit its implementation.

For point-to-point communication systems, the training transmission in training-based channel estimation is simply sending the training sequences from end to end. The CSI of the radio channel can be directly obtained by employing the optimal maximum likelihood (ML) estimator or the linear minimum-mean-square error (LMMSE) estimator. In C-RANs, the data transmission is performed by means of two communication links instead of one direct link. The first is the radio access link (ACL) between the user equipment (UE) and the remote radio head (RRH), and the second is the fronthaul link connecting the RRH and the BBU pool. The fronthaul links can be either wire or wireless, but wireless fronthaul links (WFLs) are mainly adopted owing to their low expense and flexible deployment for RRHs. The traditional training scheme in point-to-point systems can be used to obtain the CSI of cascaded channels in C-RANs, and the whole process is completed in two phases. Just knowing the cascaded channels is not sufficient to support the optimal system design in C-RANs, however, and the individual CSI for the two links should be obtained at the BBU pool, as required by certain technologies, e.g., beamforming design [6]. If the traditional training scheme is applied to obtain the individual CSI, a three-phase training strategy is imperative since the RRHs have to feedback the individual CSI estimated at the RRHs to the BBU pool. As a result, the training overhead is significant and the estimation performance is degraded owing to the CSI feedback during individual CSI acquisition. Moreover, this training scheme is incompatible with two-phase data transmission, and the training block cannot be embedded into the data frame. To tackle this problem, the superimposed training scheme in [7] is a promising solution for channel estimation in C-RANs, in which each RRH superimposes its own training sequence on the received sequence in such a way that the individual CSI can be estimated at the BBU pool. In this way, a better tradeoff between channel estimation accuracy and training overhead can be achieved with no occupation of additional training resources, in contrast with the traditional training scheme.

This superimposed training scheme has demonstrated its effectiveness in keeping a good balance between spectral efficiency and estimation accuracy. In [8] a two-phase-based superimposed training strategy in amplify-and-forward relay networks was proposed and implemented with the corresponding data transmission. In this strategy, a low complexity linear estimator is used to obtain the initial estimation, and this is followed by iterative estimation for further improvement in accuracy. The CSI for the source–relay link and for the relay–destination link can be separately obtained with high efficiency by taking advantage of superimposed training. In [9] the superimposed training scheme is extended to explore the corresponding channel estimation problem in multiple-access relay networks. The ML method is proposed for both the composite source–relay–destination-link estimation and the individual relay–destination-link estimation. Although the superimposed training scheme is an effective way to improve the spectral efficiency, the system performance will be degraded as both the training sequences and the data symbols have to share the power originally allocated for data transmission alone. Regarding this degradation in the superimposed training scheme, the segment training scheme proposed in [10] is expected to be an alternative way to acquire the individual CSI.

In this segment training scheme, two consecutive segments in the data frame are dedicated to CSI acquisition for two individual links, respectively. In [11], the segment training scheme is applied to design distributed space–time coding for amplify-and-forward one-way relay networks (AF-OWRNs) with channel estimation errors taken into consideration. Segment-training-based channel estimation in AF-OWRNs was fully exploited in [12]. The LMMSE estimator is employed to obtain an initial estimate, and an iterative maximum *a posteriori* probability estimation is then used to further improve the accuracy. By minimizing the MSE of individual channel estimation, the training-power allocation strategy for system performance enhancement is derived. The segment training scheme has also been successfully applied in C-RANs, in which a hybrid of superimposed training and segment training is proposed in order to reduce the training overhead [13]. A maximum *a posteriori* probability (MAP) estimator is developed to obtain the basis-expansion-model (BEM) coefficients of the ACLs and the channel fading of the WFLs, and the time-domain channel samples of the ACLs are recovered by maximizing the average effective signal-to-noise ratio (AESNR). In [14] the instantaneous and individual CSI of composite links combining ACLs and WFLs is obtained at the centralized BBU pool with the signal processing overhead of RRHs greatly alleviated. The sequential minimum-mean-square-error (SMMSE) estimator is developed by making use of the Kalman filter to improve the channel estimation accuracy, and the corresponding training design subject to the power constraint is derived using the MSE minimization criterion.

Training-based channel estimation is an efficient and reliable way for CSI acquisition in C-RANs. To obtain a high estimation accuracy, the training overhead needed is significant, which in turn decreases the transmission bandwidth efficiency. Furthermore, training is undesirable for certain communication systems, and the intrinsic properties of data symbols are not exploited in training-based channel estimation. Thus non-training-based channel estimation can be employed when the spectrum resource is scarce. A subspace-based method [15] using the redundant linear precoding and the noise subspace method effectively obtains accurate channel estimation in multiple-input multiple-output (MIMO) with the orthogonal frequency-division multiplexing (OFDM) modulation scheme. Using the second-order cyclostationary statistics induced by a periodic nonconstant-modulus antenna precoding, blind channel estimation and equalization for the MIMO-OFDM scheme in [16] can also achieve a good estimate with channel utilization increased. However, a large number of data symbols is required to improve the blind channel estimation accuracy. The inherent ambiguity, e.g., the phase ambiguity, also blocks the implementation of blind channel estimation in C-RANs if one merely relies on the data samples to estimate the CSI. A small amount of training is still needed to obtain an initial estimate. Semi-blind channel estimation, as a hybrid of blind and training-based approaches, is considered as a pragmatic compromise. By incorporating both the training signal and the data samples, semi-blind channel estimation can achieve higher accuracy than simply using either training-based or blind estimation.

5.1.2   The Structure of this Chapter

In C-RANs, most related technologies require an accurate knowledge of CSI to fully exploit their theoretically predicted performance. Therefore, developing efficient channel estimation schemes for CSI acquisition in C-RANs is essential. Motivated by the high efficiency of superimposed training, the latter is adopted as an effective tool to obtain the CSI in C-RANs. More specifically, the segment training scheme is also used for channel estimation in C-RANs, to improve estimation accuracy. Since the training overhead is significant if applying training-based channel estimation in C-RANs, we use blind or semi-blind channel estimation to acquire the CSI using the statistical properties of the data instead of the training sequences only.

The rest of this chapter is organized as follows. In Section 5.2 superimposed training-based channel estimation is investigated. The optimal training to improve estimation performance is derived in a closed form. Moreover, a hybrid form of the superimposed-segment training scheme, used to reduce the training overhead, is discussed. Then an iterative channel estimation algorithm using the MAP method is proposed, and its convergence condition is proved. In Section 5.3 the segment training scheme is applied to obtain the individual CSI in C-RANs. By using the Kalman filter, the sequential minimum-mean-square-error (SMMSE) estimator is developed through prior knowledge of the channel statistics and the latest estimate. The optimal training, subject to the power constraint, is derived by minimizing the MSE. In Section 5.4 we present semi-blind channel estimation for C-RANs, which takes full advantage of training and data, and channel estimation in fronthaul constrained C-RANs is discussed in Section 5.5. Finally, a summary is given in Section 5.6.

5.2   Superimposed Training Scheme in C-RANs

For obtaining individual CSI in C-RANs while remaining compatible with the data transmission, the superimposed training-based scheme is an attractive option in practical applications. In superimposed training, RRHs superimpose their own training sequences on the received sequence before forwarding the combination to the destination. Such a scheme ensures high spectrum efficiency of transmission since in this way UEs and RRHs share the same training resource. Hence, applying superimposed training in C-RANs can significantly reduce the training overhead, while maintaining a reliable estimate of the individual CSI.

5.2.1   Superimposed Training-Based Channel Estimation

First, we investigate training-based channel estimation with superimposed training used in the uplink C-RANs. The ML method is implemented to obtain the required CSI not only for composite channel estimation but also for individual channel estimation.

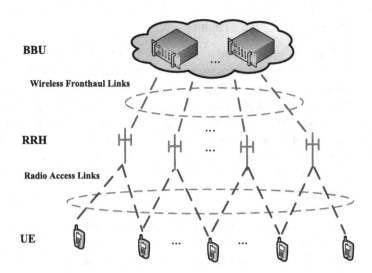

Figure 5.1 System model for a uplink C-RAN.

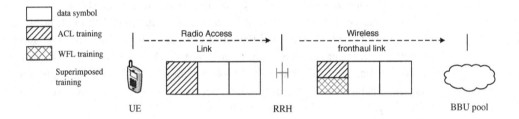

Figure 5.2 Superimposed training scheme in C-RANs.

System Model

Consider a typical uplink C-RAN structure with M UEs ($\mathbb{S}_m$, $m = 1, \ldots, M$), K RRHs ($\mathbb{R}_k$, $k = 1, \ldots, K$), and one BBU pool $\mathbb{D}$, as depicted in Fig. 5.1. All the nodes are assumed to be equipped with a single antenna, and RRHs operate in half-duplex modes. The channels are assumed to be quasi-static flat fading. The channel coefficients from $\mathbb{S}_m$ to $\mathbb{R}_k$ and from $\mathbb{R}_k$ to $\mathbb{D}$ are denoted by h_{mk} and h_{Rk}, respectively. Moreover, h_{mk} and h_{Rk} are assumed to have zero mean and variances υ_{mk} and υ_{Rk}, respectively, which are assumed to be known at the RRHs and the BBU pool. The transmit power of $\mathbb{S}_m$ is denoted by P_s and that of $\mathbb{R}_k$ is denoted by P_R. The training sequence sent from $\mathbb{S}_m$ is denoted by $\mathbf{t}_m$.

The transmission block consists of two phases as shown in Fig. 5.2. In the first phase, each UE inserts a training sequence $\mathbf{t}_m$ of length N, whose power is constrained by

$$\|\mathbf{t}_m\|^2 = NP_s. \tag{5.1}$$

Let us define $\mathbf{T} = [\mathbf{t}_1\ \mathbf{t}_2 \cdots \mathbf{t}_M]$. The received training sequence at $\mathbb{R}_k$ can be expressed as

$$\mathbf{x}_{Rk} = \mathbf{T}\mathbf{h}_k + \mathbf{w}_{Rk}, \tag{5.2}$$

where $\mathbf{h}_k$ is the kth column of an $(M \times K)$-dimensional matrix $\mathbf{H}_{SR}$, whose (m, k)th element is h_{mk}, while $\mathbf{w}_{Rk}$ represents an $N \times 1$ AWGN vector with covariance $\sigma_n^2 \mathbf{I}_N$.

In the second phase, each RRH superimposes a length-N training sequence over the received sequence before forwarding the result to the BBU pool. Since RRHs transmit in different time slots and the overall length of the second phase is NK, we could assume that the superimposed sequence sent by each RRH is identical and is denoted by $\mathbf{t}_r$. The following power constraint should be satisfied:

$$\|\mathbf{t}_r\|^2 = NP_r. \tag{5.3}$$

Therefore, the training sequence forwarded by $\mathbb{R}_k$ is written as

$$\mathbf{r}_{Rk} = \gamma_k \mathbf{x}_{Rk} + \mathbf{t}_r, \tag{5.4}$$

where

$$\gamma_k = \sqrt{\frac{P_R - P_r}{\left(\sum_{m=1}^{M} \upsilon_{mk}\right) P_s + \sigma_n^2}} \tag{5.5}$$

in order to keep the overall power of $\mathbb{R}_k$ as P_R. The received sequence during the kth time-slot of the second phase is expressed as

$$\mathbf{x}_{Dk} = \gamma_k h_{Rk} \mathbf{Th}_k + h_{Rk} \mathbf{t}_r + \gamma_k h_{Rk} \mathbf{w}_{Rk} + \mathbf{w}_{Dk}, \tag{5.6}$$

where $\mathbf{w}_{Dk}$ is an $N \times 1$ AWGN vector with covariance $\sigma_n^2 \mathbf{I}_N$. Stacking the $\mathbf{x}_{Dk}$ into a vector form yields

$$\begin{aligned}
\mathbf{x}_D &= \begin{bmatrix} \mathbf{x}_{D1}^T \ \mathbf{x}_{D2}^T \cdots \mathbf{x}_{DK}^T \end{bmatrix}^T \\
&= \left\{ \underbrace{\mathrm{diag}\left\{\gamma_1, \gamma_2, \ldots, \gamma_K\right\}}_{\boldsymbol{\Gamma}} \otimes \mathbf{T} \right\} \mathrm{vec}(\mathbf{G}) \\
&\quad + \mathrm{vecd}(\mathbf{H}_R) \otimes \mathbf{t}_r + \left\{ (\boldsymbol{\Gamma} \mathbf{H}_R) \otimes \mathbf{I}_N \right\} \mathbf{w}_R + \mathbf{w}_D,
\end{aligned} \tag{5.7}$$

where $\mathbf{w}_R = \begin{bmatrix} \mathbf{w}_{R1}^T \mathbf{w}_{R2}^T \cdots \mathbf{w}_{RK}^T \end{bmatrix}^T$ and $\mathbf{w}_D = \begin{bmatrix} \mathbf{w}_{D1}^T \mathbf{w}_{D2}^T \cdots \mathbf{w}_{DK}^T \end{bmatrix}^T$.

Channel Estimation with Superimposed Training

Let $\mathbf{g}_k$ denote the kth column of $\mathbf{G}$. The probability density function of $\mathbf{x}_D$ is given by

$$\begin{aligned}
p(\mathbf{x}_D) &= \frac{1}{\left(\pi \sigma_n^2\right)^{(K+1)N} \prod_{k=1}^{K} \left(\gamma_k^2 |h_{Rk}|^2 + 1\right)^N} \\
&\quad \times \exp\left\{ -\sum_{k=1}^{K} \frac{\left\| \mathbf{x}_{Dk} - \gamma_k \mathbf{Tg}_k - h_{Rk} \mathbf{t}_r \right\|^2}{\left(\gamma_k^2 |h_{Rk}|^2 + 1\right) \sigma_n^2} \right\}.
\end{aligned} \tag{5.8}$$

Note that the above estimation problem can be divided into K individual subproblems. In particular, the estimates of $\mathbf{g}_k$ and h_{Rk} can be obtained from

$$\begin{aligned}
\left\{\hat{\mathbf{g}}_k, \hat{h}_{Rk}\right\} = \arg\max_{\mathbf{g}_k, h_{Rk}} \Bigg\{ &-\frac{\left\| \mathbf{x}_{Dk} - \gamma_k \mathbf{Tg}_k - h_{Rk} \mathbf{t}_r \right\|^2}{\left(\gamma_k^2 |h_{Rk}|^2 + 1\right) \sigma_n^2} \\
&- N \log\left(\gamma_k^2 |h_{Rk}|^2 + 1\right) \Bigg\}.
\end{aligned} \tag{5.9}$$

For a given h_{Rk}, $\hat{\mathbf{g}}_k$ can be derived as follows:

$$\hat{\mathbf{g}}_k = \arg\min_{\mathbf{g}_k} \|\mathbf{x}_{Dk} - \gamma_k \mathbf{T}\mathbf{g}_k - h_{Rk}\mathbf{t}_r\|^2 = \frac{1}{\gamma_k} \left(\mathbf{T}^H\mathbf{T}\right)^{-1} \mathbf{T}^H \left(\mathbf{x}_{Dk} - h_{Rk}\mathbf{t}_r\right). \quad (5.10)$$

Substituting (5.10) back into (5.9), $\hat{h}_{Rk}$ can be derived as

$$\hat{h}_{Rk} = \arg\min_{h_{Rk}} \left\{ N\log\left(\gamma_k^2 |h_{Rk}|^2 + 1\right) \right.$$
$$\left. + \frac{\mathbf{x}_{Dk}^H \mathbf{M}\mathbf{x}_{Dk} - 2|h_{Rk}|\Re\{e^{j\angle h_{Rk}}\mathbf{x}_{Dk}^H \mathbf{M}\mathbf{t}_r\} + |h_{Rk}|^2 \mathbf{t}_r^H \mathbf{M}\mathbf{t}_r}{\left(\gamma_k^2 |h_{Rk}|^2 + 1\right)\sigma_n^2} \right\}, \quad (5.11)$$

where $\mathbf{M} \triangleq \mathbf{I} - \mathbf{T}\left(\mathbf{T}^H\mathbf{T}\right)^{-1}\mathbf{T}^H$. Obviously, the phase $\angle h_{Rk}$ could be separately estimated by maximizing $\Re\{e^{j\angle h_{Rk}}\mathbf{x}_{Dk}^H \mathbf{M}\mathbf{t}_r\}$, i.e.,

$$\angle \hat{h}_{Rk} = -\angle\left(\mathbf{x}_{Dk}^H \mathbf{M}\mathbf{t}_r\right). \quad (5.12)$$

For notational simplicity, we write $z = |h_{Rk}|$, whose estimate $\hat{z}$ can be obtained from

$$\hat{z} = \arg\min_z \left\{ \underbrace{\frac{\mathbf{x}_{Dk}^H \mathbf{M}\mathbf{x}_{Dk} - 2z\|\mathbf{x}_{Dk}^H \mathbf{M}\mathbf{t}_r\| + z^2\mathbf{t}_r^H \mathbf{M}\mathbf{t}_r}{\left(\gamma_k^2 z^2 + 1\right)\sigma_n^2} + N\log\left(\gamma_k^2 z^2 + 1\right)}_{f(z)} \right\}, \quad (5.13)$$

where $f(z)$ abbreviates the factor within braces. The first-order derivative of $f(z)$ is computed as

$$\dot{f}(z) = \frac{\partial f(z)}{\partial z} = \frac{\mathbf{t}_r^H \mathbf{M}\mathbf{t}_r z - 2\|\mathbf{x}_{Dk}^H \mathbf{M}\mathbf{t}_r\|}{\gamma_k^2 z^2 \sigma_n^2 + \sigma_n^2} + \frac{2N\gamma_k z}{\gamma_k^2 z^2 + 1}$$
$$- \frac{2\gamma_k^2 \sigma_n^2 z \left(\mathbf{x}_{Dk}^H \mathbf{M}\mathbf{x}_{Dk} - 2z\|\mathbf{x}_{Dk}^H \mathbf{M}\mathbf{t}_r\| + z^2\mathbf{t}_r^H \mathbf{M}\mathbf{t}_r\right)}{\left(\gamma_k^2 z^2 \sigma_n^2 + \sigma_n^2\right)^2}$$
$$= \frac{C_1 z^3 + C_2 z^2 + C_3 z + C_4}{\left(\gamma_k^2 z^2 \sigma_n^2 + \sigma_n^2\right)^2}, \quad (5.14)$$

where

$$C_1 = 2\gamma_k^2 \sigma_n^2 \mathbf{t}_r^H \mathbf{M}\mathbf{t}_r + 2N\gamma_k^4 \sigma_n^4,$$
$$C_2 = \mathbf{t}_r^H \mathbf{M}\mathbf{t}_r + N\gamma_k^2 \sigma_n^4 - 2\gamma_k^2 \sigma_n^2 |\mathbf{x}_{Dk}^H \mathbf{M}\mathbf{t}_r|,$$
$$C_3 = 2\sigma_n^2 \mathbf{t}_r^H \mathbf{M}\mathbf{t}_r + 2N\gamma_k^2 \sigma_n^4 - 2\gamma_k^2 \sigma_n^2 \mathbf{x}_{Dk}^H \mathbf{M}\mathbf{x}_{Dk} - 2|\mathbf{x}_{Dk}^H \mathbf{M}\mathbf{t}_r|,$$
$$C_4 = N\sigma_n^4 - 2\sigma_n^2 |\mathbf{x}_{Dk}^H \mathbf{M}\mathbf{t}_r|. \quad (5.15)$$

Let us define

$$\Delta_0 = C_2^2 - 3C_1 C_3,$$
$$\Delta_1 = 18C_1 C_2 C_3 C_4 - 4C_2^3 C_4 + C_2^2 C_3^2 - 4C_1 C_3^3 - 27C_1^2 C_4^2 \quad (5.16)$$

for further use. By solving the cubic equation $\dot{f}(z) = 0$, we have following results.

- **Case 1** ($\Delta_1 > 0$)   The equation $\dot{f}(z) = 0$ has three real roots z_1, z_2, and z_3. Without loss of generality, we can assume $z_1 < z_2 < z_3$. It can be checked that $\lim_{z \to -\infty} \dot{f}(z) < 0$ and $\lim_{z \to \infty} \dot{f}(z) > 0$. Hence, z_1 and z_3 are local minima while z_2 is a local maximum. Since $|h_{Rk}| \geq 0$, the estimate of $|h_{Rk}|$ is

$$
|\hat{h}_{Rk}| = \begin{cases} \arg\min\limits_{z \in \{z_1, z_3\}} \{f(z_1), f(z_3)\}, & z_1 \geq 0, & (5.17a) \\[2mm] \arg\min\limits_{z \in \{0, z_3\}} \{f(0), f(z_3)\}, & z_1 < 0 \leq z_3, & (5.17b) \\[2mm] 0, & z_3 < 0. & (5.17c) \end{cases}
$$

- **Case 2** ($\Delta_1 = 0$ and $\Delta_0 \neq 0$)   The equation $\dot{f}(z) = 0$ has three real roots,

$$
z_1 = z_2 = \frac{9C_1 C_4 - C_2 C_3}{2\Delta_0} \text{ and } z_3 = \frac{4C_1 C_2 C_3 - 9C_1^2 C_4 - C_2^3}{C_1 \Delta_0}.
$$

Since $\lim_{z \to -\infty} \dot{f}(z) < 0$ and $\lim_{z \to \infty} \dot{f}(z) > 0$, z_3 is a local minimum and z_1 (z_2) is an inflection point. Since $|h_{Rk}| \geq 0$, the estimate of $|h_{Rk}|$ is

$$
|\hat{h}_{Rk}| = \max\left\{ \frac{4C_1 C_2 C_3 - 9C_1^2 C_4 - C_2^3}{C_1 \Delta_0}, 0 \right\}. \tag{5.18}
$$

- **Case 3** ($\Delta_1 = 0$ and $\Delta_0 = 0$)   The equation $\dot{f}(z) = 0$ has three real roots $z_1 = z_2 = z_3 = -C_2/3C_1$. Since $\lim_{z \to -\infty} \dot{f}(z) < 0$ and $\lim_{z \to \infty} \dot{f}(z) > 0$, we know that z_1 is a local minimum. Thus

$$
|\hat{h}_{Rk}| = \max\left\{ -\frac{C_2}{3C_1}, 0 \right\}. \tag{5.19}
$$

- **Case 4** ($\Delta_1 < 0$)   The equation $\dot{f}(z) = 0$ has one real root z_1 and two complex roots. Since $\lim_{z \to -\infty} \dot{f}(z) < 0$ and $\lim_{z \to \infty} \dot{f}(z) > 0$, we know that z_1 is a local minimum. Thus

$$
|\hat{h}_{Rk}| = \max\{z_1, 0\}. \tag{5.20}
$$

Once $\hat{h}_{Rk}$ is derived, $\hat{g}_k$ can be obtained by substituting $\hat{h}_{Rk}$ into (5.10).

Training Design for Minimizing the Estimation MSE

- **Case 1** ($N \geq M + 1$)   In the high-SNR region, i.e., $\sigma_n^2 \to 0$ with given P_s and P_r, (5.11) can be rewritten as

$$
\hat{h}_{Rk} = \arg\min_{h_{Rk}} \left\{ |h_{Rk}|^2 \mathbf{t}_r^H \mathbf{M} \mathbf{t}_r - 2|h_{Rk}| \Re(e^{j\angle h_{Rk}} \mathbf{x}_{Dk}^H \mathbf{M} \mathbf{t}_r) \right\}. \tag{5.21}
$$

Then we have

$$
\hat{h}_{Rk} = \frac{\mathbf{t}_r^H \mathbf{M} \mathbf{x}_{Dk}}{\mathbf{t}_r^H \mathbf{M} \mathbf{t}_r}. \tag{5.22}
$$

The MSE summation of all h_{Rk} is then derived as

$$
\delta_h = \frac{\sigma_n^2}{|\mathbf{t}_r^H \mathbf{M} \mathbf{t}_r|^2} \operatorname{tr}\left(\mathbf{\Gamma} \mathbf{H}_R \mathbf{H}_R^H \mathbf{\Gamma} + \mathbf{I}_K \right). \tag{5.23}
$$

Clearly, minimizing δ_h is equivalent to maximizing $|\mathbf{t}_r^H \mathbf{M} \mathbf{t}_r|$. Hence, the optimal $\mathbf{t}_r$ to minimize δ_h can be derived from

$$\underset{\mathbf{t}_r}{\text{maximize}} \quad |\mathbf{t}_r^H \mathbf{M} \mathbf{t}_r|$$

$$\text{s.t.} \quad \|\mathbf{t}_r\|^2 = NP_r. \tag{5.24}$$

LEMMA 5.1   *The optimal training sequence $\mathbf{t}_r$ to minimize the estimation MSE of h_{Rk} should satisfy $\mathbf{T}^H \mathbf{t}_r = \mathbf{0}_N$.*

Proof   The optimization problem can be solved by using Karush–Kuhn–Tucker (KKT) conditions, and the Lagrangian function is expressed as

$$\mathcal{L} = -\mathbf{t}_r^H \mathbf{M} \mathbf{t}_r + \lambda \left(\mathbf{t}_r^H \mathbf{t}_r - NP_r \right), \tag{5.25}$$

where λ is the Lagrangian multiplier. The KKT conditions can be established as

$$\frac{\partial \mathcal{L}}{\partial \mathbf{t}_r^H} = -\mathbf{M} \mathbf{t}_r + \lambda \mathbf{t}_r = \mathbf{0}_N, \tag{5.26}$$

leading to

$$\mathbf{T} \left(\mathbf{T}^H \mathbf{T} \right)^{-1} \mathbf{T}^H \mathbf{t}_r = (\lambda - 1) \mathbf{t}_r. \tag{5.27}$$

The solution of the above equation is

$$\mathbf{T} \left(\mathbf{T}^H \mathbf{T} \right)^{-1} \mathbf{T}^H \mathbf{t}_r = \mathbf{0}_N, \tag{5.28}$$

with $\lambda = 1$. Multiplying (5.28) by $\mathbf{T}^H$ gives $\mathbf{T}^H \mathbf{t}_r = 0$, which completes the proof. $\square$

Substituting the expression for $\hat{h}_{Rk}$ into (5.10), the covariance matrix for the estimation error of $\mathbf{g}_k$ is computed as

$$\mathbf{C}_{\mathbf{g}_k} = \mathcal{E} \left\{ \left(\hat{\mathbf{g}}_k - \mathbf{g}_k \right) \left(\hat{\mathbf{g}}_k - \mathbf{g}_k \right)^H \right\} = \frac{\left(\gamma_k^2 |h_{Rk}|^2 + 1 \right) \sigma_n^2}{\gamma_k^2} \left(\mathbf{T}^H \mathbf{T} \right)^{-1}$$

$$+ \frac{\| \gamma_k \mathbf{t}_r^H \mathbf{T} \mathbf{g}_k \|^2 - \|\mathbf{t}_r\|^2 \left(\gamma_k^2 |h_{Rk}|^2 + 1 \right) \sigma_n^2}{\gamma_k^2 \|\mathbf{t}_r\|^4}$$

$$\times \left(\mathbf{T}^H \mathbf{T} \right)^{-1} \mathbf{T}^H \mathbf{t}_r \mathbf{t}_r^H \mathbf{T} \left(\mathbf{T}^H \mathbf{T} \right)^{-1}. \tag{5.29}$$

With the optimal $\mathbf{t}_r$, for which $\mathbf{T}^H \mathbf{t}_r = \mathbf{0}_N$, the estimation MSE of $\mathbf{g}_k$ simplifies to

$$\delta_{\mathbf{g}_k} = \text{Tr}(\mathbf{C}_{\mathbf{g}_k}) = \left(\gamma_k^2 |h_{Rk}|^2 + 1 \right) \sigma_n^2 \text{Tr} \left[\left(\mathbf{T}^H \mathbf{T} \right)^{-1} \right]. \tag{5.30}$$

Thus the optimal training sequences to minimize the estimation MSE can be computed from

$$\underset{\mathbf{T}}{\text{minimize}} \quad \text{Tr} \left[\left(\mathbf{T}^H \mathbf{T} \right)^{-1} \right]$$

$$\text{s.t.} \quad \|\mathbf{t}_m\|^2 = NP_s, \quad \forall m \in [1, M]. \tag{5.31}$$

LEMMA 5.2 *The optimal training matrix $\mathbf{T}$ to minimize the estimation MSE should satisfy $\mathbf{T}^H\mathbf{T} = NP_s\mathbf{I}_N$.*

Proof Assume that $\mathbf{T}_{\text{opt}}$ is the optimal training matrix and define $\mathbf{U} = \mathbf{T}_{\text{opt}}^H\mathbf{T}_{\text{opt}}$. For each sequence group $\left(\mathbf{t}_1' \cdots \mathbf{t}_M'\right)$ that satisfies the power constraint in (5.1), $\text{tr}\left(\mathbf{U}^{-1}\right) \le \text{tr}\left(\mathbf{T}_0^H\mathbf{T}_0\right)$ holds, where $\mathbf{T}_0 = \left[\mathbf{t}_1' \cdots \mathbf{t}_M'\right]$. If $\mathbf{U}$ is not a diagonal matrix, then the diagonal elements of $\mathbf{U}$ are selected to construct a diagonal matrix, denoted by $\mathbf{U}_0$.

Note that an arbitrary $k \times k$ positive-definite matrix L should satisfy

$$\text{Tr}\left(\mathbf{L}^{-1}\right) \ge \sum_{i=1}^{k}\left(L_{i,i}\right)^{-1}, \tag{5.32}$$

and the equality holds if and only if $\mathbf{L}$ is diagonal. Hence, $\text{Tr}\left(\mathbf{U}^{-1}\right) > \text{Tr}\left(\mathbf{U}_0^{-1}\right)$, which contradicts the assumption that $\mathbf{U}$ is the optimal solution. Therefore, the matrix $\left(\mathbf{T}^H\mathbf{T}\right)^{-1}$ is diagonal. Then we have

$$\text{Tr}\left[\left(\mathbf{T}^H\mathbf{T}\right)^{-1}\right] = \sum_{m=1}^{M}\frac{1}{\|\mathbf{t}_m\|^2} \ge \frac{M}{NP_s}.$$

It is easy to see that $\text{tr}\left[\left(\mathbf{T}^H\mathbf{T}\right)^{-1}\right]$ achieves a minimal value when $\mathbf{T}^H\mathbf{T} = NP_s\mathbf{I}_N$. □

- **Case 2** $(N \le M)$ The orthogonal training design cannot be achieved. Let us write $\mathbf{T}_{sr} = \left[\gamma_k\mathbf{T}\ \mathbf{t}_r\right]$. It can be checked that $\mathbf{T}_{sr}^H\mathbf{T}_{sr}$ is singular, and thus it is impossible to adopt a least-squares (LS) estimation. We then use the match-filter-based (MF) estimation, and the channel estimates for the kth subscript can be obtained from

$$\underbrace{\left[\hat{\mathbf{g}}_k^T\ \hat{h}_{Rk}\right]^T}_{\hat{\mathbf{f}}_k} = \underbrace{\left[\mathfrak{D}\left(\mathbf{T}_{sr}^H\mathbf{T}_{sr}\right)\right]^{-1}\mathbf{T}_{sr}^H}_{\mathbf{D}_{MF}}\mathbf{x}_{Dk}. \tag{5.33}$$

The covariance matrix is calculated as

$$\begin{aligned}
\mathbf{C}_{fk} &= \mathbb{E}\left\{\left[\mathbf{D}_{MF}\mathbf{x}_{Dk} - \mathbf{f}_k\right]\left[\mathbf{D}_{MF}\mathbf{x}_{Dk} - \mathbf{f}_k\right]^H\right\} \\
&= \mathbf{D}_{MF}\mathbf{R}_{wk}\mathbf{D}_{MF}^H + \mathbf{D}_{MF}\mathbf{T}_{sr}\mathbf{R}_{fk}\mathbf{T}_{sr}^H\mathbf{D}_{MF}^H + \mathbf{R}_{fk} \\
&\quad - \left(\mathbf{D}_{MF}\mathbf{T}_{sr}\mathbf{R}_{fk} + \mathbf{R}_{fk}\mathbf{T}_{sr}^H\mathbf{D}_{MF}^H\right),
\end{aligned} \tag{5.34}$$

where

$$\mathbf{R}_{fk} = \left[\begin{array}{cc} \upsilon_{Rk}\,\text{diag}\{\upsilon_{1k}, \upsilon_{2k}, \ldots, \upsilon_{Mk}\} & \mathbf{0} \\ \mathbf{0} & \upsilon_{Rk} \end{array}\right],$$

$$\mathbf{R}_{wk} = \left(\gamma_k^2|h_{Rk}|^2 + 1\right)\sigma_n^2\mathbf{I}_N, \tag{5.35}$$

and the MSE is derived as

$$\delta_{fk} = \text{Tr}\left(\mathbf{C}_{fk}\right) = \upsilon_{Rk}\sum_{m=1}^{M}\upsilon_{mk}\left(\frac{\sum_{i=1}^{M}\|\mathbf{t}_m^H\mathbf{t}_i\|^2}{\|\mathbf{t}_m\|^4} - 1\right) + \upsilon_{Rk}\sum_{m=1}^{M}\frac{\|\mathbf{t}_m^H\mathbf{t}_r\|^2}{\|\mathbf{t}_m\|^4}$$

$$+ \sum_{m=1}^{M} \frac{\sigma_n^2 \left(\gamma_k^2 |h_{Rk}|^2 + 1 \right)}{\|\mathbf{t}_m\|^2} + \upsilon_{Rk} \sum_{i=1}^{M} \frac{\upsilon_{mk} \|\mathbf{t}_r^H \mathbf{t}_i\|^2}{\|\mathbf{t}_r\|^4}$$

$$+ \frac{\sigma_n^2 \left(\gamma_k^2 |h_{Rk}|^2 + 1 \right)}{\|\mathbf{t}_r\|^2}. \tag{5.36}$$

The training sequences $\mathbf{t}_m$ and $\mathbf{t}_r$ to minimize the estimation MSE can be obtained from

$$\underset{\mathbf{t}_m, \mathbf{t}_r}{\text{minimize}} \quad \upsilon_{Rk} \sum_{m=1}^{M} \left(\upsilon_{mk} \sum_{i=1, i \neq m}^{M} \|\mathbf{t}_m^H \mathbf{t}_i\|^2 \right) + \upsilon_{Rk} \sum_{m=1}^{M} (\upsilon_{mk} + 1) \|\mathbf{t}_m^H \mathbf{t}_r\|^2$$

$$\text{s.t.} \qquad \|\mathbf{t}_m\|^2 = NP_s \quad \forall m \in [1, M],$$

$$\|\mathbf{t}_r\|^2 = NP_r. \tag{5.37}$$

It is easy to show that the above problem is convex with respect to the $\mathbf{t}_m$ and $\mathbf{t}_r$. Hence, the problem can be effectively solved from convex optimization tools, e.g. [17]. The details are omitted here.

Simulation Results

The performance of the proposed channel estimation methods obtained from Monte Carlo simulations is evaluated in this section. The parameters are set as $M = 2$ and $K = 3$, and all channel coefficients are generated by independent circularly symmetric complex Gaussian random variables with zero mean and unit variance. The average transmitting powers of the UEs and RRHs are set to be the same, i.e., $P_R = P_s$, and the AWGN at the RRHs and at the BBU pool is assumed to have unit variance. The training sequences $\mathbf{t}_1$, $\mathbf{t}_2$, and $\mathbf{t}_r$ are selected as the first, second, and third columns of the $N \times N$ DFT matrix. The average estimation MSE (EMSE) is derived from 10^5 Monte Carlo runs.

Figure 5.3, gives the average EMSE performance for the S-R-D and R-D channels with the corresponding (CRBs). It is known that the deterministic CRB depends on the instantaneous value of h_{Rk}; thus we use the accurate value of h_{Rk} to compute the corresponding CRB for comparison. In accordance with the curves, we see that the EMSE of ML estimation for both the channels quickly decreases with increasing SNR, which demonstrates the effectiveness of the proposed ML-based channel estimation in C-RANs. Moreover, it is seen that the EMSE for the S-R-D channel with ML solution coincides with the CRB, while the EMSE for the R-D channel is higher than the CRB but draws near to the CRB with increasing SNR. This is so because the ML estimation for the S-R-D channel is unbiased and that for the R-D channel is biased.

Figure 5.4 illustrates the average EMSE of the S-R-D channel versus SNR for different training sequences when $N = M$. The traditional sequence reused scheme is chosen as the baseline, i.e., $\mathbf{t}_m$ is assigned to be the $[\text{mod}\,(m, N) + 1]$th column of the $N \times N$ DFT matrix while $\mathbf{t}_r$ is randomly chosen from the columns of the $N \times N$ DFT matrix. It is observed that the optimized training design from solving (5.37) achieves

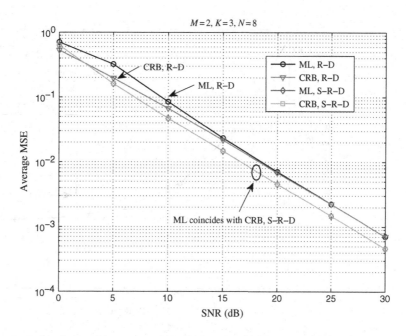

Figure 5.3 Estimation MSEs versus SNR for the channels S-R-D and R-D.

lower EMSE than that of the baseline, thus significantly improving the estimation performance. Besides, we see that error-floors always occur due to the interference between non-orthogonal training sequences.

5.2.2 Superimposed-Segment Training-Based Channel Estimation

In order to get close to the real channel situation, time-varying channels between UEs and RRHs are considered here. The superimposed-training scheme can significantly save spectral resources and is valid to perform channel estimation for time-varying environments using the complex-exponential basis-expansion model (CE-BEM). However, straightforward implementation of superimposed training in C-RANs would degrade transmission quality because superimposing both ACL and wireless FL training sequences on the data signal decreases the effective SNR. Aimed to reduce the training overhead and enhance the channel estimation performance at the BBU pool, superimposed-segment training design is proposed here; in this design superimposed-training is implemented for the ACL while segment-training is applied for the wireless FL. Moreover, a CE-BEM-based maximum *a posteriori* probability (MAP) channel estimation algorithm is developed in which the basis-expansion model (BEM) coefficients of the quasi-static wireless FL are first obtained, after which the time-domain channel samples of the ACL are restored by maximizing the average effective SNR.

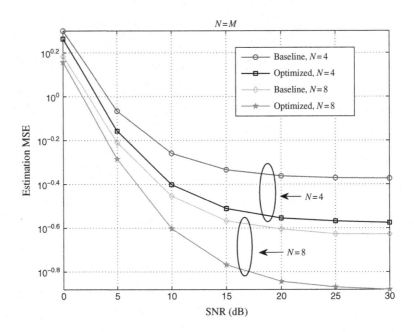

Figure 5.4 Estimation MSEs versus SNR with different training sequences when $N = M$.

System Model and Training Design

Consider a C-RAN operated in half-duplex mode and assume that different UEs served by the same RRU are allocated with a single subcarrier through the orthogonal frequency division multiplexing access (OFDMA) technique, as depicted in Fig. 5.1. It is also assumed that the UEs move continuously, while the RRUs remain fixed. Thus, the radio channels of the ACLs will vary during one transmission block, while those of wireless FLs will undergo quasi-static flat fading. Owing to the orthogonality characteristics of OFDMA for the accessing of multiple UEs, we can focus on the transmission of a single UE.

Let **b** and $\mathbf{t}_s$ denote the data vector and cyclical training sequence transmitted from the UE, respectively. The training sequence from the RRH is denoted by $\mathbf{t}_r$. The nth channel sample of the time-varying radio ACL is denoted by $h(n)$, with mean zero and variance υ_h, while the channel fading of the quasi-static flat WFL is a complex Gaussian distribution g of mean zero and variance υ_g. The transmit powers of the UEs and RRHs are denoted by P_s and P_r, respectively. The noise variance at the RRHs and at the BBU pool is denoted by σ_n^2. It is assumed that the BBU pool acquires a knowledge of $\mathbf{t}_s$, $\mathbf{t}_r$, υ_h, υ_g, P_s, P_r, and σ_n^2.

The superimposed-segment scheme is as shown in Fig. 5.5. During each transmission block, the UE transmits a signal **s** of length N_s symbols to the RRH, in which the nth entry of **s** is given by

$$s(n) = \sqrt{1 - \epsilon}\,b(n) + \sqrt{\epsilon}\,t_s(n)\,, \quad 0 \le n \le N_s - 1, \tag{5.38}$$

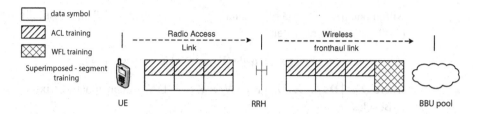

Figure 5.5 The superimposed-segment training scheme in C-RANs.

where $b(n)$ denotes the nth entry of **b** with M-ary phase shift keying (MPSK) modulation constrained by $\mathbb{E}\{|b(n)|^2\} = P_s$ and where $t_s(n)$ represents the nth entry of $\mathbf{t}_s$ with $|t_s(n)|^2 = P_s$, whose period is denoted by N_p. The parameter ϵ satisfies $0 < \epsilon < 1$. Without loss of generality, we may further assume that $K = N_s/N_p$ is an integer[18]. The nth observation at the RRH is written as

$$x_R(n) = h(n)\, s(n) + w_R(n), \quad 0 \le n \le N_s - 1, \tag{5.39}$$

where $w_R(n)$ is the AWGN at the RRH. Then the RRH scales the received signal by

$$\alpha = \sqrt{\frac{P_r}{\upsilon_h P_s + \sigma_n^2}}$$

and inserts $\mathbf{t}_r$ prior to the received signal. The sequence $\mathbf{t}_r$ is of length N_r and its nth entry satisfies $|t_r(n)|^2 = P_r$. The BBU pool receives two separate signals as

$$\mathbf{x}_s = \alpha g\, \mathrm{diag}(\mathbf{h})\ \mathbf{s} + \alpha g \mathbf{w}_R + \mathbf{w}_{Ds}, \tag{5.40}$$

$$\mathbf{x}_r = g\mathbf{t}_r + \mathbf{w}_{Dr}, \tag{5.41}$$

where $\mathbf{w}_R = [w_R(1) \cdots w_R(N_s)]^T$, $\mathbf{w}_{Ds} = [w_{Ds}(1) \cdots w_{Ds}(N_s)]^T$, and $\mathbf{w}_{Dr} = [w_{Dr}(1) \cdots w_{Dr}(N_r)]^T$ are AWGN vectors with each entry having variance σ_n^2. In order to perform coherent reception and to adopt cooperative processing techniques at the BBU pool, a knowledge of the $h(n)$ and g should be obtained; this is explained in the following section.

In [20], it is indicated that the time-domain training consumption for estimating CE-BEM parameters is much higher than that for estimating channel coefficients for the flat fading channels. Thus, we apply superimposed training for ACLs and segment training for WFLs, to ensure a high transmission spectrum efficiency.

CE-BEM-Based MAP Channel Estimation Algorithm

The CE-BEM for time-selective but frequency-flat fading channels is chosen in [19] and [20] to model the time-varying ACL as

$$h(n) = \sum_{q=-Q}^{Q} \lambda_q e^{j2\pi qn/N_s}, \quad 0 \le n \le N_s - 1, \tag{5.42}$$

where the coefficients λ_q are assumed to satisfy independent complex Gaussian distributions with mean zero and variance υ_q. Moreover, the λ_q are assumed to remain invariant within one transmission block, and $\sum_{q=-Q}^{Q} v_q = v_h$ is satisfied. Substituting (5.42) into (5.40), $\mathbf{x}_r$ can be rewritten as

$$\mathbf{x}_s = \alpha g \mathbf{D} \left(\mathbf{I}_{2Q+1} \otimes \mathbf{s} \right) \boldsymbol{\lambda} + \alpha g \mathbf{w}_R + \mathbf{w}_{Ds}, \tag{5.43}$$

where $\mathbf{D} = \left[\mathbf{D}_{-Q}, \ldots, \mathbf{D}_Q \right]$ is an $N_s \times (2Q+1)N_s$-dimensional matrix.
Besides,

$$\mathbf{D}_q = \text{diag}\left\{ 1, e^{j2\pi q/N_s}, \ldots, e^{j2\pi q(N_s-1)/N_s} \right\}. \tag{5.44}$$

By defining $\mathbf{J} = K^{-1}\mathbf{1}_K^T \otimes \mathbf{I}_{N_p}$ of dimension $N_p \times N_s$, we obtain

$$\mathbf{J}\mathbf{D}_q\mathbf{t}_s = \begin{cases} \tilde{t}_s, & q = 0, \tag{5.45a} \\ 0, & q \neq 0, \tag{5.45b} \end{cases}$$

where $\tilde{t}_s$ is an $(N_p \times 1)$-dimensional vector. Left-multiplying $\mathbf{x}_s$ by $\left(\mathbf{I}_{Q+1} \otimes \mathbf{J} \right)\mathbf{D}^H$ yields

$$\mathbf{r} = \left(\mathbf{I}_{Q+1} \otimes \mathbf{J} \right)\mathbf{D}^H \mathbf{x}_s, \tag{5.46}$$

whose $\left[(q+Q)N_p + 1 \right]$th to $\left[(q+Q+1)N_p \right]$th entries, denoted by $\mathbf{r}_q$, can be expressed as

$$\mathbf{r}_q = \alpha g \tilde{t}_s \lambda_q + \alpha g \underbrace{\sum_{l=-Q}^{Q} \mathbf{J}\mathbf{D}_{l-q}\mathbf{b}\lambda_l + \mathbf{J}\mathbf{D}_q^H \left(\alpha g \mathbf{w}_R + \mathbf{w}_{Ds} \right)}_{\mathbf{w}_q}. \tag{5.47}$$

It is can be shown that

$$\mathbb{E}\{\mathbf{w}_{q1}\mathbf{w}_{q1}^H\} = v_n \mathbf{I}_{N_p}, \quad \mathbb{E}\{\mathbf{w}_{q1}\mathbf{w}_{q2}^H\} = \mathbf{0}_{N_p} \tag{5.48}$$

for $q_1 \neq q_2$ with $v_n = k^{-1}\alpha^2 v_g v_h(1 - \epsilon)P_s + \left(\alpha^2 v_g + 1 \right)\sigma_n^2$. It is effective to use the Gaussian distribution to model the noise behavior for estimation problems, and thus we choose the complex Gaussian function to be the nominal likelihood function of $\mathbf{w}_q$. In this case, the likelihood function of $\mathbf{r}$ is written as

$$p(\mathbf{r}|\boldsymbol{\lambda}, g) = \prod_{q=-Q}^{Q} \left(\frac{1}{\pi v_n} \right)^{N_p} e^{-\|\mathbf{r}_q - \alpha g \lambda_q \tilde{t}_s\|^2 / v_n}. \tag{5.49}$$

A. *Estimation of λ_q and g* Defining $\boldsymbol{\theta} = \left[\lambda_{-Q} \cdots \lambda_Q, g \right]^T$, the MAP estimation for $\boldsymbol{\theta}$ gives

$$\hat{\boldsymbol{\theta}} = \arg \max_{\boldsymbol{\theta}} \left\{ p(\mathbf{r}|\boldsymbol{\lambda}, g)p(\mathbf{x}_r|g)p(\boldsymbol{\lambda})\,p(g) \right\}$$

$$= \arg \min_{\boldsymbol{\theta}} \underbrace{\left\{ \sum_{q=-Q}^{Q} \left\{ \frac{\|\mathbf{r}_q - \alpha g \lambda_q \tilde{t}_s\|^2}{v_n} + \frac{|\lambda_q|^2}{v_q} \right\} + \frac{\|\mathbf{x}_r - g\mathbf{t}_r\|^2}{\sigma_n^2} + \frac{|g|^2}{v_g} \right\}}_{\mathcal{L}(\boldsymbol{\theta})}, \tag{5.50}$$

where $p(\mathbf{x}_r|g), p(\lambda)$, and $p(g)$ are Gaussian distribution functions. For a given g, the estimate of λ_q can be obtained as

$$\hat{\lambda}_q = \frac{\alpha g^H \tilde{\mathbf{t}}_s^H \mathbf{r}_q}{\alpha^2 |g|^2 \|\tilde{\mathbf{t}}_s\|^2 + \upsilon_n/\upsilon_q}, \quad -Q \leq q \leq Q. \tag{5.51}$$

Substituting (5.51) into (5.50), the estimate of g is given by

$$\hat{g} = \arg\min_g \underbrace{\left\{ \frac{\|\mathbf{x}_r - g\mathbf{t}_r\|^2}{\sigma_n^2} + \frac{|g|^2}{\upsilon_g} - \frac{\alpha^2 |g|^2 \sum_{q=-Q}^{Q} |\tilde{\mathbf{t}}_s^H \mathbf{r}_q|^2}{\alpha^2 \upsilon_n \|\tilde{\mathbf{t}}_s\|^2 |g|^2 + \upsilon_n^2/\upsilon_q} \right\}}_{\mathcal{L}(g)}. \tag{5.52}$$

Note that only the first term in $\mathcal{L}(g)$ relates to the phase of g, denoted $\angle g$, and thus $\angle g$ can be directly estimated by minimizing $\|\mathbf{x}_r - g\mathbf{t}_r\|^2$ as follows:

$$\widehat{\angle g} = \arg\min\left\{ \|\mathbf{x}_r - g\mathbf{t}_r\|^2 \right\} = \angle(\mathbf{t}_r^H \mathbf{x}_r). \tag{5.53}$$

The estimate of $|g|$ must be either a local minimum of $\mathcal{L}(|g|\angle g)$ or at the boundary $|g| = 0$, which can be obtained from solving $\partial\mathcal{L}(|g|\angle g)/\partial|g| = 0$. Unfortunately, a closed-form expression for $|\hat{g}|$ is hard to derive since $\partial\mathcal{L}(|g|\angle g)/\partial|g|$ is an mth-order polynomial of $|g|$ with $m \geq 4$, and thus numerical methods such as a one-dimensional search are needed to compute the value of $|\hat{g}|$. To reduce the complexity of such approaches, an iterative approach is developed whereby $\hat{g}$ is initialized from

$$\hat{g} = \arg\max_g \left\{ p(\mathbf{x}_r|g)\, p(g) \right\} = \frac{\mathbf{t}_r^H \mathbf{x}_r}{\|\mathbf{t}_r\|^2 + \sigma_n^2/\upsilon_g}. \tag{5.54}$$

With $\hat{g}$ obtained, the $\hat{\lambda}_q$ can be estimated according to (5.51) with $\hat{g}$ in place of g. Then, $\hat{g}$ can be further updated, by substituting these $\hat{\lambda}_q$ into (5.53), as

$$\hat{g} = \frac{\sum_{q=-Q}^{Q} \alpha \hat{\lambda}_q^H \tilde{\mathbf{t}}_s^H \mathbf{r}_q/\upsilon_n + \mathbf{t}_r^H \mathbf{x}_r/\sigma_n^2}{\sum_{q=-Q}^{Q} \alpha^2 |\hat{\lambda}_q|^2 \|\tilde{\mathbf{t}}_s\|^2/\upsilon_n + \|\mathbf{t}_r\|^2/\sigma_n^2 + 1/\upsilon_g}. \tag{5.55}$$

B. Channel restoration for the samples $h(n)$ With the $\hat{\lambda}_q$ obtained, $\hat{h}(n)$ is restored as

$$\hat{h}(n) = \sum_{q=-Q}^{Q} \eta_q \hat{\lambda}_q e^{j2\pi qn/N_s}, \quad 0 \leq n \leq N_s - 1, \tag{5.56}$$

where the η_q are real factors. The vector $\boldsymbol{\eta}$ that maximizes the AESNR [21], denoted by $\boldsymbol{\eta}^*$, is obtained from

$$\boldsymbol{\eta}^* = \arg\max_{\boldsymbol{\eta}} \frac{\mathbb{E}\left\{ \mathbb{E}\left\{ |\hat{g}\hat{h}(n)|^2 \Big| \lambda, g \right\} \right\}(1 - \epsilon)}{\underbrace{\mathbb{E}\left\{ \mathbb{E}\left\{ |\hat{g}\hat{h}(n) - gh(n)|^2 \Big| \lambda, g \right\} + \left(|g|^2 + 1/\alpha^2\right)\sigma_n^2/P_s \right\}}_{\bar{\gamma}}}. \tag{5.57}$$

Define $\phi_{q,n} = e^{j2\pi qn/N_s}$, and denote by Δg and $\Delta \lambda_q$ the estimation errors of g and λ_q, respectively. We obtain

$$\mathbb{E}\left\{|\hat{g}\hat{h}(n)|^2\Big|\boldsymbol{\lambda},g\right\} = |g|^2\underbrace{\sum_{q=-Q}^{Q}\eta_q^2\mathbb{E}\left\{|\Delta\lambda_q|^2\right\}}_{\delta_{\lambda_q}} + \underbrace{\left|\sum_{q=-Q}^{Q}\eta_q\lambda_q\phi_{q,n}\right|^2\underbrace{\mathbb{E}\left\{|\Delta g|^2\right\}}_{\delta_g}}_{\tilde{h}(n)}$$

$$+ |g|^2\left|\sum_{q=-Q}^{Q}\eta_q\lambda_q\phi_{q,n}\right|^2 + \mathbb{E}\left\{\left|\Delta g\sum_{q=-Q}^{Q}\eta_q\Delta\lambda_q\phi_{q,n}\right|^2\right\}. \quad (5.58)$$

Note that $\delta_{\lambda_q}, \delta_g \sim \mathcal{O}(\sigma_n^2/P_s)$ while the last term in (5.58) is of order $\mathcal{O}\left[(\sigma_n^2/P_s)^2\right]$; thus we remove the last term for a high-SNR approximation, i.e.,

$$\mathbb{E}\left\{|\hat{g}\hat{h}(n)|^2\Big|\boldsymbol{\lambda},g\right\} = |g|^2\sum_{q=-Q}^{Q}\eta_q^2\delta_{\lambda_q} + |\tilde{h}(n)|^2\delta_g + |g|^2|\tilde{h}(n)|^2. \quad (5.59)$$

Similarly,

$$\mathbb{E}\left\{|\hat{g}\hat{h}(n) - gh(n)|^2\Big|\boldsymbol{\lambda},g\right\} \approx |g|^2\sum_{q=-Q}^{Q}\eta_q^2\delta_{\lambda_q} + |\tilde{h}(n)|^2\delta_g$$

$$+ |g|^2\left|\sum_{q=-Q}^{Q}(\eta_q - 1)\lambda_q\phi_{n,q}\right|^2. \quad (5.60)$$

Owing to the non-linearity of the MAP estimation, it is hard to derive the corresponding MSE expressions in closed forms. Moreover, it is known that the MAP-estimation MSEs converge to the Cramér–Rao bound (CRB) when the training length is sufficiently large, and thus it is effective to use CRBs in the computation of the AESNR as

$$\delta_{\lambda_q} = \mathrm{CRB}_{\lambda_q} = \frac{\upsilon_n}{\alpha^2|g|^2\|\tilde{\mathbf{t}}_s\|^2}, \quad \delta_g = \mathrm{CRB}_g = \frac{1}{\alpha^2\|\boldsymbol{\lambda}\|^2\|\tilde{\mathbf{t}}_s\|^2/\upsilon_n + \|\mathbf{t}_r\|^2/\sigma_n^2}. \quad (5.61)$$

Substituting the above expressions for δ_{λ_q} and δ_g into $\bar{\gamma}$ and taking the expectation with respect to g, the λ_q, and the noise terms, we obtain

$$\bar{\gamma} = \frac{\boldsymbol{\eta}^T\boldsymbol{\Xi}\boldsymbol{\eta}(1-\epsilon)}{\boldsymbol{\eta}^T\boldsymbol{\Xi}\boldsymbol{\eta} - 2\times\mathbf{1}_{2Q+1}^T\boldsymbol{\Upsilon}\boldsymbol{\eta} + C}, \quad (5.62)$$

where

$$\boldsymbol{\Upsilon} = \frac{\alpha^2\upsilon_g\|\tilde{\mathbf{t}}_s\|^2}{\upsilon_n}\mathbf{R}_\lambda^2 + \frac{\alpha^2\upsilon_g\|\tilde{\mathbf{t}}_s\|^2}{\upsilon_n}\mathrm{Tr}\{\mathbf{R}_\lambda\}\mathbf{R}_\lambda + \frac{\upsilon_g\|\mathbf{t}_r\|^2}{\sigma_n^2}\mathbf{R}_\lambda, \quad (5.63)$$

$$\boldsymbol{\Xi} = \boldsymbol{\Upsilon} + \mathbf{R}_\lambda + \left(\upsilon_h + \frac{\upsilon_n\|\mathbf{t}_r\|^2}{\sigma_n^2\alpha^2\|\tilde{\mathbf{t}}_s\|^2}\right)\mathbf{I}_{2Q+1}, \quad (5.64)$$

$$C = \mathbf{1}_{2Q+1}^T\boldsymbol{\Upsilon}\mathbf{1}_{2Q+1} + \left(\upsilon_g + \frac{1}{\alpha^2}\right)\left(\frac{\upsilon_h\alpha^2\|\tilde{\mathbf{t}}_s\|^2}{\upsilon_n} + \frac{\|\mathbf{t}_r\|^2}{\sigma_n^2}\right)\frac{\sigma_n^2}{P_s}, \quad (5.65)$$

Table 5.1 Iterative channel estimation algorithm

Initialize $\hat{g}$ in accordance with (5.54). For each q, obtain $\hat{\lambda}_q$ by substituting $\hat{g}$ into (5.51).
Repeat

- $g_{\text{temp}} = \hat{g}; \lambda_{q,\text{temp}} = \hat{\lambda}_q$.
- Update $\hat{g}$ by substituting $\lambda_{q,\text{temp}}$'s into (5.55).
- For each q, update $\hat{\lambda}_q$ by substituting g_{temp} into (5.51).

Until $|\hat{g} - g_{\text{temp}}|^2 + \sum_q |\hat{\lambda} - \lambda_{q,\text{temp}}|^2 \leq 0.001$ is satisfied.

Calculate the optimal η^* according to (5.71).
Restore the $\hat{h}(n)$ according to (5.56) by using the $\hat{\lambda}_q$ and η^*.
Return the $\hat{h}(n)$ and $\hat{g}$.

---

and $\mathbf{R}_\lambda = \text{diag}\{v_{-Q}, \ldots, v_Q\}$. On setting $\varphi^* = (\eta^*)^T \, \Xi \, \eta^*$, the optimization for (5.57) transforms to

$$\underset{\eta}{\text{maximize}} \quad 2 \times 1_{2Q+1}^T \Upsilon \eta \tag{5.66}$$

$$\text{s.t.} \quad \eta^T \Xi \eta = \varphi^*. \tag{5.67}$$

Clearly, the optimization problem stated in (5.66) and (5.67) is concave, and η^* can be directly obtained from the Lagrange dual function as

$$\eta^* = \frac{\sqrt{\varphi^*} \, \Xi^{-1} \Upsilon 1_{2Q+1}}{\sqrt{1_{2Q+1}^T \Upsilon \Xi^{-1} \Upsilon 1_{2Q+1}}}. \tag{5.68}$$

Substituting (5.68) back into (5.62), the optimization problem becomes

$$\underset{\varphi}{\text{maximize}} \quad \frac{\varphi}{\varphi - 2\sqrt{1_{2Q+1}^T \Upsilon \Xi^{-1} \Upsilon 1_{2Q+1}} \sqrt{\varphi} + C} \tag{5.69}$$

$$\text{s.t.} \quad \varphi > 0, \tag{5.70}$$

whose solution is

$$\varphi^* = \frac{C^2}{1_{2Q+1}^T \Upsilon \Xi^{-1} \Upsilon 1_{2Q+1}}, $$

leading to

$$\eta^* = \frac{C \Xi^{-1} \Upsilon 1_{2Q+1}}{1_{2Q+1}^T \Upsilon \Xi^{-1} \Upsilon 1_{2Q+1}}. \tag{5.71}$$

Combining the estimation for θ and channel restoration for the $h(n)$, the proposed channel estimation algorithm is summarized in Table 5.1. Moreover, the following proposition is given to demonstrate the effectiveness of the proposed algorithm.

PROPOSITION 5.3   The iterative channel estimation algorithm is convergent, and it achieves a lower MSE than that of the ML method.

Proof   Each iteration consists of $2Q + 2$ steps. Denote the ith entry of θ by θ_i; the updated estimate of θ_i, denoted by $\hat{\theta}_i^{\text{new}}$, satisfies $\mathcal{L}\left(\hat{\theta}_i^{\text{new}}\right) > \mathcal{L}\left(\hat{\theta}_i\right)$. This indicates that $\mathcal{L}(\theta)$ strictly increases after each step as well as after one round of iteration. It is known that $\mathcal{L}(\theta)$ is continuous with respect to θ_i and that $\mathcal{L}(\theta) < +\infty$. Thus, it may be concluded that the iterative algorithm is convergent.

From (5.55), the MAP estimate of g with a given λ is

$$
\hat{g}^{\text{MAP}} = \frac{\alpha^2 \|\lambda\|^2 \|\tilde{\mathbf{t}}_s\|^2 / \upsilon_n + \|\mathbf{t}_r\|^2 / \sigma_n^2}{\alpha^2 \|\lambda\|^2 \|\tilde{\mathbf{t}}_s\|^2 / \upsilon_n + \|\mathbf{t}_r\|^2 / \sigma_n^2 + 1/\upsilon_g} g
$$
$$
+ \frac{\sum_{q=-Q}^{Q} \alpha \lambda_q^H \tilde{\mathbf{t}}_s^H \mathbf{w}_q / \upsilon_n + \mathbf{t}_r^H \mathbf{w}_{Dr} / \sigma_n^2}{\alpha^2 \|\lambda\|^2 \|\tilde{\mathbf{t}}_s\|^2 / \upsilon_n + \|\mathbf{t}_r\|^2 / \sigma_n^2 + 1/\upsilon_g}, \tag{5.72}
$$

whose MSE is calculated as

$$
\delta_g^{\text{MAP}} = \mathbb{E}\left\{\left\|\hat{g}^{\text{MAP}} - g\right\|^2\right\}
$$
$$
= \frac{1}{\alpha^2 \|\lambda\|^2 \|\tilde{\mathbf{t}}_s\|^2 / \upsilon_n + \|\mathbf{t}_r\|^2 / \sigma_n^2} \frac{\upsilon_g}{\upsilon_g + \frac{1}{(\alpha^2 \|\lambda\|^2 \|\tilde{\mathbf{t}}_s\|^2 / \upsilon_n + \|\mathbf{t}_r\|^2 / \sigma_n^2)^{-1}}}. \tag{5.73}
$$

The ML estimate of g gives

$$
\hat{g}^{\text{ML}} = g + \frac{\sum_{q=-Q}^{Q} \alpha \lambda_q^H \tilde{\mathbf{t}}_s^H \mathbf{w}_q / \upsilon_n + \mathbf{t}_r^H \mathbf{w}_{Dr} / \sigma_n^2}{\alpha^2 \|\lambda\|^2 \|\tilde{\mathbf{t}}_s\|^2 / \upsilon_n + \|\mathbf{t}_r\|^2 / \sigma_n^2}, \tag{5.74}
$$

whose MSE is calculated as

$$
\delta_g^{\text{ML}} = \mathbb{E}\left\{\left\|\hat{g}^{\text{ML}} - g\right\|^2\right\} = \frac{1}{\alpha^2 \|\lambda\|^2 \|\tilde{\mathbf{t}}_s\|^2 / \upsilon_n + \|\mathbf{t}_r\|^2 / \sigma_n^2}. \tag{5.75}
$$

Clearly, $\delta_g^{\text{MAP}} < \delta_g^{\text{ML}}$ always holds, and it can be also stated that $\delta_{\lambda_q}^{\text{MAP}} < \delta_{\lambda_q}^{\text{ML}}$ is satisfied similarly.   □

Remark   According to (5.73) and (5.75) we see that $\delta_g^{\text{MAP}} < \text{CRB}_g$ and $\delta_g^{\text{ML}} = \text{CRB}_g$. This is so because the proposed MAP estimation algorithm is biased, since $\mathbb{E}\{\hat{g}^{\text{MAP}}\} \neq g$.

Simulation Results

Simulation results were obtained to evaluate the performance of the CE-BEM-based MAP (C-MAP) channel estimation algorithm. The ACL channel $\{h(n)\}$ and the WFL channel g are generated from the spatial channel model (SCM) in *3GPP TR 25.996* [22]. The parameters are set as $N_s = 800$ and $N_r = 4$. We assume binary-phase-shift-keying (BPSK) modulation for $\{b(n)\}$, while $\tilde{\mathbf{t}}_s$ is selected as the second column of the $N_p \times N_p$ discrete Fourier transform (DFT) matrix and $\mathbf{t}_r$ is selected as the third column of the $N_r \times N_r$ DFT matrix. The transmit power P_s and P_r are set to be equal, and the noise variance is set at a unit value; thus the SNR is equal to P_s.

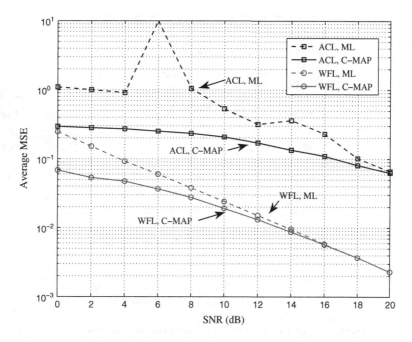

Figure 5.6 Average MSE versus SNR for different estimation methods.

In Fig. 5.6 the average MSEs of both the ACL and WFL channels for C-MAP estimation are compared with that for ML estimation. It is observed that the proposed C-MAP estimation outperforms the traditional ML method, since C-MAP achieves lower MSEs than ML for both ACL and WFL channels. Moreover, it is seen that the MSE of the ACL channel for the ML method is not convergent. The reason is that the random generation of g would result in a singularity, e.g., $g \rightarrow 0$, leading to $\Delta\lambda_q \rightarrow \infty$ for the ML method, while the proposed C-MAP algorithm is robust at the singularity. In Fig. 5.7, we evaluate the AESNR performance for the optimal weighted approach (OWA) to channel restoration. It is seen that the OWA obtains higher AESNRs than the baseline (performing restoration according to CE-BEM), especially in the low-SNR region, and it draws near to the baseline as SNR increases.

5.3   Segment Training Scheme in C-RANs

In the superimposed training scheme, the training sequences and data symbols have to share the SNR originally allocated for data transmission. As a result, inevitably the performance of both the data transmission rate and the estimation accuracy will degrade. In order to improve the data transmission rate and the estimation accuracy, the segment training scheme is used. It discussed in this section.

The segment training scheme introduced in [10] can be applied to acquire individual CSI for the ACLs and WFLs in C-RANs. In segment training scheme, two consecutive

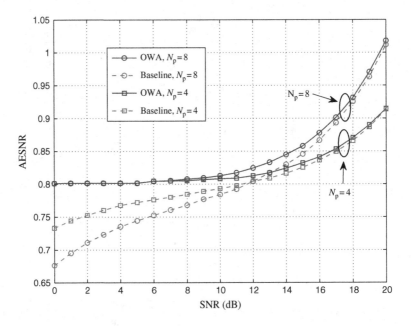

Figure 5.7 AESNR versus SNR for different channel-restoring methods.

segments are implemented for the CSI acquisition of ACLs and WFLs, respectively. In [13] the segment training scheme was utilized as a part of superimposed-segment training. However, it was applied only for the estimation of WFLs, and the corresponding optimal training design was not considered. In [14], two individual training segments were allocated for individual CSI estimation for the ACLs and WFLs, and the segment training scheme demonstrated the effectiveness of keeping the estimation accuracy high.

System Model and Training Transmission

The estimation of block-fading channels is now considered. The first T symbols of each fading block are dedicated for training. A two-phase-based segment training scheme illustrated in Fig. 5.8 is presented that acquires individual CSI for the ACLs and WFLs. During phase I, the UEs transmit the training sequence $\boldsymbol{\phi}_1$ to the RRHs. In phase II the RRHs add another training sequence $\boldsymbol{\phi}_2$ in front of $\boldsymbol{\phi}_1$ and then forward the reorganized signals to the BBU pool. The sequences $\boldsymbol{\phi}_1$ and $\boldsymbol{\phi}_2$ are assumed to occupy times τ_1 and τ_2, and the length of training satisfies $\tau_1 + \tau_2 = T$. The training signal received at the RRHs during phase I is given by

$$\mathbf{Y}_{r1} = \mathbf{G}_{r1}\boldsymbol{\Psi} + \mathbf{N}_{r1}, \tag{5.76}$$

where $\boldsymbol{\Psi}$ is a $M \times \tau_1$ transmitted training matrix stacked by $\boldsymbol{\phi}_1$, $\mathbf{G}_{r1}$ is a spatially uncorrelated channel matrix, and the elements in $\mathbf{N}_{r1}$ form an additive white Gaussian noise matrix with zero mean and variances σ_n^2. The matrix $\mathbf{G}_{r1}$ can be represented as

$$\mathbf{G}_{r1} = \boldsymbol{\Gamma}_1 \odot \mathbf{H}_{r1}, \tag{5.77}$$

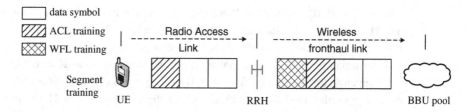

Figure 5.8 The segment training scheme in C-RANs.

where $\mathbf{H}_{r1}$ is an $K \times M$ fast-fading channel matrix, the elements of which are uncorrelated circularly symmetric complex Gaussian random variables with zero means and unit variances, and $\mathbf{\Gamma}_1$ denotes a $K \times M$ matrix of large-scale fading coefficients; $[\mathbf{\Gamma}_1]_{ij} = \sqrt{\kappa_{ij}}$ is the (i,j)th element of $\mathbf{\Gamma}_1$, which is assumed to be time-invariant over many coherence time intervals and to be known *a priori*.

By vectorizing the received training signal, we have

$$\mathbf{y}_{r1} = \left(\mathbf{\Psi}^T \otimes \mathbf{I}_M \right) \mathbf{D}^{1/2} \mathbf{h}_{r1} + \mathbf{n}_{r1}, \tag{5.78}$$

where $\mathbf{y}_{r1} = \text{vec}\,(\mathbf{Y}_{r1})$, $\mathbf{D}^{1/2} = \text{diag}\,\{\text{vec}(\mathbf{\Gamma}_1)\}$, $\mathbf{h}_{r1} = \text{vec}\,(\mathbf{H}_{r1})$, and $\mathbf{n}_{r1} = \text{vec}\,(\mathbf{N}_{r1})$.

During phase II, the segment training scheme at the RRHs can be expressed as

$$\mathbf{\Psi}_1 = [\mathbf{\Phi}, \mathbf{C}\mathbf{Y}_{r1}], \tag{5.79}$$

where $\mathbf{\Psi}_1$ is an $M \times T$ matrix, $\mathbf{C}$ is an $K \times K$ known transformation matrix serving as a co-training matrix at the RRHs, and $\mathbf{\Phi}$ is an $K \times \tau_2$ training matrix added in front of the received training signal.

The received training signal at the BBU pool forwarded by the RRHs can be partition into two parts:

$$\mathbf{Y}_r = \mathbf{F}\mathbf{G}_{r2}\mathbf{\Phi} + \mathbf{F}\mathbf{N}_r, \tag{5.80}$$

$$\mathbf{Y}_d = \mathbf{G}_{r2}\mathbf{C}\mathbf{G}_{r1}\mathbf{\Psi} + \mathbf{G}_{r2}\mathbf{C}\mathbf{N}_{r1} + \mathbf{N}_{r2}, \tag{5.81}$$

where $\mathbf{N}_r$ and $\mathbf{N}_{r2}$ are the $N \times \tau_2$ and $N \times \tau_1$ noise matrices with zero means and element-wise variances σ_n^2, respectively, and $\mathbf{F}$ is a known $N \times N$ unitary matrix applied at the BBU pool; $\mathbf{G}_{r2}$ is an $N \times M$ channel matrix of the WFLs, which is given by

$$\mathbf{G}_{r2} = \mathbf{H}_{r2}\mathbf{\Gamma}_2^{1/2}, \tag{5.82}$$

where $\mathbf{H}_{r2}$ is an $N \times K$ fast-fading channel matrix and $\mathbf{\Gamma}_2$ is a $K \times K$ diagonal large-scale fading matrix with β_i as the ith diagonal element. Here $\mathbf{H}_{r2}$ is assumed to be temporally and spatially correlated, which means that it satisfies

$$\mathbf{H}_{r2} = \mathbf{R}_c\mathbf{H}_w, \tag{5.83}$$

$$\mathbf{h}_{r2,0} = (\mathbf{I} \otimes \mathbf{R}_c)\mathbf{v}_0, \tag{5.84}$$

$$\mathbf{h}_{r2,i} = \eta\mathbf{h}_{r2,i-1} + \sqrt{1 - \eta^2}(\mathbf{I} \otimes \mathbf{R}_c)\mathbf{v}_i, \quad i \geq 1, \quad 0 \leq \eta \leq 1, \tag{5.85}$$

where $\mathbf{R} = \mathbf{R}_c\mathbf{R}_c^H$ is an $N \times N$ received correlation matrix known at the BBU pool and $\mathbf{H}_w$ is an $N \times K$ Gaussian random matrix with independent and identically distributed (i.i.d.) zero-mean and unit-variance entries. As the spatial correlation matrix $\mathbf{R}$ is a Hermitian positive definite matrix, it can be decomposed as $\mathbf{R} = \mathbf{U}\mathbf{\Lambda}\mathbf{U}^H$, where $\mathbf{U}$ is the eigenvector matrix and $\mathbf{\Lambda} = \mathrm{diag}\{[\lambda_{0,1}\lambda_{0,2}\cdots\lambda_{0,N}]\}$ is the eigenvalue matrix in descending order. In (5.84) and (5.85) $\mathbf{h}_{r2,i}$ is the vectorization of $\mathbf{H}_{r2}$ in the ith fading block ($i = 0, 1, \ldots$), η is a temporal correlation coefficient, and $\mathbf{v}_i \sim \mathcal{CN}(\mathbf{0}, \mathbf{I}_{KN})$ is an innovation process.

If $\mathbf{Y}_r$ is the training signal, denote $\mathbf{Y}_{r,i}$ as the signal received in the ith fading block. Using the vec operator, $\mathbf{Y}_{r,i}$ can be rewritten as

$$\mathbf{y}_i = \left[(\mathbf{\Phi}_i^T\mathbf{\Gamma}_2^{1/2}) \otimes \mathbf{F}\right]\mathbf{h}_{r2,i} + (\mathbf{I} \otimes \mathbf{F})\,\mathbf{n}_{r,i}, \quad i \geq 0, \tag{5.86}$$

where $\mathbf{y}_i = \mathrm{vec}\left(\mathbf{Y}_{r,i}\right)$; $\mathbf{h}_{r2,i}$ and $\mathbf{n}_{r,i}$ are the vectorizations of $\mathbf{H}_{r2}$ and $\mathbf{N}_r$ in the ith fading block, respectively.

SMMSE Channel Estimation

In this subsection the SMMSE channel estimation algorithm is developed along with the optimal training design for $\mathbf{H}_{r1}$ and $\mathbf{H}_{r2}$. The individual channel estimation of ACLs and WFLs is derived using the Kalman filter and eigenvalue decomposition (EVD). According to the MSE minimization criterion, the optimal training of the SMMSE estimator is thereby derived as well.

A. Training Design for $\boldsymbol{H}_{r2}$ Considering the temporal correlation, the estimation of $\mathbf{h}_{r2,i}$ can rely not only on the current received training signal $\mathbf{y}_i$ but also on the previously received training signals $\{\mathbf{y}_k\}_{k=0}^{i-1}$. On the basis of the channel evolution (5.85), the estimation of $\mathbf{h}_{r2,i}$ is similar to the state prediction in dynamical systems. Relying on the Kalman filter, an accurate estimate is obtained by using the entire received training signals $\{\mathbf{y}_k\}_{k=0}^{i}$ and the channel statistics η. Denote $\hat{\mathbf{h}}_{r2,i_1|i_2}$ as the estimated value of $\mathbf{h}_{r2,i_1}$, given $\{\mathbf{y}_k\}_{k=0}^{i_2}$ for $i_1 \geq i_2$. Let $\tilde{\mathbf{\Phi}}_i = (\mathbf{\Phi}_i^T\mathbf{\Gamma}_2^{1/2}) \otimes \mathbf{F}$. Then the SMMSE estimation based on the Kalman filter is given by Algorithm 3.

The optimal training design of $\mathbf{\Phi}_i$ in the ith fading block is formulated as

$$\underset{\mathbf{X}_i \geq 0}{\text{minimize}} \; \mathcal{M}_i = \mathbf{\Phi}_i$$

$$\text{s.t.} \quad \mathrm{Tr}\,(\mathbf{X}_i) \leq \tau_2 P_r, \tag{5.87}$$

where $\mathbf{X}_i = \mathbf{\Phi}_i^*\mathbf{\Phi}_i^T \geq 0$ is a positive semidefinite constraint and $\mathcal{M}_i = \mathrm{Tr}\left(\mathbf{R}_{i|i}\right)$ is the MSE in the ith fading block.

Assuming that $\mathbf{F} = \mathbf{U}^H$ and using the identity $(\mathbf{I} + \mathbf{AB})^{-1} = \mathbf{I} - \mathbf{A}\,(\mathbf{I} + \mathbf{BA})^{-1}\,\mathbf{B}$, the resulting MSE $\mathcal{M}_0$ can be directly computed as

$$\mathcal{M}_0 = \mathrm{Tr}\left\{\left(\mathbf{I}_K \otimes \mathbf{R}_c^H\mathbf{R}_c\right)\left(\mathbf{I}_{KN} + \frac{1}{\sigma_n^2}(\mathbf{\Gamma}_2^{1/2}\mathbf{\Phi}_0^*\mathbf{\Phi}_0^T\mathbf{\Gamma}_2^{1/2}) \otimes (\mathbf{R}_c^H\mathbf{R}_c)\right)^{-1}\right\}. \tag{5.88}$$

Algorithm 3 Sequential MMSE estimation

1: **Initialization:**
2: The initial estimate: $\hat{\mathbf{h}}_{r2,0|-1} = \mathbf{0}$,
3: the initial covariance matrix: $\mathbf{R}_{0|-1} = \mathbb{E}\{\mathbf{h}_{r2,0}\mathbf{h}_{r2,0}^H\} = \mathbf{I}_K \otimes \mathbf{R}$.
4: **Prediction:** $\hat{\mathbf{h}}_{r2,i|i-1} = \eta\hat{\mathbf{h}}_{r2,i-1|i-1}$
5: **Minimum prediction covariance matrix:**
6:    $\mathbf{R}_{i|i-1} = \eta^2\mathbf{R}_{i-1|i-1} + (1 - \eta^2)(\mathbf{I}_K \otimes \mathbf{R})$.
7: **Optimized estimation of the current state:**
8: The Kalman gain matrix:
9:    $\mathbf{K}_i = \mathbf{R}_{i|i-1}\tilde{\mathbf{\Phi}}_i^H \left(\sigma_n^2\mathbf{I}_{N\tau_2} + \tilde{\mathbf{\Phi}}_i\mathbf{R}_{i|i-1}\tilde{\mathbf{\Phi}}_i^H\right)^{-1}$,
10: The optimized estimation:
11:    $\hat{\mathbf{h}}_{r2,i|i} = \hat{\mathbf{h}}_{r2,i|i-1} + \mathbf{K}_i \left(\mathbf{y}_i - \tilde{\mathbf{\Phi}}_i\hat{\mathbf{h}}_{r2,i|i-1}\right)$.
12: **Minimum covariance matrix:**
13:    $\mathbf{R}_{i|i} = \left(\mathbf{I}_{KN} - \mathbf{K}_i\tilde{\mathbf{\Phi}}_i\right)\mathbf{R}_{i|i-1}$.

Then, the corresponding Lagrangian function with the diagonal constraint can be written as

$$L(\mathbf{\Phi}_0, \mu)$$

$$= \mathrm{Tr}\left\{\left(\mathbf{I}_{KN} + \frac{1}{\sigma_n^2}(\mathbf{\Gamma}_2^{1/2}\mathbf{\Phi}_0^*\mathbf{\Phi}_0^T\mathbf{\Gamma}_2^{1/2}) \otimes \mathbf{\Lambda}\right)^{-1}(\mathbf{I}_K \otimes \mathbf{\Lambda})\right\} + \mu\left(\mathrm{Tr}(\mathbf{\Phi}_0^*\mathbf{\Phi}_0^T) - \tau_2 P_r\right),$$

$$= \sum_{k}^{N} \lambda_{0,k}\mathrm{Tr}\left\{\left(\mathbf{I}_K + \frac{\lambda_{0,k}}{\sigma_n^2}(\mathbf{\Gamma}_2^{1/2}\mathbf{\Phi}_0^*\mathbf{\Phi}_0^T\mathbf{\Gamma}_2^{1/2})\right)^{-1}\right\} + \mu\left(\mathrm{Tr}(\mathbf{\Phi}_0^*\mathbf{\Phi}_0^T) - \tau_2 P_r\right). \quad (5.89)$$

The Karush–Kuhn–Tucker (KKT) conditions with respect to (5.89) are given by

$$\frac{\partial L}{\partial \mathbf{\Phi}_0^H} = \mathbf{\Phi}_0^T\left\{-\sum_{k}^{N} \frac{\lambda_{0,k}^2}{\sigma_n^2}\mathbf{\Gamma}_2^{1/2}\left(\mathbf{I}_K + \frac{\lambda_{0,k}}{\sigma_n^2}(\mathbf{\Gamma}_2^{1/2}\mathbf{\Phi}_0^*\mathbf{\Phi}_0^T\mathbf{\Gamma}_2^{1/2})\right)^{-2T}\right.$$

$$\left. \times \mathbf{\Gamma}_2^{1/2} + \mu\mathbf{I}\right\}, \quad (5.90)$$

$$\mu\left(\mathrm{Tr}(\mathbf{\Phi}_0^*\mathbf{\Phi}_0^T) - \tau_2 P_r\right) = 0. \quad (5.91)$$

Therefore, the optimal training design of $\mathbf{\Phi}_0$ can be obtained by setting (5.90) to zero, and we have

$$\sum_{k}^{N} \frac{\lambda_{0,k}^2}{\sigma_n^2}\left(\mathbf{I}_K + \frac{\lambda_{0,k}}{\sigma_n^2}(\mathbf{\Gamma}_2^{1/2}\mathbf{\Phi}_0^*\mathbf{\Phi}_0^T\mathbf{\Gamma}_2^{1/2})\right)^2 = \mu\mathbf{\Gamma}_2^{-1}. \quad (5.92)$$

It is easy to observe from (5.92) that the optimal training matrix $\mathbf{\Phi}_0$ must satisfy $\mathbf{\Phi}_0^*\mathbf{\Phi}_0^T = \tau_2 p_r\mathbf{I}_K$ for any $\mu > 0$.

When the optimal training matrix $\mathbf{\Phi}_0$ is applied, $\mathbf{R}_{1|0}$ can be calculated according to the Kalman filter:

$$\mathbf{R}_{1|0} = \eta^2\mathbf{R}_{0|0} + (1 - \eta^2)\tilde{\mathbf{R}}$$

$$= \tilde{\mathbf{U}} \operatorname{diag} \left\{ \lambda_{0,1} - \frac{\eta^2 \tau_2 p_r \beta_1 \lambda_{0,1}^2}{\sigma_n^2 + \tau_2 p_r \beta_1 \lambda_{0,1}}, \ldots, \lambda_{0,1} - \frac{\eta^2 \tau_2 p_r \beta_{\tau_2} \lambda_{0,1}^2}{\sigma_n^2 + \tau_2 p_r \beta_{\tau_2} \lambda_{0,1}}, \ldots, \right.$$

$$\left. \lambda_{0,\tau_2} - \frac{\eta^2 \tau_2 p_r \beta_{\tau_2} \lambda_{0,\tau_2}^2}{\sigma_n^2 + \tau_2 p_r \beta_{\tau_2} \lambda_{0,\tau_2}}, \lambda_{0,\tau_2+1}, \ldots, \lambda_{0,N} \right\} \tilde{\mathbf{U}}^H, \qquad (5.93)$$

where $\tilde{\mathbf{R}} = \mathbf{I} \otimes \mathbf{R}$, $\tilde{\mathbf{U}} = \mathbf{I} \otimes \mathbf{U}$, and $\tilde{\mathbf{\Lambda}} = \mathbf{I} \otimes \mathbf{\Lambda}$. The form of $\mathbf{R}_{1|0}$ is similar to $\tilde{\mathbf{R}}$ and this is true for all the following $\mathbf{R}_{i|i-1}, i > 1,$. Through the recursive derivation, the optimal training matrix $\mathbf{\Phi}_i$ for $i \geq 1$ also satisfies $\mathbf{\Phi}_i^* \mathbf{\Phi}_i^T = \tau_2 p_r \mathbf{I}_M$.

Then, applying the optimal $\mathbf{\Phi}_i$, the MSE of the ith fading block ($i \geq 1$) can be generalized as

$$\mathcal{M}_i = \operatorname{Tr} \left\{ \mathbf{R}_{i|i-1} - (\sigma_n^2 \mathbf{I}_{N\tau_2} + \tilde{\mathbf{\Phi}}_i \mathbf{R}_{i|i-1} \tilde{\mathbf{\Phi}}_i^H)^{-1} \tilde{\mathbf{\Phi}}_i \mathbf{R}_{i|i-1}^2 \tilde{\mathbf{\Phi}}_i^H \right\}$$

$$= \tau_2 \sum_{j=1}^{N} \lambda_{0,j} - \sum_{p=0}^{i} \sum_{j=1}^{\tau_2} \sum_{k=1}^{\tau_2} \frac{\eta^{2(i-p)} \tau_2 p_r \beta_j \lambda_{p,(j-1)\tau_2+k}^2}{\sigma_n^2 + \tau_2 p_r \beta_j \lambda_{p,(j-1)\tau_2+k}}, \qquad (5.94)$$

where $\lambda_{i,j}$ is the jth dominant eigenvalue of $\mathbf{R}_{i|i-1}$. The MSE in (5.94) shows that the estimation accuracy can be improved with longer training sequences. When the channel coherence time T is fixed, allocating more symbols for training degrades the data transmission rate. Hence, a satisfactory tradeoff between the lengths of training and the number of data symbols is important to improve the overall system performance.

B. Training Design for $\mathbf{H}_{r1}$ Assuming that $\mathbf{G}_{r2}$ is known at the BBU pool and that the estimate of $\mathbf{G}_{r2}$ is near perfect, the estimation error of $\mathbf{G}_{r2}$ has little impact on the estimation of $\mathbf{G}_{r1}$ and can be neglected. The vectorization of the received signals $\mathbf{Y}_d$ is given by

$$\mathbf{y}_d = \operatorname{vec}(\mathbf{Y}_d) = (\mathbf{\Psi}^T \otimes \mathbf{G}_{r2} \mathbf{C}) \mathbf{D}^{1/2} \mathbf{h}_{r1} + (\mathbf{I} \otimes \mathbf{G}_{r2} \mathbf{C}) \mathbf{n}_{r1} + \mathbf{n}_{r2}. \qquad (5.95)$$

According to the properties of the LMMSE estimator, the covariance matrix of the estimation error for $\mathbf{h}_{r1}$ is calculated as [23]

$$\mathbf{R}_{\delta \mathbf{h}_{r1}} = \mathbf{R}_{\mathbf{h}_{r1}} - \mathbf{R}_{\mathbf{h}_{r1} \mathbf{y}_d^H} \mathbf{R}_{\mathbf{y}_d \mathbf{y}_d^H}^{-1} \mathbf{R}_{\mathbf{h}_{r1} \mathbf{y}_d^H}^H$$

$$= \mathbf{I} - \mathbf{D}^{1/2} \tilde{\mathbf{\Psi}}^H (\sigma_n^2 \mathbf{I}_{N\tau_1} + \tilde{\mathbf{\Psi}} \mathbf{D} \tilde{\mathbf{\Psi}}^H + \sigma_i^2 \mathbf{I}_{\tau_1} \otimes \mathbf{G}_{r2} \mathbf{C} \mathbf{C}^H \mathbf{G}_{r2}^H)^{-1} \tilde{\mathbf{\Psi}} \mathbf{D}^{1/2}, \qquad (5.96)$$

where $\tilde{\mathbf{\Psi}} = \mathbf{\Psi}^T \otimes (\mathbf{G}_{r2} \mathbf{C})$.

By minimizing the covariance $\mathbf{R}_{\delta \mathbf{h}_{r1}}$, the optimal structure of $\mathbf{\Psi}$ is derived. Then, the optimization training design of $\mathbf{\Psi}$ and $\mathbf{C}$ subject to the power constraints at both the UEs and the RRHs is given by

$$\underset{\mathbf{\Psi}, \mathbf{C}}{\operatorname{minimize}} \quad \mathcal{M} = \operatorname{Tr}(\mathbf{R}_{\delta \mathbf{h}_{r1}})$$

$$\text{s.t.} \qquad \operatorname{Tr}(\mathbf{\Psi} \mathbf{\Psi}^H) \leq \tau_1 P_s,$$

$$\mathbb{E}\{\operatorname{Tr}(\mathbf{y}_s \mathbf{y}_s^H)\} \leq \tau_1 P_r, \qquad (5.97)$$

where $\mathbf{y}_s = \text{vec}\,(\mathbf{CY}_r)$ and $P_s = Kp_s$ is the power bound at the UEs. The power constraint at the RRHs can be calculated as

$$\mathbb{E}\{\text{Tr}(\mathbf{y}_s\mathbf{y}_s^H)\} = \text{Tr}\left(\mathbf{D}(\mathbf{\Psi}_i^*\mathbf{\Psi}_i^T) \otimes (\mathbf{C}^H\mathbf{C}) + \sigma_n^2(\mathbf{I} \otimes \mathbf{CC}^H)\right) \le \tau_1 P_r. \tag{5.98}$$

In order to solve the problem (5.97), we resort to the (EVD) with the training and transformation matrices decomposed into unitary components and diagonal components, respectively. The EVDs of $\mathbf{\Psi}^T\mathbf{\Psi}^*$ and $\mathbf{G}_{r2}\mathbf{CC}^H\mathbf{G}_{r2}^H$ are given by

$$\mathbf{\Psi}^T\mathbf{\Psi}^* = \mathbf{U}_1\mathbf{\Lambda}_1\mathbf{U}_1^H, \tag{5.99}$$

$$\mathbf{G}_{r2}\mathbf{CC}^H\mathbf{G}_{r2}^H = \mathbf{U}_2\mathbf{\Lambda}_2\mathbf{U}_2^H, \tag{5.100}$$

respectively, where $\mathbf{U}_1$ and $\mathbf{U}_2$ are unitary eigenvector matrices and $\mathbf{\Lambda}_1$ and $\mathbf{\Lambda}_2$ are diagonal eigenvalue matrices with elements in descending order. Then $\mathbf{\Psi}$ and $\mathbf{G}_{r2}\mathbf{C}$ can be separately written as

$$\mathbf{\Psi}^T = \mathbf{U}_1\mathbf{\Lambda}_1^{1/2}\mathbf{Q}_1, \tag{5.101}$$

$$\mathbf{G}_{r2}\mathbf{C} = \mathbf{U}_2\mathbf{\Lambda}_2^{1/2}\mathbf{Q}_2, \tag{5.102}$$

where $\mathbf{Q}_1$ and $\mathbf{Q}_2$ are unitary matrices. The singular value decomposition (SVD) of $\mathbf{G}_{r2}$ is $\mathbf{U}_g\mathbf{\Lambda}_g\mathbf{Q}_g^H$, with the singular values in descending order, where $\mathbf{U}_g$ and $\mathbf{Q}_g$ are unitary singular vector matrices.

Note that the minimization of $\mathbf{R}_{\delta\mathbf{h}_{r1}}$ is equivalent to maximizing the second term on the RHS of (5.96), which can be simplified as follows:

$$\tilde{\mathcal{M}} = \text{Tr}\left\{\mathbf{D}^{1/2}\tilde{\mathbf{\Psi}}^H\left(\sigma_n^2\mathbf{I}_{N\tau_1} + \sigma_n^2\mathbf{I}_{\tau_1} \otimes \mathbf{G}_{r2}\mathbf{CC}^H\mathbf{G}_{r2}^H + \tilde{\mathbf{\Psi}}\mathbf{D}\tilde{\mathbf{\Psi}}^H\right)^{-1}\tilde{\mathbf{\Psi}}\mathbf{D}^{1/2}\right\}$$

$$= \text{Tr}\left\{\left[(\mathbf{\Lambda}_1^{1/2} \otimes \mathbf{\Lambda}_2^{1/2})(\mathbf{Q}_1 \otimes \mathbf{Q}_2)\mathbf{D}(\mathbf{Q}_1^H \otimes \mathbf{Q}_2^H)(\mathbf{\Lambda}_1^{H/2} \otimes \mathbf{\Lambda}_2^{H/2})\right] \tag{5.103}\right.$$

$$\times \left[\sigma_n^2\mathbf{I}_{N\tau_1} + \sigma_n^2\mathbf{I}_{\tau_1} \otimes \mathbf{\Lambda}_2\right.$$

$$\left.\left. + (\mathbf{\Lambda}_1^{1/2} \otimes \mathbf{\Lambda}_2^{1/2})(\mathbf{Q}_1 \otimes \mathbf{Q}_2)\mathbf{D}(\mathbf{Q}_1^H \otimes \mathbf{Q}_2^H)(\mathbf{\Lambda}_1^{H/2} \otimes \mathbf{\Lambda}_2^{H/2})\right]^{-1}\right\}.$$

Since the term $\tilde{\mathcal{M}}$ is invariant under $\mathbf{U}_1$ and $\mathbf{U}_2$, the RRH power constraint can be derived according to the lemmas in [24]:

$$\text{Tr}\left\{\mathbf{D}(\mathbf{Q}_1^H\mathbf{\Lambda}_1\mathbf{Q}_1) \otimes (\mathbf{Q}_2^H\mathbf{\Lambda}_2^{H/2}\mathbf{U}_2^H\mathbf{U}_g\mathbf{\Lambda}_g^{-2}\mathbf{U}_g^H\mathbf{U}_2\mathbf{\Lambda}_2^{1/2}\mathbf{Q}_2)\right\}$$

$$+ \tau_1 \text{Tr}\left\{\mathbf{\Lambda}_g^{-2}\mathbf{U}_g^H\mathbf{U}_2\mathbf{\Lambda}_2\mathbf{U}_2^H\mathbf{U}_g\right\}$$

$$\ge \text{Tr}\left\{\mathbf{D}(\mathbf{Q}_1^H \otimes \mathbf{Q}_2^H)(\mathbf{\Lambda}_1 \otimes \mathbf{\Lambda}_2^{H/2}\mathbf{\Lambda}_g^{-2}\mathbf{\Lambda}_2^{1/2})(\mathbf{Q}_1 \otimes \mathbf{Q}_2)\right\} + \tau_1 \text{Tr}\left(\mathbf{\Lambda}_g^{-2}\mathbf{\Lambda}_2\right)$$

$$\ge \text{Tr}\left\{\mathbf{D}(\mathbf{\Lambda}_1 \otimes \mathbf{\Lambda}_2^{H/2}\mathbf{\Lambda}_g^{-2}\mathbf{\Lambda}_2^{1/2})\right\} + \tau_1 \text{Tr}\left(\mathbf{\Lambda}_g^{-2}\mathbf{\Lambda}_2\right), \tag{5.104}$$

where $\mathbf{\Lambda}_g^2 = \mathbf{\Lambda}_g\mathbf{\Lambda}_g^H$.

The power constraint above is equal to the lower bound when $\mathbf{U}_2 = \mathbf{U}_g^H$ and $\mathbf{Q}_1 \otimes \mathbf{Q}_2 = \mathbf{I}$. The maximization of (5.103) is achieved when $\mathbf{Q}_1 \otimes \mathbf{Q}_2 = \mathbf{I}$. Namely, we can choose the optimal $\mathbf{Q}_1 = \mathbf{I}$, $\mathbf{Q}_2 = \mathbf{I}$, and $\mathbf{U}_2 = \mathbf{U}_g^H$ when the RRHs consume the least amount of power and $\mathbf{R}_{\delta \mathbf{h}_{r1}}$ is minimized. Since both the cost function and the first constraint are invariant under $\mathbf{U}_1$, $\mathbf{U}_1$ must be an arbitrary unitary matrix. The optimal $\mathbf{\Psi}$ and $\mathbf{C}$ are given by

$$\mathbf{\Psi} = \mathbf{U}_1 \mathbf{\Lambda}_1^{1/2}, \tag{5.105}$$

$$\mathbf{C} = \mathbf{Q}_g \mathbf{\Lambda}_g^{-1} \mathbf{\Lambda}_2^{1/2}. \tag{5.106}$$

Therefore, the conclusion is reached and the training design problem of $\mathbf{H}_{r1}$ is simplified to a search for the optimal $\mathbf{\Lambda}_1$ and $\mathbf{\Lambda}_2$. Thus the training design problem can be converted to the following maximization problem:

$$\begin{aligned}
\underset{\mathbf{\Lambda}_1, \mathbf{\Lambda}_2}{\text{maximize}} \quad & \mathrm{Tr}\left\{\left[\sigma_n^2 \mathbf{I}_{\tau_1} \otimes \mathbf{\Lambda}_2 + \left(\mathbf{\Lambda}_1^{1/2} \otimes \mathbf{\Lambda}_2^{1/2}\right) \mathbf{D} \left(\mathbf{\Lambda}_1^{H/2} \otimes \mathbf{\Lambda}_2^{H/2}\right) + \sigma_n^2 \mathbf{I}_{N\tau_1}\right]^{-1} \right. \\
& \left. \times \left(\mathbf{\Lambda}_1^{1/2} \otimes \mathbf{\Lambda}_2^{1/2}\right) \mathbf{D} \left(\mathbf{\Lambda}_1^{H/2} \otimes \mathbf{\Lambda}_2^{H/2}\right)\right\}
\end{aligned}$$

s.t. $\mathrm{Tr}\{\mathbf{\Lambda}_1\} \leq \tau_1 P_s$,

$$\mathrm{Tr}\left\{\mathbf{D}(\mathbf{\Lambda}_1 \otimes \mathbf{\Lambda}_2^{H/2} \mathbf{\Lambda}_g^{-2} \mathbf{\Lambda}_2^{1/2})\right\} + \tau_1 \sigma_n^2 \, \mathrm{Tr}\left(\mathbf{\Lambda}_g^{-2} \mathbf{\Lambda}_2\right) \leq \tau_1 P_r. \tag{5.107}$$

Denote $\lambda_1(i)$, $\lambda_2(i)$ and $\lambda_g(i)$ as the ith diagonal elements of $\mathbf{\Lambda}_1$, $\mathbf{\Lambda}_2$, and $\mathbf{\Lambda}_g$, respectively; $\mathbf{G}_{r2}$ is assumed to be full rank. The training design problem above can then be further expressed as

$$\underset{\lambda_1(i)\geq 0, \lambda_2(j)\geq 0}{\text{maximize}} \quad \sum_{i=1}^{\tau_1} \sum_{j=1}^{N} \frac{\kappa_{ij}\lambda_1(i)\lambda_2(j)}{\sigma_n^2 + \sigma_n^2 \lambda_2(j) + \kappa_{ij}\lambda_1(i)\lambda_2(j)}$$

s.t. $$\sum_{i=1}^{\tau_1} \lambda_1(i) \leq \tau_1 P_s,$$

$$\sum_{i=1}^{\tau_1} \sum_{j=1}^{N} \frac{\kappa_{ij}\lambda_1(i)\lambda_2(j)}{\lambda_g(j)^2} + \tau_1 \sigma_n^2 \sum_{j=1}^{N} \frac{\lambda_2(j)}{\lambda_g(j)^2} \leq \tau_1 P_r. \tag{5.108}$$

To solve this non-convex problem, we can fix all the $\lambda_1(i)$ or $\lambda_2(j)$ subject to UE or RRH power constraints, and optimization over all the $\lambda_2(j)$ or $\lambda_1(i)$ becomes a convex problem. First, $\lambda_2(j)$ is optimized with $\lambda_1(i)$ fixed, $i = 1, \ldots, \tau_1$. Let the fixed $\lambda_1(i)$ satisfy $\sum_{i=1}^{\tau_1} \lambda_1(i) = \tau_1 P_s$. Then the training design problem is subject only to the RRH power constraint and can be solved by using the KKT conditions:

$$\sum_{i=1}^{\tau_1} \frac{\kappa_{ij}\lambda_1(i)}{\left[1 + \lambda_2(j) + \kappa_{ij}\lambda_1(i)\lambda_2(j)/\sigma_n^2\right]^2} = \nu \left[\sum_{i=1}^{\tau_1} \kappa_{ij}\lambda_1(i)\lambda_g(j)^{-2} + \tau_1 \sigma_n^2 \lambda_g(j)^{-2}\right], \tag{5.109}$$

$$\sum_{i=1}^{\tau_1} \sum_{j=1}^{N} \kappa_{ij}\lambda_1(i)\lambda_g(j)^{-2}\lambda_2(j) + \tau_1 \sigma_n^2 \sum_{j=1}^{N} \frac{\lambda_2(j)}{\lambda_g(j)^2} = \tau_1 P_r, \tag{5.110}$$

where $v > 0$ and $\lambda_2(j)$ is either zero or a positive value. For any given $v > 0$, $\lambda_2(j)$ is either equal to a positive solution to the first equation above or is set to zero. The optimal v can also be found from the second equation by substituting the $\lambda_2(j)$ obtained earlier.

Below the optimization of $\lambda_1(i)$ with fixed $\lambda_2(j), j = 1, 2, \ldots, N$, is discussed. The Lagrangian function is given by

$$\sum_{j=1}^{N} \frac{\kappa_{ij}\lambda_2(j)[1 + \lambda_2(j)]}{\left[1 + \lambda_2(j) + \kappa_{ij}\lambda_1(i)\lambda_2(j)/\sigma_n^2\right]^2} = v_1 + v_2 \sum_{j=1}^{N} \kappa_{ij}\frac{\lambda_2(j)}{\lambda_g(j)^2}, \qquad (5.111)$$

where $v_1 \geq 0$ and $v_2 \geq 0$ are the Lagrange multipliers for the first and second constraints, respectively. For any given pair v_1 and v_2, $\lambda_1(i)$ is obtained by solving the equation above or is simply equal to zero. To search for the optimal pair v_1 and v_2, one of the pair can be determined by a bisection search with the other held fixed when the two power constraints are active, namely, $v_1 > 0$ and $v_2 > 0$. In the other case either $v_1 = 0$ or $v_2 = 0$ is possible, and thus possibility should be examined first.

Numerical Results

The performance of our proposed training design with the SMMSE estimator under different scenarios was numerically simulated. The power bounds for the UEs and RRHs were set to be equal, i.e., $p_s = p_r = P$. The noise variance σ_n^2 was set to one, and the SNR was defined as $P/\sigma_n^2 = P$. We took $K = M = 4$ and $N = 8$. In total there were 10^5 Monte Carlo iterations. We assumed that the lengths of the training sequences ϕ_1 and ϕ_2 were equal, i.e., $\tau_1 = \tau_2 = \tau$. The power allocation between two training segments was provided and a power allocation factor $0 \leq \varepsilon \leq 1$ was adopted. In the estimation of $\mathbf{H}_{r2}$ each iteration contained 10 temporally and spatially correlated fading blocks. The elements of the correlation matrix $[\mathbf{R}]_{i,j}$ were taken to be $r^{|i-j|}, |r| = 0.2$. The temporal correlation coefficient η was set at 0.9881 [25]. The normalized MSEs of $\mathbf{H}_{r1}$ and $\mathbf{H}_{r2}$ were defined as $(N\tau)^{-1}\mathcal{M}_i$ and $(N\tau)^{-1}\mathcal{E}_{\mathbf{H}_{r2}}[\mathcal{M}]$, respectively. The expectation of the normalized MSE of $\mathbf{H}_{r1}$ was computed by taking the average over 100 realizations of $\mathbf{H}_{r2}$. The large-scale fading coefficient β_m and κ_{km} were modeled as

$$\beta_m = \frac{z_m}{(r_m/r_0)^v} \quad \text{and} \quad \kappa_{km} = \frac{z_{km}}{(r_{km}/r_0)^v},$$

where z_m and z_{mk} are lognormal random variables with standard deviation 8 dB; r_m is the distance between the mth RRH and the BBU pool, and r_{km} is the distance between the kth user and the mth RRH; r_0 was set at 100 m, and the path loss exponent v was 3.8.

Evaluation of the normalized MSE performance of $\mathbf{H}_{r2}$ under different correlation coefficient values $|\rho|$, shown in Fig. 5.9, verifies the optimality of orthogonal training ($|\rho| = 0$) in the estimation of $\mathbf{H}_{r2}$. The first fading block with $i = 0$ has SNR $= 0$ dB. The correlation coefficients

$$|\rho|_{ij} = \frac{\phi_2(i)^H \phi_2(j)}{\|\phi_2(i)\|^2},$$

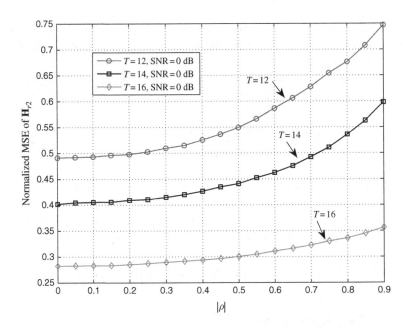

Figure 5.9 Normalized MSE of $\mathbf{H}_{r2}$ versus $|\rho|$.

and the correlation coefficients of any two training sequences were set to be equal, i.e., $|\rho_{ij}| = |\rho|$, $i \neq j$. The optimal power allocation coefficient ε was generated by maximizing the uplink ergodic capacity. As $|\rho|$ increases the normalized MSE performance degrades, and the lowest normalized MSE is achieved when $|\rho| = 0$ for the different total lengths T of the training sequences. The simulation result shows that orthogonal training is optimal, which is consistent with the theoretical analysis. For different T it can be observed that the longer T is, the better the MSE performance that can be achieved. Hence, lengthening the training sequence degrades the system's spectral efficiency in return. Therefore it is important to keep a balance between the lengths of the training sequences and the numbers of data symbols.

The normalized MSE curves versus fading-block index i under different total training sequence lengths T are depicted in Fig. 5.10. The SNR is fixed at 0 dB. Two power allocation schemes are compared. The power allocation factor ε was set to one, for equi-power allocation. Orthogonal training is assumed to be applied for both $\mathbf{\Phi}$ and $\mathbf{\Psi}$. It is easy to verify from Fig. 5.10 that the performance of SMMSE estimation becomes better as the fading block index i increases. This is reasonable since the estimation of $\mathbf{H}_{r2}$ is based on the previously received training symbols and temporal correlation is taken into consideration to improve the estimation accuracy. The MSE curves adopting the optimal power allocation are lower than those with the equi-power allocation, which demonstrates that more power is allocated to $\mathbf{\Phi}$ for the estimation of $\mathbf{H}_{r2}$. Comparing with the MSE curves for $T = 14$, a larger T can provide better MSE performance.

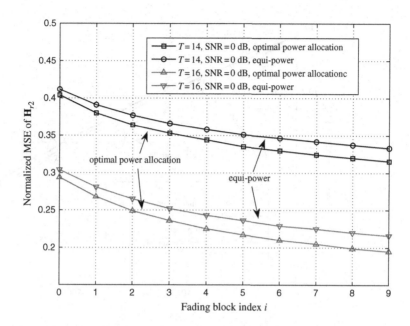

Figure 5.10 Normalized MSE of $\mathbf{H}_{r2}$ versus fading block index i.

The normalized MSE of $\mathbf{H}_{r2}$ versus SNR for the SMMSE estimator is shown in Fig. 5.11, where the CRBs are presented as benchmarks and orthogonal training is used for all MSE curves. The optimal power allocation, the equi-power allocation, and the CRBs under the optimal power allocation scheme may be compared. The MSE curves decrease as the SNRs increase or the total training sequence length T increases. It can be observed that the SMMSE estimator achieves a lower MSE under the optimal power allocation scheme than under the equi-power allocation scheme, which matches observations from Fig. 5.9. Moreover, the normalized MSEs of $\mathbf{H}_{r2}$ under optimal training are close to their corresponding CRBs when $T = 12$ and $T = 16$.

Figure 5.12 shows the normalized MSE of $\mathbf{H}_{r1}$ versus SNR under the optimal training design, the orthogonal training design, and the CRB, where the optimal power allocation scheme is applied for the estimation of $\mathbf{H}_{r1}$. The optimal training design is derived numerically from the equation. An orthogonal training design satisfies the following equations:

$$\mathbf{\Psi}^H\mathbf{\Psi} = \frac{P}{\sigma_n^2}\mathbf{I} \quad \text{and} \quad \mathbf{C}^H\mathbf{C} = \frac{P}{\bar{\kappa}P + \sigma_n^2}\mathbf{I},$$

where $\bar{\kappa}$ is the average value of the κ_{ij}. As expected, the MSE of $\mathbf{H}_{r1}$ is a decreasing function of SNR, and the optimal training design yields better MSE performance than the orthogonal training design. It can be seen that the MSE curves are relatively worse than the corresponding CRBs in the high-SNR region. This happens because the training power allocated to the estimation of $\mathbf{H}_{r1}$ is relatively small and the estimation of $\mathbf{H}_{r1}$ suffers from severe fading.

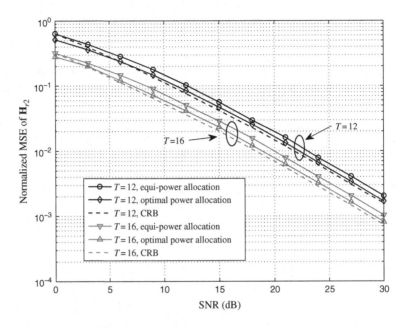

Figure 5.11 Normalized MSE of $\mathbf{H}_{r2}$ versus SNR.

5.4   Non-Training-Based Channel Estimation in C-RANs

Training-based channel estimation has demonstrated its effectiveness for acquiring a reliable estimate of the CSI in C-RANs. However, the training overhead of training-based channel estimation is significant, and the transmission bandwidth efficiency is decreased in turn. Thus non-training-based channel estimation can be regarded as an efficient method to resolve the problem of unsufficient frequency resources.

Blind channel estimation is first proposed for CSI acquisition without a training sequence. However, if only the data symbols are used for the estimation, the inherent drawbacks of blind channel estimation, e.g., phase ambiguity, slow convergence speed, and high computation complexity, can block its implementation in C-RANs. Semi-blind channel estimation has emerged as a more effective estimation method; it incorporates both the training signal and the data samples. A higher estimation accuracy can be achieved in semi-blind channel estimation by this use of a hybrid of training and data symbols than can be achieved by using purely the training or data symbols.

Inspired by the work in [26], semi-blind channel estimation is employed in C-RANs to achieve more accurate estimation without sacrificing bandwidth efficiency. The (LS) estimation is used to obtain a initial estimate and then the quasi-Newton method is applied to improve the estimation accuracy using the maximum likelihood principle. By treating the data symbols as Gaussian-distributed parameters, the joint likelihood function of the training signal $\mathbf{y}$ and the received data $\mathbf{z}$ is given by

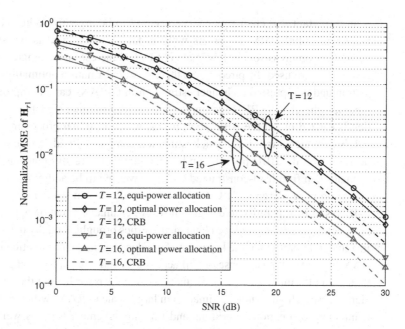

Figure 5.12 Normalized MSE of $\mathbf{H}_{r1}$ versus SNR.

$$f(\mathbf{y}, \mathbf{z}; \theta) = \frac{1}{\pi^N |\mathbf{C}|} e^{-(\mathbf{y}-\mu)^H \mathbf{C}^{-1}(\mathbf{y}-\mu)} \times \prod_{k=1}^{K} \frac{1}{\pi^N |\mathbf{Q}|} e^{-(\mathbf{y}-\mu_k)^H \mathbf{Q}^{-1}(\mathbf{y}-\mu_K)}, \qquad (5.112)$$

where θ is the estimated channel vector and $\mathbf{C}$ and $\mathbf{Q}$ are the corresponding covariance matrices for the training and data symbols, respectively. The maximization problem above is actually non-linear, and we resort to an iterative quasi-Newton method to obtain the CSI in C-RANs with a limited number of data symbols. The conventional training-based estimator, i.e., the LS estimator, is used to obtain the initial estimated values. The equivalent channel vector $\mathbf{q}$ is given, using the LS approach, by

$$\hat{\mathbf{q}} = \mathbf{\Omega}^\dagger \mathbf{y}, \qquad (5.113)$$

where $\mathbf{\Omega}$ is the aggregate training matrix. In addition, orthogonal training sequences are used to improve the channel estimation performance. In this way semi-blind channel estimation can achieve substantial improvements over the conventional training-based method in terms of estimation accuracy without a decrease in spectral efficiency. The details are omitted here.

5.5   Channel Estimation in Fronthaul Constrained and Large-Scale C-RANs

Since a constrained fronthaul is one of the major practical hurdles for C-RANs' large-scale implementation, channel estimation taking the fronthaul loading into consideration is an important and challenging issue needing to be urgently addressed. Channel estimation in uplink MIMO systems with backhaul constrained is investigated in [27], where the compress-forward estimate (CFE) and the estimate compress-forward (ECF) approaches are used. Inspired by the theoretical information optimality of separate estimation and compression, channel estimation in C-RANs can be improved by the CFE scheme, whereby the RRHs compress the received training signals and channel estimation is carried out at the BBU pool. In this way, the signal processing overhead at the RRHs is reduced and the channel estimation can take full advantage of the large-scale processing at the BBU pool. It can be seen that there are still many aspects of channel estimation in fronthaul-constrained C-RANs waiting to be addressed.

The proposed large-scale C-RANs will greatly reform the conventional structure of C-RANs. It is imperative to design innovative channel estimation approaches from new perspectives due to the unique channel properties brought about by the dramatic antenna increment. In particular, the key point of conducting channel estimation in large-scale C-RANs is to avoid the expense of CSI acquisition and signal processing overheads rather than a simple emphasis on high estimation accuracy regardless of the computation costs. Hence, research on channel estimation in large-scale C-RANs will mainly focus on low-complexity estimation algorithms and training scheme design as well as eliminating the negative effects of the antennas increasing tremendously. In general, training-based channel estimation is preferable in large-scale C-RANs on account of the high computational complexity required for the blind estimation. However, training-based channel estimation in large-scale C-RANs is troubled by the size of the training overheads. If the number of RRHs scales up to infinity as the massive MIMO, the interference caused by training reuse will turn into an overriding factor for performance degradation. Moreover, the training sequence deficiency in large-scale C-RANs means that orthogonality is no longer satisfied, and the performance of channel estimation will further degrade in this case. To alleviate the training overheads and contamination effects, semi-blind data-aided channel estimation can offer a relatively high data rate and low training overhead with reasonable estimation accuracy. The existing training scheme also needs to be adjusted to be more spectrally effective.

5.6   Summary

In this chapter we have presented several effective approaches for channel estimation in C-RANs, including training-based and non-training-based schemes. Specifically, the superimposed training scheme has been applied for individual channel estimation in C-RANs and has significantly reduced the training overhead since two training sequences share the same spectral resources. Furthermore, the segment training scheme has also been used for CSI acquisition in C-RANs by assigning two training segments dedicated for channel estimation. Estimation accuracy under the segment training scheme has been greatly increased, and the data transmission quality has also been improved without superimposing the training sequences. In terms of tradeoffs between estimation accuracy and spectral efficiency, semi-blind channel estimation has utilized both data symbols and training sequences and has achieved a significant performance improvement.

It is expected that compression techniques to overcome constrained fronthaul will definitely be applied in the large-scale implementation of C-RANs. Both the training sequences and data symbols are evidently influenced by compression disturbance. It is imperative to design jointly the channel estimation and compression schemes for overall system-performance optimization in the future.

References

[1] P. Rost *et al.*, "Cloud technologies for flexible 5G radio access networks," *IEEE Commun. Mag.*, vol. 52, no. 5, pp. 68–76, May 2014.

[2] M. Webb, "Smart 2020: enabling the low carbon economy in the information age." The Climate Group, London, 2008.

[3] M. Biguesh and A. Gershman, "Training-based MIMO channel estimation: a study of estimator tradeoffs and optimal training signals," *IEEE Trans. Signal Process.*, vol. 54, no. 3, pp. 884–893, March 2006.

[4] I. Barhumi, G. Leus, and M. Moonen, "Optimal training design for MIMO OFDM systems in mobile wireless channels," *IEEE Trans. Signal Process.*, vol. 51, no. 6, pp. 1615–1624, June 2003.

[5] J. K. Tugnait, "Blind estimation and equalization of MIMO channels via multidelay whitening," *IEEE J. Sel. Areas Commun.*, vol. 19, no. 8, pp. 1507–1519, August 2001.

[6] Z. Ding and H. V. Poor, "A general framework of precoding design for multiple two-way relaying communications," *IEEE Trans. Signal Process.*, vol. 61, no. 6, pp. 1531–1535, March 2013.

[7] J. K. Tugnait and W. Luo, "On channel estimation using superimposed training and first-order statistics," *IEEE Commun. Lett.*, vol. 7, no. 9, pp. 413–415, September 2003.

[8] F. Gao *et al.*, "Superimposed training based channel estimation for OFDM modulated amplify-and-forward relay networks," *IEEE Trans. Commun.*, vol. 59, no. 7, pp. 2029–2039, July 2011.

[9] X. Xie *et al.*, "Superimposed training based channel estimation for uplink multiple access relay networks," *IEEE Trans. Wireless Commun.*, vol. 14, no. 8, pp. 4439–4453, August 2015.

[10] T. Kong and Y. Hua, "Optimal design of source and relay pilots for MIMO relay channel estimation," *IEEE Trans. Signal Process.*, vol. 59, no. 9, pp. 4438–4446, September 2011.

[11] S. Sun and Y. Jing, "Training and decodings for cooperative network with multiple relays and receive antennas," *IEEE Trans. Commun.*, vol. 60, no. 6, pp. 1534–1544, June 2012.

[12] Z. Shun *et al.*, "Segment training based individual channel estimation in one-way relay network with power allocation," *IEEE Trans. Wireless Commun.*, vol. 12, no. 3, pp. 1300–1309, March 2013.

[13] X. Xie *et al.*, "Training design and channel estimation in uplink cloud radio access networks," *IEEE Signal Process. Lett.*, vol. 22, no. 8, pp. 1060–1064, August 2015.

[14] Q. Hu *et al.*, "Segment training based channel estimation and training design in cloud radio access networks," in *Proc. IEEE ICC.*, London, June 2015.

[15] S. Zhou, B. Muquet, and G. B. Giannakis, "Subspace-based blind channel estimation for block precoded space-time OFDM," *IEEE Trans. Signal Process.*, vol. 50, no. 5, pp. 1215–1228, May 2002.

[16] H. Böcskei, R. W. Heath, and A. J. Paulraj, "Blind channel identification and equalization in OFDM-based multiantenna systems," *IEEE Trans. Signal Process.*, vol. 50, no. 1, pp. 96–109, January 2002.

[17] M. Grant, S. Boyd and Y. Ye, CVX: Matlab software for disciplined convex programming. Online, July 2010. Available at http://cvxr.com/cvx.

[18] G. B. Giannakis and C. Tepedelenlioglu, "Basis expansion models and diversity techniques for blind identification and equalization of time-varying channels," *Proc. IEEE*, vol. 86, no. 10, pp. 1969–986, October 1998.

[19] X. Ma and G. B. Giannakis, "Maximum-diversity transmissions over doubly-selective wireless channels," *IEEE Trans. Inf. Theory.*, vol. 49, no. 7, pp. 1832–1840, July 2003.

[20] G. Wang *et al.*, "Channel estimation and training design for two-way relay networks in time-selective fading environments," *IEEE Trans. Wireless Commun.*, vol. 10, no. 8, pp. 2681–2691, August 2011.

[21] F. Gao, R. Zhang, and Y. C. Liang, "Optimal channel estimation and training design for two-way relay networks," *IEEE Trans. Commun.*, vol. 57, no. 10, pp. 3024–3033, October 2009.

[22] "Spatial channel model for multiple input multiple output simulations." 3GPP TR 25.996, release 9, December 2009.

[23] H. V. Poor, *An Introduction to Signal Detection and Estimation. Second Edition.* Springer, 1994.

[24] A. W. Marshall and I. Olkin, *Inequalities: Theory of Majorization and Its Applications.* Academic Press, 1979.

[25] J. Choi, D. J. Love, and P. Bidigare, "Downlink training techniques for FDD massive MIMO systems: open-loop and closed-loop training with memory," *IEEE J. Sel. Topics Signal Process.*, vol. 8, no. 5, pp. 802–814, October 2014.

[26] S. Abdallah and I.N. Psaromiligkos, "Semi-blind channel estimation with superimposed training for OFDM-based AF two-way relaying," *IEEE Trans. Wireless Commun.*, vol. 13, no. 5, pp. 2468–2477, May 2014.

[27] J. Kang *et al.*, "Joint signal and channel state information compression for the backhaul of uplink network MIMO systems," *IEEE Trans. Wireless Commun.*, vol. 13, no. 3, pp. 1555–1567, March 2014.

6   Massive MIMO in C-RANs

Shi Jin, Jun Zhang, Kai-Kit Wong, and Hongbo Zhu

6.1   Introduction

Recently, massive multiple-input multiple-output (MIMO) antenna systems, which promote the use of a very large antenna array, are becoming one of the hottest research topics in wireless communications and their extraordinary benefits in wireless networks have been illustrated in [1–4]. The main idea is that, as the number of collocated antennas grows very large, the effects of uncorrelated noise and fast fading vanish and intra-cellular interference can be mitigated. For cellular networks with M-antenna base stations (BSs), it has been shown that the transmit power per antenna can be scaled down as $O(1/\sqrt{M})$ if the channel state information (CSI) is estimated from the uplink pilots, while the the spectral efficiency is maintained at a very high level if multiple users are served in the same time–frequency resource [4]. Despite this preliminary success there are major bottlenecks such as geometric attenuation and pilot contamination that have cast doubts about systems with a collocated massive antenna array [5–7]. However, C-RANs, in which the BSs are replaced by a number of remote antenna units (RAUs), have been actively investigated for providing better coverage and reducing the distance between RAUs and users [8–12]. If we scale up the number of RAUs by several orders of magnitude, it is anticipated that one would possess the advantages of both collocated massive MIMO systems and C-RANs [13, 14].

In [13] the joint optimization problem of antenna selection, regularization factor, and power allocation to maximize the average weighted sum-rate in massive C-RANs was investigated. The power consumption by a massive C-RAN was optimized in [14] by deriving the optimal rate allocation among the users. In these references, only a limited or no-cooperation transmission scheme was considered. However, there are two popular transmission schemes for the traditional distributed antenna system, namely the blanket transmission scheme (BTS) and the selection transmission scheme (STS) [8]. The potential for BTS and STS when the number of users increases without limit has been studied in [15–24]. The gain in cooperation with BTS for femto- and pico-cells over independent femto-cell operation in a heterogeneous network (HetNet) was also demonstrated by system-level simulations in [15]. Earlier, in [16], spatial interference cancellation with multiple receive antennas was exploited in ad hoc networks, which bear some similarity to multicell networks; the throughput gain was found to be linear in the number of receive antennas. The upper bound of the spectral efficiency as the transmit power grows large in cooperative wireless network was given in [17]. It was argued that joint processing is not helpful in a large wireless network. A new cross-layer algorithm was conceived in [18] that achieved high performance of the coordinated multipoint (CoMP) systems at a lower feedback amount and at lower algorithmic complexity. In [19] and [20] the information-theoretic limits and the capacity scaling law for wireless networks with a large number of paired nodes was found by studying the physical geometric structure or the benefits of structure support, respectively. More recently [21, 22] characterized how the capacity of the distributed MIMO transmission scales with the number of cooperating users in several general fading environments and showed the relationship between the throughput scaling and the spatial degrees of freedom (DoFs). The papers [23, 24] addressed the energy-efficiency and spectral-efficiency optimization problems in multi-relay MIMO-OFDMA cellular networks, and they indicated that a greater number of relay nodes may degrade the energy efficiency. It is worth noting that all the results in [16–24] were given for the networks where transmitters and receivers appear in pairs. For a massive C-RAN using a few hundred RAUs to serve tens of users [14, 25, 26], which scheme is better for the uplink has not yet been established.

In this chapter we derive expressions for the achievable rate and energy efficiency of an *uplink massive C-RAN* with a fixed number of users and provide a comparative analysis for systems using a joint decoding reception scheme (JDRS) or a selection reception scheme (SRS), whose roles correspond to the downlink transmission schemes BTS and STS.

We regard SRS as an easy-to-implement reception mode in massive C-RANs, whereas JDRS is thought to be require large-scale cooperation. Therefore, in our comparison, cooperation between the RAUs is not allowed in SRS.[1] Differently from previous studies on distributed antenna systems, for dense network scenarios we adopt a two-stage channel model to obtain more realistic results. For the calculation of the energy efficiency, we consider the energy consumption for both signal transmission and signal reception.

In summary, this chapter makes the following contributions.

- We derive closed-form rate expressions for SRS and asymptotic performance scaling for the two reception schemes, in both single-cell and multi-cell cellular systems.
- By a comparative analysis we show that with JDRS the transmit power can be arbitrarily small when the number of RAUs M grows without limit, whereas with SRS a much higher energy efficiency is achieved while a comparable performance of achievable rate with JDRS is maintained. With SRS, both the energy efficiency and the achievable rate can increase proportionally to $c \log_2(M)$. Thus, full cooperation is not rewarding but a waste of energy in massive C-RANs.
- On the basis of our analytical results, we then investigate efficient joint user scheduling and RAU selection algorithms to improve the sum-rate of the system. Our results reveal that, for SRS, the RAU selection algorithm loses its benefits when M is an integer multiple of the number of users. A simple selection algorithm in which each user selects the nearest available RAU has a near-optimal excellent performance. Using

---

[1] Cooperative reception among the neighboring RAUs would further improve the system performance for SRS; studies of its performance and scheduling problems are interesting topics for future work.

this observation we then propose an efficient user-scheduling and RAU-selection scheme with which the achievable sum-rate of SRS can exceed that of JDRS even with a small number of antennas.

The rest of this chapter is organised as follows. Section 6.2 presents the network model. In Section 6.3, we analyze the achievable rate of the uplink massive C-RAN and the asymptotic performance for JDRS and SRS. The energy efficiencies and the power scaling laws for the two schemes are derived in Section 6.4. In Section 6.5 we investigate the problem of user scheduling and RAU selection and propose efficient algorithms for SRS. Numerical results are provided in Section 6.6 and we conclude the chapter in Section 6.7.

Notation
In this chapter we use column vectors denoted by lower-case boldface letters, whereas matrices are represented by upper-case boldface letters. Sets are denoted by an upper-case script font. The superscript $*$ denotes the conjugate transpose operation, while $|\cdot|$ denotes the norm of an input vector or the number of elements of an input set and $\bar{\mathcal{A}}$ denotes the set complementary to $\mathcal{A}$.

6.2 System Model

Consider the uplink of a general seven-cell C-RAN architecture. Each cell comprises M RAUs and K ($\ll M$) uniformly distributed single-antenna users (the same assumptions are made in [14, 25, 26]). We start from a simple scenario in which the RAUs in a cell form an F-tier hexagonal structure, as shown in Fig. 6.1 (e.g., if $F = 4$ then $M = 61$). We choose this scenario for the convenience of simulations, but it should be noted that the theorems presented in the subsequent sections also apply to networks with randomly and uniformly distributed RAUs. Moreover, we assume for simplicity that each RAU has one antenna, but our results can be easily extended to the case with multi-antenna RAUs. A densely deployed network is considered. When the RAUs' density increases, mobile users are more likely to be close to the RAU. Since the accuracy of the channel model will affect the analytical results, we will use different models to characterize short-range and long-range channels.

6.2.1 Short-Range Model

We use the model in [27] for users located close to the corresponding RAU. In this case we denote the complex propagation coefficient between the mth RAU in the jth cell and the kth user in the lth cell by

$$g_{mjkl} = \left(\frac{\lambda}{4\pi d_{mjkl}} \right) \sqrt{\Omega} h_{mjkl}, \quad m = 1, \ldots, M, \ k = 1, \ldots, K, \ j, l = 1, \ldots, L, \quad (6.1)$$

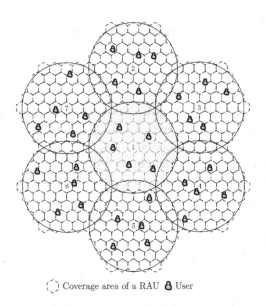

⌒⌒ Coverage area of a RAU   ⛄ User

Figure 6.1 A general seven-cell architecture for a massive C-RAN where each cell employs a few hundred RAUs to serve tens of users.

where M denotes the number of RAUs in each cell, L denotes the number of active cells, K denotes the number of terminals in each cell, λ is the carrier wavelength, d_{mjkl} is the distance between the kth user in the lth cell and the mth RAU in the jth cell, Ω is the antenna gain, taken as 10.3 dB, and h_{mjkl} is the fast-fading coefficient, which is assumed to have zero mean and unit variance.

6.2.2   Long-Range Model

For users far from the corresponding RAU, we model the complex propagation coefficient as [28]

$$g_{mjkl} = h_{mjkl}\sqrt{\Omega \beta_{mjkl}}, \tag{6.2}$$

where h_{mjkl} denotes the fast-fading coefficient and is assumed to have zero mean and unit variance and β_{mjkl} is the large-scale fading, which can be factored as follows:

$$\beta_{mjkl} = \frac{Z_{mjkl}}{d_{mjkl}^{\gamma}}, \tag{6.3}$$

where γ is the decay exponent and Z_{njkl} is a log-normal random variable with standard deviation σ_{shad}.

The transition distance in both model, d_t, is randomly chosen to be between 30 and 70 meters, which models the location of an obstacle that blocks a user's line of sight (LOS) [27]. For the fading component in the propagation models, we randomly sample a fading realization for the network and keep it fixed throughout the transmission. Here, our idealized channel model does not include a time variation in the fading.

6.2.3 Node Distributions

To describe the distributions of the RAUs and the users, we use polar coordinates (r, θ) relative to the center of a cell. Let a denote the polar radius of the fixed point A and b the polar radius of the integral variable point B. As both the users and RAUs are distributed uniformly in the cell, A and B can be either user or RAU. Then its probability density function (pdf) is given by

$$f_B(b) = \frac{b}{\int_{r \in \mathcal{B}} r dr}, \quad b \in \mathcal{B}, \tag{6.4}$$

where $\mathcal{B} = \{b \mid 0 \leq b \leq R, d_{AB} \geq R_0\}$ is the effective coverage area, in which R is the cell radius and $d_{AB} = \sqrt{a^2 + b^2 - 2ab \cos \theta_{AB}}$ is the radius between A and B. A small protective radius R_0 is specified for feasibility of the integration. Let θ_u denote the polar angle of user u with pdf

$$f_\Theta(\theta) = \frac{1}{2\pi}, \quad \theta \in [1, 2\pi]. \tag{6.5}$$

Therefore, the average value of a function G of the position of b over the cell can be calculated as

$$\mathbb{E}\{G\} = \int_{b \in \mathcal{B}}^{R} \frac{2b}{R^2} \int_{\theta=0}^{\pi} \frac{G}{\pi} d\theta db. \tag{6.6}$$

6.3 Achievable Rate

In this section, we analyze the achievable rate of the uplink massive C-RAN using JDRS and SRS in the multi-cell and single-cell scenarios, respectively. For JDRS, the CSI is estimated from the uplink pilots and the pilot sequences are of length $\tau = K$, which is the smallest amount of training that can be required. For SRS, perfect instantaneous CSI is assumed at the receiver side. Under these assumptions, expressions for the achievable rate and the asymptotic performance are derived and compared with each other.

6.3.1 Joint Decoding Reception Scheme

Multi-Cell Massive C-RAN with Imperfect CSI

In the uplink, for JDRS we consider that there are a large number of cooperating RAUs to serve a fixed number of users. As a result, in the uplink the received signal from M RAUs at the jth cell is

$$\mathbf{x}_j = \sqrt{p_s} \sum_{l=1}^{L} \mathbf{G}_{jl} \mathbf{a}_l + \boldsymbol{\omega}_j, \tag{6.7}$$

where $\mathbf{a}_l$ is the $K \times 1$ vector of the signal from the users in the lth cell, whose kth element is a_l^k with $\mathbb{E}\left\{|a_l^k|^2\right\} = 1$; $\boldsymbol{\omega}_j$ denotes an additive white Gaussian noise, whose entries are i.i.d. circular complex Gaussian random variables with zero mean; p_s is a measure of the signal-to-noise ratio (SNR), which is equal for each user without power allocation, and $\mathbf{G}_{jl}$ is the $M \times K$ propagation matrix between the M RAUs of the jth cell and the K terminals in the lth cell, with $\left[\mathbf{G}_{jl}\right]_{mk} = g_{mjkl}$ for $m = 1, \ldots, M$ and $k = 1, \ldots, K$.

The channel is estimated by pilot sequences sent from users, which are orthogonal in the same cell and reused among the cells, so that channel estimation experiences interference from the same pilot sequence sent from adjacent cells, so-called pilot contamination [29]; the worst case occurs when synchronized transmission happens from the same pilot. Thus, the estimate for the $M \times K$ propagation matrix is expressed as

$$\hat{\mathbf{G}}_{jj} = \sqrt{p_{\mathrm{p}}} \sum_{l=1}^{L} \mathbf{G}_{jl} + \mathbf{V}_j, \tag{6.8}$$

where $\mathbf{V}_j$ is an $M \times K$ matrix of the receiver noise and p_{p} is the pilot transmit power. Assuming that a total of K OFDM symbols are used for the pilots of the K users, we have $p_{\mathrm{p}} = K p_{\mathrm{s}}$. When the MF detector is used, a BS processes its received signal by multiplying it by the conjugate-transpose of the channel estimate, which yields

$$\mathbf{y}_j = \hat{\mathbf{G}}_{jj}^* \mathbf{x}_j$$
$$= \left(\sqrt{p_{\mathrm{p}}} \sum_{l_1=1}^{L} \mathbf{G}_{jl_1} + \mathbf{V}_j \right)^* \left(\sqrt{p_{\mathrm{s}}} \sum_{l=1}^{L} \mathbf{G}_{jl} \mathbf{a}_l + \boldsymbol{\omega}_j \right). \tag{6.9}$$

As a consequence, we have

$$\mathbf{y}_j = \sqrt{p_{\mathrm{p}} p_{\mathrm{s}}} \sum_{l_1=1}^{L} \sum_{l=1}^{L} \mathbf{G}_{jl_1}^* \mathbf{G}_{jl} \mathbf{a}_l + \sqrt{p_{\mathrm{p}}} \sum_{l=1}^{L} \mathbf{G}_{jl}^* \boldsymbol{\omega}_j$$
$$+ \sqrt{\rho_{\mathrm{s}}} \sum_{l=1}^{L} \mathbf{V}_j^* \mathbf{G}_{jl} \mathbf{a}_l + \mathbf{V}_j^* \boldsymbol{\omega}_j. \tag{6.10}$$

The noise effect here is taken to be

$$\mathbf{z}_j = \sqrt{p_{\mathrm{s}}} \sum_{u=1}^{K} \mathbf{v}_{uj}^* \mathbf{g}_{uj} \mathbf{a}_{uj} + \sqrt{p_{\mathrm{p}} p_{\mathrm{s}}} \sum_{u \neq k}^{K} \sum_{l=1}^{L} \mathbf{g}_{ul}^* \mathbf{g}_{uj} \mathbf{a}_{uj}$$
$$+ \sqrt{p_{\mathrm{s}}} \sum_{l \neq j}^{L} \left(\sum_{l_1=1}^{L} \sqrt{p_{\mathrm{p}}} \mathbf{G}_{jl_1} + \mathbf{V}_j \right)^* \mathbf{G}_{jl} \mathbf{a}_l + \sqrt{p_{\mathrm{p}}} \sum_{l=1}^{L} \mathbf{G}_{jl}^* \boldsymbol{\omega}_j + \mathbf{V}_j^* \boldsymbol{\omega}_j, \tag{6.11}$$

where $\mathbf{g}_{ul}$ and $\mathbf{v}_{ul}$ are the channel propagation vector and the thermal noise vector from user u in the lth cell to the jth cell, respectively. Then the uplink achievable rate for user k in the jth cell is given by

$$C_{\mathrm{JDRS}kj} = \log_2 \left(1 + \frac{p_{\mathrm{p}} p_{\mathrm{s}} \left| \sum_{l=1}^{L} \mathbf{g}_{kl}^* \mathbf{g}_{kj} \right|^2}{p_{\mathrm{s}} \left| \mathbf{v}_{kj}^* \mathbf{g}_{kj} \right|^2 + \left| \left(\sqrt{p_{\mathrm{p}}} \sum_{l=1}^{L} \mathbf{g}_{kl} + \mathbf{v}_{kj} \right)^* \boldsymbol{\omega}_j \right|^2 + \mathrm{ICI}} \right), \tag{6.12}$$

where the inter-cell interference is given by

$$\text{ICI} = p_s \sum_{(l,u) \neq (j,k)} \left| \left(\sqrt{P_p} \sum_{l_1=1}^{L} \mathbf{g}_{kl_1} + \mathbf{v}_{kj} \right)^* \mathbf{g}_{ul} \right|^2. \tag{6.13}$$

As M grows without limit, the L2-norms of the above vectors grow proportionally to M, while the inner products of uncorrelated vectors, by assumption, grow at a lesser rate. Hence, we have

$$\mathbf{G}_{jl_1}^* \mathbf{G}_{jl_2} \to \mathbf{D}_{\boldsymbol{\beta}_{jl}} \delta_{l_1 l_2}, \tag{6.14}$$

where $\mathbf{D}_{\boldsymbol{\beta}_{jl}}$ denotes a $K \times K$ diagonal matrix whose diagonal elements are $[\boldsymbol{\beta}_{jl}]_k = \sum_{m=1}^{M} \beta_{mjkl}$ when $l_1 = l_2 = l$. Thus, the kth component of the processed signal becomes

$$\frac{1}{\sqrt{P_p P_s}} y_{kj} \to \sum_{m=1}^{M} \beta_{mjkj} a_{kj} + \sum_{l \neq j}^{L} \sum_{m=1}^{M} \beta_{mjkl} a_{kl}. \tag{6.15}$$

As $M \to \infty$, the signal-to-interference ratio (SIR) for the kth user in the jth cell is given by

$$\text{SIR}_{kj} \to \frac{\left(\sum_{m=1}^{M} \beta_{mjkj} \right)^2}{\sum_{l \neq j}^{L} \left(\sum_{m=1}^{M} \beta_{mjkl} \right)^2}. \tag{6.16}$$

Then the asymptotic uplink rate for JDRS is given by

$$C_{\text{JDRS}} \to \log_2 \left(1 + \frac{\left(\sum_{m=1}^{M} \beta_{mjkj} \right)^2}{\sum_{l \neq j}^{L} \left(\sum_{m=1}^{M} \beta_{mjkl} \right)^2} \right). \tag{6.17}$$

For brevity, in (6.17) we have used C_{JDRS} to denote $C_{\text{JDRS}_{kj}}$. For ease of comparison, we now analyze the scaling law for the asymptotic rate for JDRS. When M grows without limit, from the law of large numbers the arithmetic average of β_{mjkj} is given by

$$\frac{1}{M} \sum_{m=1}^{M} \beta_{mjkj} = \mathbb{E}\{\beta_{mjkj} | B = b\}. \tag{6.18}$$

It is difficult to express the expectation in an explicit form for an arbitrary value of γ. However, we can derive an explicit result for a few specific values, e.g., the typical values $\gamma = 2$ or $\gamma = 4$.

THEOREM 6.1   *When $M \to \infty$ with K fixed, for $2 \leq \gamma \leq 4$ and the channel propagation coefficient given in Section 6.2, the scaling law of the achievable uplink rate for JDRS can be given as*

$$C_{\text{JDRS}} = O(c), \tag{6.19}$$

where c is independent of M.

Proof See Appendix section 6.8.1.   □

From Theorem 6.1 we can see that, when M grows large, the uplink rate is limited by pilot contamination and is independent of the transmit powers. This means that the transmitted power can be arbitrarily small and yet a fixed rate can still be maintained. Nevertheless, if the wireless channel environment is different from the model used in this chapter then the rate-scaling law may change.

Remark 6.1   If there is no LOS channel in any distance range and $\beta_{mjkl} = Z_{mjkl}/d_{mjkl}^4$ then we may have $C_{JDRS} = O\left(2\log_2(M)\right)$. However, considering the near-field effect and the high chance of having an LOS, this result is hardly likely to occur. In this chapter we focus on the more likely models.

Intuitively, if we increase the radius of the cell while the density of RAUs remains the same, the desired signal benefits from more cooperative RAUs while both the intra-cell interference and the inter-cell interference suffer from the longer distance. From the proof of Theorem 6.1, we have the following proposition.

PROPOSITION 6.2   Assuming $R \gg d_t$, when $M \to \infty$ with K fixed the achievable rate of a multi-cell massive C-RAN with JDRS almost surely increases with the cell radius R.

Proof See Appendix section 6.8.2.   □

Single-Cell Massive C-RAN with Imperfect CSI

In the above analysis we derived expressions for the achievable rate and the asymptotic rate for JDRS in a multi-cell massive C-RAN, where the system performance is limited by inter-cell interference. For a single-cell scenario, the inter-cell interference disappears and so does pilot contamination. When the number of RAUs grows without limit, the system is noise-limited. Intuitively, the achievable rate will grow with a power of M even with imperfect CSI at the receiver side. Following similar derivations as those for the multi-cell scenarios, we obtain the following rate expression and scaling law.

PROPOSITION 6.3   Assuming imperfect CSI at the BS, when $M \to \infty$ the achievable uplink rate for a given user with JDRS in the single-cell scenario is given by

$$C_{JDRS}^{s\text{-}cell} \to \log_2\left(1 + p\sum_{m=1}^{M}\beta_{mjkj}\right), \qquad (6.20)$$

where $p = \sqrt{p_p p_s}$ is fixed. When $2 \le \gamma \le 4$, the rate-scaling law can be written as

$$C_{JDRS}^{s\text{-}cell} = O(2\log_2 M). \qquad (6.21)$$

Proof   In the single-cell scenario, inter-cell interference can be ignored. All RAUs of the jth cell receive an $M \times 1$ vector

$$\mathbf{x}_j = \sqrt{p_s}\mathbf{G}_j\mathbf{a}_j + \boldsymbol{\omega}_j. \qquad (6.22)$$

Then, with the assumption of imperfect CSI, channel estimation is obtained by the uplink pilots and is given by $\hat{\mathbf{G}}_j = \sqrt{p_p}\mathbf{G}_j + \mathbf{V}_j$, where $\mathbf{V}_j$ is the receiver noise.

Multiplying the received signal by the conjugate-transpose of the channel estimate, we have

$$\mathbf{y}_j = \left(\sqrt{p_p}\mathbf{G}_j + \mathbf{V}_j\right)^* \left(\sqrt{p_s}\mathbf{G}_j\mathbf{a}_j + \boldsymbol{\omega}_j\right). \tag{6.23}$$

Using the assumption in (6.14), we obtain (6.20) when $M \rightarrow \infty$. From the analysis in Appendix section 6.8.1, it is easily seen that when $2 \leq \gamma \leq 4$ we have

$$\sum_{m=1}^{M} \beta_{mjkj} = O(M). \tag{6.24}$$

Substituting (6.24) into (6.20) yields (6.21), which completes the proof. □

Proposition 6.3 provides the asymptotic rate with imperfect CSI at the BS, whereas the result with perfect CSI can be easily obtained by replacing $p = \sqrt{K}p_s$ in (6.20) by p_k. Compared with the single-input single-output (SISO) system, it is possible to reduce p proportionally to $1/M$ in a single-cell massive C-RAN. Meanwhile the uplink sum-rate increases K times since K users are served with the same time–frequency resource. If we scale down the transmit power proportionally to $1/M^a$, where $a > 1$, then the signal-to-interference plus noise ratio (SINR) of the uplink will go to zero when $M \rightarrow \infty$. Therefore, the largest proportion by which the transmit power can be cut in the single-cell JDRS is $1/M$, conditioned on a fixed rate.

Compared with the result derived in Theorem 6.1, the single-cell massive C-RAN with JDRS improves the scaling law for the achievable rate because inter-cell interference disappears. This suggests that mitigating pilot contamination or using other interference-cancellation techniques would enlarge the achievable rate of massive C-RANs with JDRS by a factor $O(c \log_2 M)$ at most, where c is a constant independent of M.

Single-Cell Massive C-RAN with Perfect CSI

In this subsection, to obtain the theoretical maximum achievable rate we further study the single-cell scenario with perfect CSI at the receiver side. Here, we consider the case with multi-antenna RAUs, equipped with $N_1, \ldots, N_M$ antennas, respectively, and with the spatial correlation of channels, which characterizes the Kronecker model. Specifically, the channel from the kth user to the mth RAU the jth cell[2] can be written as

$$\mathbf{g}_{m,k} = \mathbf{R}_{m,k}^{1/2}\mathbf{x}_{m,k}, \tag{6.25}$$

where $\mathbf{R}_{m,k}$ is an $N_m \times N_m$ deterministic nonnegative-definite matrix, which characterizes the spatial correlation at the mth RAU and the fast-fading coefficient between the kth user and the mth RAU, and $\mathbf{x}_{m,k}$ is an $N_m \times 1$ complex Gaussian random vector, in which elements have zero mean and unit variance.

All RAUs of the jth cell receive an $N \times 1$ vector

$$\mathbf{y} = \sqrt{p_s}\mathbf{G}\mathbf{a} + \boldsymbol{\omega}, \tag{6.26}$$

[2] We will drop the index j as we are considering a single cell.

where $N = \sum_{m=1}^{M} N_m$, $\mathbf{G} = [\mathbf{G}_1^T \cdots \mathbf{G}_M^T]^T$, $\mathbf{G}_m = [\mathbf{g}_{m,1} \cdots \mathbf{g}_{m,K}]$, $\mathbf{a}$ is the $K \times 1$ vector of the signal from the users, $\boldsymbol{\omega}$ denotes an $N \times 1$ additive white Gaussian noise, whose entries are i.i.d. circular complex Gaussian random variables with zero mean, and p_s is a measure of the SNR.

By employing JDRS, the ergodic achievable rate of the above system model can be expressed as

$$\mathcal{V}_{\mathbf{G}} = \frac{1}{N}\mathbb{E}\left\{\log\det\left(\mathbf{I}_N + p_s\mathbf{G}\mathbf{G}^H\right)\right\}. \tag{6.27}$$

Specifically, $\mathcal{V}_{\mathbf{G}}$ provides a performance metric regarding the number of bits per second per hertz per antenna that can be transmitted reliably over the channel matrix $\mathbf{G}$.

Here, we are interested in understanding the ergodic achievable rate of the channel of interest in the asymptotic regime where M is fixed and $N_1, \ldots, N_M, K$ grow to infinity with ratios $\beta_m = N_m/K$ such that $0 < \beta_m < \infty$. We assume that the deterministic matrices $\{\mathbf{R}_{m,k}\}_{\forall m,k}$ are deterministic nonnegative-definite and that the spectral norm of $\mathbf{R}_{m,k}$ is bounded by a constant, i.e., $\max_{k,l}\|\mathbf{R}_{m,k}\| \leq C_{\max}$.

THEOREM 6.4   *Under the assumption that perfect CSI is available at the BS, as $M, K \to \infty$ the ergodic achievable rate satisfies*

$$\mathbb{E}\{\mathcal{V}_G\} - \mathcal{V}_M = O\left(\frac{1}{N}\right), \tag{6.28}$$

where

$$\mathcal{V}_M = \frac{1}{N}\sum_{m=1}^{M}\log\det\left(\mathbf{I}_{N_m} + p_s\sum_{k=1}^{K}\frac{1}{1 + \sum_{m=1}^{M}N_m e_{m,k}}\mathbf{R}_{m,k}\right)$$
$$+ \frac{1}{N}\sum_{k=1}^{K}\log\left(1 + \sum_{m=1}^{M}N_m e_{m,k}\right) - \frac{1}{N}\sum_{m,k}\frac{e_{m,k}}{1 + \sum_{m=1}^{M}N_m e_{m,k}} \tag{6.29}$$

and $\{e_{m,k}\}$ is a unique solution of the following $M \times K$ equations:

$$e_{m,k} = \frac{1}{N_m}\mathrm{Tr}\left\{\mathbf{R}_{m,k}\left(\frac{1}{p_s}\mathbf{I}_{N_m} + \sum_{k=1}^{K}\frac{1}{1 + \sum_{m=1}^{M}N_m e_{m,k}}\mathbf{R}_{m,k}\right)^{-1}\right\}, \tag{6.30}$$

for $m = 1, \ldots, M$ and $k = 1, \ldots, K$.

Proof   Using the same approach as in the proof of [30, Theorem 2], we obtain this result.   □

Theorem 6.4 provides the deterministic equivalent of the ergodic achievable rate in a single-cell massive C-RAN with perfect CSI by employing the approach of large-dimension random matrix theory, which is different from the approach used previously.

6.3.2   Selection Reception Scheme

In SRS, we assume that each user is only associated with one home RAU. There are K RAUs selected in each cell and the other $M - K$ RAUs switch to sleep mode. We denote the RAU selected by the kth user as $\mathcal{R}_k$, and the channel propagation coefficient between $\mathcal{R}_k$ in the jth cell and the uth user in the lth cell is g_{kjul}. The received signal at $\mathcal{R}_k$ in the jth cell can be written as

$$y_k = g_{kjkj}a_{kj} + \sum_{\substack{u=1 \\ u \neq k}}^{K} g_{kjuj}a_{uj} + \sum_{\substack{l=1 \\ l \neq j}}^{L} \sum_{u=1}^{K} g_{kjul}a_{ul} + n_{kj}, \tag{6.31}$$

where $|a_{ul}| = \sqrt{p_{ul}}$. On the right-hand side of the above equation the first term is the desired signal from the given user, the second term is the intra-cell interference, the third term is the inter-cellular interference, and n_{kj} is the additive noise with unit variance. The ergodic achievable rate of the kth user for SRS can be written as

$$C_{SRS} = \mathbb{E} \left\{ \log_2 \left(1 + \frac{p_{kj}g_{kj}^2}{\sum_{\{l,u\} \neq \{j,k\}} p_{ul}g_{kjul}^2 + 1} \right) \right\}, \tag{6.32}$$

where we use g_{kj} instead of g_{kjkj} for simplicity, and the expectation is taken over the small-scale fading factors. As for JDRS, we use C_{SRS} to denote $C_{SRS_{kj}}$. If we approximate the sum of the interference plus noise as a complex Gaussian random variable with variance $\sigma^2 = \sum_{\{u,l\} \neq \{k,j\}} p_{ul}\beta_{kjul} + 1$, where $\beta_{kjul} = (\lambda/4\pi d_{kjul})^2$ when $d_{kjul} < d_t$, a lower bound of the ergodic rate can be obtained. Since g_{kj} follows an exponential distribution, the lower bound of the ergodic achievable rate for SRS can be obtained from [31] as

$$C_{SRS} = \log_2 e \, e^{\alpha_{kj}} \int_{\alpha_{kj}}^{\infty} \frac{e^{-t}}{t} dt, \tag{6.33}$$

where

$$\alpha_{kj} = \frac{\sum_{\{u,l\} \neq \{k,j\}} p_{ul}\beta_{kjul} + 1}{p_{kj}\beta_{kjkj}}. \tag{6.34}$$

Furthermore, when the number of RAUs goes to infinity, with the number of users and the selected RAUs fixed, we can derive the asymptotic performance for SRS using the following theorem.

THEOREM 6.5   When $M \to \infty$ with K fixed, for any $\epsilon > 0$ the scaling law of the achievable uplink rate for SRS is given by

$$C_{SRS} \geq O \left((1 - \epsilon) \log_2 M \right). \tag{6.35}$$

Proof See Appendix section 6.8.3.   □

The above theorem is applicable for both multi-cell and single-cell scenarios. When $M \to \infty$, the ergodic rate of SRS increases proportionally to $\log_2 M$ at least. Compared with JDRS, SRS can provide lower backhaul loads and a higher achievable rate when the number of RAUs is sufficiently large. It will be shown that the energy efficiency is also improved by using SRS because most RAUs are sleeping.

To study the power-scaling law for SRS, we first scale down the transmit power proportionally to $1/M^a$, $a > 0$. Note that here we assume that the BS has perfect CSI for the reason that the distance between the user and the selective antenna of the same cell is quite close; therefore the power allocated to the pilot is small, with the result that pilot contamination to the inter-cell can be omitted. Thus, when $M \to \infty$ with K fixed, the achievable rate follows from (6.68):

$$C_{SRS} \geq \log_2 \left(1 + \frac{P\mathbb{E}\left\{g_{kj}^2\right\}}{M^a + \left(M^{1/2-\epsilon} - 1\right)^{-2} P\mathbb{E}\left\{g_{kj}^2\right\}} \right), \tag{6.36}$$

where P is the power of the transmitted signals. When M grows large, for any $\epsilon > 0$ we have the probability

$$P[d_{kj} \leq R/M^{(1+\epsilon)/2}] = 1/M^\epsilon \to 0. \tag{6.37}$$

From (6.37) and (6.62), for SRS it follows that $d_{kj} \overset{a.s.}{\to} O(1/\sqrt{M})$. Then, using the model in (6.1), we have

$$\mathbb{E}\left\{g_{kj}^2\right\} \overset{a.s.}{\to} O(M). \tag{6.38}$$

Substituting (6.38) into (6.36), as $a > 0$ we can calculate the lower bound of the rate-scaling law as $C_{SRS} \geq O\left((1 - a)\log_2 M\right)$, which is lower than the result in (6.35). The reason is that when the transmit power is scaled down by M^a ($a > 0$), the system is limited by thermal noise. This implies that we cannot scale down the transmit power by M in a massive C-RAN with SRS while maintaining an increasing sum-rate as in (6.35). In the same way, it can be found that to keep a fixed rate for SRS ($C_{SRS} \geq O(1)$), the largest fraction by which the transmit power can be cut is $1/M$.

Compared with the results for JDRS in the previous subsection, it is found that in multi-cell networks the rate-scaling law of for SRS is better than that for JDRS. Moreover, from (6.31) we have found that the achievable rate for SRS can be further improved by the optimization of user scheduling and RAU selection, whereas for JDRS the scaling law can hardly be enhanced by scheduling because of the channel-hardening effect. We will study the scheduling problem in SRS in Section 6.5.

6.4   Energy Efficiency

In this section we define the energy efficiency (EE) of a system as the ratio of the sum-rate and the sum of the consumed power in a cell. The power consumed by both the signal transmission and the reception is taken into account. Typically, increasing the spectral efficiency is associated with decreasing the energy efficiency; this is caused by the increase in the transmit power. However, by studying the energy efficiency and the power-scaling law in massive C-RANs, it is possible to jointly increase the achievable rate and the energy efficiency.

6.4.1   Joint Decoding Reception Scheme

Multi-Cell Massive C-RAN

In order to reveal the relationship between the energy consumption and the energy efficiency, and for ease of analysis, we consider only the power consumption for signal transmission and reception, with the energy expenditure of processing and backhaul left for future research. We define the energy efficiency as the ratio of the sum-rate and the sum of the consumed power in a cell. Then the energy efficiency of a given cell for JDRS can be given as

$$EE_{JDRS} = \frac{\sum_k C^k_{JDRS}}{\eta \sum_k p_k + \sum_m q_m},$$ (6.39)

where η denotes the power amplifier factor, p_k denotes the signal transmission power consumed by user k, and q_m represents the reception power consumed by RAU m. First, we consider a multi-cell scenario. The energy efficiency expression for JDRS can be obtained by substituting (6.12) into (6.39). When M grows large, from (6.17) we know that C_{JDRS} is independent of the transmitted powers and, according to (6.19), C_{JDRS} is independent of the number of RAUs. Therefore, the transmission power can be arbitrarily small, i.e., $p_k \rightarrow 0$ as $M \rightarrow \infty$. From the law of large numbers, we have

$$\sum_{m=1}^{M} q_m = M \int_{b \in \mathcal{B}} q(b) f_B(b) db.$$ (6.40)

Since the RAUs are uniformly placed in the cell, we assume that both $q(b)$ and $f_B(b)$ are independent of M. Thus, we have $\sum_m q_m = c_1 M$, where c_1 is a constant that is independent of M. Without loss of generality, we can assume that the transmit powers of all the users are of the same magnitude.

THEOREM 6.6 *When $M \rightarrow \infty$ with K fixed, the scaling law of the energy efficiency of multi-cell C-RAN with JDRS is given by*

$$EE_{JDRS} \rightarrow \begin{cases} O(c/M) & \text{if } p_k \rightarrow 0, \\ O(c/M) & \text{if } p_k = PM^a, a < 1, \\ O(c_2/(\eta PM^a)) & \text{if } p_k = PM^a, a \geq 1, \end{cases}$$ (6.41)

where c_2 and c are constants satisfying $C_{JDRS} \xrightarrow{a.s.} O(c_2)$ and $c = Kc_2/c_1$, respectively, and a is the power scaling factor.

Proof From (6.17), C^k_{JDRS} is independent of p_k. If $p_k \rightarrow 0$, the energy efficiency for JDRS is given by $EE_{JDRS} \rightarrow \sum_k C^k_{JDRS}/(c_1 M)$. If we scale p_k by M as $p_k = PM^a$, $EE_{JDRS} = \sum_k C^k_{JDRS}/(\eta KPM^a + c_1 M)$. Therefore, when $a < 1$, $EE_{JDRS} \rightarrow \sum_k C^k_{JDRS}/(c_1 M)$, when $a \geq 1$, $EE_{JDRS} \rightarrow \sum_k C^k_{JDRS}/(\eta KPM^a)$. Since we have $C_{JDRS} \rightarrow O(c_2)$ (c_2 is a constant) in a multi-cell massive C-RAN with JDRS, (6.41) is therefore obtained. $\square$

As proved in the above theorem, the system energy efficiency of a multi-cell massive C-RAN with JDRS decreases heavily when the number of RAUs grows large; this is caused by the increasing cost of energy for packet reception. In this case controlling the transmit power will be pointless, owing to severe inter-cell interference caused by pilot contamination.

Single-Cell Massive C-RAN

In the single-cell massive C-RAN, with the assumption that K OFDM symbols are used for uplink pilots we have $p_p = Kp_s$. As a consequence, substituting $p = \sqrt{K}p_k$ into (6.20) the energy efficiency as M grows can be rewritten as

$$EE_{JDRS}^{\text{s-cell}} = \frac{\sum_k \log_2 \left[1 + Kp_k^2 \left(\sum_{m=1}^{M} \beta_{mjkj}\right)^2\right]}{\eta \sum_k p_k + \sum_m q_m}. \tag{6.42}$$

Similarly to the derivation of Theorem 6.6 for multi-cell networks, we obtain the following proposition.

PROPOSITION 6.7   When $M \to \infty$ with K fixed, the energy efficiency of single-cell C-RAN with JDRS satisfies

$$EE_{JDRS}^{\text{s-cell}} \to \begin{cases} 0 & \text{if } p_k \to 0, \\ 0 & \text{if } p_k = PM^a,\ a < -1, \\ O(c/M) & \text{if } p_k = P/M, \\ O\left(c(a+1)\log_2 M/M\right) & \text{if } p_k = PM^a,\ -1 < a \le 1, \\ O\left(c(a+1)\log_2 M/M^a\right) & \text{if } p_k = PM^a,\ a > 1, \end{cases} \tag{6.43}$$

where c is a factor that depends on the power scaling factor a but is independent of M.

Proof   When $p_k = 0$, we have $C_{JDRS}^{\text{s-cell}} = 0$. According to (6.42), $EE_{JDRS}^{\text{s-cell}} = 0$ is obtained. When p_k scales with M according to $p_k = PM^a$, the energy efficiency for a single-cell massive C-RAN is given by

$$EE_{JDRS}^{\text{s-cell}} = \frac{\sum_k \log_2 \left[1 + KP^2 M^{2a} \left(\sum_{m=1}^{M} \beta_{mjkj}\right)^2\right]}{\eta KPM^a + c_1 M}. \tag{6.44}$$

Since $\sum_{m=1}^{M} \beta_{mjkj} \overset{a.s.}{\to} O(M)$ when $M \to \infty$, we have

$$EE_{JDRS}^{\text{s-cell}} \to \frac{K\log_2\left(1 + KP^2 G^2 M^{2a+2}\right)}{\eta KPM^a + c_1 M}, \tag{6.45}$$

where G is independent of M. We distinguish among the four cases $a < -1$, $a = -1$, $-1 < a < 1$, and $a \ge 1$. Then the results in (6.43) can be derived from (6.45), which completes the proof.   □

When the power consumed by reception is taken into consideration, the energy efficiency decreases proportionally to at least $1/M$ in the multi-cell scenario, and is obtained as $p_k = M^a(a < 1)$ or even $p_k \to 0$. In single-cell cases, the energy efficiency decreases as at least $O\left(c\log_2 M/M\right)$ and is obtained as $p_k = MP$. If we scale down the transmit power proportionally to $1/M$, the energy efficiency is still comparable with that in multi-cell cases. As proved in the above theorems, when the energy consumed by packet reception is taken into consideration an increasing number of RAUs is always harmful to the system energy efficiency in JDRS. The improvement in the rate cannot make up for the loss of energy. Comparing with the results in Theorem 6.6, energy efficiency benefits the single-cell system for the reason that pilot contamination disappears. It means that using a pilot decontamination technique can enlarge the energy efficiency of a massive C-RAN with JDRS by a factor of at most $O(\log_2(M))$.

6.4.2   Selection Reception Scheme

The energy efficiency of a massive C-RAN with SRS is given as the ratio of the sum-rates and the sum of the consumed power in a cell. The achievable rate of a given user for SRS is defined in (6.31). Then the energy efficiency over a cell is given by

$$\text{EE}_{\text{SRS}} = \left[\sum_k (\eta p_k + q_k)\right]^{-1} \sum_k \log_2\left(1 + \frac{p_{kj}g_{kj}^2}{\sum_{\{l,u\}\neq\{j,k\}} p_{ul}g_{kjul}^2 + 1}\right). \tag{6.46}$$

As the number of RAUs grows large, for SRS it can be seen that the average distance from the selected RAU to the corresponding user will be very small. We normally use low-power devices in this situation and q_k can therefore be considered as a constant, c_1. Then we can derive the relation between the asymptotic energy efficiency and the power scaling for SRS.

THEOREM 6.8   *When $M \to \infty$ with K fixed, the scaling law of the energy efficiency of SRS can be described as*

$$\text{EE}_{\text{SRS}} \to \begin{cases} 0 & \text{if } p_k \to 0, \\ 0 & \text{if } p_k = PM^a, \ a < -1, \\ c & \text{if } p_k = PM^a, \ a = -1, \\ O\left(c\log_2(M)\right) & \text{if } p_k = PM^a, \ -1 < a \leq 0, \\ O\left(c\log_2(M)/M^a\right) & \text{if } p_k = M^a, \ a > 0, \end{cases} \tag{6.47}$$

where c is a factor that depends on the power scaling factor a but is independent of M.

Proof In the case $p_k \to 0$, the result is the same as that for JDRS. When p_k scales with M as $p_k = PM^a$, the energy efficiency for SRS becomes

$$\text{EE}_{\text{SRS}} = \left[K\left(\eta PM^a + c_1\right)\right]^{-1} \sum_k \log_2\left(1 + \frac{PM^a g_{kj}^2}{PM^a \sum_{\{l,u\}\neq\{j,k\}} g_{kjul}^2 + 1}\right). \tag{6.48}$$

When $M \to \infty$, (6.48) can be written according to (6.38) and (6.68) as

$$\text{EE}_{\text{SRS}} \geq [\eta PM^a + c_1]^{-1} \log_2\left(1 + \frac{PM^{a+1}}{PM^{a+1}(KL-1)\left(M^{1/2-\epsilon}-1\right)^{-2} + c_2}\right), \tag{6.49}$$

where ϵ is any positive constant, and c_2 is independent of M. We use the lower bound when $\epsilon \to 0$ as an approximation for the energy efficiency. Then the result depends on the value of a. We divide the value of a into four cases, $a < -1$, $a = -1$, $-1 < a \leq 0$, and $a > 0$, and obtain (6.47).   □

If we scale down p_k faster than $1/M$, the energy efficiency will go to zero as $M \to \infty$. A constant energy efficiency is obtained when the transmit power is scaled down as

$p_k = P/M$. The maximum energy efficiency is achieved when we scale p_k according to $p_k = PM^a$ $(-1 \le a \le 0)$. With the above analysis of the two transmission schemes for a massive C-RAN, we can see that SRS can provide much higher energy efficiency than JDRS if the users' transmission power is reasonably allocated. The achievable rate can still scale up with M as (6.35) and the transmission power can be scaled down with M as fast as $1/M^a$ $(0 < a < 1)$. The reason is that in SRS the number of RAUs selected to receive the users' transmission is limited, and a majority of the RAUs go to sleep. With SRS, massive C-RAN not only has the advantages of both massive MIMO and C-RAN but also avoids pilot contamination and energy wastage.

6.5   Joint User Scheduling and RAU Selection Algorithms

As discussed, the RAUs should be optimally selected to maximize the average achievable rate for SRS. Moreover, since the channel-hardening effect exists only in JDRS, a remarkable gain can be achieved in SRS through effective scheduling. Owing to the correlation between closely spaced users in JDRS, as illustrated in [7], a user scheduling algorithm is also required for JDRS to avoid severe intra-cell interference. Thus, we first propose an efficient joint user scheduling and an RAU selection algorithm utilizing long-term CSI. As $M \to \infty$, the impact of the RAU selection on the performance weakens greatly. Therefore, a simple yet effective RAU selection scheme where the users choose closer RAUs is considered, this gives a good performance when it is applied together with our proposed user scheduling algorithm.

6.5.1   RAU Selection Algorithm

For known long-term CSI between a user and any RAU, we propose an efficient RAU selection algorithm for a given scheduled user set. We consider that all the given K users must be accommodated in the same time–frequency resource, and the RAUs should be optimally selected to maximize the average achievable rate. The optimal solution of this combinatorial problem requires a high complexity exhaustive search, which is undesirable. We therefore propose an efficient algorithm A1 to reduce its complexity.

Algorithm A1 contains three main steps. In step 1 we identify the candidate RAUs according to a threshold, where $\mathcal{M}$ denotes the set of all RAUs of the jth cell, $\mathcal{N}_k$ consists of the alternative RAUs for user k, $\mathcal{R}$ is the set of all selected RAUs of jth cell, and R_k is the RAU selected for user k. The value of the threshold β_{th} can be determined by extensive offline simulations. In step 2, we use an efficient way to select those RAUs with strong links to the users. The variable F_k is used as a flag, and $F_k = 1$ in step 2 means that the uplink from k to R_k is not the strongest uplink. In step 3, we have $\bar{\mathcal{N}}_k = \mathcal{M}/\mathcal{N}_k$, and $A_m = |\{k : \beta_{mjkl} > \beta_{th}\}|$ shows whether the chosen RAU experiences heavy interference from other users. The condition $A_m = 1$ means that the selected RAU m has a strong uplink only with its own user and has weak links with other users, while the condition $A_m > 1$ means that the RAU m has strong links with more than one user. In the latter case, if m is selected then it may suffer from heavy intra-cell interference.

Algorithm A1

Initialization:

    Let $\mathcal{M} = \{1, \ldots, M\}$, $\mathcal{R} = \emptyset$, $\mathcal{N}_k = \emptyset$, $F_k = 0$ and $R_k = 0$

    Step 1: Determining the candidate RAU set $\mathcal{N}_k$.

      For $k = 1$ to K, $m = 1$ to M

        If $\beta_{mjkj} \geq \beta_{\text{th}}$, then $\mathcal{N}_k = \mathcal{N}_k \cup \{m\}$

      End

    Step 2: Selecting those RAUs with strong uplinks.

      For $k = 1$ to K, let $\mathcal{D} = \mathcal{N}_k$

        While 1 do

          $n^* = \arg\max_{n \in \mathcal{D}} \{\beta_{njkj}\}$

          If $\mathcal{R} \cap \{n^*\} = \emptyset$, then $\mathcal{R} = \mathcal{R} \cup \{n^*\}$, $R_k = n^*$, break

          Else then $\mathcal{D} = \mathcal{D}/\{n^*\}$, $F_k = 1$

      End

    Step 3: Greedy search for re-examining the chosen RAUs.

      For $k = 1$ to K

        If $A_{R_k} > 1$ or $F_k = 1$,

          then $m^* = \arg\max_{m \in \mathcal{N}_k \cap \{m : A_m = 1\}} \{\beta_{mjkj}\}$

        $\mathcal{D} = \mathcal{N}_k \cup m^*$

        $n^* = \arg\max_{n \in \mathcal{D}} C_{\text{SRS}} ((\mathcal{R}/R_k) \cup n)$

        If $C_{\text{SRS}} ((\mathcal{R}/R_k) \cup n^*) > C_{\text{SRS}} (\mathcal{R})$,

          then $\mathcal{R} = \mathcal{R}/R_k \cup \{n^*\}$, $R_k = n^*$

    End

By carefully determining the threshold value β_{th}, the number of elements in $\mathcal{N}_k$ in step 1 and $\mathcal{D}$ in step 3 can be decreased greatly and the complexity of the algorithm can be reduced while it maintains a good performance. Since the calculation of C_{SRS} has complexity $O(M)$, and assuming that $|\mathcal{N}_k| \approx K$, the computational complexity of step 1 is $O(KM)$ whereas the worst-case complexities of steps 2 and 3 are $O(K^2)$ and $O(K^2 M)$, respectively. Therefore, the worst-case complexity of Algorithm A1 is dominated by step 3 and is $O(K^2 M)$. As seen, the complexity is often lower than a greedy selection algorithm, because the sum-rate is calculated only when the RAUs meet the discriminant conditions in step 3.

When M grows without limit with K and $\mathcal{R}$ fixed, β_{th} will be set very close to $\max_{n \in \mathcal{M}} \{\beta_{njkj}\}$. It is almost certain that the RAUs selected in step 2 in Algorithm A1 will have one strong link with their corresponding user and weak links with other users. Each selected RAU serves the nearest user. Thus, for any user k in the jth cell, the distance between user k and the selected RAUs satisfies (6.62) and (6.64). Following the proof in Appendix section 6.8.3, the lower bound of the ergodic achievable rate for user k is given by (6.35), which proves that Algorithm A1 is asymptotically optimal for large M. From the above consideration, we can see that a simple selection scheme, which we will refer to as nearest selection (NS), in which the users simply choose the RAUs with the best large-scale fading, is asymptotically optimal when M grows large with K fixed. As a consequence, we will employ both the two schemes in our proposed joint scheduling algorithms and compare their performances to find out the most efficient scheme for massive C-RANs.

6.5.2 Joint Scheduling and RAU Selection

Here, we assume that there are U users uniformly randomly distributed over each cell, and K of them are scheduled at a given time. As demonstrated analytically in [32], in multi-cell single-hop networks where the access points cooperate in terms of joint resource allocation, the sum-rate scales as $\log U$ with simple distributed user scheduling algorithms. Unfortunately, owing to the channel hardening effect, user scheduling provides only a very limited gain in JDRS. As a consequence, for a massive C-RAN with a fixed number of users, it is advisable to use SRS together with a well-designed scheduling algorithm; this can achieve an even higher scaling behavior of the achievable sum-rate than the result given in Theorem 6.5.

First, since the complexity of the optimal solution is extremely high when $\mathcal{U}$ is large, we propose a suboptimal greedy scheduling Algorithm A2, using Algorithm A1 for RAU selection. In Algorithm A2, $\mathcal{U}$ denotes the set of all users of the jth cell; $\mathcal{K}$ denotes the set of scheduled users. The user interference index $B_k = |\{u : d_{uk} \leq d_{th}\}|$ shows whether the user k experiences heavy interference. The value of d_{th} is determined by extensive offline simulations. For example, when $R = 1000$ m and $K = 10$, choosing $d_{th} = 100$ m obtains a good performance. Since U may grow large in our considerations, the worst-case complexity of Algorithm A2 is $O(K^3 UM + U^2)$, which is inefficient. We then propose a low-complexity greedy Algorithm A3, where (NS) is used for RAU selection. As can be seen from the algorithm, when U and M both grow large the complexity will be dominated by step 2, which has complexity $O(KUM)$. As a result, when $K = 10$, Algorithm A3 reduces its complexity by 100 times compared with that of Algorithm A2. In Section 6.6 we will show that this algorithm is not only asymptotically optimal but is also promising when the number of RAUs is not very large.

6.5.3 Complexity Summary

The worst-case computational complexities of Algorithms A1, A2 and A3 are summarized in Table 6.1.

6.6 Numerical Results

In this section, we evaluate the seven-cell massive C-RAN for various numbers of single-antenna RAUs. Our numerical results were obtained using the parameters specified in Table 6.2 and based on [27].

Table 6.1 Worst-case complexity comparison

	Proposed algorithms	Worst-case complexity
	Step 1	$O(KM)$
A1	Step 2	$O(K^2)$
	Step 3	$O(K^2M)$
	Total	$O(K^2M)$
	Step 1	$O(U^2)$
A2	Step 2	$O(K^3UM + KUM)$
	Total	$O(K^3UM + U^2)$
	Step 1	$O(UM)$
A3	Step 2	$O(K^2U + KUM)$
	Total	$O(KUM)$

Algorithm A2

Initialization:

    Let $\mathcal{U} = \{1, \ldots, U\}$ and $B_k = 0$

Step 1: Determining the user interference index B_k.

    For $u_1 = 1$ to U, $u_2 = 1$ to U

        If $d_{u_1 u_2} \le d_{\text{th}}$, then $B_{u_1} = B_{u_1} + 1$

    End

Step 2: Scheduling users and their home RAUs.

    Let $(n^*, k^*) = \arg\max_{k \in \mathcal{U} \cap \{B_u = 1\}, n \in \mathcal{M}} \{\beta_{njkj}\}$, $\mathcal{K} = \{k^*\}$, and $\mathcal{R} = \{n^*\}$

    While $|\mathcal{K}| < K$ do

        For $k = 1$ to U

            Let the user set be $\mathcal{K} \cup \{k\}$

            Then employ Algorithm A1 to obtain the selected RAU set $\mathcal{D}_k$ from $\mathcal{M}/\mathcal{R}$

        End for

        $(k^*, \mathcal{D}_k^*) = \arg\max_{k \in \mathcal{U}/\mathcal{K}} C_{\text{SRS}}(\mathcal{K} \cup \{k\}, \mathcal{D}_k)$,

        $\mathcal{K} = \mathcal{K} \cup \{k^*\}$, and $\mathcal{R} = \{\mathcal{D}_k^*\}$

    End

Algorithm A3

Step 1: Same as Step 1 in Algorithm A1.

    Step 2: Scheduling users with a strong link.

    While $|\mathcal{K}| < K$ do

        For $k = 1$ to U

            Let $n_k = \arg\max_{n \in \mathcal{N}_k} \{\beta_{njkj}\}$

        End for

        Let $k^* = \arg\max_{k \notin \mathcal{K}, n_k \notin \mathcal{R}} C_{\text{SRS}}(\mathcal{K} \cup \{k\}, \mathcal{R} \cup \{n_k\})$

        $\mathcal{K} = \mathcal{K} \cup \{k^*\}$, and $\mathcal{R} = \mathcal{R} \cup \{n_{k^*}\}$

    End

Table 6.2 System parameters

Carrier frequency f	2 GHz
Carrier wavelength λ	0.15 m
Bandwidth per subcarrier	15 kHz
Thermal noise	-174 dBm/Hz
Standard deviation of shadow-fading σ	6 dB
Path loss factor γ	4
Protect distance R_0	1 m
Total number of users U	40
Scheduled number of users K	10
Length of pilot sequences τ	10
Power amplifier factor α	2.85
Power for signal reception q	0.1 W

6.6.1   Achievable Rate

We first examine our proposed analytical bound and scalings of the achievable rate. Figure 6.2 shows the average rates per user with transmit power $p_t = P$ and $P/N_0 = 120$ dB, for JDRS and SRS, respectively. For the CSI estimation from the uplink pilots, we chose pilot sequences of length $\tau = K$, which is the smallest amount of training that can be used. As predicted by the analysis, the asymptotic result for multi-cell JDRS does not change with M, whereas the asymptotic rate scaling for single-cell JDRS increases without bound. Although the gap between the simulated results and the asymptotic results for JDRS can be huge because (a) M is not large enough and (b) the rate of convergence is slow. The gap will decrease as M gets greater. The analytical result for SRS is very close to the simulation and increases with M. Clearly, single-cell systems with JDRS achieve the best spectral efficiency; however, the performances of SRS and multi-cell JDRS are relatively close. Even when M is finite, the gap between the rates for SRS and JDRS in multi-cell systems is less than 15%.

In Fig. 6.3, the results were obtained by choosing $p_t = P/M$ with the same setting as in Fig. 6.2. The results in this figure illustrate the power scaling laws. In addition, they show that for JDRS in single-cell and multi-cell systems, when the transmitted power is proportional to $1/M$ both the simulated and the asymptotic rates approach constants and so does the ergodic rate of SRS. If the power is lower than P/M, it can be easily found that the spectral efficiency of the three cases will decrease as M increases.

Figure 6.4 shows the average rate results against the cell radius with a fixed RAU density $M = 2R/250$, which verifies Proposition 6.2. When the RAU density remains unchanged, both the asymptotic curve and the simulated curve for multi-cell systems increase with R. As shown in the picture below, the same phenomenon can be observed in single-cell systems.

6.6.2   Energy Efficiency

We next examine the scaling law of energy efficiency for a massive C-RAN. The energy efficiency with SNR at the transmitter $P/N_0 = 150$ dB is illustrated in Fig. 6.5. The solid

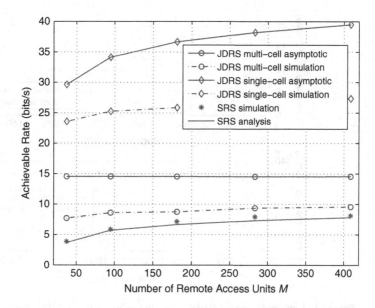

Figure 6.2 Achievable rates for JDRS and SRS versus the number of RAUs M with a fixed transmit power $p_k = P$. In this example there are $K = 10$ users in each cell, the cell radius is $R = 1000$ m, and the reference SNR is $P/N_0 = 120$ dB.

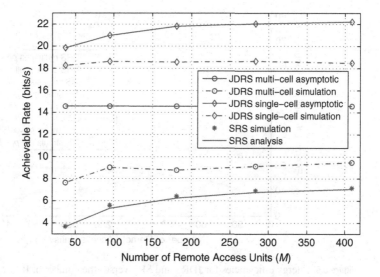

Figure 6.3 Achievable rates for JDRS and SRS versus the number of RAUs M with the transmit power decreasing with the number of RAUs as $p_k = P/M$. We chose the same parameter setting as in Fig. 6.2.

lines represent the results for transmit power $p_t = P$. As expected, when M increases the energy efficiency for JDRS decreases to 0 whereas the results for SRS increase slowly. In the case $p_t = P/M$, the energy efficiency for JDRS in multi-cell networks performs

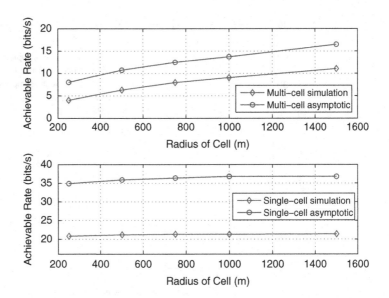

Figure 6.4 Achievable rate for JDRS versus the radius of the cell. The density of RAUs is fixed to be $M = 2R/250$, the number of users in each cell is $K = 10$, and the SNR at the transmitter is 120 dB.

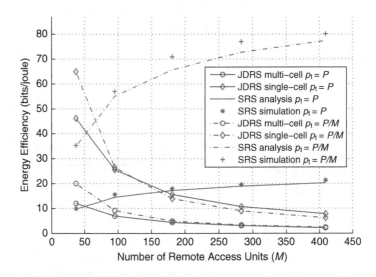

Figure 6.5 Energy efficiencies for JDRS and SRS versus the number of RAUs, taking into account power consumption by signal transmission and reception. We chose the SNR at the transmitter as $P/N_0 = 150$ dB, and the reception power was 0.1 W.

the same as for the case $p_t = P$ when M grows large, whereas the result for single-cell JDRS decreases at a much higher speed than the result with constant transmit power. For SRS, when the transmit power is proportional to $1/M$, with increasing M the growth of the energy efficiency slows down and will approach a constant value, confirming the

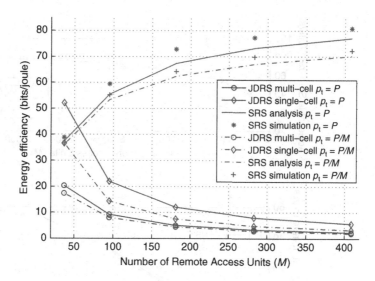

Figure 6.6 Energy efficiency for JDRS and SRS versus the number of RAUs, with $P/N_0 = 120$ dB.

result reported in Theorem 6.8. Also, the energy efficiency of SRS is higher than that of JDRS by more than 10 times when M is greater than 100. With SRS, a majority of RAUs go to sleep and the system achieves a very good result.

In Fig. 6.6, the results were obtained by choosing $P/N_0 = 120$ dB with the same setting as in Fig. 6.2. When compared with Fig. 6.5, the results give the same insight but the gap between the performance of SRS with $p_t = P$ and that with $p_t = P/M$ is reduced. This is so because the transmit power is lower when $P/N_0 = 120$ dB and the power reduction with $p_t = P/M$ is less. Furthermore, as the SNR is high in Fig. 6.5, for SRS, when the transmit power is reduced to $1/M$ the energy efficiency increases owing to energy saving. However, when the SNR is low the reduction in the transmit power leads to a decrease in the ergodic rate and leads to reduction in the energy efficiency. There is no significant change in the gap between the system performance with JDRS in the two cases because, for JDRS, the performance is limited by inter-cell interference and does not change the transmit power much when SNR is high, as reported in [17].

6.6.3 User Scheduling and RAU Selection Algorithms

As illustrated previously, for SRS the analytical bound is very tight. Therefore, we will use this bound for all the numerical results here. Figure 6.7 shows the average sum-rates for three RAU selection algorithms versus the numbers of RAUs and the SNR at the transmitter; NS corresponds to a system in which the RAUs with the best large-scale fading are chosen for the corresponding users. However, with GSA, by which we mean a greedy selection algorithm, the RAUs are selected by a low-complexity greedy algorithm, where the users choose the RAUs that maximize the sum-rate one

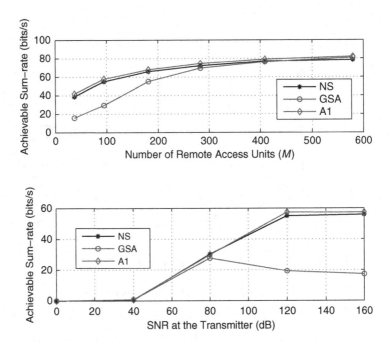

Figure 6.7 Achievable sum-rates for SRS for various RAU selection algorithms versus the number of RAUs M and the SNR at the transmitter, respectively.

by one, and if the selection decreases the sum-rate then the user will not be selected. The results in this figure demonstrate that the proposed Algorithm A1 is always the best for various values of M and SNR at the transmitter, but NS has a performance very close to it when $M > 30$ with $K = 10$. Therefore, we can simply choose the RAUs with the best large-scale fading even when M is only a multiple of the number of users.

Figure 6.8 shows the average sum-rates for SRS and JDRS with different scheduling algorithms versus the number of RAUs. For JDRS, we set the distance from the user to a scatterer as $r = 30$ m. According to [33], in order to keep the independence of the channels, we can schedule only users with distance greater than $2r$ from the other users. Because of the channel-hardening effect, all the users satisfying the above condition are scheduled in a round-robin manner for the sake of fairness. The radius of a cell is $R = 1000$ m and other simulation parameters are chosen as in Table 6.2. Additionally, it can be shown that, when $M \to \infty$, the difference between Algorithm A2 and Algorithm A3 is negligible. With our proposed schemes, SRS achieves much better performance than JDRS when M grows sufficiently large.

Figure 6.9 illustrates the average sum-rate results for different scheduling algorithms versus the SNR at the transmitter. A better performance is achieved by our proposed algorithms for a range of values of the SNR. Altogether, extraordinary spectral and energy efficiencies are achieved by SRS with our proposed schemes, and we have proved that full cooperation is useless and a waste of energy in a massive C-RAN.

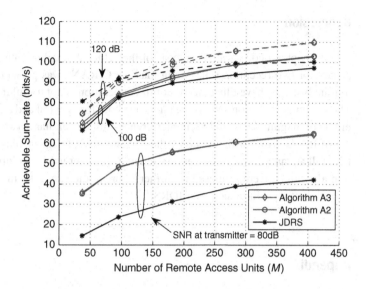

Figure 6.8 Achievable sum-rates for SRS using joint user scheduling and RAU selection algorithms and JDRS with user scheduling versus the number of RAUs.

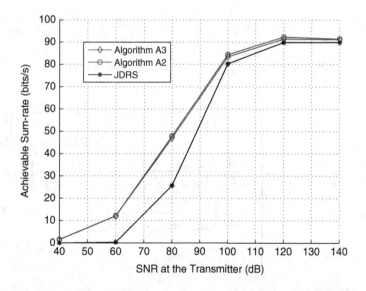

Figure 6.9 Achievable sum-rates for SRS using joint user scheduling and RAU selection algorithms and for JDRS with user scheduling, versus the SNR at the transmitter when $M = 95$.

6.7   Conclusion

In this chapter a comparative analysis of the achievable uplink rate and the energy efficiency for two reception schemes in massive C-RANs was presented. On the basis of
this analysis, we have illustrated that full cooperation is not useful in a massive C-RAN.
The selection reception scheme can improve both the spectral and energy efficiencies by
orders of magnitude in multi-cell interference-limited cellular networks. When the number of RAUs grows without bound, the performance of SRS increases without limit. To
take full advantages of this reception scheme, efficient user scheduling and RAU selection algorithms were proposed to further improve the sum-rate; these schemes achieve
a much better performance than JDRS.

6.8   Appendix

6.8.1   Proof of Theorem 6.1

When M grows without limit, $\mathcal{B}$ can be simplified to $\{b \,|\, 0 \le b \le R, |b - a| \ge R_0\}$.
When $\gamma = 4$, from [31] using the integer result

$$\frac{1}{\pi} \int_0^\pi \left(a^2 + b^2 - 2ab\cos\theta\right)^{-2} d\theta = \frac{a^2 + b^2}{|a^2 - b^2|^3} \tag{6.50}$$

in (6.6), the value of $\mathbb{E}_B\{d_{AB}^{-4}\}$ at a fixed point A is given by

$$\mathbb{E}_B\{d_{AB}^{-4}\} \to \begin{cases} \dfrac{1}{R^2}\left(\dfrac{(a-R_0)^2}{\left[(a-R_0)^2 - a^2\right]^2} + \dfrac{(a+R_0)^2}{\left[(a+R_0)^2 - a^2\right]^2} - \dfrac{R^2}{\left(R^2 - a^2\right)^2}\right) \\ \qquad\qquad\qquad\qquad \text{if } R_0 < a < R - R_0, \\ \dfrac{1}{R^2}\left(\dfrac{(a+R_0)^2}{\left[(a+R_0)^2 - a^2\right]^2} - \dfrac{R^2}{\left(R^2 - a^2\right)^2}\right) \quad \text{if } a < R_0, \\ \dfrac{1}{R^2}\left(\dfrac{(a-R_0)^2}{\left[(a-R_0)^2 - a^2\right]^2}\right) \\ \qquad\qquad\qquad\qquad \text{if } R - R_0 < a < R + R_0, \\ \dfrac{1}{\left(R^2 - a^2\right)^2} \qquad\qquad \text{if } a > R + R_0. \end{cases} \tag{6.51}$$

If the transmission distance $d_t \le R_0$, from the law of large numbers we have
$M^{-1}\sum_{m=1}^{M} \beta_{mjkj} = \mathbb{E}_B[d_{AB}^{-4}]$. Otherwise, the difference between $M^{-1}\sum_{m=1}^{M} \beta_{mjkj}$
and $\mathbb{E}_B\{d_{AB}^{-4}\}$ is non-negligible when M grows without limit. According to (6.51), an
upper bound and a lower bound for the achievable rate can be obtained.

First, we ignore the transmission-distance obstacle modeled by d_t and use the short-
range model to compute the channel power gain for all the RAUs with $a - d_t < b <
a + d_t$. With this method, an upper bound is given by

$$M^{-1} \sum_{m=1}^{M} \beta_{mjkj} \le$$

$$
\begin{cases}
\dfrac{\mathbb{E}Z}{R^2} \left(\dfrac{(a+d_t)^2}{\left[(a+d_t)^2 - a^2\right]^2} - \dfrac{R^2}{(R^2 - a^2)^2} \right) + \left(\dfrac{\lambda}{4\pi R} \right)^2 \ln\left(4a^2 d_t^2 - d_t^4\right), \\
\qquad\qquad\qquad\qquad\qquad\qquad \text{if } 0 \le a < d_t, \\[2pt]
\dfrac{\mathbb{E}Z}{R^2} \left(\dfrac{(a-d_t)^2}{\left[(a-d_t)^2 - a^2\right]^2} + \dfrac{(a+d_t)^2}{\left[(a+d_t)^2 - a^2\right]^2} - \dfrac{R^2}{(R^2 - a^2)^2} \right) + \left(\dfrac{\lambda}{4\pi R} \right)^2 \ln\left(4a^2 d_t^2 - d_t^4\right), \\
\qquad\qquad\qquad\qquad\qquad\qquad \text{if } d_t \le a \le R - d_t, \\[2pt]
\dfrac{\mathbb{E}Z}{R^2} \dfrac{(a-d_t)^2}{\left[(a-d_t)^2 - a^2\right]^2} + \left(\dfrac{\lambda}{4\pi R} \right)^2 \ln\left(4a^2 d_t^2 - d_t^4\right), \\
\qquad\qquad\qquad\qquad\qquad\qquad \text{if } R - d_t < a \le R.
\end{cases}
\tag{6.52}
$$

Then we give the lower bound using the following inequality. For RAUs in the range $d_t \le a \le R - d_t$, we have

$$\int_{a-d_t}^{a+d_t} \frac{2b}{R^2} \int_{\theta=0}^{\pi} \frac{g_{AB}}{\pi} d\theta\, db \ge \int_{1}^{d_t} \frac{2d_{AB}}{R^2} \int_{\theta=0}^{\pi} \frac{g_{AB}}{\pi} d\theta\, d(d_{AB}). \tag{6.53}$$

Using (6.51) and (6.53), for users with $d_t \le a \le R - d_t$, we obtain the lower bound as

$$M^{-1} \sum_{m=1}^{M} \beta_{mjkj} \ge \frac{\mathbb{E}Z}{R^2} \left(\frac{(a-d_t)^2}{\left[(a-d_t)^2 - a^2\right]^2} + \frac{(a+d_t)^2}{\left[(a+d_t)^2 - a^2\right]^2} - \frac{R^2}{(R^2 - a^2)^2} \right)$$
$$+ \left(\frac{\lambda}{4\pi R} \right)^2 \ln d_t^2. \tag{6.54}$$

According to (6.51), the lower bounds for users with $0 \le a < d_t$ and with $R - d_t < a \le R$ can be obtained in a similar way, the derivation is omitted here. The interference can also be calculated from (6.51). For $a \ge R + d_t$, we obtain

$$\frac{z_{kj}}{M^2} = \frac{\sum_{l \ne j}^{L} \left(\sum_{m=1}^{M} \beta_{mjkl} \right)^2}{M^2} \to \sum_{l \ne j}^{L} \frac{(\mathbb{E}Z)^2}{\left(R^2 - a_l^2\right)^4}. \tag{6.55}$$

For $R < a < R + d_t$, computation of an accurate result becomes very complicated, but (6.55) turns out to be a very good approximation. Thus, we will apply it for all users in neighboring cells. Substituting (6.52), (6.54), and (6.55) into (6.17), we find that both the upper bound and the lower bound of the asymptotic rate are independent of M, and we have $\left(\sum_{m=1}^{M} \beta_{mjkj} \right)^2 = O(M^2)$ and $z_{kj} = O(M^2)$. Hence, when M grows without limit and $\gamma = 4$, the achievable rate of a massive C-RAN settles down to a finite value.

When $\gamma = 2$, substituting

$$\frac{1}{\pi} \int_0^{\pi} \left(a^2 + b^2 - 2ab\cos\theta \right)^{-1} d\theta = \frac{1}{|a^2 - b^2|}$$

into (6.6) and following a similar analysis to that in the above case of $\gamma = 4$, we have $C_{\text{JDRS}} = O(c)$. Since for $2 < \gamma < 4$ we have

$$\left(\sum_{m=1}^{M} \beta_{mjkj}\right)^2_{\gamma=4} < \left(\sum_{m=1}^{M} \beta_{mjkj}\right)^2_{2<\gamma<4} < \left(\sum_{m=1}^{M} \beta_{mjkj}\right)^2_{\gamma=2} \tag{6.56}$$

and

$$\left(z_{kj}\right)_{\gamma=4} < \left(z_{kj}\right)_{2<\gamma<4} < \left(z_{kj}\right)_{\gamma=2}, \tag{6.57}$$

therefore, for $2 \le \gamma \le 4$ it follows that $\left(\sum_{m=1}^{M} \beta_{mjkj}\right)^2 = O(M^2)$ and $z_{kj} = O(M^2)$ and hence $C_{\text{JDRS}} = O(c)$.

6.8.2   Proof of Proposition 6.2

In the case of $R \gg d_t$, the upper and lower bound of the asymptotic rate calculated in Appendix 6.8.1 are very close to the accurate result. With this assumption, we make the following proof. For users with $d_t \le a \le R - d_t$, from (6.52) and (6.55), we have

$$\text{SIR}_{kj} \le \text{SIR}_{kj}^{ub}$$

$$= \left[\sum_{l \ne j}^{L} \frac{(\mathbb{E}Z)^2}{\left(1 - a_l^2/R^2\right)^4}\right]^{-1} R^4 \left(\mathbb{E}Z\left(\frac{(a - d_t)^2}{\left[(a - d_t)^2 - a^2\right]^2} + \frac{(a + d_t)^2}{\left[(a + d_t)^2 - a^2\right]^2}\right)\right.$$

$$\left. - \frac{R^2}{\left(R^2 - a^2\right)^2}\right) + \left(\frac{\lambda}{4\pi}\right)^2 \ln\left(4a^2 d_t^2 - d_t^4\right)\right)^2. \tag{6.58}$$

To analyze the above formula, we write $a_l = \xi_l R$, where ξ_l is independent with R. Obviously, the first factor is unchanged with an increase in R. For the second factor, we write

$$a = \xi R, \Delta_1 = R^2 \left[\frac{(a - d_t)^2}{\left((a - d_t)^2 - a^2\right)^2} + \frac{(a + d_t)^2}{\left((a + d_t)^2 - a^2\right)^2}\right],$$

$$\Delta_2 = R^2 \frac{R^2}{\left(R^2 - a^2\right)^2},$$

$$\Delta_3 = R^2 \ln\left(4a^2 d_t^2 - d_t^4\right),$$

where ξ is a coefficient independent of R. Second factor in (6.58) is then equal to $(\Delta_1 + \Delta_2 + \Delta_3)^2$, where

$$\frac{d\Delta_1}{dR} = \frac{d}{dR}\left(\frac{R^2 a^2 d_t^4 + 2R^2 d_t^2\left(2a^2 - d_t^2\right)^2}{d_t^4 - 4a^2 d_t^2}\right) > 0, \tag{6.59}$$

$$\frac{d\Delta_2}{dR} = \frac{d}{dR}\left(\frac{1}{\left(1 - a^2/R^2\right)^2}\right) = 0, \tag{6.60}$$

$$\frac{d\Delta_3}{dR} = \frac{d}{dR}\left(R^2 \ln\left(4\alpha^2 d_t^2 R^2 - d_t^4\right)\right) > 0. \tag{6.61}$$

Hence, the second factor in (6.58) increases with R. Overall, for $d_t \leq a \leq R - d_t$ we have

$$\frac{d\,\mathrm{SIR}_{kj}^{ub}}{dR} > 0 \tag{6.62}$$

Using a similar method, from (6.52) and (6.54) we can derive that, for all users in a cell, we have $d\,\mathrm{SIR}_{kj}^{ub}/dR > 0$ and $d\,\mathrm{SIR}_{kj}^{lb}/dR > 0$. Since $\log_2(\cdot)$ is an increasing function of the input factor, $dC_{JDRS}dR/ \overset{a.s.}{>} 0$ is obtained, which completes the proof.

6.8.3   Proof of Theorem 6.5

For any user k in the jth cell, we have

$$\min_{l,m} d_{mlkj} \leq \frac{R}{\sqrt{M}}. \tag{6.63}$$

For the distance between user k and the RAU of any other user, it follows that

$$\min_{\{l,u\}\neq\{k,j\}} d_{kjul} \geq d_{\min} - \frac{R}{\sqrt{M}}, \tag{6.64}$$

where $d_{\min}$ is the minimum distance between any two users. Since in a circular cell structure $d_{\min}$ satisfies

$$P\left(d_{\min} \geq d_0\right) \geq 1 - K\left[1 - \left(1 - \frac{d_0^2}{R^2}\right)^{K-1}\right], \tag{6.65}$$

when $M \to \infty$, for any $\epsilon_1 > 0$, it satisfies $P\left(d_{\min} \geq R/M^{\epsilon_1}\right) \to 1$, i.e., $d_{\min} \overset{a.s.}{\geq} R/M^{\epsilon_1}$. We then obtain

$$\min_{\{l,u\}\neq\{k,j\}} d_{kjul} \overset{a.s.}{\geq} \frac{R}{M^{\epsilon_1}} - \frac{R}{\sqrt{M}}. \tag{6.66}$$

As $M \to \infty$, we have $d_{kjkj} < d_t$. It follows that $g_{kj} = \left(\lambda/4\pi d_{kjkj}\right)h_{kjkj}$. Hence, for any u, l, it follows from (6.62) and (6.66) that

$$\frac{\mathbb{E}\left\{g_{kj}^2\right\}}{\mathbb{E}\left\{g_{kjul}^2\right\}} \geq \frac{d_{kj}^{-2}\,\mathbb{E}\left\{|h_{kj}|^2\right\}}{d_{kjul}^{-2}\,\mathbb{E}\left\{|h_{kjul}|^2\right\}}$$

$$\geq \left(M^{1/2-\epsilon_1} - 1\right)^2. \tag{6.67}$$

Then using (6.31) and Jensen's inequality, the ergodic uplink rate is lower bounded by

$$C_{\mathrm{SRS}} \geq \log_2\left(1 + \frac{P\,\mathbb{E}\left\{g_{kj}^2\right\}}{1 + P\left(KL - 1\right)\left(M^{1/2-\epsilon_1} - 1\right)^{-2}\mathbb{E}\left[g_{kj}^2\right]}\right), \tag{6.68}$$

in which the power of the transmitted signals is assumed to be a fixed value P for simplicity. It follows from (6.62) that $\mathbb{E}\left\{g_{kj}^2\right\} \geq \left(\frac{\lambda}{4\pi R}\right)^2 M\mathbb{E}\left\{\left|h_{kj}\right|^2\right\}$. Thus for any $\epsilon > 0$ we have $C_{\text{SRS}} \geq O\left((1-\epsilon)\log_2 M\right)$.

References

[1] T. L. Marzetta, "Noncooperative cellular wireless with unlimited numbers of base station antennas," *IEEE Trans. Wireless Commun.*, vol. 9, no. 11, pp. 3590–3600, November 2010.

[2] F. Rusek, D. Persson, B. K. Lau, E. G. Larsson, T. L. Marzetta, O. Edfors *et al.*, "Scaling up MIMO: opportunities and challenges with very large arrays," *IEEE Signal Process Mag.*, vol. 30, no. 1, pp. 40–60, January 2013.

[3] T. Bai and R. W. Heath, Jr., "Asymptotic coverage and rate in massive MIMO networks," in *Proc. IEEE Global Conf. on Signaling and Information Processing*, Atlanta, GA, December 2014, pp. 602–606.

[4] H. Q. Ngo, E. G. Larsson, and T. L. Marzetta, "Energy and spectral efficiency of very large multiuser MIMO systems," *IEEE Trans. Commun.*, vol. 61, no. 4, pp. 1436–1449, April 2013.

[5] J. Jose, A. Ashikhmin, T. L. Marzetta, and S. Vishwanath, "Pilot contamination and precoding in multi-cell TDD systems," *IEEE Trans. Wireless Commun.*, vol. 10, no. 8, pp. 2640–2651, August 2011.

[6] E. G. Larsson, F. Tufvesson, O. Edfors, and T. L. Marzetta, "Massive MIMO for next generation wireless systems," *IEEE Commun. Mag.*, vol. 52, no. 2, pp. 186–195, February 2014.

[7] H. Yin, D. Gesbert, M. Filippou, and Y. Liu, "A coordinated approach to channel estimation in large-scale multiple-antenna systems," *IEEE J. Sel. Areas Commun.*, vol. 31, no. 2, pp. 264–273, February 2013.

[8] W. Choi and J. G. Andrews, "Downlink performance and capacity of distributed antenna systems in a multicell environment," *IEEE Trans. Wireless Commun.*, vol. 6, no. 1, pp. 69–73, January 2007.

[9] J. Zhang and J. G. Andrews, "Distributed antenna systems with randomness," *IEEE Trans. Wireless Commun.*, vol. 7, no. 9, pp. 3636–3646, September 2008.

[10] China Mobile Research Institute, "C-RAN: the road towards green RAN," October 2011.

[11] R. Heath Jr., T. Wu, Y. H. Kwon, and A. Soong, "Multiuser MIMO in distributed antenna systems with out-of-cell interference," *IEEE Trans. Sig. Proc.*, vol. 59, no. 10, pp. 4885–4899, October 2011.

[12] M. Peng, Y. Li, Z. Zhao, and C. Wang, "System architecture and key technologies for 5G heterogeneous cloud radio access networks," *IEEE Network*, vol. 29, no. 2, pp. 6–14, March 2015.

[13] A. Liu and V. K. N. Lau, "Joint power and antenna selection optimization for energy-efficient large distributed MIMO networks," in *Proc. IEEE Int. Conf. on Communication Systems (ICCS)*, Singapore, November 2012, pp. 230–234.

[14] S. Mahboob, C. Mahapatra, and V. C. M. Leung, "Energy-efficient multiuser MIMO downlink transmissions in massively distributed antenna systems with predefined capacity constraints," in *Proc. Int. Conf. on Broadband, Wireless Computing, Commununications and Applications*, Victoria, BC, November 2012, pp. 208–211.

[15] H. Li, J. Hajipour, A. Attar, and V. Leung, "Efficient HetNet implementation using broadband wireless access with fiber-connected massively distributed antennas architecture," *IEEE Wireless Commun.*, vol. 18, no. 3, pp. 72–78, June 2011.

[16] N. Jindal, J. G. Andrews, and S. Weber, "Rethinking MIMO for wireless networks: linear throughput increases with multiple receive antennas," in *Proc. IEEE Int. Conf. Commun.*, Dresden, June 2009, pp. 1–6.

[17] A. Lozano, R. Heath, and J. G. Andrews, "Fundamental limits of cooperation," *IEEE Trans. Inf. Theory*, vol. 59, no. 9, pp. 5213–5226, September 2013.

[18] Y. Yang, B. Bai, W. Chen, and L. Hanzo, "A low-complexity cross-layer algorithm for coordinated downlink scheduling and robust beamforming under a limited feedback constraint," *IEEE Trans. Veh. Technol.*, vol. 63, no. 1, pp. 107–118, January 2014.

[19] M. Franceschetti, D. Migliore, and P. Minero, "The capacity of wireless networks: information-theoretic and physical limits," *IEEE Trans. Inf. Theory*, vol. 55, no. 8, pp. 3413–3424, August 2009.

[20] W.-Y. Shin, S.-W. Jeon, N. Devroye, M. H. Vu, S.-Y. Chung, Y. H. Lee, and V. Tarokh, "Improved capacity scaling in wireless networks with infrastructure," *IEEE Trans. Inf. Theory*, vol. 57, no. 8, pp. 5088–5102, August 2011.

[21] A. Ozgur, O. Leveque, and D. Tse, "Spatial degrees of freedom of large distributed MIMO systems and wireless ad hoc networks," *IEEE J. Sel. Areas Commun.*, vol. 31, no. 2, pp. 202–214, February 2013.

[22] S. P. Shariatpanahi, B. H. Khalaj, K. Alishahi, and H. Shah-Mansouri, "Throughput of large one-hop ad-hoc wireless networks with general fading," *IEEE Trans. Veh. Technol.*, vol. 64, no. 7, pp. 3304–3310, July 2015.

[23] K. T. K. Cheung, S. Yang, and L. Hanzo, "Achieving maximum energy-efficiency in multi-relay OFDMA cellular networks: a fractional programming approach," *IEEE Trans. Commun.*, vol. 61, no. 7, pp. 2746–2757, July 2013.

[24] K. T. K. Cheung, S. Yang, and L. Hanzo, "Spectral and energy spectral efficiency optimization of joint transmit and receive beamforming based multi-relay MIMO-OFDMA cellular networks," *IEEE Trans. Wireless Commun.*, vol. 13, no. 11, pp. 6147–6165, November 2014.

[25] K. T. Truong and R. W. Heath Jr., "The viability of distributed antennas for massive MIMO systems," in *Proc. Asilomar Conf. on Signals, Systems, and Computers*, Pacific Grove, CA, November 2013, pp. 1318–1323.

[26] J. Joung, Y. K. Chia, and S. Sun, "Energy-efficient, large-scale distributed-antenna system (L-DAS) for multiple users," *IEEE J. Sel. Topics Signal Process.*, vol. 8, no. 5, pp. 954–965, March 2014.

[27] Y. Liang, A. J. Goldsmith, G. Foschini, R. Valenzuela, and D. Chizhik, "Evolution of base stations in cellular networks: denser deployment versus coordination," in *Proc. IEEE Int. Conf. on Communications*, Beijing, May 2008, pp. 4128–4132.

[28] J. B. Andersen, T. S. Rappaport, and S. Yoshida, "Propagation measurements and models for wireless communication channels," *IEEE Commun. Mag.*, vol. 33, no. 1, pp. 42–49, January 1995.

[29] L. Lu, G. Li, A. Swindlehurst, A. Ashikhmin, and R. Zhang, "An overview of massive MIMO: benefits and challenges," *IEEE J. Sel. Topics Signal Process.*, vol. 8, no. 5, pp. 742–758, April 2014.

[30] J. Zhang, C.-K. Wen, S. Jin, X. Q. Gao, and K.-K. Wong, "On capacity of large-scale MIMO multiple access channels with distributed sets of correlated antennas," *IEEE J. Sel. Areas Commun.*, vol. 31, no. 2, pp. 133–148, February 2013.

[31] I. Gradshteyn and I. Ryzhik, *Tables of Integrals, Series, and Products*. Seventh Edition. London: Academic Press, 2003.

[32] D. Gesbert and M. Kountouris, "Rate scaling laws in multicell networks under distributed power control and user scheduling," *IEEE Trans. Inf. Theory*, vol. 57, no. 1, pp. 234–244, January 2011.

[33] H. Yin and D. Gesbert, "Dealing with interference in distributed large-scale MIMO systems: a statistical approach," *IEEE J. Sel. Topics Signal Process.*, vol. 8, no. 5, pp. 942–953, May 2014.

7 Large-Scale Convex Optimization for C-RANs

Yuanming Shi, Jun Zhang, Khaled B. Letaief, Bo Bai, and Wei Chen

7.1 Introduction

7.1.1 C-RANs

The proliferation of "smart" mobile devices, coupled with new types of wireless applications, has led to an exponential growth in wireless and mobile data traffic. In order to provide high-volume and diversified data services, C-RAN [1, 2] has been proposed; it enables efficient interference management and resource allocation by shifting all the baseband units (BBUs) to a single cloud data center, i.e., by forming a BBU pool with powerful shared computing resources. Therefore, with efficient hardware utilization at the BBU pool, a substantial reduction can be obtained in both the CAPEX (e.g., via low-cost site construction) and the OPEX (e.g., via centralized cooling). Furthermore, the powerful conventional base stations are replaced by light and low-cost remote radio heads (RRHs), with the basic functionalities of signal transmission and reception, which are then connected to the BBU pool by high-capacity and low-latency optical fronthaul links. The capacity of C-RANs can thus be significantly improved through network densification and large-scale centralized signal processing at the BBU pool. By further pushing a substantial amount of data, storage, and computing resources (e.g., the radio access units and end-user devices) to the edge of the network, using the principle of mobile edge computing (i.e., fog computing) [3], heterogeneous C-RANs [4], as well as Fog-RANs and MENG-RANs [5] can be formed. These evolved architectures will further improve user experience by offering on-demand and personalized services and location-aware and content-aware applications. In this chapter we investigate the computation aspects of this new network paradigm, and in particular focus on the large-scale convex optimization for signal processing and resource allocation in C-RANs.

7.1.2 Large-Scale Convex Optimization: Challenges and Previous Work

Convex optimization serves as an indispensable tool for resource allocation and signal processing in wireless networks [6–9]. For instance, coordinated beamforming [10] often yields a convex optimization formulation, i.e., second-order cone programming (SOCP) [11]. The network max–min fairness-rate optimization [12] can be solved through the bisection method [11], in polynomial time; in this method a sequence of convex subproblems needs to be solved. Furthermore, convex relaxation provides a principled way to develop polynomial-time algorithms for non-convex or NP-hard problems, e.g., group-sparsity penalty relaxation for NP-hard mixed-integer non-linear programming problems [2], semidefinite relaxation [7] for NP-hard robust beamforming [13, 14] and multicast beamforming [15], and a sequential convex approximation to the highly intractable stochastic coordinated beamforming problem [16].

Nevertheless, in dense C-RANs [5], which may possibly need to handle hundreds of RRHs simultaneously, resource allocation and signal processing problems will be dramatically scaled up. The underlying optimization problems will have a high dimension and/or a large number of constraints, e.g., per-RRH transmit power constraints and per-MU (mobile user) QoS constraints. For instance, for a C-RAN with 100 single-antenna RRHs and 100 single-antenna MUs, the dimension of the aggregative coordinated beamforming vector of the optimization variables will be 10^4. Most advanced off-the-shelf solvers (e.g., SeDuMi [17], SDPT3 [18], and MOSEK [19]) are based on the interior-point method. However, the computational burden of such a second-order method makes it inapplicable for large-scale problems. For instance, solving convex quadratic programs has cubic complexity [20]. Furthermore, to use these solvers the original problems need to be transformed to the standard forms supported by the solvers. Although parser/solver modeling frameworks such as CVX [21] and YALMIP [22] can automatically transform original problem instances into standard forms, they may require substantial time to perform such a transformation [23], especially for problems with a large number of constraints [24].

One may also develop custom algorithms to enable efficient computation by exploiting the structures of specific problems. For instance, the uplink–downlink duality [10] can be exploited to extract the structures of optimal beamformers [25] and enable efficient algorithms. However, such an approach still has cubic complexity since it performs matrix inversion at each iteration [26]. First-order methods, e.g., the alternating-direction method of multipliers (ADMM) algorithm [27], have recently attracted attention for their distributed and parallelizable implementation as well as for their capability of scaling to large problem sizes. However, most existing ADMM-based algorithms cannot provide the certificates of infeasibility [13, 26, 28] which are needed for such problems as max–min rate maximization [24] and group sparse beamforming [2]. Furthermore, some of these algorithms may still fail to scale to large problem sizes, owing to SOCP subproblems [28] or semidefinite programming (SDP) subproblems [13] needed to be solved at each iteration.

Without efficient and scalable algorithms, previous studies of wireless cooperative networks either only demonstrate performance in small-size networks, typically with less than 10 RRHs, or resort to suboptimal algorithms, e.g., zero-forcing-based approaches [29, 30]. Meanwhile, from the above discussion, we see that the large-scale optimization algorithms to be developed should possess the following two features:

- they should scale well to large problem sizes with parallel computing capability;
- they should detect problem infeasibility effectively, i.e., provide certificates of infeasibility.

Figure 7.1 The proposed two-stage approach for large-scale convex optimization. The optimal solution or the certificate of infeasibility can be extracted from $\mathbf{x}^\star$ by the ADMM solver.

7.1.3   A Two-Stage Approach for Large-Scale Convex Optimization

To address the above two requirements in a unified way, in this chapter we shall present a two-stage approach, as shown in Fig. 7.1. The proposed framework [31] is capable of solving large-scale convex optimization problems in parallel, as well as providing certificates of infeasibility. Specifically, the original problem $\mathscr{P}$ is first transformed into a standard cone programming form $\mathscr{P}_{\text{cone}}$ [20] based on the Smith-form reformulation [32], which involves introducing a new variable for each subexpression in the disciplined convex programming form [33] of the original problem. This will eventually transform the coupled constraints in the original problem into a structured constraint consisting only of two convex sets: a subspace, and a convex set formed by a Cartesian product of a finite number of standard convex cones. Such a structure helps to develop efficient parallelizable algorithms and enable infeasibility detection capability *simultaneously* via solving the homogeneous self-dual embedding [34] of the primal–dual pair of standard form by the ADMM algorithm.

As the mapping between the standard cone program and the original problem depends only on the network size (i.e., the numbers of RRHs, MUs and antennas at each RRH), we can pre-generate and store the structures of the standard forms with different candidate network sizes. Then for each problem instance, i.e., given the channel coefficients, QoS requirements, and maximum RRH transmit powers, we only need to copy the original problem parameters to the standard cone-programming data. Thus, the transformation procedure will be very efficient and can avoid repeatedly parsing and re-generating problems [21, 22]. This technique is called *matrix stuffing* [23, 24] and is essential for the proposed framework to scale well to large problem sizes. It may also help rapid prototyping and testing in practical equipment development.

7.1.4   Outline

In Section 7.2 we demonstrate that typical signal processing and resource allocation problems in C-RANs can be solved essentially through addressing one or a sequence of large-scale convex optimization or convex feasibility problems. In Section 7.3, a systematic cone-programming form-transformation procedure is developed. The operator splitting method with detailed discussions is presented in Section 7.4. Numerical results will be reported in Section 7.5. We give a summary and conclusions in Section 7.6.

7.2   Large-Scale Convex Optimization in Dense C-RANs

Consider the following parametric family $\mathscr{P}$ of convex optimization problems:

$$\mathscr{P} : \underset{\mathbf{x}}{\text{minimize}} \quad f_0(\mathbf{x}; \boldsymbol{\alpha})$$

$$\text{s. t.} \quad f_i(\mathbf{x}; \boldsymbol{\alpha}) \leq g_i(\mathbf{x}; \boldsymbol{\alpha}), \quad i = 1, \dots, m, \tag{7.1}$$

$$u_i(\mathbf{x}; \boldsymbol{\alpha}) = v_i(\mathbf{x}; \boldsymbol{\alpha}), \quad i = 1, \dots, p, \tag{7.2}$$

where $\mathbf{x} \in \mathbb{R}^n$ is the vector of optimization variables and $\boldsymbol{\alpha} \in \mathcal{A}$ is the problem parameter vector; $\mathcal{A}$ denotes the parameter space. For each fixed $\boldsymbol{\alpha} \in \mathcal{A}$, the problem instance $\mathscr{P}(\boldsymbol{\alpha})$ is convex if the functions f_i and g_i are convex and concave, respectively, and the functions u_i and v_i are affine. The reader should refer to [11] for an introduction to the basics of convex optimization. In this section we will first illustrate that typical optimization problems in C-RANs can be formulated in this parametric form of convex programming, and then the proposed framework for large-scale convex optimization will be introduced.

7.2.1   Convex Optimization Examples in C-RANs

To illustrate the wide-ranging applications of convex optimization in C-RANs, we mainly focus on the generic scenario for downlink transmission with full cooperation among the RRHs. The proposed methodology in this chapter can be easily applied to uplink transmission and more general cooperation scenarios in heterogeneous C-RANs [4, 8], Fog-RANs, and MENG-RANs [5] as we need only exploit the convexity of the resulting optimization problems.

Signal Model
Consider a downlink fully cooperative C-RAN with L multi-antenna RRHs and K single-antenna MUs, where the lth RRH is equipped with N_l antennas. The wireless propagation channel from the lth RRH to the kth MU is denoted as $\mathbf{h}_{kl} \in \mathbb{C}^{N_l}, \forall k, l$. The received signal $y_k \in \mathbb{C}$ at MU k is given by

$$y_k = \sum_{l=1}^{L} \mathbf{h}_{kl}^H \mathbf{v}_{lk} s_k + \sum_{i \neq k} \sum_{l=1}^{L} \mathbf{h}_{kl}^H \mathbf{v}_{li} s_i + n_k, \quad \forall k, \tag{7.3}$$

where s_k is the encoded information symbol for MU k with $\mathbb{E}\{|s_k|^2\} = 1$, $\mathbf{v}_{lk} \in \mathbb{C}^{N_l}$ is the transmit beamforming vector from the lth RRH to the kth MU, and $n_k \sim \mathcal{CN}(0, \sigma_k^2)$ is the additive Gaussian noise at MU k.

We assume that the s_k and n_k are mutually independent and that all the users apply single-user detection. Therefore, the signal-to-interference-plus-noise ratio (SINR) of MU k is given by

$$\text{SINR}_k(\mathbf{v}_1, \dots, \mathbf{v}_K) = \frac{|\mathbf{h}_k^H \mathbf{v}_k|^2}{\sum_{i \neq k} |\mathbf{h}_k^H \mathbf{v}_i|^2 + \sigma_k^2}, \quad \forall k, \tag{7.4}$$

where $\mathbf{h}_k \triangleq [\mathbf{h}_{k1}^T \cdots \mathbf{h}_{kL}^T]^T \in \mathbb{C}^N$, with $N = \sum_{l=1}^{L} N_l$, is the channel vector consisting of the channel coefficients from all the RRHs to the kth MU and $\mathbf{v}_k \triangleq [\mathbf{v}_{1k}^T \mathbf{v}_{2k}^T \cdots \mathbf{v}_{Lk}^T]^T \in \mathbb{C}^N$ is the beamforming vector consisting of the beamforming coefficients from all the RRHs to MU k.

Coordinated beamforming is an efficient way to design energy-efficient and spectrally efficient systems [10] in which the beamforming vectors $\mathbf{v}_{lk}$ are designed to minimize the network power consumption and maximize the network utility, respectively. Three representative examples are given below to illustrate the power of convex optimization to design efficient transmit strategies to optimize the system performance of C-RANs, for which coordinated beamforming is the basic building block.

Example 7.1   Coordinated beamforming via second-order cone programming Consider the following coordinated beamforming problem to, that of minimizing the total transmit power while satisfying the QoS requirements and the transmit power constraints for RRHs [14]:

$$\underset{\mathbf{v}_1,\ldots,\mathbf{v}_K}{\text{minimize}} \quad \sum_{l=1}^{L}\sum_{k=1}^{K} \|\mathbf{v}_{lk}\|_2^2$$

$$\text{s. t.} \quad \frac{|\mathbf{h}_k^H \mathbf{v}_k|^2}{\sum_{i \neq k} |\mathbf{h}_k^H \mathbf{v}_i|^2 + \sigma_k^2} \geq \gamma_k, \quad k = 1,\ldots,K, \tag{7.5a}$$

$$\sum_{k=1}^{K} \|\mathbf{v}_{lk}\|_2^2 \leq P_l, \quad l = 1,\ldots,L, \tag{7.5b}$$

where $\gamma_k > 0$ is the target SINR for MU k and $P_l > 0$ is the maximum transmit power of the lth RRH.

Since the phases of $\mathbf{v}_k$ will not change the objective function or constraints of problem (7.5) [35], the SINR constraints (7.5a) are equivalent to the following second-order cone constraints:

$$\sqrt{\sum_{i \neq k} |\mathbf{h}_k^H \mathbf{v}_i|^2 + \sigma_k^2} \leq \frac{1}{\sqrt{\gamma_k}} \mathfrak{R}(\mathbf{h}_k^H \mathbf{v}_k), \quad k = 1,\ldots,K, \tag{7.6}$$

where $\mathfrak{R}(\cdot)$ denotes the real part. Therefore, problem (7.5) can be reformulated as the following (SOCP) problem:

$$\underset{\mathbf{v}_1,\ldots,\mathbf{v}_K}{\text{minimize}} \quad \sum_{l=1}^{L}\sum_{k=1}^{K} \|\mathbf{v}_{lk}\|_2^2$$

$$\text{s. t.} \quad \sqrt{\sum_{i \neq k} |\mathbf{h}_k^H \mathbf{v}_i|^2 + \sigma_k^2} \leq \frac{1}{\sqrt{\gamma_k}} \mathfrak{R}(\mathbf{h}_k^H \mathbf{v}_k), \quad k = 1,\ldots,K, \tag{7.7a}$$

$$\sqrt{\sum_{k=1}^{K} \|\mathbf{v}_{lk}\|_2^2} \leq P_l, \quad l = 1,\ldots,L. \tag{7.7b}$$

Example 7.2 **Network power minimization via group sparse beamforming** To design a green C-RAN the network power consumption, including the power consumption of each RRH and of each associated fronthaul link, needs to be minimized while satisfying the QoS requirements for all the MUs. Mathematically, we need to solve the following network-power minimization problem [2]:

$$
\underset{\mathbf{v}_1,\ldots,\mathbf{v}_K}{\text{minimize}} \quad \sum_{l=1}^{L}\sum_{k=1}^{K}\frac{1}{\eta_l}\|\mathbf{v}_{lk}\|_2^2 + \sum_{l=1}^{L}P_l^c I(\text{Supp}(\mathbf{v}) \cap \mathcal{V}_l \neq \emptyset)
$$

$$
\text{s. t.} \quad (7.7a), (7.7b), \tag{7.8}
$$

where $\eta_l > 0$ is the drain-inefficiency coefficient of the radio frequency power amplifier and $P_l^c \geq 0$ is the relative fronthaul-link power consumption [2], representing the static power saving when both the RRH and the corresponding fronthaul link are switched off. Here, $I(\text{Supp}(\mathbf{v}) \cap \mathcal{V}_l \neq \emptyset)$ is an indicator function that takes the value 0 if $\text{Supp}(\mathbf{v}) \cap \mathcal{V}_l = \emptyset$ (i.e., all the beamforming coefficients at the lth RRH are zeros, indicating that the corresponding RRH is switched off) and 1 otherwise, where $\mathcal{V}_l$ is defined as $\mathcal{V}_l := \{K\sum_{l=1}^{L-1}N_l + 1, \ldots, K\sum_{l=1}^{L}N_l\}$; $\text{Supp}(\mathbf{v})$ is the support of the beamforming vector $\mathbf{v} = [\tilde{\mathbf{v}}_l] \in \mathbb{C}^{KN}$ with $\tilde{\mathbf{v}}_l = [\mathbf{v}_{l1}^T, \ldots, \mathbf{v}_{lK}^T]^T \in \mathbb{C}^{N_l K}$ as the aggregated beamforming vector at RRH l. Problem (7.8) is a mixed combinatorial optimization problem and is NP-hard in general.

Observing that all the beamforming coefficients in the vector $\tilde{\mathbf{v}}_l$ will be set to zero simultaneously when the lth RRH is switched off, the aggregate beamforming vector $\mathbf{v}$ has the group-sparsity structure if multiple RRHs need to be switched off to reduce the network power consumption [36]. A three-stage group sparse beamforming algorithm with polynomial time complexity was thus proposed to minimize the network power consumption by adaptively selecting active RRHs via controlling the group-sparsity structure of the aggregative beamforming vector $\mathbf{v}$. Specifically, in the first stage, the group-sparsity structure of the aggregated beamformer $\mathbf{v}$ is induced by minimizing the following weighted mixed ℓ_1/ℓ_2-norm of $\mathbf{v}$:

$$
\mathscr{P}_{\text{SOCP}} : \underset{\mathbf{v}_1,\ldots,\mathbf{v}_K}{\text{minimize}} \quad \sum_{l=1}^{L}\omega_l\|\tilde{\mathbf{v}}_l\|_2
$$

$$
\text{s. t.} \quad (7.7a), (7.7b), \tag{7.9}
$$

where $\omega_l > 0$ is the corresponding weight for the beamformer coefficient group $\tilde{\mathbf{v}}_l$. On the basis of the (approximate) group sparse beamformer $\mathbf{v}^\star$, which is the optimal solution to the convex SOCP problem $\mathscr{P}_{\text{SOCP}}$ (7.9), in the second stage an RRH selection procedure is performed to switch off some RRHs so as to minimize the network power consumption. In this procedure we need to check whether the remaining RRHs can support the QoS requirements for all the MUs, i.e., to check the feasibility of problem (7.9) given the active RRHs. In the third stage, we need to further minimize the total transmit power with those RRHs determined as active while satisfying the QoS requirements for all the MUs, which is amounts to solving a coordinated beamforming problem, as in Example 7.1. More details on the group sparse beamforming algorithm can be found in [2]. Extensions on semidefinite programming for network power minimization with imperfect channel state information in multicast C-RANs can be found in [14].

Example 7.3 Stochastic coordinated beamforming via sequential convex approximations In practice, inevitably there will be uncertainty in the available channel state information (CSI). Such uncertainty may originate from various sources, e.g., training-based channel estimation [37], limited feedback [38], delays [39, 40], hardware deficiencies [41], or partial CSI acquisition [16]. The uncertainty in the available CSI brings a new technical challenge for system design. To guarantee performance, we impose a probabilistic QoS system constraint by assuming that the distribution information of the channel knowledge is available. Considering a unicast transmission scenario, the stochastic coordinated beamforming problem is formulated to minimize the total transmit power while satisfying a probabilistic QoS system constraint, as follows [16]:

$$\underset{\mathbf{v}_1,\dots,\mathbf{v}_K}{\text{minimize}} \quad \sum_{l=1}^{L}\sum_{k=1}^{K} \|\mathbf{v}_{lk}\|^2$$

$$\text{s. t.} \quad \Pr\left\{ \frac{|\mathbf{h}_k^H \mathbf{v}_k|^2}{\sum_{i\neq k}|\mathbf{h}_k^H \mathbf{v}_i|^2 + \sigma_k^2} \geq \gamma_k, \quad k=1,\dots,K \right\} \geq 1-\epsilon, \tag{7.10}$$

where the distribution information for the $\mathbf{h}_k$ is known and $0 < \epsilon < 1$ indicates that the QoS requirements should be guaranteed for all the MUs simultaneously with probability at least $1 - \epsilon$. However, problem (7.10) is a joint chance-constrained program [42] and is known to be intractable in general.

In [16] a stochastic difference-of-convex (DC) programming algorithm was proposed for finding a KKT point by solving the following stochastic convex programming problems iteratively:

$$\underset{\mathbf{v}_1,\dots,\mathbf{v}_K,\kappa>0}{\text{minimize}} \quad \sum_{l=1}^{L}\sum_{k=1}^{K} \|\mathbf{v}_{lk}\|^2$$

$$\text{subject to} \quad u(\mathbf{v},\kappa) - u(\mathbf{v}^{[j]},0) - 2\langle \nabla_{\mathbf{v}^*}u(\mathbf{v}^{[j]},0), \mathbf{v} - \mathbf{v}^{[j]}\rangle \leq \kappa\epsilon, \tag{7.11}$$

where $u(\mathbf{v},\nu) = \mathbb{E}\left\{\max_{1\leq k\leq K+1} s_k(\mathbf{v},\mathbf{h},\nu)\right\}$ is a convex function with $s_k(\mathbf{v},\mathbf{h},\nu) = \sum_{i\neq k}\mathbf{v}_i^H \mathbf{h}_k \mathbf{h}_k^H \mathbf{v}_i + \sigma_k^2 + \sum_{i\neq k}\gamma_i^{-1}\mathbf{v}_i^H \mathbf{h}_i \mathbf{h}_i^H \mathbf{v}_i, k = 1,\dots,K$, and $s_{K+1}(\mathbf{v},\mathbf{h},\nu) = \sum_{i=1}^{K}\gamma_i^{-1}\mathbf{v}_i^H \mathbf{h}_i \mathbf{h}_i^H \mathbf{v}_i$ are convex quadratic functions in $\mathbf{v}$. Here, the complex gradient of $u(\mathbf{v},0)$ with respect to $\mathbf{v}^*$ (the complex conjugate of $\mathbf{v}$) is given by

$$\nabla_{\mathbf{v}^*}u(\mathbf{v},0) = \mathbb{E}\{\nabla_{\mathbf{v}^*}s_{k^*}(\mathbf{v},\mathbf{h},0)\}, \tag{7.12}$$

where $k^* = \arg\max_{1\leq k\leq K+1} s_k(\mathbf{v},\mathbf{h},0)$, and $\nabla_{\mathbf{v}^*}s_k(\mathbf{v},\mathbf{h},0) = [\nu_{k,i}]_{1\leq i\leq K} (1 \leq k \leq K)$, with $\nu_{k,i} \in \mathbb{C}^N$ given by

$$\nu_{k,i} = \begin{cases} \left(\mathbf{h}_k\mathbf{h}_k^H + \gamma_i^{-1}\mathbf{h}_i\mathbf{h}_i^H\right)\mathbf{v}_i & \text{if } i \neq k, 1 \leq k \leq K, \\ 0, & \text{otherwise,} \end{cases} \tag{7.13}$$

and $\nabla_{\mathbf{v}^*}s_{K+1}(\mathbf{v},\mathbf{h},\kappa) = [\mathbf{v}_{K+1,i}]_{1\leq i\leq K}$ with $\mathbf{v}_{K+1,i} = \gamma_i^{-1}\mathbf{h}_i\mathbf{h}_i^H\mathbf{v}_i, \forall i$. Furthermore, the gradient of $u(\mathbf{v},0)$ with respect to κ is zero, as $\kappa = 0$ is a constant in the function $u(\mathbf{v},0)$.

To solve the stochastic convex programming problem (7.11) efficiently at each iteration, the sample-average approximation method is further adopted; this involves solving the following convex quadratically constrained quadratic program (QCQP):

$$\mathscr{P}_{\text{QCQP}} : \underset{\mathbf{v}_1,\ldots,\mathbf{v}_K,\kappa,\mathbf{x}}{\text{minimize}} \quad \sum_{l=1}^{L}\sum_{k=1}^{K}\|\mathbf{v}_{lk}\|^2$$

$$\text{s. t.} \quad \frac{1}{M}\sum_{m=1}^{M}x_m - \bar{u}(\mathbf{v}^{[j]},0) - 2\langle\bar{\nabla}_{\mathbf{v}^*}u(\mathbf{v}^{[j]},0),\mathbf{v}-\mathbf{v}^{[j]}\rangle \leq \kappa\epsilon,$$

$$s_k(\mathbf{v},\mathbf{h}^m,\kappa) \leq x_m, \quad x_m \geq 0, \quad \kappa > 0, \quad \forall k,m. \qquad (7.14)$$

Here, $\mathbf{x} = [x_m]_{1\leq m\leq M} \in \mathbb{R}^M$ is the collection of slack variables with M independent realizations of the random vector $\mathbf{h} \in \mathbb{C}^{NK}$.

More examples on applying convex optimization for resource allocation and signal processing in wireless cooperative networks can be found in [8, 9, 31]. In summary, the above examples illustrate that the new architecture of C-RANs brings up new design challenges, while convex optimization can serve as a powerful tool to formulate and solve these problems. Meanwhile, as the problem sizes scale up in C-RANs, it becomes critical to solve the resulting convex optimization problems efficiently.

7.2.2   A Unified Framework for Large-Scale Convex Optimization

As presented previously, a sequence of convex optimization problems $\mathscr{P}$ needs to be solved for typical signal processing and resource allocation problems in C-RANs. In dense C-RANs the BBU pool can support hundreds of RRHs for simultaneous transmission and reception [1]. Therefore, all the resulting convex optimization problems $\mathscr{P}$ are shifted into a new domain with a high problem dimension and a large number of constraints. Although the convex programs $\mathscr{P}$ can be solved in polynomial time using the interior-point method, which is implemented in most advanced off-the-shelf solvers (e.g., public software packages like SeDuMi [17] and SDPT3 [18] and commercial software packages like MOSEK [19]), the computational cost of such second-order methods will be prohibitive for large-scale problems. However, most custom algorithms, e.g., the uplink–downlink approach [10] and the ADMM-based algorithms [13, 26, 28], fail either to scale well to large problem sizes or to detect infeasibility effectively.

To overcome the limitations of the scalability of state-of-art solvers and the capability of infeasibility detection of custom algorithms, in this chapter we propose a two-stage large-scale optimization framework as shown in Fig. 7.1. Specifically, in the first stage the original problem will be transformed into a standard cone programming, thereby providing the capability of infeasibility detection and parallel computing. This will be presented in Section 7.3. In the second stage, the first-order alternating-direction method of multipliers (ADMM) algorithm [27], i.e., the operator splitting method, will be adopted to solve the large-scale homogeneous self-dual embedding (HSD) system. This will be presented in Section 7.4.

7.3 Matrix Stuffing for Fast Cone-Programming Transformation

Consider the following parametric family $\mathscr{P}_{\text{cone}}$ of primal conic optimization problems:

$$\mathscr{P}_{\text{cone}} : \underset{v,\mu}{\text{minimize}} \quad \mathbf{c}^T v$$

$$\text{s. t.} \qquad \mathbf{A}v + \mu = \mathbf{b}, \tag{7.15a}$$

$$(v, \mu) \in \mathbb{R}^n \times \mathcal{K}, \tag{7.15b}$$

where $v \in \mathbb{R}^n$ and $\mu \in \mathbb{R}^m$ (with $n \le m$) are the optimization variables, $\mathbf{A} \in \mathbb{R}^{m \times n}$, $\mathbf{b} \in \mathbb{R}^m$, $\mathbf{c} \in \mathbb{R}^n$, and $\mathcal{K} = \mathcal{K}_1 \times \cdots \times \mathcal{K}_q \in \mathbb{R}^m$ is a Cartesian product of q closed convex cones. Here, each $\mathcal{K}_i$ has dimension m_i such that $\sum_{i=1}^q m_i = m$. Let $\beta = \{\mathbf{A}, \mathbf{b}, \mathbf{c}\} \in \mathcal{D}$ be the problem data with $\mathcal{D}$ as the data space.

Although $\mathcal{K}_i$ is allowed to be any closed convex cone, we are primarily interested in the following symmetric cones:

- the nonnegative reals, $\mathbb{R}_+ = \{x \in \mathbb{R} | x \ge 0\}$;
- a second-order cone, $\mathcal{Q}^d = \{(y, \mathbf{x}) \in \mathbb{R} \times \mathbb{R}^{d-1} | \|\mathbf{x}\| \le y\}$;
- a positive semidefinite cone, $\mathbf{S}_+^n = \{\mathbf{X} \in \mathbb{R}^{n \times n} | \mathbf{X} = \mathbf{X}^T, \mathbf{X} \succeq \mathbf{0}\}$.

In particular, a problem instance $\mathscr{P}_{\text{cone}}(\beta)$ is known as a linear program (LP), SOCP, or SDP if all the cones $\mathcal{K}_i$ are restricted to $\mathbb{R}_+$, $\mathcal{Q}^d$, or $\mathbf{S}_+^n$, respectively. Therefore, almost all the original convex optimization problem family $\mathscr{P}$ can be equivalently transformed into the standard conic optimization problem family $\mathscr{P}_{\text{cone}}$ on the basis of the principle of disciplined convex programming [33]. This equivalence means that the optimal solution or the certificate of infeasibility of the original problem instance $\mathscr{P}(\alpha)$ can be extracted from the solution to the corresponding equivalent cone-program instance $\mathscr{P}_{\text{cone}}(\beta)$.

To develop a generic large-scale convex optimization framework, instead of exploiting the special structures in the original specific problems $\mathscr{P}$, we propose to work with the transformed equivalent standard conic optimization problems $\mathscr{P}_{\text{cone}}$. This approach bears the following advantages.

1. The convex programs $\mathscr{P}$ can be equivalently transformed into the standard cone programs $\mathscr{P}_{\text{cone}}$. This will be presented in Section 7.3.1.
2. The homogeneous self-dual embedding of the primal–dual pair of the standard cone program $\mathscr{P}_{\text{cone}}$ can be induced, thereby providing certificates of infeasibility. This will be presented in Section 7.4.1.
3. The feasible set in $\mathscr{P}_{\text{cone}}$ is formed by two sets: a subspace constraint (7.15a) and a convex cone $\mathcal{K}$, which is formed by the Cartesian product of smaller convex cones.

This salient feature will be exploited to enable parallel and scalable computing in Section 7.4.2.

7.3.1   Standard Conic Optimization Transformation

Our goal is to map the parameters $\boldsymbol{\alpha}$ in the original convex optimization problem family $\mathscr{P}$ to the problem data $\boldsymbol{\beta}$ in the equivalent standard conic optimization problem family $\mathscr{P}_{\text{cone}}$ with the description of the convex cone $\mathcal{K}$. The general idea of such a transformation is to rewrite the original problem family $\mathscr{P}$ in a Smith form by introducing a new variable for each subexpression in the disciplined convex programming form [33] of the problem family $\mathscr{P}$. In the following we will take the coordinated beamforming problem as an example to illustrate such a transformation.

Example 7.4   Standard cone program for coordinated beamforming Consider the coordinated beamforming problem (7.7) with the objective function as $\|\mathbf{v}\|_2$, which can be rewritten as the following disciplined convex programming form [33]:

$$\text{minimize} \quad \|\mathbf{v}\|_2$$
$$\text{s. t.} \quad \|\mathbf{D}_l\mathbf{v}\|_2 \leq \sqrt{P_l}, \quad l = 1, \ldots, L, \tag{7.16a}$$
$$\|\mathbf{C}_k\mathbf{v} + \mathbf{g}_k\|_2 \leq \beta_k \mathbf{r}_k^T\mathbf{v}, \quad k = 1, \ldots, K, \tag{7.16b}$$

where $\beta_k = \sqrt{1 + 1/\gamma_k}$,

$$\mathbf{D}_l = \text{blk diag}\{\mathbf{D}_l^1, \ldots, \mathbf{D}_l^K\} \in \mathbb{R}^{N_l K \times NK}$$

with $\mathbf{D}_l^k = \begin{bmatrix} \mathbf{0}_{N_l \times \sum_{i=1}^{l-1} N_i} & \mathbf{I}_{N_l \times N_l} & \mathbf{0}_{N_l \times \sum_{i=l+1}^{L} N_i} \end{bmatrix} \in \mathbb{R}^{N_l \times N}$,

$$\mathbf{g}_k = [\mathbf{0}_K^T \; \sigma_k]^T \in \mathbb{R}^{K+1},$$

$$\mathbf{r}_k = \begin{bmatrix} \mathbf{0}_{(k-1)N}^T, & \mathbf{h}_k^T & \mathbf{0}_{(K-k)N}^T \end{bmatrix}^T \in \mathbb{R}^{NK},$$

and

$$\mathbf{C}_k = [\tilde{\mathbf{C}}_k \; \mathbf{0}_{NK}]^T \in \mathbb{R}^{(K+1) \times NK}$$

with $\tilde{\mathbf{C}}_k = \text{blk diag}\{\mathbf{h}_k, \ldots, \mathbf{h}_k\} \in \mathbb{R}^{NK \times K}$.

Smith form reformulation To arrive at the standard cone program $\mathscr{P}_{\text{cone}}$, we rewrite problem (7.16) in the following Smith form [32] by introducing a new variable for each subexpression in (7.16):

$$\text{minimize} \quad x_0$$
$$\text{s. t.} \quad \|\mathbf{x}_1\| = x_0, \quad \mathbf{x}_1 = \mathbf{v},$$
$$\mathcal{G}_1(l), \mathcal{G}_2(k), \quad \forall k, l, \tag{7.17}$$

where $\mathcal{G}_1(l)$ is the Smith-form reformulation for the transmit power constraint for RRH l (7.16a), given as follows:

$$\mathcal{G}_1(l): \begin{cases} (y_0^l, \mathbf{y}_1^l) \in \mathcal{Q}^{KN_l+1}, \\ y_0^l = \sqrt{P_l} \in \mathbb{R}, \\ \mathbf{y}_1^l = \mathbf{D}_l\mathbf{v} \in \mathbb{R}^{KN_l}, \end{cases} \tag{7.18}$$

and $\mathcal{G}_2(k)$ is the Smith-form reformulation for the QoS constraint for MU k (7.16b), given as follows:

$$
\mathcal{G}_2(k) : \begin{cases}
(t_0^k, t_1^k) \in \mathcal{Q}^{K+1}, \\
t_0^k = \beta_k \mathbf{r}_k^T \mathbf{v} \in \mathbb{R}, \\
t_1^k = t_2^k + t_3^k \in \mathbb{R}^{K+1}, \\
t_2^k = \mathbf{C}_k \mathbf{v} \in \mathbb{R}^{K+1}, \\
t_3^k = \mathbf{g}_k \in \mathbb{R}^{K+1}.
\end{cases}
\tag{7.19}
$$

Nevertheless, the Smith form reformulation (7.17) is not convex owing to the non-convex constraint $\|\mathbf{x}_1\| = x_0$. We thus relax this non-convex constraint to $\|\mathbf{x}_1\| \le x_0$, yielding the following relaxed Smith form:

$$
\begin{aligned}
&\text{minimize} \quad x_0 \\
&\text{s. t.} \qquad \mathcal{G}_0, \mathcal{G}_1(l), \mathcal{G}_2(k), \quad \forall k, l,
\end{aligned}
\tag{7.20}
$$

where

$$
\mathcal{G}_0 : \begin{cases}
(x_0, \mathbf{x}_1) \in \mathcal{Q}^{NK+1}, \\
\mathbf{x}_1 = \mathbf{v} \in \mathbb{R}^{NK}.
\end{cases}
\tag{7.21}
$$

It can be easily proved that the constraint $\|\mathbf{x}_1\| \le x_0$ has to be active at the optimal solution, otherwise we could always scale down x_0 in such a way that the cost function is further minimized while still satisfying the constraints. Therefore, we conclude that the relaxed Smith form (7.20) is equivalent to the original problem (7.16).

Conic reformulation Now the relaxed Smith-form reformulation (7.20) is readily reformulated as the standard cone-programming form $\mathscr{P}_{\text{cone}}$. Specifically, define optimization variables $[x_0; \mathbf{v}]$ with the same type of equations as in $\mathcal{G}_0$; then $\mathcal{G}_0$ can be rewritten as

$$
\mathbf{M}[x_0; \mathbf{v}] + \boldsymbol{\mu}_0 = \mathbf{m},
\tag{7.22}
$$

where the slack variables $\boldsymbol{\mu}_0$ belong to the following convex set,

$$
\boldsymbol{\mu}_0 \in \mathcal{Q}^{NK+1},
\tag{7.23}
$$

and $\mathbf{M} \in \mathbb{R}^{(NK+1) \times (NK+1)}$ and $\mathbf{m} \in \mathbb{R}^{NK+1}$ are given as follows:

$$
\mathbf{M} = \begin{bmatrix} -1 & \\ \hline & -\mathbf{I}_{NK} \end{bmatrix}, \quad \mathbf{m} = \begin{bmatrix} 0 \\ \mathbf{0}_{NK} \end{bmatrix},
\tag{7.24}
$$

respectively. Now define optimization variables $[y_0^l; \mathbf{v}]$ with the same type of equations as in $\mathcal{G}_1(l)$; then $\mathcal{G}_1(l)$ can be rewritten as

$$
\mathbf{P}^l[y_0^l; \mathbf{v}] + \boldsymbol{\mu}_1^l = \mathbf{p}^l,
\tag{7.25}
$$

where the slack variables $\boldsymbol{\mu}_1^l \in \mathbb{R}^{KN_l+2}$ belongs to the following convex set formed by the Cartesian product of two convex sets,

$$
\boldsymbol{\mu}_1^l \in \mathcal{Q}^1 \times \mathcal{Q}^{KN_l+1},
\tag{7.26}
$$

and $\mathbf{P}^l \in \mathbb{R}^{(KN_l+2)\times(NK+1)}$ and $\mathbf{p}^l \in \mathbb{R}^{KN_l+2}$ are given as follows:

$$\mathbf{P}^l = \begin{bmatrix} 1 \\ \hline -1 \\ \hline -\mathbf{D}_l \end{bmatrix}, \quad \mathbf{p}^l = \begin{bmatrix} \sqrt{P_l} \\ 0 \\ \mathbf{0}_{KN_l} \end{bmatrix}, \tag{7.27}$$

respectively. Next, define optimization variables $[t_0^k; \mathbf{v}]$ with the same type of equations as in $\mathcal{G}_2(k)$; then $\mathcal{G}_2(k)$ can be rewritten as

$$\mathbf{Q}^k[t_0^k; \mathbf{v}] + \boldsymbol{\mu}_2^k = \mathbf{q}^k, \tag{7.28}$$

where the slack variables $\boldsymbol{\mu}_2^k \in \mathbb{R}^{K+3}$ belong to the following convex set formed by the Cartesian product of two convex sets,

$$\boldsymbol{\mu}_2^k \in \mathcal{Q}^1 \times \mathcal{Q}^{K+2}, \tag{7.29}$$

and $\mathbf{Q}^k \in \mathbb{R}^{(K+3)\times(NK+1)}$ and $\mathbf{q}^k \in \mathbb{R}^{K+3}$ are given as follows:

$$\mathbf{Q}^k = \begin{bmatrix} 1 & -\beta_k \mathbf{r}_k^T \\ \hline -1 & \\ \hline & -\mathbf{C}_k \end{bmatrix}, \quad \mathbf{q}^k = \begin{bmatrix} 0 \\ 0 \\ \mathbf{g}_k \end{bmatrix}, \tag{7.30}$$

respectively.

Therefore, we arrive at the standard form $\mathscr{P}_{\text{cone}}$ by writing the optimization variables $\boldsymbol{v} \in \mathbb{R}^n$ as follows:

$$\boldsymbol{v} = [x_0; y_0^1; \ldots; y_0^L; t_0^1; \ldots; t_0^K; \mathbf{v}], \tag{7.31}$$

and $\mathbf{c} = [1; \mathbf{0}_{n-1}]$. The structure of the standard cone-programming $\mathscr{P}_{\text{cone}}$ is characterized by the following data:

$$n = 1 + L + K + NK, \tag{7.32}$$

$$m = (L+K) + (NK+1) + \sum_{l=1}^{L}(KN_l+1) + K(K+2), \tag{7.33}$$

$$\mathcal{K} = \underbrace{\mathcal{Q}^1 \times \cdots \times \mathcal{Q}^1}_{L+K} \times \mathcal{Q}^{NK+1} \times \underbrace{\mathcal{Q}^{KN_1+1} \times \cdots \times \mathcal{Q}^{KN_L+1}}_{L}$$

$$\times \underbrace{\mathcal{Q}^{K+2} \times \cdots \times \mathcal{Q}^{K+2}}_{K}, \tag{7.34}$$

where $\mathcal{K}$ is the Cartesian product of $2(L+K)+1$ second-order cones, and $\mathbf{A}$ and $\mathbf{b}$ are given as follows:

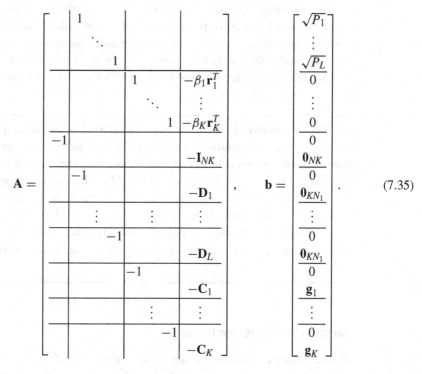

$$(7.35)$$

Extension to the complex case For $\mathbf{h}_k \in \mathbb{C}^N, \mathbf{v}_i \in \mathbb{C}^N$, we have

$$\mathbf{h}_k^H \mathbf{v}_i \Longrightarrow \underbrace{\begin{bmatrix} \Re(\mathbf{h}_k) & -\Im(\mathbf{h}_k) \\ \Im(\mathbf{h}_k) & \Re(\mathbf{h}_k) \end{bmatrix}}_{\tilde{\mathbf{h}}_k}^T \underbrace{\begin{bmatrix} \Re(\mathbf{v}_i) \\ \Im(\mathbf{v}_i) \end{bmatrix}}_{\tilde{\mathbf{v}}_i}, \qquad (7.36)$$

where $\tilde{\mathbf{h}}_k \in \mathbb{R}^{2N \times 2}$ and $\tilde{\mathbf{v}}_i \in \mathbb{R}^{2N}$. Therefore, the complex-field problem can be changed into a real-field problem by the transformations $\mathbf{h}_k \Rightarrow \tilde{\mathbf{h}}_k$ and $\mathbf{v}_i \Rightarrow \tilde{\mathbf{v}}_i$.

7.3.2 Matrix Stuffing for Fast Transformation

Inspired by the recent work [23] on fast-optimization code deployment for embedding a second-order cone program, we propose to use the matrix stuffing technique [23, 24] to transform the original problem instance $\mathscr{P}(\boldsymbol{\alpha})$ into the standard cone-program instance $\mathscr{P}_{\text{cone}}(\boldsymbol{\beta})$ quickly. Specifically, for any given network size we first generate and store the structure that maps the original problem family $\mathscr{P}$ to the standard conic optimization problem family $\mathscr{P}_{\text{cone}}$. Thus, the pre-stored standard-form structure includes the problem dimensions (i.e., m and n), the description of $\mathcal{V}$ (i.e., the array of the cone sizes $[m_1, m_2, \ldots, m_q]$), and the symbolic problem parameters $\boldsymbol{\beta} = \{\mathbf{A}, \mathbf{b}, \mathbf{c}\}$. This procedure can be done offline. Furthermore, to reduce the storage and memory overhead we store the problem data β in a sparse form [43] by storing only the non-zero entries.

Using the pre-stored structure, for a given problem instance $\mathscr{P}(\alpha)$ we need only copy its parameters α to the corresponding problem data β in the standard conic-optimization problem family $\mathscr{P}_{\text{cone}}$. As the procedure for transformation needs only to copy memory, it thus is suitable for fast transformation and can avoid repeated parsing and generating as in parser/solver modeling frameworks like CVX [21] and YALMIP [22].

Example 7.5 **Matrix stuffing for the coordinated beamforming problem** For the convex coordinated beamforming problem (7.16), to arrive at the standard cone program form $\mathscr{P}_{\text{cone}}(\beta)$, we need only copy the parameters of the transmit power constraints P_l to the data of the standard form, i.e., for the $\sqrt{P_l}$ in $\mathbf{b}$, copy the parameters of the SINR thresholds γ to the data of the standard form, i.e., the β_k in $\mathbf{A}$, and copy the parameters of the channel realizations $\mathbf{h}_k$ to the data of the standard form, i.e., the $\mathbf{r}_k$ and $\mathbf{C}_k$ in $\mathbf{A}$. As we need to copy the memory only for the transformation, this procedure can be very efficient compared with state-of-the-art numerical-based modeling frameworks such as CVX.

7.3.3   Practical Implementation Issues

We have presented a systematic way to equivalently transform the original problems $\mathscr{P}$ to standard conic optimization problems $\mathscr{P}_{\text{cone}}$. The resultant structure that maps the original problem to the standard form can be stored and reused for fast transforming via matrix stuffing. This can significantly reduce the modeling overhead compared with the parser/solver modeling frameworks such as CVX. However, it requires tedious manual work to find the mapping, and it may not be easy to verify its correctness. Chu *et al.* [23] made an attempt that was intended to automatically generate the code for matrix stuffing. However, so far the corresponding software package QCML [23] is far from complete and may not be suitable for our applications. Extending numerically based transformation modeling frameworks like CVX to symbolically based transformation modeling frameworks like QCML is non-trivial and requires tremendous mathematical and technical effort.

7.4   Operator Splitting for Large-Scale Homogeneous Self-Dual Embedding

Although the standard cone program $\mathscr{P}_{\text{cone}}$ itself is suitable for parallel computing via the operator splitting method [44], directly working on this problem may fail to provide certificates of infeasibility. To address this limitation, on the basis of the recent work by O'Donoghue *et al.* [45] we propose to solve the homogeneous self-dual embedding [34] of the primal–dual pair of the cone program $\mathscr{P}_{\text{cone}}$. The resultant homogeneous self-dual embedding is further solved via the operator splitting method, also known as the ADMM algorithm [27].

7.4.1 Homogeneous Self-Dual Embedding of Cone Programming

The basic idea of homogeneous self-dual embedding is to embed the primal and dual problems of the cone program $\mathscr{P}_{\text{cone}}$ into a single feasibility problem (i.e., finding a feasible point of the intersection of a subspace and a convex set) such that either the optimal solution or the certificate of infeasibility of the original cone program $\mathscr{P}_{\text{cone}}$ can be extracted from the solution of the embedded problem.

The dual problem of $\mathscr{P}_{\text{cone}}$ is given by [45]

$$\mathscr{D}_{\text{cone}} : \underset{\eta, \lambda}{\text{maximize}} \quad -\mathbf{b}^T \eta$$
$$\text{s. t.} \quad -\mathbf{A}^T \eta + \lambda = \mathbf{c},$$
$$(\lambda, \eta) \in \{0\}^n \times \mathcal{K}^*, \tag{7.37}$$

where $\lambda \in \mathbb{R}^n$ and $\eta \in \mathbb{R}^m$ are the dual variables and $\mathcal{K}^*$ is the dual cone of the convex cone $\mathcal{K}$. Define the optimal values of the primal program $\mathscr{P}_{\text{cone}}$ and dual program $\mathscr{D}_{\text{cone}}$ as $p^\star$ and $d^\star$, respectively. Let $p^\star = +\infty$ and $p^\star = -\infty$ indicate primal infeasibility and unboundedness, respectively. Similarly, let $d^\star = -\infty$ and $d^\star = +\infty$ indicate dual infeasibility and unboundedness, respectively. We assume strong duality for the convex cone program $\mathscr{P}_{\text{cone}}$, i.e., $p^\star = d^\star$, including the cases when they are infinite. This is a standard assumption made when practically designing solvers for conic programs, e.g., it is assumed in [17–19, 34, 45]. Besides this, we do not make any regularity assumption on the feasibility and boundedness assumptions on the primal and dual problems.

Certificates of Infeasibility

Given the cone program $\mathscr{P}_{\text{cone}}$, a main task is to detect feasibility. However, most existing custom algorithms assume that the original problem $\mathscr{P}$ is feasible [10] or provide heuristic ways to handle infeasibility [26]. Nevertheless, the only way to detect infeasibility effectively is to provide a certificate or proof of infeasibility, as presented in the following proposition.

PROPOSITION 7.1 (Certificates of infeasibility)   The following system,

$$\mathbf{A}\mathbf{v} + \mu = \mathbf{b}, \quad \mu \in \mathcal{K}, \tag{7.38}$$

is infeasible if and only if the following system is feasible

$$\mathbf{A}^T \eta = \mathbf{0}, \quad \eta \in \mathcal{K}^\star, \quad \mathbf{b}^T \eta < 0. \tag{7.39}$$

Therefore, any dual variable η satisfying the system (7.39) provides a certificate or proof that the primal program $\mathscr{P}_{\text{cone}}$ (equivalently, the original problem $\mathscr{P}$) is infeasible. Similarly, any primal variable $\mathbf{v}$ satisfying the system

$$-\mathbf{A}\mathbf{v} \in \mathcal{K}, \quad \mathbf{c}^T \mathbf{v} < 0, \tag{7.40}$$

is a certificate of the infeasibility of the dual program $\mathscr{D}_{\text{cone}}$.

Proof This result follows directly from the theorem of strong alternatives [11, Section 5.8.2]. □

Optimality Conditions

If the transformed standard cone program $\mathscr{P}_{\text{cone}}$ is feasible then $(v^\star, \mu^\star, \lambda^\star, \eta^\star)$ are optimal if and only if they satisfy the following Karush–Kuhn–Tucker (KKT) conditions:

$$\mathbf{A}v^\star + \mu^\star - \mathbf{b} = \mathbf{0}, \tag{7.41}$$

$$\mathbf{A}^T \eta^\star - \lambda^\star + \mathbf{c} = \mathbf{0}, \tag{7.42}$$

$$(\eta^\star)^T \mu^\star = 0, \tag{7.43}$$

$$(v^\star, \mu^\star, \lambda^\star, \eta^\star) \in \mathbb{R}^n \times \mathcal{K} \times \{0\}^n \times \mathcal{K}^\ast. \tag{7.44}$$

In particular, the complementary slackness condition (7.43) can be rewritten as

$$\mathbf{c}^T v^\star + \mathbf{b}^T \eta^\star = 0, \tag{7.45}$$

which explicitly forces the duality gap to be zero.

Homogeneous Self-Dual Embedding

We can first detect feasibility by Proposition 7.1 and then solve the KKT system if the problem is feasible and bounded. However, the disadvantage of such a two-phase method is that two related problems (i.e., checking feasibility and solving KKT conditions) need to be solved sequentially [34]. To avoid such inefficiency, we propose to solve the following homogeneous self-dual embedding [34]:

$$\mathbf{A}v + \mu - \mathbf{b}\tau = \mathbf{0}, \tag{7.46}$$

$$\mathbf{A}^T \eta - \lambda + \mathbf{c}\tau = \mathbf{0}, \tag{7.47}$$

$$\mathbf{c}^T v + \mathbf{b}^T \eta + \kappa = 0, \tag{7.48}$$

$$(v, \mu, \lambda, \eta, \tau, \kappa) \in \mathbb{R}^n \times \mathcal{K} \times \{0\}^n \times \mathcal{K}^\ast \times \mathbb{R}_+ \times \mathbb{R}_+, \tag{7.49}$$

which embeds all the information on infeasibility and optimality into a single system by introducing two new nonnegative variables τ and κ, which encode different outcomes. Thus the homogeneous self-dual embedding can be rewritten in the following compact form:

$$\mathscr{F}_{\text{HSD}} : \text{find} \quad (\mathbf{x}, \mathbf{y})$$

$$\text{s. t.} \quad \mathbf{y} = \mathbf{Q}\mathbf{x},$$

$$\mathbf{x} \in \mathcal{C}, \quad \mathbf{y} \in \mathcal{C}^\ast, \tag{7.50}$$

where

$$\underbrace{\begin{bmatrix} \lambda \\ \mu \\ \kappa \end{bmatrix}}_{\mathbf{y}} = \underbrace{\begin{bmatrix} \mathbf{0} & \mathbf{A}^T & \mathbf{c} \\ -\mathbf{A} & \mathbf{0} & \mathbf{b} \\ -\mathbf{c}^T & -\mathbf{b}^T & 0 \end{bmatrix}}_{\mathbf{Q}} \underbrace{\begin{bmatrix} v \\ \eta \\ \tau \end{bmatrix}}_{\mathbf{x}}. \tag{7.51}$$

In (7.51), $\mathbf{x} \in \mathbb{R}^{m+n+1}$, $\mathbf{y} \in \mathbb{R}^{m+n+1}$, $\mathbf{Q} \in \mathbb{R}^{(m+n+1)\times(m+n+1)}$, $C = \mathbb{R}^n \times \mathcal{K}^* \times \mathbb{R}_+$, and $C^* = \{0\}^n \times \mathcal{K} \times \mathbb{R}_+$. This system has a trivial solution with all variables as zeros.

The homogeneous self-dual embedding problem $\mathscr{F}_{\mathrm{HSD}}$ is thus a feasibility problem finding a non-zero solution in the intersection of a subspace and a convex cone. Let $(\nu, \mu, \lambda, \eta, \tau, \kappa)$ be a non-zero solution of the homogeneous self-dual embedding. We then have the following remarkable trichotomy, derived in [34]:

- **Case 1,** $\tau > 0$, $\kappa = 0$, then

$$\hat{\nu} = \nu/\tau, \quad \hat{\eta} = \eta/\tau, \quad \hat{\mu} = \mu/\tau \tag{7.52}$$

are the primal and dual solutions to the cone program $\mathscr{P}_{\mathrm{cone}}$.
- **Case 2,** $\tau = 0$, $\kappa > 0$; this implies that $\mathbf{c}^T \nu + \mathbf{b}^T \eta < 0$. Then
  1. If $\mathbf{b}^T \eta < 0$ then $\hat{\eta} = \eta/(-\mathbf{b}^T \eta)$ is a certificate of primal infeasibility, as

$$\mathbf{A}^T \hat{\eta} = \mathbf{0}, \quad \hat{\eta} \in \mathcal{V}^*, \quad \mathbf{b}^T \hat{\eta} = -1. \tag{7.53}$$

  2. If $\mathbf{c}^T \nu < 0$ then $\hat{\nu} = \nu/(-\mathbf{c}^T \hat{\nu})$ is a certificate of dual infeasibility, as

$$-\mathbf{A}\hat{\nu} \in \mathcal{V}, \quad \mathbf{c}^T \hat{\nu} = -1. \tag{7.54}$$

- **Case 3,** $\tau = \kappa = 0$; no conclusion can be made about the cone problem $\mathscr{P}_{\mathrm{cone}}$.

Therefore, from the solution to the homogeneous self-dual embedding, we can extract either the optimal solution (based on (7.31)) or the certificate of infeasibility for the original problem. Furthermore, as the set (7.49) is a Cartesian product of a finite number of sets, this will enable parallelizable algorithm design. With these distinct advantages of homogeneous self-dual embedding, in the sequel we focus on developing efficient algorithms to solve the large-scale feasibility problem $\mathscr{F}_{\mathrm{HSD}}$ via the operator splitting method.

7.4.2   The Operator Splitting Method

Conventionally, the convex homogeneous self-dual embedding $\mathscr{F}_{\mathrm{HSD}}$ can be solved via the interior-point method, e.g. [17–19, 34]. However, the computational cost of such a second-order method can still be prohibitive for large-scale problems. Instead, O'Donoghue et al. [45] developed a first-order optimization algorithm based on the operator splitting method, i.e., the ADMM algorithm [27], to solve a large-scale homogeneous self-dual embedding. The key observation is that the convex cone constraint in $\mathscr{F}_{\mathrm{HSD}}$ is the Cartesian product of smaller standard convex cones (i.e., second-order cones, semidefinite cones, and nonnegative reals), which enables parallelizable computing.

Specifically, the homogeneous self-dual embedding problem $\mathscr{F}_{\mathrm{HSD}}$ can be rewritten as

$$\text{minimize } I_{C \times C^*}(\mathbf{x}, \mathbf{y}) + I_{Q\mathbf{x}=\mathbf{y}}(\mathbf{x}, \mathbf{y}), \tag{7.55}$$

where I_S is the indicator function of the set S, i.e., $I_S(z)$ is zero for $z \in S$ and $+\infty$ otherwise. By replicating the variables $\mathbf{x}$ and $\mathbf{y}$, problem (7.55) can be transformed into the following consensus form [27, Section 7.1]:

$$\mathscr{P}_{\text{ADMM}} : \text{minimize} \quad I_{\mathcal{C} \times \mathcal{C}^*}(\mathbf{x}, \mathbf{y}) + I_{\mathbf{Q}\tilde{\mathbf{x}} = \tilde{\mathbf{y}}}(\tilde{\mathbf{x}}, \tilde{\mathbf{y}})$$
$$\text{s. t.} \quad (\mathbf{x}, \mathbf{y}) = (\tilde{\mathbf{x}}, \tilde{\mathbf{y}}), \tag{7.56}$$

which is readily solved by the operator splitting method.

Applying the ADMM algorithm [27, Section 3.1] to the problem $\mathscr{P}_{\text{ADMM}}$ and eliminating the dual variables by exploiting the self-dual property of the problem $\mathscr{F}_{\text{HSD}}$ (refer to [45, Section 3] on how to simplify the ADMM algorithm), the final algorithm is obtained as follows:

$$\mathcal{OS}_{\text{ADMM}} : \begin{cases} \tilde{\mathbf{x}}^{[i+1]} = (\mathbf{I} + \mathbf{Q})^{-1}(\mathbf{x}^{[i]} + \mathbf{y}^{[i]}), \\ \mathbf{x}^{[i+1]} = \Pi_{\mathcal{C}}(\tilde{\mathbf{x}}^{[i+1]} - \mathbf{y}^{[i]}), \\ \mathbf{y}^{[i+1]} = \mathbf{y}^{[i]} - \tilde{\mathbf{x}}^{[i+1]} + \mathbf{x}^{[i+1]}, \end{cases} \tag{7.57}$$

where $\Pi_{\mathcal{C}}(\mathbf{x})$ denotes the Euclidean projection of $\mathbf{x}$ onto the set $\mathcal{C}$. This algorithm has an $O(1/k)$ convergence rate [46] with k as the iteration counter (i.e., ϵ-accuracy can be achieved in $O(1/\epsilon)$ iterations) and will not converge to zero if a non-zero solution exists [45, Section 3.4]. Empirically, this algorithm can converge to modest accuracy within a reasonable amount of time. As the last step is computationally trivial, in the sequel we will focus on how to solve the first two steps efficiently.

7.4.3   Subspace Projection Algorithms

The first step in the algorithm $\mathcal{OS}_{\text{ADMM}}$ is a subspace projection. After simplification [45, Section 4], we essentially need to solve the following linear equation at each iteration, i.e.,

$$\underbrace{\begin{bmatrix} \mathbf{I} & -\mathbf{A}^T \\ -\mathbf{A} & -\mathbf{I} \end{bmatrix}}_{\mathbf{S}} \underbrace{\begin{bmatrix} v \\ -\eta \end{bmatrix}}_{\mathbf{x}} = \underbrace{\begin{bmatrix} v^{[i]} \\ \eta^{[i]} \end{bmatrix}}_{\mathbf{b}}, \tag{7.58}$$

for the given $v^{[i]}$ and $\eta^{[i]}$ at iteration i, where $\mathbf{S} \in \mathbb{R}^{d \times d}$ with $d = m + n$ is a *symmetric quasidefinite matrix* [47]. Several approaches will be presented to solve the large-scale linear system (7.58) efficiently, i.e., so as to trade off the solving time and accuracy.

Factorization Caching Approach

To enable quicker inversions and reduce memory overhead via exploiting the sparsity of the matrix $\mathbf{S}$, the method of sparse permuted LDL^T factorization [43] can be adopted. Specifically, such a factor-solve method can be carried out by first computing the sparse permuted LDL^T factorization as follows:

$$\mathbf{S} = \mathbf{P}\mathbf{L}\mathbf{D}\mathbf{L}^T\mathbf{P}^T, \tag{7.59}$$

where $\mathbf{L}$ is a lower-triangular matrix, $\mathbf{D}$ is a diagonal matrix [44], and $\mathbf{P}$ with $\mathbf{P}^{-1} = \mathbf{P}^T$ is a permutation matrix to fill in the factorization [43], i.e., the non-zero entries in $\mathbf{L}$.

Such a factorization exists for any permutation $\mathbf{P}$, as the matrix $\mathbf{S}$ is symmetric quasidefinite [47, Theorem 2.1]. Computing the factorization costs much less than $\mathcal{O}(1/3d^3)$ flops, while the exact value depends on d and the sparsity pattern of $\mathbf{S}$ in a complicated way. Note that this factorization needs to be computed only once, in the first iteration, and can be cached for reusing in the sequent iterations for subspace projections. This is called the *factorization caching* technique [45].

Given the cached factorization (7.59), solving subsequent projections $\mathbf{x} = \mathbf{S}^{-1}\mathbf{b}$ (7.58) can be carried out by solving the following much easier equations,

$$\mathbf{Px}_1 = \mathbf{b}, \quad \mathbf{Lx}_2 = \mathbf{x}_1, \quad \mathbf{Dx}_3 = \mathbf{x}_2, \quad \mathbf{L}^T\mathbf{x}_4 = \mathbf{x}_3, \quad \mathbf{P}^T\mathbf{x} = \mathbf{x}_4, \tag{7.60}$$

which cost respectively zero flops, $\mathcal{O}(sd)$ flops by forward substitution with s as the number of non-zero entries in $\mathbf{L}$, $\mathcal{O}(d)$ flops, $\mathcal{O}(sd)$ flops by backward substitution, and zero flops, respectively [11, Appendix C].

Approximate Approaches

To scale the linear system (7.58) to large problem sizes for, approximate algorithms can be adopted to trade off the accuracy of the solution and the solving time. We first rewrite (7.58) as follows:

$$v = (\mathbf{I} + \mathbf{A}^T\mathbf{A})^{-1}(v^{[i]} - \mathbf{A}^T\eta^{[i]}), \tag{7.61}$$

$$\eta = \eta^{[i]} + \mathbf{A}v. \tag{7.62}$$

The conjugate gradient method [48] is then applied to find an approximation to the above linear system. Specifically, let $\mathbf{G} = \mathbf{I} + \mathbf{A}^T\mathbf{A}$. The conjugate gradient algorithm to find $\mathbf{x}$ such that $\mathbf{Gx} = \mathbf{b}$ is given by [48, Section 10.2]

$$\beta_k = \mathbf{r}_{k-1}^T\mathbf{r}_{k-1} / \left(\mathbf{r}_{k-2}^T\mathbf{r}_{k-2}\right), \tag{7.63}$$

$$\mathbf{p}_k = \mathbf{r}_{k-1} + \beta_k\mathbf{p}_{k-1}, \tag{7.64}$$

$$\alpha_k = \mathbf{r}_{k-1}^T\mathbf{r}_{k-1} / \left(\mathbf{p}_k^T\mathbf{Gp}_k\right), \tag{7.65}$$

$$\mathbf{x}_k = \mathbf{x}_{k-1} + \alpha_k\mathbf{p}_k, \tag{7.66}$$

$$\mathbf{r}_k = \mathbf{r}_{k-1} - \alpha_k\mathbf{Gp}_k. \tag{7.67}$$

This is used until the norm of the residual $\|\mathbf{r}_k\|_2$ is sufficiently small. As the iterations only require a matrix–vector multiplication operation, the conjugate gradient method can be very efficient. Other approximate approaches to solving the linear system (7.61), (7.62) can be found in [48]. The applicability of the approximate algorithms method is based on the fact that if the subspace projection error is bounded by a summable sequence then the ADMM algorithm $\mathcal{OS}_{\text{ADMM}}$ will converge [45, 49].

7.4.4 Cone Projection

Proximal Algorithm

The second step in the algorithm $\mathcal{OS}_{\text{ADMM}}$ is to project a point ω onto the cone $\mathcal{C}$. As $\mathcal{C}$ is the Cartesian product of a finite number of smaller convex cones $\mathcal{C}_i$, we can perform projection onto $\mathcal{C}$ by projecting onto $\mathcal{C}_i$ separately and in parallel. Furthermore, the projection onto each convex cone can be done with closed forms. For example, for nonnegative real $C_i = \mathbb{R}_+$, we have that [50, Section 6.3.1]

$$\Pi_{C_i}(\boldsymbol{\omega}) = \boldsymbol{\omega}_+, \tag{7.68}$$

where the nonnegative-part operator $(\cdot)_+$ is taken elementwise. For the second-order cone $C_i = \{(y, \mathbf{x}) \in \mathbb{R} \times \mathbb{R}^{p-1} | \|\mathbf{x}\| \le y\}$, we have that [50, Section 6.3.2]

$$\Pi_{C_i}(\boldsymbol{\omega}, \tau) = \begin{cases} 0, & \|\boldsymbol{\omega}\|_2 \le -\tau, \\ (\boldsymbol{\omega}, \tau), & \|\boldsymbol{\omega}\|_2 \le \tau, \\ (1/2)(1 + \tau/\|\boldsymbol{\omega}\|_2)(\boldsymbol{\omega}, \|\boldsymbol{\omega}\|_2), & \|\boldsymbol{\omega}\|_2 \ge |\tau|. \end{cases} \tag{7.69}$$

For the semidefinite cone $C_i = \{\mathbf{X} \in \mathbb{R}^{n \times n} | \mathbf{X} = \mathbf{X}^T, \mathbf{X} \succeq \mathbf{0}\}$, we have that [50, Section 6.3.3]

$$\Pi_{C_i}(\boldsymbol{\Omega}) = \sum_{i=1}^{n}(\lambda_i)_+ \mathbf{u}_i \mathbf{u}_i^T, \tag{7.70}$$

where $\sum_{i=1}^{n} \lambda_i \mathbf{u}_i \mathbf{u}_i^T$ is the eigenvalue decomposition of $\boldsymbol{\Omega}$. More examples on the cone projection (e.g., the exponential cone projection) can be found in [50].

Randomized Algorithms

Although the cone projection can be performed in parallel with closed forms, the scaling may be prohibitive on the semidefinite cone projection via eigenvalue decomposition. Therefore, to scale well to large problem sizes for SDP problems, it is of great interest to develop efficient algorithms to solve approximately the semidefinite cone projection problem with rigorous performance and convergence guarantees for the resulting ADMM algorithm $\mathcal{OS}_{\text{ADMM}}$.

Randomized sketching provides powerful randomized and sampling techniques for large-scale matrices by compressing them into much smaller matrices, thereby saving solving time and memory by reducing the problem dimensions. For semidefinite cone projection, to project a symmetric matrix $\mathbf{A} \in \mathbb{R}^{n \times n}$ onto the positive semidefinite cone, we need to first perform its eigenvalue expansion and then drop the terms associated with negative eigenvalues. The randomized algorithms for the eigenvalue decomposition of the symmetric matrix $\mathbf{A} \in \mathbb{R}^{n \times n}$ generally consist of the following simple steps [51]:

1. generate the orthonormal matrix $\mathbf{Q} \in \mathbb{R}^{m \times n}$ ($m < n$) such that $\|\mathbf{A} - \mathbf{Q}\mathbf{Q}^T\mathbf{A}\| \le \epsilon$ with ϵ as the computational tolerance;
2. form the smaller matrix $\mathbf{B} = \mathbf{Q}\mathbf{A}\mathbf{Q}^T$;
3. compute an eigenvalue decomposition $\mathbf{B} = \mathbf{V}\boldsymbol{\Lambda}\mathbf{V}^T$;
4. form the orthonormal matrix $\mathbf{U} = \mathbf{Q}\mathbf{V}$ such that $\mathbf{A} \approx \mathbf{U}\boldsymbol{\Lambda}\mathbf{U}^T$.

For step 1, several efficient randomized schemes were discussed in [51, Section 4] to minimize the sampling size and computational cost for producing the matrix $\mathbf{Q}$. A simple scheme is based on the Gaussian random matrix. In step 3, once we have $\mathbf{B}$ we can adopt any of the standard deterministic factorization techniques in [51, Section 3.3] to produce eigenvalue decomposition. More efficient algorithms for factorization can be found in [51, Section 5.2] by exploiting the information in $\mathbf{Q}$. Typically, randomized algorithms only require $\mathcal{O}(n^2 \log(k))$ flops while classic algorithms require $\mathcal{O}(n^2 k)$ flops to do eigenvalue decomposition for a rank-k matrix $\mathbf{A}$. More recent progress on randomized sketching methods for numerical linear algebra can be found in [52, 53].

Although randomized algorithms can exploit randomness as a source for speedup, it is non-trivial to apply these algorithms directly for approximate cone projections with performance and convergence guarantees for the resulting operator splitting algorithm $\mathcal{OS}_{\text{ADMM}}$. It is thus of great interest to establish theoretical guarantees for the randomized cone projection methods in the algorithm $\mathcal{OS}_{\text{ADMM}}$.

7.4.5 Practical Implementation Issues

We have thus far presented the two-stage parallel computing framework for large-scale convex optimization in dense C-RANs. Here, we will discuss implementation issues of the proposed framework in C-RANs, thereby exploiting the computational architectures to obtain further speed gains.

Parallel and Distributed Implementation

The operator splitting algorithm $\mathcal{OS}_{\text{ADMM}}$ that we have presented is compact and parameter-free, with parallelizable computing and linear convergence. In particular, each iteration of the algorithm is simple and easy for parallel and distributed computing. This allows the algorithm $\mathcal{OS}_{\text{ADMM}}$ to utilize the cloud computing environments in C-RANs with shared computing and memory resources in a single BBU pool. Specifically, the parallel algorithms can be leveraged in the subspace projection for LDL^T factorization and sparse matrix–vector multiplication [54]. The cone projection can be parallelized easily by projecting onto $\mathcal{K}_i$ separately and in parallel. However, it is challenging to accommodate the operator splitting algorithm to the distributed environments in heterogeneous C-RANs [4, 8], Fog-RAN, and MENG-RAN [5]. In particular, message updates across the network (e.g., backhaul network) and synchronization among heterogenous computation units will significantly increase the communication complexity in the distributed algorithms, which may result in delay and loss of performance. Therefore, it is critical to design large-scale distributed optimization algorithms with the minimal requirements of synchronization and communication.

Real-Time Implementation

In dense C-RANs, to satisfy the strict low-latency demands, e.g., in Tactile Internet the end-to-end latency is constrained to one millisecond [55], we need to solve large-scale optimization problems in a millisecond. This brings significant challenges compared with large-scale optimization problems in machine learning and big data, where latency is not a big issue but the problem dimension is often in the order of millions.

To solve a large-scale optimization problem in a real-time way, one promising approach is to leverage the *symbolic* subspace and cone projections. The general idea is to generate and store all the structures and descriptions of the algorithm for the specific problem family $\mathcal{P}_{\text{cone}}$. Eventually, the ADMM solver can be symbolically based so as to provide numerical solutions for each problem instance $\mathscr{P}_{\mathrm{cone}}(\boldsymbol{\beta})$ extremely quickly within a hard real-time deadline. This idea has already been successfully applied in the code generation system CVXGEN [56] for real-time convex quadratic optimization [57] and in the interior-point based SOCP solver [58] for embedded systems. It is of great interest to implement this idea for the operator splitting algorithm $\mathcal{OS}_{\mathrm{ADMM}}$ for general real-time conic optimization.

7.5   Numerical Results

In this section we simulate the proposed two-stage large-scale convex optimization framework for performance optimization in dense C-RANs. *The corresponding MAT-LAB code that can reproduce all the simulation results using the proposed large-scale convex optimization algorithm is available online.*[1]

We considered the following channel model for the link between the kth MU and the lth RRH:

$$\mathbf{h}_{kl} = 10^{-L(d_{kl})/20}\sqrt{\varphi_{kl}s_{kl}}\mathbf{f}_{kl}, \quad \forall k, l, \tag{7.71}$$

where $L(d_{kl})$ is the path loss in dB at distance d_{kl}, as in [2, Table I], s_{kl} is the shadowing coefficient, φ_{kl} is the antenna gain, and $\mathbf{f}_{kl}$ is the small-scale fading coefficient. We used the standard cellular network parameters as in [2, Table I]. All the simulations were carried out on a personal computer with a 3.2 GHz quad-core Intel Core i5 processor and 8 GB of RAM running Linux. The reference implementation of the operator splitting algorithm SCS is available online;[2] it is a general software package for solving large-scale convex cone problems based on [45] and can be called by the modeling frameworks CVX and CVXPY [59]. The settings (e.g., the stopping criteria) of SCS can be found in [45].

The proposed two-stage approach framework, termed "Matrix Stuffing+SCS", is compared with the following state-of-the-art frameworks:

- CVX+SeDuMi/SDPT3/MOSEK   This category adopts second-order methods. The modeling framework CVX will first automatically transform the original problem instance (e.g., the problem $\mathscr{P}$ written in disciplined convex programming form) into the standard cone-programming form and then call an interior-point solver, e.g., SeDuMi [17], SDPT3 [18], or MOSEK [19].
- CVX+SCS   In this framework based on first-order methods, CVX first transforms the original problem instance into the standard form and then calls the operator splitting solver SCS.

We define the *modeling time* as the transformation time for the first stage, the *solving time* as the time spent on the second stage, and the *total time* as the time for the two stages to solve one problem instance. As the large-scale convex optimization algorithm

---

[1] https://github.com/ShiYuanming/large-scale-convex-optimization
[2] https://github.com/cvxgrp/scs

should scale well to both the modeling part and the solving part simultaneously, the time comparison of each individual stage will demonstrate the effectiveness of the proposed two-stage approach.

Given the network size, we first generate and store the problem structure of the standard conic optimization problem family $\mathscr{P}_{\text{cone}}$, i.e., the structure of $\mathbf{A}$, $\mathbf{b}$, $\mathbf{c}$, and the descriptions of $\mathcal{K}$. As this procedure can be done offline for all the candidate network sizes, we thus ignore this step for time comparison. The following procedures to solve the large-scale convex optimization problem instances $\mathscr{P}(\boldsymbol{\alpha})$ are repeated with different parameters $\boldsymbol{\alpha}$ and sizes using the proposed framework Matrix Stuffing+SCS:

1. Copy the parameters in the problem instance $\mathscr{P}(\boldsymbol{\alpha})$ to the data in the pre-stored structure of the standard cone program $\mathscr{P}_{\text{cone}}$.
2. Solve the resultant standard conic optimization problem instance $\mathscr{P}_{\text{cone}}(\boldsymbol{\beta})$ using the solver SCS.
3. Extract the optimal solutions of $\mathscr{P}(\boldsymbol{\alpha})$ from the solutions to $\mathscr{P}_{\text{cone}}(\boldsymbol{\beta})$ produced by the solver SCS.

Finally, note that all the interior-point solvers are multiple threaded (i.e., they can utilize multiple threads to gain extra speedups), while the operator splitting algorithm solver SCS is single threaded. Nevertheless, we will show that SCS performs much faster than the interior-point solvers. We also emphasize that the operator splitting method is aimed to scale well to large problem sizes and thus to provide solutions to modest accuracy within a reasonable time, while the interior-point method's intended to provide highly accurate solutions. Furthermore, the modeling framework CVX provides rapid prototyping and a user-friendly tool for automatic transformations for general problems, while the matrix-stuffing technique targets large-scale problems for the specific problem family $\mathscr{P}$. Therefore, these frameworks and solvers are not really comparable in view of their different purposes and application capabilities. We mainly use them to verify the effectiveness and reliability of our proposed framework in terms of solution time and solution quality.

7.5.1   Effectiveness and Reliability of the Large-Scale Optimization Framework

Consider a network with L two-antenna RRHs, K single-antenna MUs, and $L = K$, where all the RRHs and MUs are uniformly and independently distributed in the square region $[-3000, 3000] \times [-3000, 3000]$ meters. We consider the total transmit-power minimization problem $\mathscr{P}_{\text{SOCP}}$ with the objective function as $\|\mathbf{v}\|_2^2$ and the QoS requirements for each MU as $\gamma_k = 5$ dB, $\forall k$. Table 7.1 demonstrates for comparison the running time and solutions using different convex optimization frameworks. Each point of the simulation results is averaged over 100 randomly generated network realizations (i.e., one small-scale fading realization for each large-scale fading realization).

For the modeling time comparisons, this table shows that the time value of the proposed matrix-stuffing technique ranges between 0.01 and 30 seconds for different network sizes and can bring a speedup of about 15 to 60 times compared with the parser/solver modeling framework CVX. In particular, for large-scale problems, the

Table 7.1 Time and solution results for different convex optimization frameworks

Network Size ($L = K$)		20	50	100	200
CVX+SeDuMi	**Total time** (s)	**8.1164**	N/A	N/A	N/A
	Objective (W)	12.2488	N/A	N/A	N/A
CVX+SDPT3	**Total time** (s)	**5.0398**	**330.6814**	N/A	N/A
	Objective (W)	12.2488	6.5216	N/A	N/A
CVX+MOSEK	**Total time** (s)	**1.2072**	**51.6351**	N/A	N/A
	Objective (W)	12.2488	6.5216	N/A	N/A
CVX+SCS	**Total time** (s)	**0.8501**	**5.6432**	**51.0472**	**725.6173**
	Modeling time (s)	0.7563	4.4301	38.6921	534.7723
	Objective [W]	12.2505	6.5215	3.1303	1.5404
Matrix Stuffing+SCS	**Total time** (s)	**0.1137**	**2.7222**	**26.2242**	**328.2037**
	Modeling time (s)	0.0128	0.2401	2.4154	29.5813
	Objective (W)	12.2523	6.5193	3.1296	1.5403

transformation using CVX is time consuming and becomes the bottleneck, as the modeling time is comparable with and even larger than the solving time. For example, when $L = 150$, the modeling time using CVX is about 3 minutes, while matrix stuffing only requires about 10 seconds. Therefore, matrix stuffing for fast transformation is essential for solving large-scale convex optimization problems quickly.

For the solving time (which can be easily calculated by subtracting the modeling time from the total time) using different solvers, this table shows that the operator splitting solver can provide a speedup of several orders of magnitude over the interior-point solvers. For example, for $L = 50$, the speedup is about 20 and 130 times over MOSEK and SDPT3, respectively, while SeDuMi is inapplicable for this problem size as the running time exceeds the predefined maximum value, i.e., one hour. In particular, all the interior-point solvers fail to solve large-scale problems (i.e., $L = 100, 150, 200$), which is indicated as N/A, while the operator splitting solver SCS can scale well to large problem sizes. Regarding the largest problems, with $L = 200$, the operator splitting solver can solve them in about 5 minutes.

Regarding the quality of the solutions, Table 7.1 shows that the proposed framework can provide a solution to modest accuracy within much less time. For the two problem sizes, i.e., $L = 20$ and $L = 50$, which can be solved by the interior-point frameworks, the optimal values attained by the proposed framework are within 0.03% of that obtained via the second-order-method frameworks.

In summary, the proposed two-stage large-scale convex optimization framework scales well to simultaneous large-scale problem modeling and solving. Therefore it could provide an effective way to evaluate the system performance via large-scale optimization in dense wireless networks. However, its implementation and performance in practical systems still needs further investigation. In particular, this set of results indicates that the scale of cooperation in dense wireless networks may be fundamentally constrained by the computation complexity on time.

7.5.2 Max–Min Fairness Rate Optimization

We will simulate the minimum network-wide achievable rate maximization problem using the max–min fairness optimization algorithm in [24, Algorithm 1] via the bisection method, which requires solving a sequence of convex feasibility problems. We will not only show the quality of the solutions and speedups provided by the proposed framework but also demonstrate that the optimal coordinated beamformers significantly outperform low-complexity and heuristic transmission strategies, i.e., zero-forcing beamforming (ZFBF) [30, 60], regularized zero-forcing beamforming (RZF) [61], and maximum ratio transmission (MRT) [62].

Consider a network with $L = 55$ single-antenna RRHs and $K = 50$ single-antenna MUs uniformly and independently distributed in the square region $[-5000, 5000] \times [-5000, 5000]$ meters. Figure 7.2 demonstrates the minimum network-wide achievable rate (which is defined as the transmit power at all the RRHs divided by the receive noise power at all the MUs) versus SNR using different algorithms. Each point of the simulation results is averaged over 50 randomly generated network realizations. For optimal beamforming, this figure shows the accuracy of the solutions obtained by the proposed framework compared with the first-order method framework CVX+SCS. The average solving time and modeling time for obtaining a single point for the optimal beamforming with CVX+SCS and Matrix Stuffing+SCS are (176.3410, 55.1542) seconds and (82.0180, 1.2012) seconds, respectively. This shows that the proposed framework can reduce both the solving time and modeling time via warm-starting and matrix stuffing, respectively.

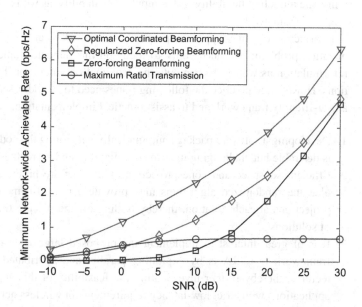

Figure 7.2 The minimum network-wide achievable rate versus transmit SNR with 55 single-antenna RRHs and 50 single-antenna MUs.

Furthermore, this figure also shows that optimal beamforming can achieve quite an improvement for the per-user rate compared with the suboptimal transmission strategies RZF, ZFBF, and MRT; this clearly shows the importance of developing optimal beamforming algorithms for such networks. The average solving time and modeling time for a single point using CVX+SDPT3 for the RZF, ZFBF, and MRT are (2.6210, 30.2053) seconds, (2.4592, 30.2098) seconds, and (2.5966, 30.2161) seconds, respectively. Note that the solving time is very small, because we only need to solve a sequence of linear programming problems for power control when the directions of the beamformers are fixed during the bisection search procedure. The main time-consuming part is the transformation using CVX.

7.6   Summary and Discussion

In this chapter we have presented a unified two-stage framework for large-scale optimization in dense C-RANs. We showed that various performance optimization problems in C-RANs can be essentially solved by solving one, or a sequence of, convex optimization or convex feasibility problems. The proposed framework requires only the convexity of the underlying problems without any other structural assumptions, e.g., smooth or separable functions. This is achieved by first transforming the original convex problem to a standard form via matrix stuffing and then using the ADMM algorithm to solve the homogeneous self-dual embedding of the primal–dual pair of the transformed standard cone program. Simulation results demonstrate the infeasibility detection capability, the modeling flexibility and computing scalability, as well as the reliability of the proposed framework.

In principle one may apply the proposed framework to any large-scale convex optimization problem; one needs only to focus on the standard conic optimization form reformulation as well as to compute the proximal operators for different cone projections. However, in practice the following issues need to be addressed in order to provide a user-friendly framework and to assist practical implementation.

1. Developing a software package automatically generating the code for matrix stuffing is desirable but challenging in terms of reliability and correctness verification.
2. Efficient subspace and cone projection algorithms are highly desirable. In particular, the randomized algorithms may provide a powerful method to scale up the projections at each iteration, thereby trading off the solving time and the accuracy of solutions.
3. It is of great interest to implement the proposed large-scale convex optimization framework in C-RANs by exploiting parallel and distributed computation architectures, thereby further investigating the feasibility of this approach for real-time applications with strict low-latency requirements in wireless networks.
4. It is also of interest apply the proposed framework to various non-convex optimization problems, e.g., optimization on manifolds [63, 64].

References

[1] China Mobile, "C-RAN: the road towards green RAN," White Paper, ver. 3.0, December 2013.

[2] Y. Shi, J. Zhang, and K. B. Letaief, "Group sparse beamforming for green Cloud-RAN," *IEEE Trans. Wireless Commun.*, vol. 13, no. 5, pp. 2809–2823, May 2014.

[3] F. Bonomi, R. Milito, P. Natarajan, and J. Zhu, "Fog computing: a platform for internet of things and analytics," in *Big Data and Internet of Things: A Roadmap for Smart Environments*, Springer International Publishing, 2014, pp. 169–186.

[4] M. Peng, Y. Li, J. Jiang, J. Li, and C. Wang, "Heterogeneous cloud radio access networks: a new perspective for enhancing spectral and energy efficiencies," *IEEE Wireless Commun. Mag.*, vol. 21, no. 6, pp. 126–135, December 2014.

[5] Y. Shi, J. Zhang, K. Letaief, B. Bai, and W. Chen, "Large-scale convex optimization for ultra-dense cloud-ran," *IEEE Wireless Commun. Mag.*, vol. 22, no. 3, pp. 84–91, June 2015.

[6] A. B. Gershman, N. D. Sidiropoulos, S. Shahbazpanahi, M. Bengtsson, and B. Ottersten, "Convex optimization-based beamforming: from receive to transmit and network designs," *IEEE Signal Process. Mag*, vol. 27, no. 3, pp. 62–75, May 2010.

[7] Z.-Q. Luo, W.-K. Ma, A.-C. So, Y. Ye, and S. Zhang, "Semidefinite relaxation of quadratic optimization problems," *IEEE Signal Process. Mag.*, vol. 27, no. 3, pp. 20–34, May 2010.

[8] E. Björnson and E. Jorswieck, "Optimal resource allocation in coordinated multi-cell systems," *Found. Trends Commun. Inf. Theory*, vol. 9, nos. 2–3, pp. 113–381, January 2013.

[9] D. P. Palomar and Y. C. Eldar, *Convex Optimization in Signal Processing and Communications*. Cambridge University Press, 2010.

[10] H. Dahrouj and W. Yu, "Coordinated beamforming for the multicell multi-antenna wireless system," *IEEE Trans. Wireless Commun.*, vol. 9, no. 5, pp. 1748–1759, September 2010.

[11] S. P. Boyd and L. Vandenberghe, *Convex Optimization*. Cambridge University Press, 2004.

[12] R. Zakhour and S. V. Hanly, "Base station cooperation on the downlink: large system analysis," *IEEE Trans. Inf. Theory*, vol. 58, no. 4, pp. 2079–2106, April 2012.

[13] C. Shen, T.-H. Chang, K.-Y. Wang, Z. Qiu, and C.-Y. Chi, "Distributed robust multicell coordinated beamforming with imperfect CSI: an ADMM approach," *IEEE Trans. Signal Process.*, vol. 60, no. 6, pp. 2988–3003, June 2012.

[14] Y. Shi, J. Zhang, and K. Letaief, "Robust group sparse beamforming for multicast green cloud-ran with imperfect csi," *IEEE Trans. Signal Process.*, vol. 63, no. 17, pp. 4647–4659, September 2015.

[15] J. Cheng, Y. Shi, B. Bai, W. Chen, J. Zhang, and K. Letaief, "Group sparse beamforming for multicast green Cloud-RAN via parallel semidefinite programming," *IEEE Int. Conf. Communications*, London, 2015.

[16] Y. Shi, J. Zhang, and K. Letaief, "Optimal stochastic coordinated beamforming for wireless cooperative networks with CSI uncertainty," *IEEE Trans. Signal Process.*, vol. 63, no. 4, pp. 960–973, February 2015.

[17] J. F. Sturm, "Using SeDuMi 1.02, a MATLAB toolbox for optimization over symmetric cones," *Optim. Methods Softw.*, vol. 11, no. 1–4, pp. 625–653, 1999.

[18] K.-C. Toh, M. J. Todd, and R. H. Tütüncü, "SDPT3 – a MATLAB software package for semidefinite programming, version 1.3," *Optim. Methods Softw.*, vol. 11, nos. 1–4, pp. 545–581, 1999.

[19] E. D. Andersen and K. D. Andersen, "The mosek interior point optimizer for linear programming: an implementation of the homogeneous algorithm," in *High Performance Optimization*, Springer, 2000, pp. 197–232.

[20] Y. Nesterov, A. Nemirovskii, and Y. Ye, *Interior-Point Polynomial Algorithms in Convex Programming*. SIAM, 1994, vol. 13.

[21] CVX Research, Inc., "CVX: Matlab software for disciplined convex programming, version 2.0 (beta)," 2013. Online. Available at http://cvxr.com/cvx/

[22] J. Lofberg, "YALMIP: a toolbox for modeling and optimization in MATLAB," in *Proc. IEEE Int. Symp. Computer-Aided Control Systems Design*, Taipei, September 2004, pp. 284–289.

[23] E. Chu, N. Parikh, A. Domahidi, and S. Boyd, "Code generation for embedded second-order cone programming," in *Proc. 2013 European Control Conf.*, July 2013, pp. 1547–1552.

[24] Y. Shi, J. Zhang, and K. Letaief, "Scalable coordinated beamforming for dense wireless cooperative networks," in *Proc. IEEE Global Communications Conf.*, Austin, TX, December 2014, pp. 3603–3608.

[25] E. Bjornson, M. Bengtsson, and B. Ottersten, "Optimal multiuser transmit beamforming: a difficult problem with a simple solution structure [lecture notes]," *IEEE Signal Process. Mag.*, vol. 31, no. 4, pp. 142–148, July 2014.

[26] W.-C. Liao, M. Hong, Y.-F. Liu, and Z.-Q. Luo, "Base station activation and linear transceiver design for optimal resource management in heterogeneous networks," *IEEE Trans. Signal Process.*, vol. 62, no. 15, pp. 3939–3952, August 2014. Online. Available at http://arxiv.org/abs/1309.4138

[27] S. Boyd, N. Parikh, E. Chu, B. Peleato, and J. Eckstein, "Distributed optimization and statistical learning via the alternating direction method of multipliers," *Found. Trends Mach. Learn.*, vol. 3, no. 1, pp. 1–122, July 2011.

[28] S. K. Joshi, M. Codreanu, and M. Latva-aho, "Distributed resource allocation for MISO downlink systems via the alternating direction method of multipliers," *EURASIP J. Wireless Commun. Netw.*, vol. 2014, no. 1, pp. 1–19, January 2014.

[29] J. Zhang, R. Chen, J. G. Andrews, A. Ghosh, and R. W. Heath, "Networked MIMO with clustered linear precoding," *IEEE Trans. Wireless Commun.*, vol. 8, no. 4, pp. 1910–1921, April 2009.

[30] T. Yoo and A. Goldsmith, "On the optimality of multiantenna broadcast scheduling using zero-forcing beamforming," *IEEE J. Sel. Areas Commun.*, vol. 24, no. 3, pp. 528–541, March 2006.

[31] Y. Shi, J. Zhang, B. O'Donoghue, and K. Letaief, "Large-scale convex optimization for dense wireless cooperative networks," *IEEE Trans. Signal Process.*, vol. 63, no. 18, pp. 4729–4743, September 2015.

[32] E. Smith, "On the optimal design of continuous processes," Ph.D. thesis, Imperial College London (University of London), 1996.

[33] M. C. Grant and S. P. Boyd, "Graph implementations for nonsmooth convex programs," in *Recent Advances in Learning and Control*, Springer, 2008, pp. 95–110.

[34] Y. Ye, M. J. Todd, and S. Mizuno, "An $\mathcal{O}(\sqrt{n}L)$-iteration homogeneous and self-dual linear programming algorithm," *Math. Oper. Res.*, vol. 19, no. 1, pp. 53–67, 1994.

[35] A. Wiesel, Y. Eldar, and S. Shamai, "Linear precoding via conic optimization for fixed MIMO receivers," *IEEE Trans. Signal Process.*, vol. 54, no. 1, pp. 161–176, January 2006.

[36] Y. Shi, J. Zhang, and K. Letaief, "Group sparse beamforming for green cloud radio access networks," in *Proc. IEEE Global Communications Conf.*, Atlanta, GA, December 2013, pp. 4635–4640.

[37] N. Jindal and A. Lozano, "A unified treatment of optimum pilot overhead in multipath fading channels," *IEEE Trans. Commun.*, vol. 58, no. 10, pp. 2939–2948, October 2010.

[38] D. J. Love, R. W. Heath, V. K. Lau, D. Gesbert, B. D. Rao, and M. Andrews, "An overview of limited feedback in wireless communication systems," *IEEE J. Sel. Areas Commun.*, vol. 26, no. 8, pp. 1341–1365, October 2008.

[39] M. A. Maddah-Ali and D. Tse, "Completely stale transmitter channel state information is still very useful," *IEEE Trans. Inf. Theory*, vol. 58, no. 7, pp. 4418–4431, July 2012.

[40] J. Zhang, R. W. Heath, M. Kountouris, and J. G. Andrews, "Mode switching for the multi-antenna broadcast channel based on delay and channel quantization," *Proc. EURASIP Conf., J. Adv. Signal Process, Special Issue on Multiuser Lim. Feedback*, vol. 2009, Article ID 802548, 15 pp., 2009.

[41] E. Björnson, G. Zheng, M. Bengtsson, and B. Ottersten, "Robust monotonic optimization framework for multicell MISO systems," *IEEE Trans. Signal Process.*, vol. 60, no. 5, pp. 2508–2523, May 2012.

[42] L. J. Hong, Y. Yang, and L. Zhang, "Sequential convex approximations to joint chance constrained programs: a Monte Carlo approach," *Oper. Res.*, vol. 59, no. 3, pp. 617–630, May–June 2011.

[43] T. A. Davis, *Direct Methods for Sparse Linear Systems*. Society for Industrial and Applied Mathematics, 2006. Online. Available at http://epubs.siam.org/doi/abs/10.1137/1.9780898718881.

[44] E. Chu, B. O'Donoghue, N. Parikh, and S. Boyd, "A primal–dual operator splitting method for conic optimization," 2013. Online. Available at www.stanford.edu/ boyd/papers/pdos.html.

[45] B. O'Donoghue, E. Chu, N. Parikh, and S. Boyd, "Conic optimization via operator splitting and homogeneous self-dual embedding," 2013. Online. Available at arxiv.org/abs/1312.3039.

[46] T. Goldstein, B. O'Donoghue, S. Setzer, and R. Baraniuk, "Fast alternating direction optimization methods," *SIAM J. Imaging Sci.*, vol. 7, no. 3, pp. 1588–1623, 2014. Online. Available at dx.doi.org/10.1137/120896219.

[47] R. Vanderbei, "Symmetric quasidefinite matrices," *SIAM J. Optim.*, vol. 5, no. 1, pp. 100–113, 1995. Online. Available at dx.doi.org/10.1137/0805005.

[48] G. H. Golub and C. F. Van Loan, *Matrix Computations*. Johns Hopkins University Press, 2012, vol. 3.

[49] J. Eckstein and D. P. Bertsekas, "On the douglas–rachford splitting method and the proximal point algorithm for maximal monotone operators," *Math. Progr.*, vol. 55, no. 1-3, pp. 293–318, 1992.

[50] N. Parikh and S. Boyd, "Proximal algorithms," *Found. Trends Optim.*, vol. 1, no. 3, January 2014. Online. Available at www.stanford.edu/ boyd/papers.

[51] N. Halko, P.-G. Martinsson, and J. A. Tropp, "Finding structure with randomness: probabilistic algorithms for constructing approximate matrix decompositions," *SIAM Review*, vol. 53, no. 2, pp. 217–288, 2011.

[52] M. W. Mahoney, "Randomized algorithms for matrices and data," *Found. Trends Mach. Learn.*, vol. 3, no. 2, pp. 123–224, 2011. Online. Available at dx.doi.org/10.1561/2200000035.

[53] D. P. Woodruff, "Sketching as a tool for numerical linear algebra," *Found. Trends Theoret. Computer Sci.*, vol. 10, no. 1-2, pp. 1–157, 2014. Online. Available at dx.doi.org/10.1561/0400000060.

[54] J. Poulson, B. Marker, R. A. van de Geijn, J. R. Hammond, and N. A. Romero, "Elemental: a new framework for distributed memory dense matrix computations," *ACM Trans. Math. Softw.*, vol. 39, no. 2, p. 13, February 2013.

[55] G. Fettweis, "The tactile internet: applications and challenges," *IEEE Veh. Technol. Mag.*, vol. 9, no. 1, pp. 64–70, March 2014.

[56] J. Mattingley and S. Boyd, "CVXGEN: a code generator for embedded convex optimization," *Optim. Engineering*, vol. 13, no. 1, pp. 1–27, 2012.

[57] J. Mattingley and S. Boyd, "Real-time convex optimization in signal processing," *IEEE Signal Process. Mag*, vol. 27, no. 3, pp. 50–61, May 2010.

[58] A. Domahidi, E. Chu, and S. Boyd, "ECOS: an SOCP solver for embedded systems," in *Proc. 2013 European Control Conf.*, IEEE, 2013, pp. 3071–3076.

[59] S. Diamond, E. Chu, and S. Boyd, "CVXPY: A Python-embedded modeling language for convex optimization, version 0.2," May 2014.

[60] J. Zhang and J. G. Andrews, "Adaptive spatial intercell interference cancellation in multicell wireless networks," *IEEE J. Sel. Areas Commun.*, vol. 28, no. 9, pp. 1455–1468, December 2010.

[61] C. B. Peel, B. M. Hochwald, and A. L. Swindlehurst, "A vector-perturbation technique for near-capacity multiantenna multiuser communication – Part I: channel inversion and regularization," *IEEE Trans. Commun.*, vol. 53, no. 1, pp. 195–202, January 2005.

[62] F. Rusek, D. Persson, B. K. Lau, E. Larsson, T. Marzetta, O. Edfors *et al.*, "Scaling up MIMO: opportunities and challenges with very large arrays," *IEEE Signal Process. Mag.*, vol. 30, no. 1, pp. 40–60, January 2013.

[63] P.-A. Absil, R. Mahony, and R. Sepulchre, *Optimization Algorithms on Matrix Manifolds*. Princeton University Press, 2009.

[64] Y. Shi, J. Zhang, and K. B. Letaief, "Low-rank matrix completion via Riemannian pursuit for topological interference management," in *Proc. IEEE Int. Symp. on Information Theory*, Hong Kong, 2015.

8 Fronthaul Compression in C-RANs

Osvaldo Simeone, Seok-Hwan Park, Onur Sahin, and Shlomo Shamai Shitz

8.1 Introduction

The C-RAN architecture relies on fronthaul links to connect each remote radio head (RRH) to the managing baseband unit (BBU). In particular, for the uplink, the fronthaul links allow the RRHs to convey their respective received signals, either in analog format or in the form of digitized baseband samples, to the BBU. For the downlink, the BBU transfers the radio signal that each RRH is to transmit on the radio interface, in analog or digital format, on the fronthaul links to the RRHs. It is this transfer of radio or baseband signals that makes possible the virtualization of the baseband and higher-layer functions of the (RRHs) at the BBU, which defines the C-RAN architecture. The analog transport solution is typically implemented by means of radio-over-fiber (see, e.g., [1]) but solutions based on copper LAN cables are also available [2]. In contrast, the digital transmission of baseband, or IQ, samples is currently carried out by following the common public radio interface (CPRI) specification [3]. This ideally requires fiber optic fronthaul links, although practical constraints motivate the development of wireless-based digital fronthauling [4]. The digital approach seems to have attracted the most interest owing to the traditional advantages of digital solutions, their including resilience to noise and to hardware impairments as well as flexibility in the transport options (see, e.g., [5]). Furthermore, the connection between an RRH and the BBU may be direct, i.e., single-hop, or it may take place over a cascade of fronthaul links, i.e., be multi-hop, as illustrated in Fig. 8.1.

In this chapter we provide an overview of the state of the art on the problem of transporting digitized IQ baseband signals on the fronthaul links. As mentioned, the current de facto standard that defines analog-to-digital processing and transport options is provided by the common public radio interface (CPRI) specification [3]. This specification is widely understood to be unsuitable for the large-scale implementation of C-RAN owing to its significant fronthaul bit rate requirements under common operating conditions. As an example, as reported in [5], the bit rate needed for an LTE base station that serves three cell sectors with carrier aggregation over five carriers and two receive antennas exceeds even the 10 Gbits/s provided by standard fiber optic links. The large bit rate is a consequence of the simple scalar quantization approach taken by CPRI, whereby each IQ sample is quantized using a given number – typically around 15 – of bits per I and Q sample. The rate requirements are even more problematic for network deployments in which fiber optic links are not available – a common occurrence, owing

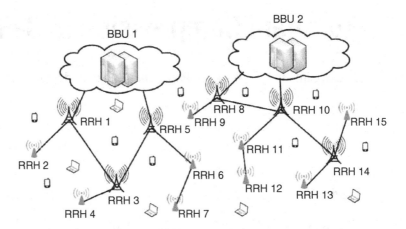

Figure 8.1 A heterogeneous cellular network based on C-RAN with two clusters of RRHs managed by two BBUs and a multi-hop fronthaul network (fronthaul links are shown as solid lines).

to the cost of installing or leasing fiber optic connections. Typical examples are heterogeneous dense networks with RRHs having small coverage, such as pico-base stations or home-base stations, for which wireless fronthauling is under study over mm-wave channels [4].

Motivated by the above mentioned shortcomings of the CPRI specification in the presence of practical fronthaul capacity limitations, this chapter provides a review of current and advanced solutions for the compression of baseband signals to be transmitted over digital fronthaul links. We observe that fronthaul links also impose constraints on the latency entailed by the transfer of information between BBU and RRHs, which have important consequences for the performance of protocols such as HARQ and random access; we refer to [5, 6] for discussions and references.

The content and organization of the chapter is as follows.

- **Point-to-Point Fronthaul Processing:** The state of the art on the fronthaul quantization and compression of baseband signals is reviewed in Section 8.2. In particular, in this section we discuss the CPRI specification and various improvements thereof that apply solutions such as filtering, scaling, or lossless compression to each fronthaul link.
- **Network-Aware Fronthaul Processing:** The point-to-point, or per-fronthaul link, quantization and compression solutions reviewed in Section 8.2 are generally oblivious to the network topology and state, e.g., the density and channel conditions. As a result, they are generally suboptimal from a network information-theoretic standpoint. On the basis of this observation, we then overview advanced fronthaul processing methods that follow network information-theoretic principles and leverage the joint processing capabilities of the BBU along with information about network topology and state. We refer to this class of techniques as being *network-aware*. Specifically, in Section 8.3, *distributed quantization and compression* is discussed for the uplink

of C-RAN systems; in Section 8.4, *multivariate quantization and compression* is presented for the downlink; and Section 8.5 elaborates on the use of *in-network processing* for multi-hop network topologies. In each section the information-theoretic principles underlying each solution are explained by the use of intuitive arguments and illustrations.

Network-aware fronthaul processing techniques operate across multiple fronthaul links and RRHs, and hence their benefits should be measured at a system level rather than merely in terms of rate reduction on each fronthaul link. Therefore, numerical results are reported throughout the chapter in order to illustrate the relative merits of different solutions in terms of network-wide criteria such as the sum-rate or edge-cell rate.

We end this introduction by emphasizing two important themes that recurr in the chapter. The first is the fact that, in a C-RAN, significant gains can be accrued by joint optimization of the operation of the system across the wireless channels and the fronthaul network. The second, broader, theme is the important role that network information theory can play in guiding the design of practical solutions for complex systems such as C-RAN and, more generally, 5G systems and beyond.

8.2 State of the Art: Point-to-Point Fronthaul Processing

In this section we first review the basics of the CPRI specification, in Section 8.2.1. Then, having identified the limitations of the scalar quantization approach prescribed by CPRI, Section 8.2.2 presents techniques that have been proposed to reduce the fronthaul bit rate by means of more advanced quantization and compression solutions applied separately on each fronthaul link, i.e., via *point-to-point* fronthaul processing.

8.2.1 Scalar Quantization: CPRI

The CPRI specification was issued by a consortium of radio equipment manufacturers with the aim of standardizing the communication interface between BBU and RRHs[1] on the fronthaul network. This specification prescribes, on the one hand, the use of sampling and scalar quantization for the digitization of the baseband signals and, on the other, a constant bit rate serial interface for the transmission of the resulting bit rate. Note that the baseband signals are either obtained from downconversion in the uplink or produced by the BBU after baseband processing in the downlink.

The CPRI interface specifies a frame structure that is designed to carry user-plane data, namely the quantized IQ samples, along with the control and management plane, for, e.g., error detection and correction, and synchronization plane data. It supports 3GPP GSM/EDGE, 3GPP UTRA, and LTE and allows for star, chain, tree, ring, and multi-hop fronthaul topologies. CPRI signals are defined at different bit rates up to

---

[1] The terminology used in CPRI is radio equipment control (REC) and radio equipment (RE), respectively.

9.8 Gbps and are constrained by strict requirements in terms of error probability (10^{-12}), timing accuracy (0.002 ppm), and delay (5 μs excluding propagation).

The line rates are proportional to the bandwidth of the signal to be digitized, to the number of receive antennas, and to the number of bits per sample. Specifically, the bit rate can be calculated as [6]

$$R_{CPRI} = 2N_{ant}R_sN_{res}N_{ov}, \qquad (8.1)$$

where N_{ant} is the number of receive antennas at the RRH; R_s is the sampling rate, which depends on the signal bandwidth according to a specified table [3]; N_{res} is the number of bits per I or Q sample, so that $2N_{res}$ is the number of bits for each complex sample; and N_{ov} accounts for the overhead of the management, control plane, and synchronization planes. The parameter N_{res} ranges in the interval from 8 to 20 bits for LTE in both the uplink and the downlink. It is noted that using (8.1) it is easy to identify common scenarios, such as the one discussed at the beginning of this section, in which the maximum CPRI rate of 9.8 Gbs is violated, particularly in the presence of carrier aggregation and/or large-array MIMO systems [5].

As discussed above, the basic approach prescribed by CPRI, which is based on sampling and scalar quantization, is bound to produce bit rates that are difficult to accommodate within the available fronthaul capacities – most notably for small cells with wireless fronthauling and for larger cells with optical fronthaul links in the presence of carrier aggregation and large-array MIMO transceivers. This has motivated the design of strategies that reduce the bit rate of the CPRI data stream while limiting the distortion incurred on the quantized signal. In the following we provide an overview of these schemes, differentiating between techniques that adhere to the standard C-RAN implementation, characterized by the full migration of baseband processing to the BBU, and solutions that explore different functional splits between RRHs and BBU. We refer to the former class as compressed CPRI and review both classes separately in the next subsections.

8.2.2   Compressed CPRI

The full separation of baseband processing at the BBU from the radio functionalities implemented at the RRHs is made possible by the fact that CPRI performs quantization of the time-domain baseband signals. We recall that these signals are either received by the RRHs in the uplink or produced by means of baseband processing at the BBU for the downlink. The separation at hand can be maintained, while reducing the required fronthaul rate, by compressing the time-domain baseband samples rather than simply performing scalar quantization. We refer to this class of approaches as compressed CPRI. Compressed CPRI is based on a number of principles, which are briefly discussed in the following.

(1) *Filtering* [7, 8]:   As per the CPRI standard, the time-domain signal is oversampled. For instance, for a 10 MHz LTE signal a sampling frequency of 15.36 MHz is adopted.

Therefore, a low-pass filter can be applied to the signal without affecting the information content.

(2) *Per-block scaling* [7, 8]:   The dynamic range of the quantizer needs to be selected to accommodate the peak-to-peak variations of the time-domain signal. Given the generally large peak-to-average power ratio (PAPR), this calls for quantization with a large number of bits in order to maintain a small quantization noise over the entire dynamic range (e.g., typically 15 bits in CPRI). In LTE, the problem is particularly relevant for the OFDM-based downlink owing to the large PAPR of OFDM signals. This limitation can be mitigated by dividing the signal into subblocks of small size (e.g., 32 samples in [7]) and rescaling the signal in each subblock so that the peak-to-peak variations within the block fit the dynamic range of the quantizer. In this fashion, the relevant peak-to-peak amplitude is that measured not across the entire block of samples but only within each subblock. Note that this approach entails some overhead, as the receiver needs to be informed regarding the scaling factor applied to each block – there is hence a tension between the effectiveness of this solution and the required fronthaul overhead.

(3) *Optimized non-uniform quantization* [7, 8]:   Rather than adopting uniform scalar quantization, the quantization levels can be optimized as a function of the statistics of the baseband signal by means of standard strategies such as the Lloyd–Max algorithm.

(4) *Noise shaping* [9]:   Owing to the correlation of successive baseband samples, predictive, or noise-shaping, quantization techniques based on a feedback filter can be beneficial to reduce the rate of optimized quantization.

(5) *Lossless compression* [10]:[2]   Any residual correlation among successive quantized baseband samples, possibly after predictive quantization, can be further leveraged by entropy coding techniques that aim at reducing the rate down to the entropy of the digitized signal.

As a rule of thumb, compressed CPRI techniques are seen to reduce the fronthaul rate by a factor around 2–3 [6].

8.2.3   Alternative Functional Splits

In order to obtain further fronthaul rate reductions by means of point-to-point compression techniques, alternative functional splits to the conventional C-RAN implementation need to be explored [6, 11, 12]. Accordingly, some baseband functionalities are implemented at the RRH, such as frame synchronization, FFT/ IFFT, or resource demapping. The rationale is that, by keeping some baseband functionalities at the RRHs, one can potentially reduce the fronthaul overhead.

A first solution prescribes the implementation of frame synchronization and FFT in the uplink and of the IFFT in the downlink at the RRH (see demarcation point A in

---

[2] Reference [10] in fact considers time-domain modulation and not OFDM but the principle is the same as that discussed here.

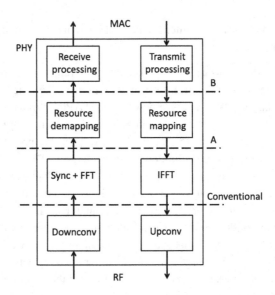

Figure 8.2 Alternative functional splits of the physical layer between BBU and RRH.

Fig. 8.2). The rest of the baseband functionalities, such as channel decoding on encoding, are instead performed at the BBU. This functional split enables the signal to be quantized in the frequency domain, i.e., after the FFT in the uplink and prior to the IFFT in the downlink. Given that the signal has a lower PAPR in the frequency domain, particularly in the LTE downlink, the number of bits per sample can be reduced at a minor cost in terms of signal-to-quantization-noise ratio. The experiments in [6] do not demonstrate, however, very significant fronthaul rate gains with this approach.

A more promising approach implements also resource demapping for the uplink and resource mapping for the downlink at the RRH (see demarcation point B in Fig. 8.2). For the uplink, this implies that the RRH can deconstruct the frame structure and distinguish among the different physical channels multiplexed in the resource blocks. As a result, the RRH can apply different quantization strategies to distinct physical channels, e.g., by quantizing more finely channels carrying higher-order modulations. More importantly, in the case of lightly loaded frames, unused resource blocks can be neglected. This approach was shown in [6, 13] to lead to compression ratios of order up to 30 – an order of magnitude larger than with compressed CPRI – in the regime of small system loads. A similar approach was also implemented in the field trials and reported in [14].

8.3   Network-Aware Fronthaul Processing: Uplink

The solutions explored so far to reduce the fronthaul capacity requirements of the C-RAN architecture have been based on point-to-point quantization and compression algorithms. Here we revisit the problem of fronthaul compression, taking a more fundamental viewpoint grounded in network information theory. Accordingly, we look at the problem at the network level rather than at the granularity of each individual fronthaul link. As the rest of this chapter illustrates, this network-aware perspective on the design of fronthaul processing has the potential to move significantly beyond the limitations of point-to-point approaches towards the network information-theoretic optimal performance. We start by analyzing the uplink with a single-hop fronthaul topology in this section, while the downlink, under the same fronthaul topology, is treated in the next section. Finally, multi-hop fronthauling is considered in Section 8.5.

8.3.1   Problem Setting

The block diagram of an uplink C-RAN system characterized by a single cluster of RRHs with a single-hop fronthaul topology is shown in Fig. 8.3. Note that the fronthaul links between the RRHs and the BBU may have significantly different capacity limitations, as indicated by the parameters C_i in the figure. For instance, some RRHs may be endowed with optical fronthaul links while others with wireless fronthauling. Depending on the available fronthaul link budget, each RRH quantizes and compresses the locally received baseband signal to convey it on the corresponding fronthaul link to the BBU. Note that we mark the fronthaul processing block at the RRHs as "Compression", although this block also includes quantization, and, furthermore, no compression may take place following quantization as in a standard CPRI implementation. An analogous discussion applies to the "Decompression" block at the BBU.

In a conventional implementation based on the point-to-point solutions reviewed in the previous section, the decompression block at the BBU involves separate decompression operations for each fronthaul link. In contrast, with a network-aware solution based on distributed source coding, joint decompression across all connected RRHs is performed. The rationale for joint decompression is that the signals received by different RRHs are correlated, as they represent noisy versions of the same transmitted signals. This correlation is expected to be particularly significant for dense networks – an important use for the C-RAN architecture. To realize the performance advantages of

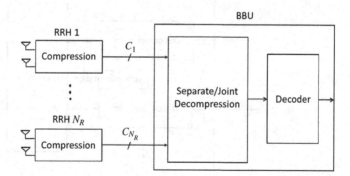

Figure 8.3 Block diagram detailing baseband processing for the uplink. With conventional point-to-point solutions, the decompression block at the BBU performs separate decompression on each fronthaul link, while, with a network-aware solution based on distributed quantization and compression, joint decompression across all connected RRHs is performed.

joint decompression, the RRHs implement the network information-theoretic technique of *distributed source coding* (see, e.g., [15] for an introduction), as discussed next.

8.3.2 Distributed Quantization and Compression (Wyner–Ziv Coding)

The impact of distributed quantization and compression, or source coding, on the performance of C-RAN systems has been investigated from an information-theoretic viewpoint in a number of papers starting from the original work [16]; these include including [17–20]. The key idea of distributed source coding can be easily explained with reference to the problem of quantization or compression with side information at the receiver's side. Specifically, given that the signals received by different RRHs are correlated, once the BBU has recovered the signal of one RRH, that signal can be used as *side information* for decompression of the signal of another RRH. As illustrated in Fig. 8.4, this process can be iterated in a decision-feedback-type loop, whereby signals that have been already decompressed can be used as side information to alleviate the fronthaul requirements for the RRHs whose signals have yet to be decompressed. As we will see, the availability of side information at the decompressor allows the required fronthaul rate to be reduced with no penalty on the accuracy of the quantized signal, or, in a dual fashion, the quantization accuracy to be enhanced at the same fronthaul rate.

As a practical remark, we note that the process at hand requires some decompression order across the RRHs to be established. In particular, as argued in [21], a choice that is generally sensible, and close to optimal, is that of decompressing first the signals coming from macro-base stations (BSs) and then those from pico- or femto-BSs in their vicinity. The rationale for this approach is that macro-BSs tend to have a larger fronthaul capacity and hence their decompressed signals provide relevant side information for the

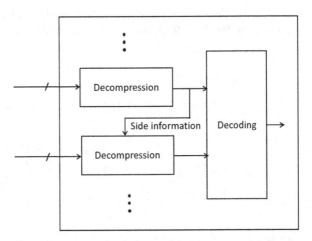

Figure 8.4 Block diagram detailing the process of network-aware decompression, where signals that have been already decompressed can be used as side information to alleviate the fronthaul requirements for another RRH.

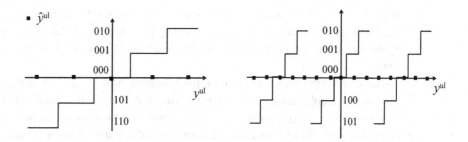

Figure 8.5 Left: Standard uplink quantization. Right: Quantization with side information (Wyner–Ziv coding).

signals coming from smaller cells, which are typically connected with lower-capacity fronthaul links.

The coding strategy to be implemented at the RRHs in order to leverage the side information at the receiver is known in information theory as *Wyner–Ziv coding* [22]. Note that Wyner–Ziv coding does not require the RRHs to be aware of the side information available at the BBU but only of the correlation between the received signal and the side information. More discussion on this point will be provided below. Rather than offering a technical description of Wyner–Ziv coding, we provide here a simple example of the most basic case of scalar quantization. Improvements based on compression are possible and follow in a manner similar to the discussion above.

Consider Fig. 8.5. On the left, a portion of a standard uniform quantizer with five levels is shown. Note that we focus for simplicity of illustration on real samples. The black boxes along the horizontal axis represent the quantization levels and the vertical axis reports the corresponding binary labels.[3] Now, assume that the receiver has some side information that is correlated with the sample to be quantized. Wyner–Ziv quantization enables the RRH to use a finer quantizer and hence to achieve an enhanced resolution without increasing the fronthaul rate or, conversely, to keep the same resolution while reducing the fronthaul rate. The first effect is illustrated on the right in Fig. 8.5, where the same number of binary labels, and hence the same fronthaul rate, is used to support a finer subdivision of the dynamic range of the received uplink signal. Note that, with Wyner–Ziv quantization, the same binary label is assigned to multiple quantization levels.

The BBU can distinguish between quantization levels that are assigned the same binary label – known collectively as a "bin" in information theory – by leveraging the side information. The quantization level that is "closer" to the side information sample is likely to be the correct one. "Closeness" generally depends on the correlation between the signal to be decompressed and the side information. For instance, a standard minimum-distance decoder may be adapted to decode within a bin under the assumption that the received signals can be approximately described as jointly Gaussian random variables – a typical assumption in the presence of channel state information

---

[3] An odd number of levels is considered here for simplicity of illustration.

at the BBU. It is emphasized that decompressing a Wyner–Ziv quantized sample hence entails the additional decoding step of selecting the correct level within the bin indicated by the quantizer. Alternatively, one can design the quantizers jointly with the demapping function, as in [23].

The description above refers to an implementation of Wyner–Ziv coding based on sample-by-sample scalar quantization. In order to reduce the probability of error for the in-bin decoding step described above and/or to complement quantization with compression, in practice block processing that leverages the mature state of the art on modern source and channel coding is desirable. A first solution is to follow standard scalar quantization, as used in CPRI, with a block that computes the syndrome of a binary linear block code on the resulting bit stream [24, 25]. In this fashion, bins are implemented as cosets of a linear binary block code. Hence, by selecting codes that admit efficient decoding, such as LDPC or turbo codes, in-bin decoding becomes feasible, particularly considering that the computational burden is on the BBU. A second alternative is that of using nested codes with the property that the finer code is a good source code or quantization codebook while the coarser code, and its cosets, are good channel codes that play the role of bins. Examples of such codes include trellis codes [26], polar codes [27], and compound LDGM-LDPC codes [28].

Another key practical issue is the need to inform each RRH about the correlation between the received signal and the side information corresponding to the signals that are decompressed by the BBU before that of the RRH at hand. This correlation is essential for the RRH to select a quantization and compression scheme that may allow the BBU to decode within a bin based on the side information with acceptable reliability. Moreover, the correlation depends on the channel state information of the involved RRHs and amounts to a covariance matrix of the size of the number of receive antennas at the RRH (see, e.g., [18]). Therefore, the BBU may convey this correlation matrix on the fronthaul link to the RRH or, more conventionally, the BBU may inform the RRH about which particular quantizer–compressor to apply among the available algorithms in a codebook of possible choices. The design of such a codebook and of rules for the selection of specific quantizers–compressors is an interesting open problem.

We finally observe that the discussion above assumes that the BBU first decompresses the quantized signals and then decodes the UEs' messages on the basis of the decompressed signals. It is known that the performance may be potentially improved by performing joint decompression and decoding at the cost of increased computational complexity [29].

8.3.3   Examples

Distributed quantization and compression, or Wyner–Ziv coding, has been demonstrated in a number of theoretical papers, including [16, 18–20], to offer significant potential performance gains for a C-RAN uplink. Here we first describe one such result and then provide some discussion about performance under a more complete scenario, of relevance for LTE systems. Throughout, achievable rates are computed by using standard information-theoretic characterizations (i.e., by evaluating appropriate mutual information terms). The relationship between the fronthaul overhead and the accuracy of the quantized and compressed baseband signals is also modeled by using information-theoretic arguments, namely from rate-distortion theory. The use of information-theoretic metrics implies that the displayed performance of point-to-point solutions corresponds to the maximum theoretically achievable rate with optimal point-to-point techniques that leverage all the ideas described in the previous section. The performance of distributed source coding also reflects that of an optimal block-based implementation that uses state-of-the-art codes and perfect information about the correlation between RRHs, as introduced above. In this regard we note that such information, in this example, reduces to a scalar since we consider each RRH to have a single antenna.

Figure 8.6 plots the achievable per-cell uplink sum-rate for point-to-point and distributed compression versus the uplink signal-to-noise ratio (SNR) in a standard three-cell circulant Wyner model (see, e.g., [30]), where each cell contains a single-antenna UE and single-antenna RRH, and inter-cell interference takes place only between adjacent cells (the first and third cells are considered to be adjacent). Note that the sum-rate is calculated here under the assumption of joint decoding of the signals of all users at the CU for both point-to-point and distributed compression. The inter-cell channel gains are set to be 5 dB smaller than the intra-cell channel gain, and every RRH has the same fronthaul capacity of 3 bits/s per hertz, where normalization is with respect to the uplink bandwidth, or, equivalently, 3 bits for each sample of the received signal if sampled at the baud rate. For reference, we also show the per-cell sum-rate achievable with single-cell processing, whereby each RRH decodes the signal of the in-cell UE by treating all other UE signals as noise, and the cut-set upper bound [30]. It can

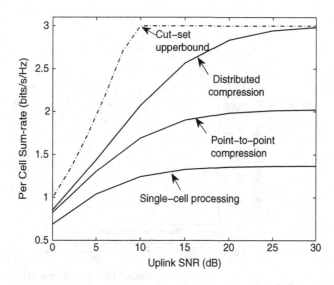

Figure 8.6 Per-cell uplink sum-rate versus the uplink SNR for a circulant Wyner model with fronthaul capacity of 3 bits/s per Hz and inter-cell channel 5 dB smaller than the intra-cell channel gain.

be seen that the performance advantage of distributed compression over point-to-point compression increases as the SNR grows larger, since the correlation of the received signals at the RRHs becomes more pronounced when the effect of noise is less relevant. We also note that with distributed compression the uplink SNR at which the quantization noise becomes the dominant factor limiting performance, hence causing a floor on the achievable sum-rate, increases significantly.

We now provide a performance evaluation using the standard cellular topology and channel models described in the LTE document [31]. We focus on the performance of the macro cell located at the center of a two-dimensional 19-cell hexagonal cellular layout. In each macro cell, there are K randomly and uniformly located single-antenna UEs and a number of RRHs as follows: a macro-base station (BS) with three sectorized antennas placed in the center and N randomly and uniformly located single-antenna pico-BSs. A single-hop fronthaul topology is assumed, where a BBU is connected directly to the macro BS and the pico-BS in the macro cell. The fronthaul links to each macro-BS antenna and to each pico-BS have capacities C_{macro} and C_{pico}, respectively. All interference signals from other macro cells are treated as independent noise signals. More details on the system parameters can be found in [32].

We adopt the conventional metric of cell-edge throughput versus the average per-UE spectral efficiency (see, e.g., [33, Fig. 5]). The achievable rates are evaluated using the rate functions in [31] that account for the smallest and largest spectral efficiencies allowed by existing modulation and coding schemes. Moreover, the rates are computed by running a proportional fairness scheduler on a sequence of T time slots with independent fading realizations, and by then evaluating the cell-edge throughput as the five percentile rate and the average spectral efficiency as the average sum-rate normalized by

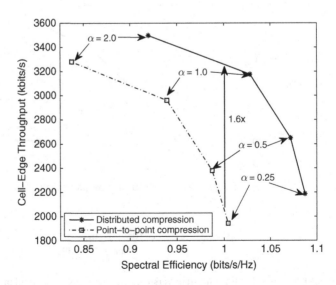

Figure 8.7 Cell-edge throughput (five percentile rate) versus the average per-UE spectral efficiency for various fairness constants α in the uplink of a C-RAN with $N = 3$ pico-BSs, $K = 5$ UEs, $(C_{\text{macro}}, C_{\text{pico}}) = (9, 3)$ bits/s/Hz, $T = 10$, $\beta = 0.5$ and a bandwidth of 10 MHz.

the number of UEs. We recall that the proportional fairness scheduler maximizes at each time slot the weighted sum $R_{\text{sum}}^{\text{fair}} = \sum_{k=1}^{K} R_k^{\text{dl}}/\bar{R}_k^{\alpha}$ of per-UE rates R_k^{dl}, where $\alpha \geq 0$ is a fairness constant and $\bar{R}_k$ is the average data rate accrued by UE k so far. After each time slot the rate $\bar{R}_k$ is updated as $\bar{R}_k \leftarrow \beta \bar{R}_k + (1 - \beta) R_k^{\text{dl}}$, where $\beta \in [0, 1]$ is a forgetting factor. Increasing α leads to a fairer rate allocation among the UEs.

Figure 8.7 plots the cell-edge throughput versus the average spectral efficiency for $N = 3$ pico-BSs, $K = 5$ UEs, $(C_{\text{macro}}, C_{\text{pico}}) = (9, 3)$ bits/s per hertz, $T = 10$, $\beta = 0.5$, and a bandwidth of 10 MHz. The curve is obtained by varying the fairness constant α in the utility function $R_{\text{sum}}^{\text{fair}}$. It is observed that spectral efficiencies larger than 1.01 bit/s per hertz are not achievable with point-to-point compression, while they can be obtained with multivariate compression. Moreover, it is seen that distributed compression provides a gain of 1.6× in terms of cell-edge throughput for a spectral efficiency of 1 bit/s per hertz.

8.4 Network-Aware Fronthaul Processing: Downlink

In this section we consider the downlink of a C-RAN with a single-hop fronthaul topology. As seen in the previous section, in the uplink the traditional solution consisting of separate fronthaul quantizers and compressors is suboptimal, from a network information-theoretic viewpoint based on the principle of distributed source coding. In this section, we will see that a different principle, namely that of multivariate quantization and compression (see, e.g., [15]), is relevant for the downlink.

The block diagram of a downlink C-RAN system with a single cluster of RRHs connected by means of a single-hop fronthaul topology to a BBU is shown in Fig. 8.8. The BBU performs baseband processing to encode the downlink data streams to be delivered to the UEs. Specifically, the BBU first carries out channel encoding on each data stream and then applies linear precoding, which is computed from the available channel state information, in order to enable multi-user MIMO transmission. The resulting baseband signals are then quantized and compressed before being transmitted on the fronthaul links to the RRHs. As for the uplink, we mark as "Compression" and "Decompression" blocks that also include quantization.

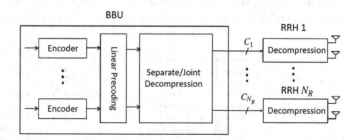

Figure 8.8 Block diagram detailing baseband processing for the downlink. With conventional point-to-point solutions, the compression block at the BBU performs separate decompression on each fronthaul link, while, with a network-aware solution based on multivariate quantization/compression, joint compression across all connected RRHs is performed.

In a conventional implementation based on the point-to-point solutions reviewed in Section 8.2, the compression block at the BBU involves separate compression for each fronthaul link. In contrast, with a network-aware solution based on multivariate quantization and compression, joint compression across all connected RRHs is performed. As discussed below, joint compression allows the BBU to shape the distribution of the quantization noise across nearby RRHs and, in so doing, to control the impact of the quantization noise on the downlink reception at the UEs.

8.4.1   Multivariate Quantization–Compression

To understand the key ideas in the simplest terms, let us start by assessing the impact of scalar quantization on the downlink performance. We focus, as above, on the quantization of real samples for simplicity of illustration. On the left-hand side of Fig. 8.9 we illustrate the quantization regions resulting from standard scalar quantization of the signals $\tilde{x}_i^{dl}$ with $i = 1, 2$ to be transmitted by two RRHs. The small black squares represent the quantization levels x_i^{dl} with $i = 1, 2$ used by the two quantizers and the small black disks denote the corresponding quantization levels on the plane. Given the current channel state information, an UE may be more sensitive to quantization noise in a certain spatial dimension. In particular, if the UE has a single antenna, the channel vector from the two RRHs to the UE defines this signal direction. However, as seen in the figure, with traditional separate quantization, no control of the shape of the quantization regions is possible. Therefore, one cannot leverage the channel state information at BBU to implement a more effective quantizer that reduces the impact of quantization noise along the signal direction.

As proposed in [34], the limitation identified above can be alleviated by multivariate compression, whereby the vector $\tilde{\mathbf{x}}^{dl} = [\tilde{x}_1^{dl}\ \tilde{x}_2^{dl}]^T$ is jointly, rather than separately, quantized at the BBU. Note that, unlike vector quantization, multivariate compression is constrained by the numbers of levels for each axis, which yield the fronthaul rates for each RRH, rather than on the total number of levels on the plane. As illustrated on the right-hand side of Fig. 8.9, multivariate compression enables the shaping of the quantization regions on the plane, hence allowing a finer control of the impact of the quantization noise on the received signals.

As mentioned above, the implementation of multivariate compression hinges on the availability of channel state information at the BBU, which is to be expected. Moreover,

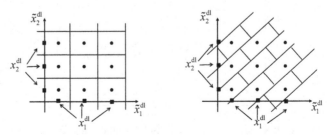

Figure 8.9 Left: standard downlink quantization for two RRHs. Right: multivariate quantization.

it requires the BBU to inform each RRH about the quantization levels, i.e., the quantization codebook, to be used. Importantly, however, the RRHs need not be informed about the specific mapping carried out by joint compression (e.g., about the shapes of the quantization regions on the plane in Fig. 8.9) as a function of the channel state information. In practice, therefore, the selection of quantization codebooks should be performed on a coarse time scale, based only on long-term channel state information, while the specific mapping carried out by joint compression should be adapted on the basis of current channel state information at the BBU.

8.4.2 Example

This section provides a performance evaluation of the fronthaul compression techniques discussed above, by considering the downlink of the LTE-based scenario adopted for Fig. 8.7. Figure 8.10 plots the cell-edge throughput versus the average spectral efficiency for $N = 1$ pico-BS, $K = 4$ UEs, $(C_{macro}, C_{pico}) = (6, 2)$ bits/s per hertz, $T = 5$, and $\beta = 0.5$. We recall that the plot is obtained by varying the fairness constant α in the utility function R_{sum}^{fair}. The rates are evaluated here under the assumption of an optimized linear precoder based on channel state information at the CU for both point-to-point and multivariate compression. As a side remark, we note that in [34] the performance is also evaluated for non-linear precoders following the "dirty paper coding" principle. As for the uplink, it is seen that spectral efficiencies larger than 3.12 bps/Hz are not achievable with point-to-point compression, while they can be obtained with multivariate compression. Furthermore multivariate compression provides about a twofold gain in terms of cell-edge throughput for a spectral efficiency of 3 bps/Hz.

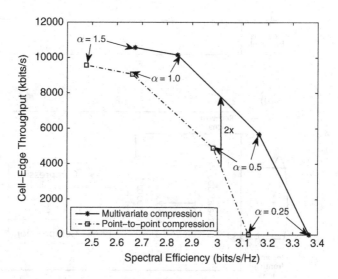

Figure 8.10 Cell-edge throughput, (five percentile rate) versus the average per-UE spectral efficiency for various fairness constants α in the downlink of a C-RAN with $N = 1$ pico-BS, $K = 4$ UEs, $(C_{macro}, C_{pico}) = (6, 2)$ bits/s per hertz, $T = 5$, and $\beta = 0.5$.

8.5   Network-Aware Fronthaul Processing: In-Network Processing

In this section we study the case in which the fronthaul network has a general multi-hop topology. As an example, in Fig. 8.1, RRH 7 communicates to the BBU via a two-hop fronthaul connection that passes through RRH 6 and RRH 5. Note that each RRH may have multiple incoming and outgoing fronthaul links. As will be discussed, the information-theoretic idea of in-network processing plays a key role in this scenario.

8.5.1   Problem Setting

In order to convey the quantized IQ samples from the RRHs to the BBU through multiple hops, each RRH forwards, on each outgoing fronthaul link, some information about the signals received on the wireless channel and the incoming fronthaul links. A first standard option based on point-to-point fronthaul processing is to use routing: the bits received on the incoming links are simply forwarded, along with the bit stream produced by the local quantizer/compressor, on the outgoing links without any additional processing, as illustrated in Fig. 8.11(a). This approach requires the optimization of

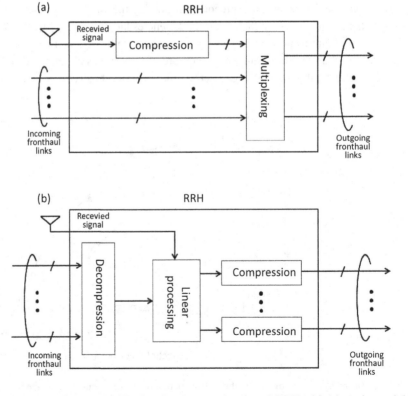

Figure 8.11 Block diagram detailing baseband processing at each RRH with (a) routing and (b) in-network processing strategies.

standard flow variables that define the allocation of fronthaul capacity to the different bit streams. The problem is formulated and addressed in [35].

Routing may be highly inefficient in the presence of a dense deployment of RRHs. In fact, under this assumption, an RRH may be close to a large number of other RRHs, all of which receive correlated baseband signals. In this case it is wasteful of the available fronthaul capacity merely to forward all the bit streams received from the connected RRHs. Instead, it is possible to combine the correlated baseband signals at the RRH prior to forwarding in order to reduce redundancy. We refer to this processing of incoming signals as *in-network processing*.

8.5.2   In-Network Fronthaul Processing

A possible implementation of in-network processing based on linear operations is shown in Fig. 8.11(b). Accordingly, in order to allow for in-network processing, each RRH first decompresses the received bit streams from the connected RRHs so as to recover the baseband signals. The decompressed baseband signals are then linearly processed, along with the IQ signal received locally by the RRH. After in-network processing, the obtained signals must be recompressed before they can be sent on the outgoing fronthaul links. Thus the effect of the quantization noise resulting from this second quantization step must be counterbalanced by the advantages of in-network processing in order to make the strategy preferable to routing. The optimal design of in-network processing is addressed in [35].

Example

We now compare the sum-rates achievable with routing and with in-network processing for the uplink of a C-RAN with a two-hop fronthaul network. Specifically, there are N RRHs in the first layer and two RRHs in the second layer. The RRHs in the first layer do not have direct fronthaul links to the BBU, while the RRHs in the second layer do. Half the RRHs in the first layer are connected to one RRH in the second layer, and half to the other RRH in the second layer. We assume that all fronthaul links have capacity equal to 2–4 bits/s per hertz and all channel matrices have identically and independently distributed (i.i.d.) complex Gaussian entries with unit power (Rayleigh fading). Figure 8.12 shows the average sum-rate versus the number N of RRHs in the first layer, with $N_M = 4$ UEs and an average received per-antenna SNR of 20 dB at all RRHs. It is first observed that the performance gain of in-network processing over routing becomes more pronounced as the number N of RRHs in the first layer increases. This suggests that, as the RRHs' deployment becomes more dense, it is desirable for each RRH in the second layer to perform in-network processing of the signals received from the first layer. Moreover, in-network processing is more advantageous when the fronthaul links have a larger capacity, as the distortion introduced by the recompression step discussed above becomes smaller.

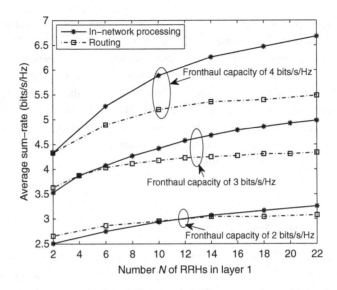

Figure 8.12 Average sum-rate versus the number of RRHs in the first layer of a two-hop topology with $N_M = 4$ UEs, average received per-antenna SNR of 20 dB, and fronthaul capacity of 2, 3, 4 bits/s per hertz.

8.6   Concluding Remarks

In this chapter we have provided an overview of the state of the art on fronthaul quantization and compression for C-RANs. We have differentiated between point-to-point, or per-fronthaul-link, quantization–compression solutions, which are generally oblivious to the network topology and state, and network-aware approaches, which instead follow network information-theoretic principles and leverage the joint processing capabilities of the BBU. It was demonstrated, via various examples, that the information-theoretic concepts of distributed quantization and compression, multivariate quantization and compression, and in-network processing provide useful frameworks on which to base the design of fronthaul processing techniques that are able to significantly outperform point-to-point solutions in terms of network-wide performance criteria. Interesting open problems concerning the implementation of network-aware fronthaul compression include the design of efficient feedback mechanisms on the fronthaul network aimed at satisfying the channel state information requirements of this class of techniques, as discussed in this chapter.

In closing we would like to mention a related technique that is also inspired by network information theory and that may play a role in the design of next-generation cellular systems based on generalizations of the C-RAN architecture, namely compute-and-forward [36]. Compute-and-forward relies on the use at the UEs of nested lattice codes, whose structure guarantees that any integer (modulo-) sum of codewords is a codeword in the same lattice codebook. Thanks to this property, in the uplink the RRHs can decode a linear function of the uplink codewords with the aim of providing the CU with enough linear equations to recover all transmitted messages. Since the codebook size of the possible functions to be decoded can be adapted to the fronthaul capacity, compute-and-forward does not require any quantization at the RRHs. Drawbacks of the method include an increased complexity of the RRHs, which need to operate as fully fledged base stations. A version of this technique also exists for the downlink, as proposed in [37]. A discussion on compute-and-forward in the context of the C-RAN architecture can be also found in [21].

Finally, we observe that many network information-theoretic problems underlying the design of C-RANs are still open (see, e.g., [38]) and hence advances in this domain may lead to progress in the C-RAN technology.

8.7 Acknowledgments

The work of O. Simeone was partly supported by US NSF under grant CCF-1525629. The work of S.-H. Park was supported by the National Research of Korea, Funded by the Korea Government (MSIP) under grant NF-2015R1C1A1A01051825. The work of S. Shamai was supported by the European Union's Horizon, 2020 Research and Innovation programme, grant agreement no. 694630.

References

[1] H. Al-Raweshidy and S. Komaki, *Radio over Fiber Technologies for Mobile Communications Networks*, Artech House, 2002.

[2] C. Lu, H. Almeida, E. Trojer, K. Laraqui, M. Berg, O. V. Tidblad *et al.*, "Connecting the dots: small cells shape up for high-performance indoor radio," *Ericsson Review*, December 2014.

[3] Ericsson AB, Huawei Technologies, NEC Corporation, Alcatel Lucent, and Nokia Siemens Networks, "Common public radio interface (CPRI); interface specification," CPRI specification v5.0, September 2011.

[4] Fujitsu, "The benefits of cloud-RAN architecture in mobile network expansion," White Paper, 2015.

[5] A. Checko, H. L. Christiansen, Y. Yan, L. Scolari, G. Kardaras, M. S. Berger *et al.*, "Cloud RAN for mobile networks – a technology overview," *IEEE Commun. Surveys Tutorials*, vol. 17, no. 1, pp. 405–426, first quarter 2015.

[6] U. Dotsch, M. Doll, H. P. Mayer, F. Schaich, J. Segel, and P. Sehier, "Quantitative analysis of split base station processing and determination of advantageous architectures for LTE," *Bell Labs. Technical J.*, vol. 18, no. 1, pp. 105–128, September 2013.

[7] D. Samardzija, J. Pastalan, M. MacDonald, S. Walker, and R. Valenzuela, "Compressed transport of baseband signals in radio access networks," *IEEE Trans. Wireless Commun.*, vol. 11, no. 9, pp. 3216–3225, 2012.

[8] B. Guo, W. Cao, A. Tao, and D. Samardzija, "LTE/LTE-A signal compression on the cpri interface," *Bell Labs. Technical J.*, vol. 18, no. 2, pp. 117–133, 2013.

[9] K. F. Nieman and B. L. Evans, "Time-domain compression of complex-baseband lte signals for cloud radio access networks," in *Proc. IEEE Glob. Conf. on Signaling and Information Processing*, December 2013, pp. 1198–1201,

[10] M. W. A. Vosoughi and J. R. Cavallaro, "Baseband signal compression in wireless base stations," in *Proc. IEEE Glob. Comm. Conf.*, December 2012.

[11] D. Wubben, P. Rost, J. Bartelt, M. Lalam, V. Savin, M. Gorgoglione *et al.*, "Benefits and impact of cloud computing on 5G signal processing: flexible centralization through cloud-RAN," *IEEE Signal Process Mag.*, vol. 31, no. 6, pp. 35–44, November 2014.

[12] J. Liu, S. Xu, S. Zhou, and Z. Niu, "Redesigning fronthaul for next-generation networks: beyond baseband samples and point-to-point links," ArXiv e-prints, August 2015.

[13] J. Lorca and L. Cucala, "Lossless compression technique for the fronthaul of lte/lte-advanced cloud-RAN architectures," in *Proc. IEEE Int. Symp. World of Wireless, Mobile and Multimedia Networks*, June 2013.

[14] S. B. S. Grieger and G. Fettweis, "Large scale field trial results on frequency domain compression for uplink joint detection," in *Proc. IEEE Glob. Communication Conf.*, December 2012.

[15] A. E. Gamal and Y.-H. Kim, *Network Information Theory*. Cambridge University Press, 2011.

[16] A. Sanderovich, O. Somekh, H. V. Poor, and S. Shamai, "Uplink macro diversity of limited backhaul cellular network," *IEEE Trans. Inf. Theory*, vol. 55, no. 8, pp. 3457–3478, August 2009.

[17] A. Sanderovich, S. Shamai, and Y. Steinberg, "Distributed MIMO receiver – achievable rates and upper bounds," *IEEE Trans. Inf. Theory*, vol. 55, no. 10, pp. 4419–4438, October 2009.

[18] A. del Coso and S. Simoens, "Distributed compression for MIMO coordinated networks with a backhaul constraint," *IEEE Trans. Wireless Commun.*, vol. 8, no. 9, pp. 4698–4709, September 2009.

[19] S.-H. Park, O. Simeone, O. Sahin, and S. Shamai, "Robust and efficient distributed compression for cloud radio access networks," *IEEE Trans. on Veh. Technol.*, vol. 62, no. 2, pp. 692–703, February 2013.

[20] L. Zhou and W. Yu, "Uplink multicell processing with limited backhaul via per-base-station successive interference cancellation," *IEEE J. Sel. Areas Commun.*, vol. 31, no. 10, pp. 2246–2254, October 2013.

[21] S.-H. Park, O. Simeone, O. Sahin, and S. Shamai, "Fronthaul compression for cloud radio access networks," *IEEE Signal Proc. Mag., Special Issue on Signal Processing for the 5G Revolution*, vol. 31, no. 6, pp. 69–79, November 2014.

[22] A. D. Wyner and J. Ziv, "The rate-distortion function for source coding with side information at the decoder," *IEEE Trans. Inf. Theory*, vol. 22, no. 1, pp. 1–10, January 1976.

[23] A. Saxena, J. Nayak, and K. Rose, "On efficient quantizer design for robust distributed source coding," in *Proc. Data Compression Conf., 2006,* IEEE, 2006, pp. 63–72.

[24] Z. Liu, S. Cheng, A. D. Liveris, and Z. Xiong, "Slepian-wolf coded nested lattice quantization for Wyner-Ziv coding: High-rate performance analysis and code design," *IEEE Trans. Inf. Theory*, vol. 52, no. 10, pp. 4358–4379, October 2006.

[25] A. Aaron and B. Girod, "Compression with side information using turbo codes," *Proc. IEEE Data Compression Conf.*, April 2002.

[26] S. S. Pradhan and K. Ramchandran, "Distributed source coding using syndromes (DISCUS): design and construction," *IEEE Trans. Inf. Theory*, vol. 49, no. 3, pp. 626–643, March 2010.

[27] S. B. Korada and R. L. Urbanke, "Polar codes are optimal for lossy source coding," *IEEE Trans. Inf. Theory*, vol. 56, no. 4, pp. 1751–1768, April 2010.

[28] E. Martinian and M. J. Wainwright, "Analysis of LDGM and compound codes for lossy compression and binning." Available at arXiv:0602.046.

[29] S.-H. Park, O. Simeone, O. Sahin, and S. Shamai, "Joint decompression and decoding for cloud radio access networks," *IEEE Signal Process Lett.*, vol. 20, no. 5, pp. 503–506, May 2013.

[30] O. Simeone, N. Levy, A. Sanderovich, O. Somekh, B. M. Zaidel, H. V. Poor *et al.* "Cooperative wireless cellular systems: an information-theoretic view," *Found. Trends Commun. Inf. Theory*, 2011.

[31] "3GPP tr 36.931 ver. 9.0.0 rel. 9," no. 6, May 2011.

[32] S.-H. Park, O. Simeone, O. Sahin, and S. Shamai, "Performance evaluation of multiterminal backhaul compression for cloud radio access networks," in *Proc. Conf. on Inf. Science and Systems*, March 2014.

[33] R. Irmer, H. Droste, P. March, M. Grieger, G. Fettweis, S. Brueck *et al.* "Coordinated multipoint: concepts, performance, and field trial results," *IEEE Commun. Mag.*, vol. 49, no. 2, pp. 102–111, February 2011.

[34] S.-H. Park, O. Simeone, O. Sahin, and S. Shamai, "Joint precoding and multivariate backhaul compression for the downlink of cloud radio access networks," *IEEE Trans. Signal Process*, vol. 61, no. 22, pp. 5646–5658, November 2013.

[35] S.-H. Park, O. Simeone, O. Sahin, and S. Shamai, "Multihop backhaul compression for the uplink of cloud radio access networks," Available at arXiv:1312.7135.

[36] B. Nazer and M. Gastpar, "Compute-and-forward: harnessing interference through structured codes," *IEEE Trans. Inf. Theory*, vol. 57, no. 10, pp. 6463–6486, October 2011.

[37] S.-N. Hong and G. Caire, "Compute-and-forward strategies for cooperative distributed antenna systems," *IEEE Trans. Inf. Theory*, vol. 59, no. 9, pp. 5227–5243, September 2013.

[38] N. Liu and W. Kang, "A new achievability scheme for downlink multicell processing with finite backhaul capacity," in *Proc. 2014 IEEE Int. Symp. Information Theory*, pp. 1006–1010.

9   Adaptive Compression in C-RANs

Thang X. Vu, Tony Q. S. Quek, and Hieu D. Nguyen

Cloud radio access networks (C-RANs) provide a promising architecture for the future mobile networks needed to sustain the exponential growth of the data rate. In C-RAN, one data processing center or baseband unit communicates with users through distributed remote radio heads, which are connected to the baseband unit (BBU) via high-capacity low-latency so-called fronthaul links. The architecture of C-RAN, however, imposes a burden of fronthaul bandwidth because raw I/Q samples are exchanged between the RRHs and the BBU. Therefore, signal compression is required on fronthaul links owing to their limited capacity. This chapter exploits the advance of joint signal processing to reduce the transmission rate on fronthaul uplinks. In particular, we first propose a joint decompression and detection (JDD) algorithm which exploits the correlation among RRHs and jointly performs decompressing and detecting. The JDD algorithm takes into consideration both fading and quantization effects in a single decoding step. Second, the block error rate of the JDD algorithm is analyzed in a closed form by using pairwise error probability analysis under both deterministic and Rayleigh fading channel models. Third, on the basis of the analyzed block error rate (BLER), we introduce adaptive compression schemes subject to quality of service constraints to minimize the fronthaul transmission rate while satisfying the predefined target QoS. The premise of the proposed compression methods originates from practical scenarios, where most applications tolerate a non-zero BLER. As a dual problem, we also develop a scheme to minimize the signal distortion subject to the fronthaul rate constraint. We finally consider the counterparts of these two adaptive compression schemes for Rayleigh-fading channels and analyze their asymptotic behavior as the constraints approach extremes.

9.1   Introduction

Cloud radio access networks have been widely accepted as a new architecture for future mobile networks to sustain the ever increasing demand in the data rate [1]. In a C-RAN, one centralized processor or BBU communicates with users distributed in a graphical area via a number of remote radio heads (RRHs), which act as "soft" relaying nodes and are connected to the BBU via high-capacity and low-latency fronthaul links. By moving all baseband processing functions from RRHs to a centralized processor, the C-RAN enables adaptive load balancing via a virtual base station pool [2] and effective network-wide inter-cell interference management thanks to multi-cell processing [3, 4]. The promise of C-RANs over traditional mobile networks includes system throughput improvement, high power efficiency, and dynamic resource management, which eventually result in a cost saving on capital expenditure (CAPEX) and operating expenditure (OPEX) [1, 5]. Because the baseband processing functions are executed at the BBU, the in-phase/quadrature-phase (I/Q) samples which represent the physical signal obtained through the sampling of the complex baseband signals are exchanged between the RRHs and the BBU, resulting in an enormous transmission rate on the fronthaul links. Reducing this rate is extremely important in the implementation of C-RANs since the fronthaul links' capacity is limited in practice.

Numerous research efforts have recently investigated the compression of C-RANs, mostly from the information-theoretic perspective; this work has aimed to design and optimize the quantization noise in order to maximize the achievable sum-rate [6–12]. This problem can be seen as a network multiple-input multiple-output (MIMO) problem with limited backhaul capacity [13–15]. The compression process is implemented via a test channel and the quantization noise is modeled as an independent Gaussian random variable, whose variance is linked to the capacity of the test channel. It is shown, in general, that a joint design of the precoding and quantization noise matrix can significantly improve the system sum-rate over a separate design [7]. Such an improvement results from the correlation among the RRHs when distributed source coding is applied [3]. The quality of the received signal at one RRH can be enhanced by exploiting the signal at other RRHs as side information. In [7] a robust distributed compression for uplink baseband signal was proposed based on the Karhunen–Loeve transform. In that work the correlated data at base stations were assumed to be imperfect and were modeled as deterministic additive errors on the bound of the eigenvalue of the error matrix. A further performance gain can be achieved by optimizing the test channel [16]. The authors in [8] proposed a hybrid compression and message-sharing strategy that allows a BS to perform a mix of compression and data sharing on the downlinks. It is shown that the hybrid solution achieves a better rate region than the pure methods of compression or data sharing. In [10] an optimum compression method was derived for sensor networks to compress noisy sensor measurements by minimizing the trace or determinant of the error covariance matrix. Further reviews on C-RAN are presented in [17, 18].

From the practical system point of view, various compression techniques have been studied in both the time and frequency domains (subcarrier compression) [1]. The key idea in those techniques is to minimize the redundancy of control information in common public radio interface (CPRI) package structure [19]. Lossless compression is proposed to achieve a good compression ratio due to two added nodes at the ends of the fronthaul links that optimize the redundancy in both the time and frequency domains [20, 21]. Statistical multiplexing gain is achieved as follows: (i) the information data of only active users is transmitted via the fronthaul links; (ii) the information needed for the reconstruction of the control information is minimized, since a large amount of control information, which is completely or semistatic, is locally generated; and (iii) a reduced set of the precoding matrix is transferred. A similar time-domain compression technique was proposed in [22]. Note that most compression techniques proposed in the time domain are for a single base station, and so cannot exploit the residential gain from the correlation among the RRHs.

In this chapter we focus on the compression on C-RAN uplinks and the receiver design at the BBU. In practical systems the received signal at an RRH is first uniformly quantized into bit sequences by an analog-to-digital converter (ADC).[1] These bits are then transmitted to the BBU via ideal fronthaul links. The compression ratio can be managed by changing the resolution of the ADC. The proposed compression method is not limited to the time domain and can be applied to the frequency domain with little modification. From the observation that treating the decompression and demodulation separately leads to a very suboptimal solution [9], we thus narrow this chapter to consideration of a joint decompression and detection (JDD) algorithm that jointly performs decompressing and detecting in a single step and effectively exploits the correlation among the RRHs; it achieves a significant improvement in the information-theoretic sense [3, 16]. The first goal is to minimize the transmission rate on the fronthaul links with an acceptable distortion of the decompressed signal so that the BBU can support a maximum number of RRHs. This design criterion is different from that in [3, 16], which aims to fully occupy the fronhaul link capacity. This objective comes from practical situations where most applications can tolerate an acceptable non-zero BLER. The second goal is to minimize the signal distortion given a fronthaul rate constraint. An analytical closed-form expression for the BLER is derived using pairwise error probability (PEP) analysis under both deterministic and Rayleigh-fading channel models, where the BLER is shown as a function of the channel fading, thermal noise, and quantization noise. From the analyzed BLER expression, two adaptive compression schemes with a quality of service (QoS) constraint are proposed to maximize the compression ratio while satisfying a given BLER target. We also present a JDD scheme which aims to minimize the distortion given a predetermined fronthaul rate. The counterparts of these two adaptive compression schemes are also studied for Rayleigh-fading channels, and their asymptotic behavior as the constraints approach extremes, is analyzed.

The rest of the chapter is organized as follows. Section 9.2 describes in detail the compression scheme and the proposed JDD algorithm. In Section 9.3 the performance of the proposed algorithm is analysed. Adaptive compression schemes are proposed in Section 9.4. Section 9.5 gives numerical results. Finally, Section 9.6 concludes the chapter.

9.2   System Model

We consider a C-RAN system consisting of M users denoted by $U_1, \ldots, U_M$, N RRHs denoted by $R_1, \ldots, R_N$, and one BBU, as shown in Fig. 9.1. The users communicate with the RRHs via a wireless medium, while the RRHs connect to the BBU by high-speed low-latency optical fiber (or wireless) links, which are known as fronthaul links [1]. A distinguished feature of an RRH compared with the classical base station (BS) is that

---

[1] Other non-linear quantization methods can also be applied.

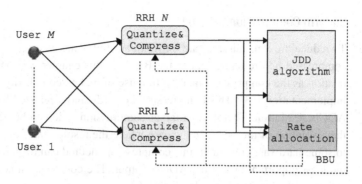

Figure 9.1 Block diagram of uplinks in a C-RAN with joint decompression and detection algorithm at the BBU. The adaptive compression scheme employs a rate allocation block to feedback the optimal sampling rate to the RRHs.[2]

the RRH's function is much simpler than that of a traditional BS because all baseband processing functions are immigrated to the BBU. Therefore, an RRH can be seen as a "soft" relaying node that forwards an I/Q signal to the BBU. The users and RRHs are equipped with a single antenna. In practical system, a multiple-antenna RRH can be seen as a band of single-antenna RRHs which are subject to a sum-rate constraint, because all baseband processing functions are performed at the BBU. Owing to the to limited capacity on the fronthaul links, the I/Q signal needs to be compressed before being sent to the processing center [3]. The BBU decompresses the received signal from the RRHs and then performs further processing. In the following, we focus on the compression and decompression on the fronthaul uplinks.

We assume that all nodes are synchronous and all wireless channels are block Rayleigh fading. The BBU is assumed to know all the channel state information (CSI) in the network. Denote by c_m a modulated symbol emitted by user U_m. The modulated symbol $c_m, 1 \leq m \leq M$, belongs to the source codebook $\mathcal{S} = \{s_1, \ldots, s_{|\mathcal{S}|}\}$, where $|\cdot|$ denotes the cardinality of a set. The source codebook satisfies a unit power constraint, e.g., $\mathbb{E}_{s \in \mathcal{S}} |s|^2 = 1$. Denote $\mathbf{c} = [c_1, \ldots, c_M]^T$ as a codeword transmitted by the users, where $(\cdot)^T$ represents the vector on matrix transpose. The signal received at R_n is given by

$$y_n = \sum_{m=1}^{M} h_{nm} \sqrt{P_m} c_m + z_n = \mathbf{h}_n \mathbf{P} \mathbf{c} + z_n, \qquad (9.1)$$

where $\mathbf{P} = \text{diag}\{\sqrt{P_1}, \ldots, \sqrt{P_M}\}$, P_m is the average transmit power of user U_m; h_{nm} is the channel-fading coefficient between U_m and R_n, including the path loss, which is a complex Gaussian random variable with zero mean and variance $\sigma_{h_{nm}}^2$; $\mathbf{h}_n = [h_{n1} \cdots h_{nM}]$ is the channel vector from all users to $\mathbf{R}_n$; and z_n is i.i.d. Gaussian noise with zero mean and variance σ^2.

Upon receiving analog signals from the users, the RRHs quantize and compress them into digital bits and then forward these bits to the BBU.

9.2.1   Uniform Compression Scheme

To reduce the transmission rate on fronthaul links, the received signal at each RRH is compressed before being sent to the BBU. In this chapter we consider uniform quantization as the compression method because of its low complexity and practicability of implementation [23]. This compression method can be realized by flexibly tuning the ADC's resolution. Therefore, a target compression ratio can be achieved by changing the resolution of the ADC. In the case where the resolution of the ADC is fixed owing to some hardware constraint, this compression method can be performed by truncating the least important bits in the ADC's output. The compression is executed on the real and imaginary parts separately [1].

First, the received signal at every RRH is normalized by a scaling factor to restrict the normalized signal to lie within $[-1, 1]$ with high probability. In the literature there are two methods for choosing the scaling factor: static and dynamic. In static method the scaling factor is computed only on the basis of the source codebook and the transmitted power. This method thus requires a minimum overhead signal at the RRHs. The dynamic method, however, calculates the scaling factor using the channel-fading coefficients $\mathbf{h}_n$ as well. As such, it requires more overhead information. Denote η_n^s and η_n^d as the scaling factor in the static and dynamic methods, respectively. Here, we assume the "three-sigma" rule [26] in which the scaling factor is equal to three times the square root of the power of y_n. For a given $\mathbf{h}_n$, it is straightforward to compute the power of y_n as $||\mathbf{h}_n\mathbf{P}||^2 + \sigma^2$. Applying the three-sigma rule, the BBU computes $\eta_n^d = 3\sqrt{||\mathbf{h}_n\mathbf{P}||^2 + \sigma^2}$, which is assumed to be known for free at the nth RRH because its overhead is negligible compared with that of the data. For the static method, $\eta_n^s = 3\sqrt{||\mathbf{P}||^2 + \sigma^2}$. For ease of representation we use $\eta_n \in \{\eta_n^s, \eta_n^d\}$ for both methods.

Let y_n^R and y_n^I be the real and imaginary parts of y_n, respectively. The normalized signal at the nth RRH is given by

$$\bar{y}_n = \frac{y_n^R}{\eta_n} + i\frac{y_n^I}{\eta_n} = \bar{y}_n^R + i\bar{y}_n^I.$$

In the next step the normalized signal $\bar{y}_n$ is quantized into $\tilde{y}_n = \tilde{y}_n^R + i\tilde{y}_n^I$ by an uniform quantizer whose resolution equals Q_n bits. The compressed signal can be calculated from the normalized signal as follows:

$$\tilde{y}_n^a = \eta_n \frac{\text{round}(\bar{y}_n^a \times 2^{Q_n})}{2^{Q_n}},$$

where the superscript a represents either R or I and the function round(x) denotes the closest integer to x. The quantization error at R_n is given as $q_n = y_n - \tilde{y}_n = q_n^R + iq_n^I$. When the absolute value of y_n is large compared with the quantization step, q_n^R and q_n^I can be well modeled as uniform random variables with support $[-\delta_n, \delta_n]$, where $\delta_n = \eta_n 2^{-Q_n-1}$ with $\eta_n \in \{\eta_n^s, \eta_n^d\}$. We observe via intensive simulations that, with the three-sigma rule, such an assumption is still feasible even with a small number of quantization bits (see Fig. 9.2). After compression, $\tilde{y}_n$ is converted into a bit sequence which is later sent to the BBU via error-free fronthaul links.

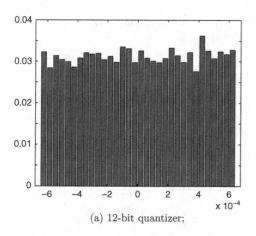

(a) 12-bit quantizer;

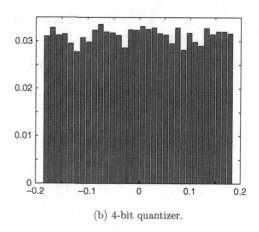

(b) 4-bit quantizer.

Figure 9.2 Quantization noise modeled as a uniform random variable: (a) 12-bit quantizer; (b) 4-bit quantizer.

9.2.2 Joint Decompression and Detection Algorithm

In this section we propose a JDD algorithm that performs decompressing and detecting for the source codeword simultaneously, by exploiting the structure of the quantizer and the codebook. The BBU is assumed to know the CSI of all wireless links. The CSI can be obtained via, e.g., channel estimation with pilot transmission in a training period. Given the compressed bit sequences, the BBU optimally estimates the source codeword by using the maximum a posteriori (MAP) receiver as follows:

$$\hat{\mathbf{c}} = \arg \max_{\mathbf{c}} P\{\mathbf{c}|\tilde{y}_1, \ldots, \tilde{y}_N\} \overset{(a)}{=} \arg \max_{\mathbf{c}} P\{\mathbf{c}\} \prod_{n=1}^{N} P\{\tilde{y}_n|\mathbf{c}\}, \qquad (9.2)$$

where the equality (a) results from the fact that $P\{\tilde{y}_1, \ldots, \tilde{y}_N\}$ is constant for any codeword, and the noise z_n and compressed signals are independent given the source codeword.

In (9.2), $P\{\tilde{y}_n|\mathbf{c}\}$ is the probability that the quantizer outputs $\tilde{y}_n$ from the observation $y_n = \mathbf{h}_n \mathbf{Pc} + z_n$. It is worth mentioning that for a real signal the linear quantizer outputs y if the distance between the input and y is less than or equal to the quantization error. For a complex signal y_n, the quantizer outputs $\tilde{y}_n$ if both $|y_n^R - \tilde{y}_n^R|$ and $|y_n^I - \tilde{y}_n^I|$ are less than the quantization error. Because the quantization is performed independently for the real and the imaginary parts, we have

$$P\{\tilde{y}_n|\mathbf{c}\} = P\{y_n^R \in [\tilde{y}_n^R - \delta_n, \tilde{y}_n^R + \delta_n] \cap y_n^I \in [\tilde{y}_n^I - \delta_n, \tilde{y}_n^I + \delta_n]\}$$
$$= P\{y_n^R \in [\tilde{y}_n^R - \delta_n, \tilde{y}_n^R + \delta_n]\} \times P\{y_n^I \in [\tilde{y}_n^I - \delta_n, \tilde{y}_n^I + \delta_n]\}.$$

To derive the above probability we recall that, for a given the codeword and fading channels, y_n^R and y_n^I are Gaussian distributed with the same variance $\sigma^2/2$ and means $\mathcal{R}(\mathbf{h}_n \mathbf{Pc})$ and $\mathcal{I}(\mathbf{h}_n \mathbf{Pc})$, respectively, where $\mathcal{R}(x)$ and $\mathcal{I}(x)$ are the real and imaginary parts of x. Therefore, the conditional probability density functions (PDFs) $f(y_n^R|\mathbf{c})$ and $f(y_n^I|\mathbf{c})$ of y_n^R and y_n^I, respectively, are given as [24]

$$\frac{1}{\sqrt{\pi}\sigma}\exp\left(-\frac{|y_n^R - \mathcal{R}(\mathbf{h}_n\mathbf{Pc})|^2}{\sigma^2}\right) \quad \text{and} \quad \frac{1}{\sqrt{\pi}\sigma}\exp\left(-\frac{|y_n^I - \mathcal{I}(\mathbf{h}_n\mathbf{Pc})|^2}{\sigma^2}\right).$$

By substituting the above PDFs into $P\{\tilde{y}_n|\mathbf{c}\}$ we obtain

$$P\{\tilde{y}_n|\mathbf{c}\} = \int_{\tilde{y}_n^R - \delta_n}^{\tilde{y}_n^R + \delta_n} f(y_n^R|\mathbf{c})dy_n^R \times \int_{\tilde{y}_n^I - \delta_n}^{\tilde{y}_n^I + \delta_n} f(y_n^I|\mathbf{c})dy_n^I$$

$$= \frac{1}{4} \times \left[\text{erfc}\left(\frac{\tilde{y}_n^R - \mathcal{R}(\mathbf{h}_n\mathbf{Pc}) - \delta_n}{\sigma}\right) - \text{erfc}\left(\frac{\tilde{y}_n^R - \mathcal{R}(\mathbf{h}_n\mathbf{Pc}) + \delta_n}{\sigma}\right)\right]$$

$$\times \left[\text{erfc}\left(\frac{\tilde{y}_n^I - \mathcal{I}(\mathbf{h}_n\mathbf{Pc}) - \delta_n}{\sigma}\right) - \text{erfc}\left(\frac{\tilde{y}_n^I - \mathcal{I}(\mathbf{h}_n\mathbf{Pc}) + \delta_n}{\sigma}\right)\right], \quad (9.3)$$

where $\text{erfc}(\cdot)$ denotes the complementary error function. Substituting (9.3) into (9.2), we obtain a decoding rule for codeword $\hat{\mathbf{c}}$.

9.3   Block Error Rate Analysis

9.3.1   Deterministic Channel Models

This section analyzes the BLER of the JDD algorithm introduced in Section 9.2.2. The BLER is defined as the probability of receiving codeword $\hat{\mathbf{c}}$ when a codeword $\mathbf{c} \neq \hat{\mathbf{c}}$ was transmitted. A block error event occurs when at least one of M symbols c_m, $1 \leq m \leq M$, is decoded with error. Since the BLER is difficult to investigate, we instead resort to the union bound on the BLER and consider instead the average pairwise error probability (APEP), as follows:

$$\text{BLER} \leq \text{APEP} = \frac{1}{|\mathcal{S}|^M} \sum_{\mathbf{c} \in \mathcal{S}^M} \sum_{\mathbf{c} \neq \tilde{\mathbf{c}} \in \mathcal{S}^M} P\{\mathbf{c} \to \tilde{\mathbf{c}}\}. \quad (9.4)$$

where $P\{\mathbf{c} \to \tilde{\mathbf{c}}\}$ is the instantaneous PEP of receiving $\tilde{\mathbf{c}}$ when $\mathbf{c}$ was transmitted; it depends on the channel fading coefficients, and $\tilde{\mathbf{c}}$ is the only candidate.

To evaluate the PEP, we model the quantization effect by an uniformly distributed random variable that is independent of the input. This assumption can be well justified when the absolute value of the input is much larger than the quantization step. Under such an assumption, the compressed signal from the nth RRH is modeled as

$$\tilde{y}_n = \mathbf{h}_n\mathbf{Pc} + z_n + q_n, \quad (9.5)$$

where $q_n = q_n^R + iq_n^I$ is the quantization noise at R_n. Since both q_n^R and q_n^I are uniformly distributed in $[-\delta_n, \delta_n]$ (see Section 9.2 for more details), it is straightforward to verify that q_n has zero mean and a variance which is computed as

$$\sigma_{q_n}^2 = \text{Var}(q_n) = \frac{1}{2\delta_n}\int_{-\delta_n}^{\delta_n} |q_n^R|^2 dq_n^R + \frac{1}{2\delta_n}\int_{-\delta_n}^{\delta_n} |q_n^I|^2 dq_n^I = \frac{2\delta_n^2}{3}.$$

Denote $\mathbb{M}(\mathbf{c}) = \prod_{n=1}^{N} P\{\tilde{\mathbf{y}}_n | \mathbf{c}\}$ as the detection metric of codeword $\mathbf{c}$, where $P\{\tilde{\mathbf{y}}_n | \mathbf{c}\}$ is computed in (9.3). A pairwise error occurs if the metric of the transmitted codeword is smaller than that of another candidate:

$$P\{\mathbf{c} \rightarrow \tilde{\mathbf{c}}\} = P\{\mathbb{M}(\mathbf{c}) < \mathbb{M}(\tilde{\mathbf{c}})\}. \tag{9.6}$$

A computation of (9.6) based on the exact expression in (9.3) is very complicated owing to the multi-fold product of $\text{erfc}(\cdot)$ functions. As an alternative, we use the first-order Taylor approximation $f(x) \simeq f(x_0) + f'(x_0)(x - x_0)$, where x_0 is any feasible point. Applying this to the function $\text{erfc}(\cdot)$ in (9.3) with $x_0 = [\tilde{y}_n^R - \mathcal{R}(\mathbf{h}_n \mathbf{P}\mathbf{c})]/\sigma$ for the real part and $x_0 = [\tilde{y}_n^I - \mathcal{I}(\mathbf{h}_n \mathbf{P}\mathbf{c})]/\sigma$ for the imaginary part, $P\{\tilde{\mathbf{y}}_n | \mathbf{c}\}$ can be written as [24]

$$P\{\tilde{\mathbf{y}}_n | \mathbf{c}\} \simeq \frac{\delta_n^2}{\pi \sigma^2} \exp\left(-\frac{|\tilde{\mathbf{y}}_n - \mathbf{h}_n \mathbf{P}\mathbf{c}|^2}{\sigma^2}\right). \tag{9.7}$$

Remark 9.1 The derivation of $P\{\tilde{\mathbf{y}}_n | \mathbf{c}\}$ in (9.3) is exact and (9.3) can be used as the (exact) decoding metric. However, under a high-SNR regime and fading channel, the argument of the $\text{erfc}(\cdot)$ function in (9.3) can be very large, resulting in over-buffer and wrong decoding. For a practical implementation of our scheme, an approximation using the first-order Taylor approximation (9.7) can be used instead to avoid such problems.

Substituting (9.7) into $\mathbb{M}(\mathbf{c})$, we obtain $\mathbb{M}(\mathbf{c}) = K \exp(-\mathbb{D}(\mathbf{c}))$, where $K = \prod_{n=1}^{N} \delta_n^2 / (\pi \sigma^2)^N$ is a constant and $\mathbb{D}(\mathbf{c}) = \sum_{n=1}^{N} |\tilde{\mathbf{y}}_n - \mathbf{h}_n \mathbf{P}\mathbf{c}|^2$. Then, the PEP is derived as $P\{\mathbf{c} \rightarrow \tilde{\mathbf{c}}\} = P\{\mathcal{I}(\mathbf{c}, \tilde{\mathbf{c}}) > 0\}$, where

$$\mathcal{I}(\mathbf{c}, \tilde{\mathbf{c}}) = \sum_{n=1}^{N} \left[\tilde{\mathbf{y}}_n^T \mathbf{h}_n \mathbf{P}(\tilde{\mathbf{c}} - \mathbf{c}) + (\tilde{\mathbf{c}} - \mathbf{c})^T \mathbf{P}\mathbf{h}_n^T \tilde{\mathbf{y}}_n + |\mathbf{h}_n \mathbf{P}\mathbf{c}|^2 - |\mathbf{h}_n \mathbf{P}\tilde{\mathbf{c}}|^2\right].$$

Substituting (9.5) into $\mathcal{I}(\mathbf{c}, \tilde{\mathbf{c}})$, we obtain

$$\mathcal{I}(\mathbf{c}, \tilde{\mathbf{c}}) = Z_1 + Z_2 - \psi,$$

where $Z_1 = \sum_{n=1}^{N}[z_n^T \mathbf{h}_n \mathbf{P}(\tilde{\mathbf{c}} - \mathbf{c}) + (\tilde{\mathbf{c}} - \mathbf{c})^T \mathbf{P}\mathbf{h}_n^T z_n]$, $Z_2 = \sum_{n=1}^{N}[q_n^T \mathbf{h}_n \mathbf{P}(\tilde{\mathbf{c}} - \mathbf{c}) + (\tilde{\mathbf{c}} - \mathbf{c})^T \mathbf{P}\mathbf{h}_n^T q_n]$, and $\psi = \sum_{n=1}^{N} |\mathbf{h}_n \mathbf{P}(\tilde{\mathbf{c}} - \mathbf{c})|^2$.

Because each z_n is a complex Gaussian random variable with zero mean and variance σ^2, and the z_n are mutually independent, Z_1 is also a Gaussian random variable, with zero mean and variance

$$\sigma_{Z_1}^2 = 2\sigma^2 \sum_{n=1}^{N} |\mathbf{h}_n \mathbf{P}(\tilde{\mathbf{c}} - \mathbf{c})|^2.$$

However, because q_n is uniformly distributed, to compute the exact joint PDF of Z_2 is complicated. For ease of analysis we model Z_2 by a Gaussian variable $\bar{Z}_2$ that has a similar mean and variance to Z_2, i.e., $\bar{Z}_2 \sim \mathcal{N}(\mu_{Z_2}, \sigma_{Z_2}^2)$, where $\mu_{\bar{Z}_2} = \mathbb{E}\{Z_2\} = 0$ and

$$\sigma_{\bar{Z}_2}^2 = \mathbb{E}\left\{|Z_2|^2\right\} = \frac{4}{3} \sum_{n=1}^{N} \delta_n^2 |\mathbf{h}_n \mathbf{P}(\tilde{\mathbf{c}} - \mathbf{c})|^2.$$

Then the sum $Z = Z_1 + Z_2$ is also a Gaussian random variable with zero mean and variance $\sigma_Z^2 = \sigma_{Z_1}^2 + \sigma_{Z_2}^2$. Therefore we can compute the PEP as follows:

$$P\{\mathbf{c} \to \tilde{\mathbf{c}}\} = P\{Z > \psi\} \tag{9.8}$$

$$= \frac{1}{2} \mathrm{erfc} \left(\frac{\sum_{n=1}^{N} |\mathbf{h}_n \mathbf{P}(\tilde{\mathbf{c}} - \mathbf{c})|^2}{\sqrt{4\sigma^2 \sum_{n=1}^{N} |\mathbf{h}_n \mathbf{P}(\tilde{\mathbf{c}} - \mathbf{c})|^2 + \frac{8}{3} \sum_{n=1}^{N} \delta_n^2 |\mathbf{h}_n \mathbf{P}(\tilde{\mathbf{c}} - \mathbf{c})|^2}} \right).$$

It can be observed from (9.8) that the PEP depends on how the relative distance between $\mathbf{c}$ and $\tilde{\mathbf{c}}$ is distorted by the fading channels, the thermal noise power σ^2, and the compression noise δ_n. Substituting (9.8) into (9.4), we obtain an upper bound for the BLER. Note that (9.8) is valid for both static and dynamic compression methods.

9.3.2 Rayleigh-Fading Channel Models

In this section, we analyse the BLER of the JDD algorithm in Rayleigh-fading channels. We restrict the analysis to the static compression method only, i.e., $\eta_n = \eta_n^s$ which is independent of the fading channels. The BLER over the fading channels is computed as follows:

$$\mathrm{BLER}_{\mathrm{fading}} \leq \frac{1}{|\mathcal{S}|^M} \sum_{\mathbf{c} \in \mathcal{S}^M} \sum_{\tilde{\mathbf{c}} \in \mathcal{S}^M, \tilde{\mathbf{c}} \neq \mathbf{c}} \overline{\mathrm{PEP}}_{\mathbf{c} \to \tilde{\mathbf{c}}}, \tag{9.9}$$

where $\overline{\mathrm{PEP}}_{\mathbf{c} \to \tilde{\mathbf{c}}} = \mathbb{E}\{\mathrm{PEP}_{\mathbf{c} \to \tilde{\mathbf{c}}}\}$ denotes the expectation over the fading channels of $\mathrm{PEP}_{\mathbf{c} \to \tilde{\mathbf{c}}}$, which was computed in (9.8).

Before deriving $\overline{\mathrm{PEP}}_{\mathbf{c} \to \tilde{\mathbf{c}}}$, we observe that $\{\mathbf{h}_n \mathbf{P}(\tilde{\mathbf{c}} - \mathbf{c})\}_{n=1}^{N}$ are random variables with a $\mathcal{CN}(0, |\mathbf{\Lambda}_n(\tilde{\mathbf{c}} - \mathbf{c})|^2)$ distribution, where $\mathbf{\Lambda}_n = [\sigma_{h_{n1}}^2 \sqrt{P_1}, \ldots, \sigma_{h_{nM}}^2 \sqrt{P_M}]$. Define $G_n^{\mathbf{c} \to \tilde{\mathbf{c}}} t_n \triangleq |\mathbf{h}_n \mathbf{P}(\tilde{\mathbf{c}} - \mathbf{c})|^2$, where $G_n^{\mathbf{c} \to \tilde{\mathbf{c}}} \triangleq |\mathbf{\Lambda}_n(\tilde{\mathbf{c}} - \mathbf{c})|^2$. Thus, $\{t_n\}_{n=1}^{N}$ are i.i.d. exponential random variables each with distribution $e^{-t}, 0 < t < \infty$. We can re-express the term inside (9.8) as

$$\frac{\sum_{n=1}^{N} G_n^{\mathbf{c} \to \tilde{\mathbf{c}}} t_n}{\sqrt{\sum_{n=1}^{N} \left(4\sigma^2 + \frac{8}{3}\delta_n^2\right) G_n^{\mathbf{c} \to \tilde{\mathbf{c}}} t_n}} = \frac{\sum_{n=1}^{N} G_n^{\mathbf{c} \to \tilde{\mathbf{c}}} t_n}{\sqrt{\sum_{n=1}^{N} \beta_n G_n^{\mathbf{c} \to \tilde{\mathbf{c}}} t_n}},$$

where $\beta_n \triangleq 4\sigma^2 + \frac{8}{3}\delta_n^2$.

Replacing the variable t_n by $v_n = \beta_n G_n^{\mathbf{c} \to \tilde{\mathbf{c}}} t_n$, the average PEP over the fading channels is evaluated as follows:

$$\overline{\mathrm{PEP}}_{\mathbf{c} \to \tilde{\mathbf{c}}} = \tag{9.10}$$

$$\frac{1}{2 \prod_{n=1}^{N} \beta_n G_n^{\mathbf{c} \to \tilde{\mathbf{c}}}} \int_0^{\infty} \cdots \int_0^{\infty} \mathrm{erfc} \left(\frac{\sum_{n=1}^{N} v_n / \beta_n}{\sqrt{\sum_{n=1}^{N} v_n}} \right) \exp \left(-\sum_{n=1}^{N} \frac{v_n}{\beta_n G_n^{\mathbf{c} \to \tilde{\mathbf{c}}}} \right) dv_1 \cdots dv_N.$$

The exact computation of (9.10) for an arbitrary set $\{\beta_n\}_{n=1}^N$ is challenging, especially over the set $\{v_n\}_{n=1}^N$. Thus, we seek for a lower bound and upper bound of (9.10). We will show later that (9.10) has the exact expression for identical $\{\beta_n\}$, i.e., $\beta_1 = \cdots = \beta_N$.

Lower Bound

Let $\mathcal{P}_{\text{ind}}$ denote the set of all permutations of $\{1, 2, \ldots, N\}$, i.e.,

$$\mathcal{P}_{\text{ind}} = \{\{i_1, \ldots, i_N\} : i_m \neq i_q, \forall m \neq q; \{i_1, \ldots, i_N\} \equiv \{1, \ldots, N\}\}. \tag{9.11}$$

Furthermore, the function $f(x) \triangleq \text{erfc}(\alpha x) \exp(-x)$ can be shown to be convex in $x > 0$ for $\alpha > 0$. Considering any given value set $\{v_n\}_{n=1}^N$, we thus have

$$\sum_{\mathcal{P}_{\text{ind}}} \text{erfc}\left(\frac{\sum_{n=1}^N v_n/\beta_n}{\sqrt{\sum_{n=1}^N v_n}}\right) \exp\left(-\sum_{n=1}^N \frac{v_n}{\beta_n G_n^{c \to \tilde{c}}}\right)$$

$$\geq N! \, \text{erfc}\left(\frac{1}{N!}\frac{\sum_{\{i_n\}_n \in \mathcal{P}_{\text{ind}}} \sum_{n=1}^N v_{i_n}/\beta_n}{\sqrt{\sum_{n=1}^N v_n}}\right) \exp\left(-\frac{1}{N!}\sum_{\{i_n\}_n \in \mathcal{P}_{\text{ind}}} \sum_{n=1}^N \frac{1}{\beta_n G_n^{c \to \tilde{c}}} v_{i_n}\right)$$

$$= N! \, \text{erfc}\left[\left(\frac{1}{N}\sum_{n=1}^N \frac{1}{\beta_n}\right)\sqrt{\sum_{n=1}^N v_n}\right] \exp\left[-\left(\frac{1}{N}\sum_{n=1}^N \frac{1}{\beta_n G_n^{c \to \tilde{c}}}\right)\sum_{n=1}^N v_n\right]. \tag{9.12}$$

Combining (9.10) and (9.12), we can bound the average PEP below as

$$\overline{\text{PEP}}_{c \to \tilde{c}} \geq \frac{1}{2\prod_{n=1}^N \beta_n G_n^{c \to \tilde{c}}} \int_0^\infty \cdots \int_0^\infty \text{erfc}\left[\left(\frac{1}{N}\sum_{n=1}^N \frac{1}{\beta_n}\right)\sqrt{\sum_{n=1}^N v_n}\right]$$

$$\times \exp\left[-\left(\frac{1}{N}\sum_{n=1}^N \frac{1}{\beta_n G_n^{c \to \tilde{c}}}\right)\sum_{n=1}^N v_n\right] dv_1 \cdots dv_N. \tag{9.13}$$

To facilitate the computation of (9.13), we use a tight approximation of the error complementary function [28]:

$$\text{erfc}(x) \approx \frac{1}{6}e^{-x^2} + \frac{1}{2}e^{-4x^2/3}. \tag{9.14}$$

After some algebraic manipulations, we obtain

$$\overline{\text{PEP}}_{c \to \tilde{c}} \geq \frac{1}{12\prod_{n=1}^N \beta_n G_n^{c \to \tilde{c}}}\left[\left(\frac{1}{N}\sum_{n=1}^N \frac{1}{\beta_n}\right)^2 + \frac{1}{N}\sum_{n=1}^N \frac{1}{\beta_n G_n^{c \to \tilde{c}}}\right]^{-N} \tag{9.15}$$

$$+ \frac{1}{4\prod_{n=1}^N \beta_n G_n^{c \to \tilde{c}}}\left[\frac{4}{3}\left(\frac{1}{N}\sum_{n=1}^N \frac{1}{\beta_n}\right)^2 + \frac{1}{N}\sum_{n=1}^N \frac{1}{\beta_n G_n^{c \to \tilde{c}}}\right]^{-N}.$$

Substituting (9.15) into (9.9), we obtain the lower of the union bounds in fading channels.

Upper Bound

Let us define $\beta_{\max} = \max_{n \in \{1,\dots,N\}} \beta_n$; then we have

$$\sum_{n=1}^{N} \frac{v_n}{\beta_n} \geq \sqrt{\sum_{n=1}^{N} \frac{v_n}{\beta_n}} \sqrt{\sum_{n=1}^{N} \frac{v_n}{\beta_{\max}}} = \sqrt{\sum_{n=1}^{N} \frac{v_n}{\beta_n}} \sqrt{\frac{\sum_{n=1}^{N} v_n}{\beta_{\max}}}. \tag{9.16}$$

Applying (9.16) to (9.10) while noting that erfc(x) is a decreasing function, we obtain

$$\overline{\text{PEP}}_{\mathbf{c} \to \tilde{\mathbf{c}}} \leq \frac{1}{2 \prod_{n=1}^{N} \beta_n G_n^{\mathbf{c} \to \tilde{\mathbf{c}}}} \int_0^\infty \cdots \int_0^\infty \text{erfc}\left(\frac{1}{\sqrt{\beta_{\max}}} \sqrt{\sum_{n=1}^{N} \frac{v_n}{\beta_n}} \right)$$

$$\times \exp\left(-\sum_{n=1}^{N} \frac{v_n}{\beta_n G_n^{\mathbf{c} \to \tilde{\mathbf{c}}}} \right) dv_1 \cdots dv_N$$

$$\overset{(a)}{\approx} \frac{1}{2 \prod_{n=1}^{N} \beta_n G_n^{\mathbf{c} \to \tilde{\mathbf{c}}}} \int_0^\infty \cdots \int_0^\infty \left\{ \frac{1}{6} \exp\left[-\sum_{n=1}^{N} \left(\frac{1}{\beta_{\max}} + \frac{1}{G_n^{\mathbf{c} \to \tilde{\mathbf{c}}}} \right) \frac{v_n}{\beta_n} \right] \right.$$

$$\left. + \frac{1}{2} \exp\left[-\sum_{n=1}^{N} \left(\frac{4}{3\beta_{\max}} + \frac{1}{G_n^{\mathbf{c} \to \tilde{\mathbf{c}}}} \right) \frac{v_n}{\beta_n} \right] \right\} dv_1 \cdots dv_N$$

$$= \frac{1}{12 \prod_{n=1}^{N} \left(G_n^{\mathbf{c} \to \tilde{\mathbf{c}}} / \beta_{\max} + 1 \right)} + \frac{1}{4 \prod_{n=1}^{N} \left(4 G_n^{\mathbf{c} \to \tilde{\mathbf{c}}} / 3\beta_{\max} + 1 \right)}. \tag{9.17}$$

where (a) results from the approximation (9.14).

Substituting (9.17) into (9.9), we obtain an upper bound for the BLER in fading channels. It is observed that C-RAN achieves full diversity of order N with respect to the signal-to-compression-plus-Gaussian-noise ratio. The accuracy of the bounds obviously depends on how diverse $\{\beta_n\}_{n=1}^{N}$ is. When the quantization noise powers at the RRHs are identical, i.e., $\delta_1 = \cdots = \delta_N$ and therefore $\beta_1 = \cdots = \beta_N$, the lower bound and upper bound completely coincide.

9.4   Adaptive Compression under QoS Constraint

In practical systems, various applications might require different QoS depending on specific contexts. For example, a voice message usually requires a lower QoS than a video call. A flexible compression scheme should be capable of adapting the compression ratio to satisfy a predefined QoS and maximize the compression efficiency. In this section we first propose two adaptive compression schemes to maximize the compression efficiency under a certain target BLER so that a fronthaul link can support a maximal number of antennas. Such schemes are desirable for systems which support large fronthaul feedback and/or require a stringent QoS. Furthermore, we also consider an adaptive compression design which minimizes the BLER, specifically the PEP as a proxy of the BLER, given a compression efficiency. In contrast with the previous two counterparts, this design focuses on systems with a stricter constraint on the fronthaul bandwidth. Note that the proposed compression schemes in this section are valid for both static and dynamic compression methods.

Sections 9.4.1 and 9.4.2 will introduce optimization problems for deterministic channel models. In Section 9.4.3 we will analyze the counterparts of such optimizations assuming Rayleigh-fading channels, in particular the optimal solutions under extreme cases.

9.4.1 Minimization of the Number of Bits Given the BLER

In this subsection we consider systems which require a certain BLER QoS while tolerating a possible large fronthaul bandwidth. In particular, we would like to minimize the number of bits for quantization under a QoS constraint ζ_{tag}, as follows:

$$\underset{\{Q_n : Q_n \geq 1\}_{n=1}^{N}}{\text{minimize}} \quad \sum_{n=1}^{N} Q_n \tag{9.18}$$

$$\text{s.t.} \quad \frac{1}{|S|^M} \sum_{\tilde{c} \neq c \in S^M} \sum P\{c \rightarrow \tilde{c}\} \leq \zeta_{tag},$$

where ζ_{tag} is the predefined BLER target, and $P\{c \rightarrow \tilde{c}\}$ is given in (9.8).

PEP-Based Algorithm

The problem in (9.18) is difficult to solve due to its non-convexity. Consider an alternative approach which gives us an upper bound of (9.18), as follows:

$$\underset{\{Q_n : Q_n \geq 1\}_{n=1}^{N}}{\text{minimize}} \quad \sum_{n=1}^{N} Q_n$$

$$\text{s.t.} \quad \frac{1}{2} \text{erfc}\left(\sqrt{\Phi_{\tilde{c},c}}\right) \leq \zeta_{tag}, \quad \forall \tilde{c} \neq c, \tag{9.19}$$

where

$$\Phi_{\tilde{c},c} = \frac{\left(\sum_{n=1}^{N} |h_n P(\tilde{c} - c)|^2\right)^2}{4\sigma^2 \sum_{n=1}^{N} |h_n P(\tilde{c} - c)|^2 + \frac{8}{3} \sum_{n=1}^{N} \delta_n^2 |h_n P(\tilde{c} - c)|^2},$$

and the first constraint in (9.19) is obtained by using $P\{c \rightarrow \tilde{c}\}$ in (9.8).

We note that the optimal solution of (9.19) always satisfies (9.18), i.e., the optimal objective value of (9.19) is an upper bound for that of (9.18). The proof is as follows. Let $Pe(c)$ be the error probability for the transmission of c and the reception of $\hat{c} \neq c$, i.e., $Pe(c) = P\{\hat{c} \in S_{\sim c}^M | c\}$, where $S_{\sim c}^M$ denotes the set of codewords, c. Obviously, $P\{\hat{c} \in S_{\sim c}^M | c\} \leq \sum_{\hat{c} \neq c} PEP\{c \rightarrow \hat{c}\}$. This confirms that the optimal objective value of (9.19) is an upper bound for that of (9.18). Note that (9.19) is an integer programming problem, which is difficult to solve. A conventional approach to obtaining a convex formulation of (9.19) is to relax the integer constraint of Q_n.

By introducing $\mu_n = 2^{-2(Q_n+1)}$, we can reformulate (9.19) as:

$$\underset{\{\mu_n : \mu_n \le \frac{1}{4}\}_{n=1}^N}{\text{minimize}} \quad \sum_{n=1}^N -\frac{1}{2}\log_2(\mu_n)$$

$$\text{s.t.} \qquad (9.21) \text{ holds}, \quad \forall \tilde{\mathbf{c}} \ne \mathbf{c}, \qquad (9.20)$$

where

$$4\sigma^2 \sum_{n=1}^N |\mathbf{h}_n \mathbf{P}(\tilde{\mathbf{c}} - \mathbf{c})|^2 + \frac{8}{3}\sum_{n=1}^N \eta_n^2 |\mathbf{h}_n \mathbf{P}(\tilde{\mathbf{c}} - \mathbf{c})|^2 \mu_n \le \frac{\left(\sum_{n=1}^N |\mathbf{h}_n \mathbf{P}(\tilde{\mathbf{c}} - \mathbf{c})|^2\right)^2}{\alpha}. \qquad (9.21)$$

In (9.21) α is an auxiliary variable satisfying $\frac{1}{2}\mathrm{erfc}(\sqrt{\alpha}) = \frac{\zeta_{\text{tag}}}{|\mathcal{S}|^M - 1}$. Note that we can consider α as the maximum PEP such that the ζ_{tag} constraint can still be satisfied. Thus the problem (9.18) with a BLER constraint has been effectively transformed to a PEP-based counterpart. We denote the scheme that solves (9.20) as MinBits-PEP.

It can be proved that (9.20) is a convex optimization problem and thus can be solved efficiently by using, e.g., the primal–dual interior point method [27]. Furthermore, (9.19) is substantially simpler than (9.18) and is preferable for systems requiring low complexity. The integer quantization bit Q_n can be obtained from μ_n simply by choosing the smallest integer following $\hat{Q}_n = 1 - \frac{1}{2}\log_2 \mu_n$, i.e., $\lfloor\hat{Q}_n\rfloor$. In general, there is no bound for the optimality loss of such an approximation. However, as the constraint threshold ζ_{tag} tends to 0, the loss also tends to 0. The reason is that each Q_n becomes large in such a case, which leads to a small $(\hat{Q}_n - \lfloor\hat{Q}_n\rfloor)\hat{Q}_n$.

SDR-Based Algorithm

An approximate solution for problem (9.18) can also be obtained by using semidefinite programming relaxation (SDR). The problem (9.18) is first rewritten as

$$\underset{\{Q_n : Q_n \ge 1\}_{n=1}^N}{\text{minimize}} \quad \sum_{n=1}^N Q_n$$

$$\text{s.t.} \qquad \frac{1}{|\mathcal{S}|^M} \sum_{\tilde{\mathbf{c}} \ne \mathbf{c} \in \mathcal{S}^M} \frac{1}{2}\mathrm{erfc}(\sqrt{\alpha_{\tilde{\mathbf{c}},\mathbf{c}}}) \le \zeta_{\text{tag}},$$

$$\alpha_{\tilde{\mathbf{c}},\mathbf{c}} \le \Phi_{\tilde{\mathbf{c}},\mathbf{c}}, \quad \forall \tilde{\mathbf{c}} \ne \mathbf{c}. \qquad (9.22)$$

Let us introduce μ_n as in Section 9.4.1, $\mathbf{x} = [\alpha_1 \cdots \alpha_L \; \mu_1 \cdots \mu_N \; 1]^T$, and $\mathbf{A}_{\tilde{\mathbf{c}},\mathbf{c}} \in \mathbb{R}^{(L+N+1)\times(L+N+1)}$, where $L = |\mathcal{S}|^M \times (|\mathcal{S}|^M - 1)$, which is given by (9.23):

$$\begin{bmatrix} & & & \mathbf{0} & & & \\ & & & \vdots & & & \\ & & & \mathbf{0} & & & \\ 0 \cdots 0 & \frac{8}{3}\eta_1^2|\mathbf{h}_1\mathbf{P}(\tilde{\mathbf{c}}-\mathbf{c})|^2 & \ldots & \frac{8}{3}\eta_N^2|\mathbf{h}_N\mathbf{P}(\tilde{\mathbf{c}}-\mathbf{c})|^2 & 4\sigma^2 \sum_{n=1}^N |\mathbf{h}_n\mathbf{P}(\tilde{\mathbf{c}}-\mathbf{c})|^2 \\ & & & \mathbf{0} & & & \\ & & & \vdots & & & \\ & & & \mathbf{0} & & & \end{bmatrix}. \qquad (9.23)$$

Problem (9.22) can be expressed as

$$\underset{\{\mu_n:\mu_n\leq\frac{1}{4}\}_{n=1}^N,\mathbf{x}}{\text{minimize}} \quad \sum_{n=1}^N -\frac{1}{2}\log_2(\mu_n)$$

$$\text{s.t.} \quad \frac{1}{|\mathcal{S}|^M}\sum_{\tilde{\mathbf{c}}\neq\mathbf{c}\in\mathcal{S}^M}\frac{1}{2}\,\text{erfc}(\sqrt{\alpha_{\tilde{\mathbf{c}},\mathbf{c}}})\leq\zeta_{\text{tag}},$$

$$\text{Tr}(\mathbf{A}_{\tilde{\mathbf{c}},\mathbf{c}}\mathbf{x}\mathbf{x}^T)\leq\left(\sum_{n=1}^N|\mathbf{h}_n\mathbf{P}(\tilde{\mathbf{c}}-\mathbf{c})|^2\right)^2. \tag{9.24}$$

By defining $\mathbf{X}=\mathbf{x}\mathbf{x}^T$, problem (9.24) becomes equivalent to the following:

$$\underset{\mathbf{X}}{\text{minimize}} \quad \sum_{n=L+1}^{L+N} -\frac{1}{4}\log_2([\mathbf{X}]_{n,n})$$

$$\text{s.t.} \quad \frac{1}{|\mathcal{S}|^M}\sum_{l=1}^L\frac{1}{2}\,\text{erfc}([\mathbf{X}]_{l,l}^{1/4})\leq\zeta_{\text{tag}},$$

$$\text{Tr}(\mathbf{A}_{\tilde{\mathbf{c}},\mathbf{c}}\mathbf{X})\leq\left(\sum_{n=1}^N|\mathbf{h}_n\mathbf{P}(\tilde{\mathbf{c}}-\mathbf{c})|^2\right)^2,$$

$$[\mathbf{X}]_{n,n}\leq\frac{1}{16}, \quad n=L+1,\dots,L+N,$$

$$\text{rank}(\mathbf{X})=1. \tag{9.25}$$

The SDR of (9.25) is obtained by ignoring the rank constraint. It can be shown that the SDR of (9.25) is a convex optimization problem and is solvable by using, e.g., the primal–dual interior point method [27]. We denote the scheme that solves (9.25) without the rank constraint as MinBits-SDR.

Compared with PEP-based minimization, BLER-based minimization is expected to achieve higher compression ratios, the tradeoff beging higher computing complexity. This is so because the PEP-based solution guarantees that all PEP satisfy the target QoS, which can result in a smaller BLER than necessary. Consequently, the PEP-based solution requires a greater fronthaul rate to achieve a better BLER.

Remark 9.2 In general, a SDR solution of (9.25) might violate the rank-1 constraint, which is, in fact, a generic problem of SDR. To obtain an approximate (vector) solution $\mathbf{x}^*$ for (9.25) from a SDR counterpart $\mathbf{X}^*$, we implement the Gaussian randomization procedure [29], where more details of such procedure and its approximation accuracies under several setups can be found.

Remark 9.3 To facilitate the computation of the first constraint in (9.25), we use the tight approximation (9.14). The resulting problem (9.26) is still convex and solvable as follows:

$$\underset{\mathbf{X}}{\text{minimize}} \quad \sum_{n=L+1}^{L+N} -\frac{1}{4}\log_2([\mathbf{X}]_{n,n})$$

$$\text{s.t.} \quad \frac{1}{|\mathcal{S}|^M}\sum_{n=1}^L\frac{1}{12}\exp\left(-[\mathbf{X}]_{n,n}^{1/2}\right)+\frac{1}{4}\exp\left(-\frac{4}{3}[\mathbf{X}]_{n,n}^{1/2}\right)\leq\zeta_{\text{tag}},$$

$$\text{Tr}(\mathbf{A}_{\tilde{\mathbf{c}},\mathbf{c}}\mathbf{X}) \leq \left(\sum_{n=1}^{N} |\mathbf{h}_n\mathbf{P}(\tilde{\mathbf{c}} - \mathbf{c})|^2\right)^2, \quad \forall\tilde{\mathbf{c}} \neq \mathbf{c},$$

$$[\mathbf{X}]_{n,n} \leq \frac{1}{16}, \quad n = L+1,\ldots,L+N. \tag{9.26}$$

9.4.2   Minimization of the Maximum PEP Given the Number of Bits Q_{sum}

In this section we investigate the dual problem of (9.18), which is solved in Section 9.4.1 using PEP- and SDR-based algorithms. In particular we want to minimize the BLER given that $\sum_{n=1}^{N} Q_n \leq Q_{\text{sum}}$. This problem arises under systems with limited fronthaul bandwidth but a less stringent BLER constraint. The problem, however, is difficult to solve. Therefore, as in Section 9.4.1, we will consider an alternative problem based on PEP which gives an upper-bound solution for the original optimization. The alternative problem is expressed mathematically as follows:

$$\underset{\{Q_n:Q_n\geq 1\}_{n=1}^{N}}{\text{minimize}} \quad \underset{\tilde{\mathbf{c}}\neq\mathbf{c}}{\max} \ \ P\{\mathbf{c} \rightarrow \tilde{\mathbf{c}}\} = \frac{1}{2}\,\text{erfc}(\sqrt{\Phi_{\tilde{\mathbf{c}},\mathbf{c}}})$$

$$\text{s.t.} \quad \sum_{n=1}^{N} Q_n \leq Q_{\text{sum}}. \tag{9.27}$$

Because erfc$(\cdot)$ is a monotonic function, by introducing an auxiliary varable α we can reformulate (9.27) as:

$$\underset{\{Q_n:Q_n\geq 1\}_{n=1}^{N}}{\text{maximize}} \quad \alpha$$

$$\text{s.t.} \quad \alpha \leq \Phi_{\tilde{\mathbf{c}},\mathbf{c}}, \ \ \forall\tilde{\mathbf{c}} \neq \mathbf{c},$$

$$\sum_{n=1}^{N} Q_n \leq Q_{\text{sum}}, \tag{9.28}$$

where $\Phi_{\tilde{\mathbf{c}},\mathbf{c}}$ was defined in Section V.A. In another form, (9.28) is identical to the following problem:

$$\underset{\{Q_n:Q_n\geq 1\}_{n=1}^{N}}{\text{maximize}} \quad \alpha$$

$$\text{s.t.} \quad \text{(9.30) holds} \ \ \forall\tilde{\mathbf{c}} \neq \mathbf{c},$$

$$\sum_{n=1}^{N} Q_n \leq Q_{\text{sum}}, \tag{9.29}$$

where the first constraint in problem (9.29) is given as follows:

$$\frac{8}{3}\sum_{n=1}^{N} \eta_n^2|\mathbf{h}_n\mathbf{P}(\tilde{\mathbf{c}} - \mathbf{c})|^2 2^{-2(Q_n+1)} \leq \frac{\left(\sum_{n=1}^{N} |\mathbf{h}_n\mathbf{P}(\tilde{\mathbf{c}} - \mathbf{c})|^2\right)^2}{\alpha}$$

$$-4\sigma^2\sum_{n=1}^{N} |\mathbf{h}_n\mathbf{P}(\tilde{\mathbf{c}} - \mathbf{c})|^2. \tag{9.30}$$

Table 9.1 MinBLER-PEP Algorithm

1:	Initialize α_H, α_L, and ϵ.
2:	$\alpha_M = (\alpha_H + \alpha_L)/2$.
3:	If (9.31) is feasible, then $\alpha_L := \alpha_M$.
	Otherwise $\alpha_H := \alpha_M$.
4:	Repeat steps 2 and 3 until $|\alpha_H - \alpha_L| \leq \epsilon$.

As in Section 9.4.1, the maximum PEP has been used as a proxy for the BLER minimization. The auxiliary variable α effectively represents the minimum SNR across all users which is achievable under the Q_{sum} constraint. Note that, given α, (9.29) is a convex optimization problem and thus solvable using standard methods. A standard approach is to apply the bisection technique to solve (9.29). The resulting optimization problem, whose steps are given in Table 9.1, is given as follows:

$$
\begin{aligned}
\underset{\{Q_n : Q_n \geq 1\}_{n=1}^{N}}{\text{maximize}} \quad & \alpha_M \\
\text{s.t.} \quad & \Gamma(\alpha_M), \ \forall \tilde{c} \neq c, \\
& \sum_{n=1}^{N} Q_n \leq Q_{\text{sum}},
\end{aligned}
\tag{9.31}
$$

where $\Gamma(\alpha_M)$ is obtained by replacing α in (9.30) by α_M. We denote this scheme as MinBLER-PEP.

Remark 9.4   The decoder and optimization derived here are based on the ML receiver and thus have a complexity that increases exponentially with the number of users. In order to reduce the computing complexity one can consider a sphere decoder, which significantly reduces the computing complexity by reducing the search circle but yet can achieve a performance close to that of the ML decoder.

9.4.3   Rayleigh-Fading Channel Models

Assuming Rayleigh-fading channels, the optimization counterparts of (9.19) and (9.27) are given as in (9.32) and (9.33), respectively:

$$
\begin{aligned}
\underset{\{Q_n : Q_n \geq 1\}_{n=1}^{N}}{\text{minimize}} \quad & \sum_{n=1}^{N} Q_n \\
\text{s.t.} \quad & \overline{\text{PEP}}_{c \to \tilde{c}} \leq \frac{\zeta_{\text{tag}}}{|S|^M - 1}, \ \forall c \neq \tilde{c},
\end{aligned}
\tag{9.32}
$$

$$
\begin{aligned}
\underset{\{Q_n : Q_n \geq 1\}_{n=1}^{N}}{\text{minimize}} \quad & \underset{c,\tilde{c}}{\max} \ \overline{\text{PEP}}_{c \to \tilde{c}} \\
\text{s.t.} \quad & \sum_{n=1}^{N} Q_n \leq Q_{\text{sum}},
\end{aligned}
\tag{9.33}
$$

where $\overline{\text{PEP}}_{c \to \tilde{c}}$ is computed as in Section 9.3.2. Here, the optimization of $\{Q_n\}_{n=1}^N$ is based only on the statistics of the channel since the PEP and BLER are averaged over a Rayleigh-fading channel.

It is difficult to obtain solutions for (9.32) and (9.33) in general. In the following, we observe the asymptotic behaviors of the solutions instead. The propositions below investigate (9.32) and (9.33) as the optimization constraints approach extreme cases [25].

PROPOSITION 9.1   As the QoS threshold $\zeta_{\text{tag}} \to 0$, the solution of the problem (9.32) approaches that for uniform quantization noise, i.e., $\delta_1 = \cdots = \delta_N$.

PROPOSITION 9.2   If there exist $\{q_n : q_n \geq 1\}_{n=1}^N$ such that $\eta_1 2^{-q_1} = \eta_2 2^{-q_2} = \cdots = \eta_N 2^{-q_N}$, and $\sum_{n=1}^N q_n = Q_{\text{sum}}$, then the solution of problem (9.33) satisfies uniform quantization noise and is given as

$$Q_n^\star = \frac{1}{N} Q_{\text{sum}} + \frac{1}{2} \log_2 \left(\frac{2}{3} \eta_n^2 \right) - \frac{1}{2N} \sum_{n=1}^N \log_2 \left(\frac{2}{3} \eta_n^2 \right), \quad \forall n.$$

Propositions 9.1 and 9.2 are strong enough to state that the optimal solution of (9.32) and (9.33) approaches that of uniform compression noise as the BLER threshold decreases, since they provide an upper bound for the original problems. Nevertheless, Propositions 9.1 and 9.2 provide a justification for implementing the sampling that imposes uniform compression noise, especially under delay-constrained system where sophisticated adaptive sampling might not be appropriate.

9.5   Simulation Results

To demonstrate the effectiveness of the proposed algorithms, simulations were carried out on a network that consisted of $M = 3$ users and $N = 3$ RRHs. We assumed block Rayleigh-fading channels and a symmetric network with equal user transmit energy, e.g. $P_1 = \cdots = P_M = P$. The average SNR is defined as P/σ^2. In addition, $\sigma_{h_{nn}}^2 = 1 \forall n$ and $\sigma_{h_{nm}}^2 = 0.5$ for $n \neq m$. During each channel realization, the users emit a message comprising $K = 1000$ data symbols that belong to a QPSK codebook with normalized power, i.e., $S = \{-1 - i, -1 + i, 1 - i, 1 + i\}/\sqrt{2}$. All the RRHs apply uniform quantization. The BBU is assumed to know CSI for the entire network.

Figure 9.3 presents the BLER performance for the dynamic compression method, in which the scaling factor depends on the instantaneous fading channels. Here we assume that the RRHs are subject to identical compression noise. Similar results as for the static compression method are observed. In particular, at low SNRs the contribution of quantization noise is small. For example, for SNRs between 0 dB and 8 dB, sampling at 4 bits per sample or 12 bits per sample yields an almost similar BLER. This is due to the fact that under a small- or medium-SNR regime, the thermal noise is large and therefore is dominant compared with the quantization noise. In contrast, under a high-SNR regime, the thermal noise is comparable with or even smaller than the quantization

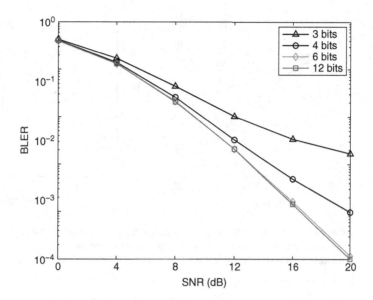

Figure 9.3 BLER of the dynamic compression method versus fronthaul sampling resolution for $Q = 3, 4, 6, 12$ bits.

noise. Decreasing Q in this case can result in severe loss in BLER. It is also observed that a 6-bit quantizer achieves an almost similar performance to a 12-bit counterpart under the setting considered and the SNR range of observation. This interesting observation suggests an adaptive compression scheme to minimize the fronthaul rate while maintaining the BLER under a given QoS.

Figure 9.4(a) shows the performance of C-RAN systems with the static compression method and symmetric network topology, i.e., $\eta_1 = \eta_2 = \cdots = \eta_N$. The sampling rate is equally allocated, i.e., $Q_1 = Q_2 = Q_3$ bits. As a result, the quantization noise at every RRH is identical. Therefore, the upper bound and lower bound of the PEP coincide. For all cases, the derived bounds closely match the simulation results, which validates our analysis. It is observed that the sampling rate Q has an effect on both the BLER and the diversity order. In general, a larger Q results in a better BLER, and the BLER will saturate as Q decreases. Figure 9.4(b) shows simulation results and the corresponding bounds under a non-uniform quantization noise scenario. As SNR increase, the analysed BLERs based on the upper bound of the PEP and on the lower bound of the PEP diverge. Although the lower bound underestimates the non-uniform compression noise, it is shown in [25] that the optimal sampling allocation which achieve, the minimum BLER satisfies the uniform compression noise scenario.

Figure 9.5 presents for BLER of adaptive compression versus SNR under a given QoS constraint. The premise is that different applications require different QoS, e.g., BLER, levels. For a given BLER, we want to maximize the compression efficiency or equivalently to minimize the fronthaul transmission rate. Two adaptive compression schemes, based on the PEP constraint (Section 9.4.1, the MinBits-PEP) or the BLER constraint (Section 9.4.1, the MinBits-SDR) are presented. In addition, results for the

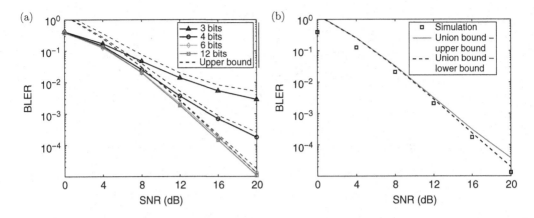

Figure 9.4 Validation of the BLER bounds in the static compression method. (a) Uniform compression noise, i.e., $\beta_1 = \cdots = \beta_N$, for different sampling rates Q_n; (b) non-uniform compression noise with $(Q_1, Q_2, Q_3) = (5, 6, 7)$.[3]

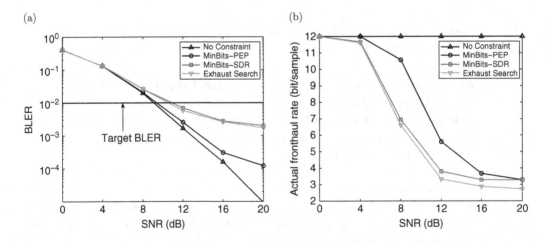

Figure 9.5 (a) BLER performance and (b) compression efficiency of the adaptive compression schemes with target $\zeta_{tag} = 10^{-2}$ for different SNR. The fronthaul bandwidth $Q = 12$ bits.[3]

scheme without a QoS constraint which fully occupies the fronthaul bandwidth (named No Constraint) and for the Exhaustive Search optimization are also plotted. For Exhaustive Search we checked every combination of the rate allocation $[Q_1, Q_2, \ldots, Q_N]$ and found the one which gives the minimum fronthaul rate while satisfying the BLER target. Such a scheme yields the optimal rate allocation but is hindered by its NP-complete complexity. Three hundred channel realizations were conducted in the simulations. The threshold ζ_{tag} is equal to 10^{-2}. The BLER performance is shown in Fig. 9.5(a) and the actual fronthaul rate is presented in Fig. 9.5(b). At very low SNRs the BLER does not satisfy the target QoS because the channel is so poor. Even using all 12 bits for quantization does not satisfy the target BLER. One good thing is that, while not satisfying the target BLER, adaptive compressions achieve a smaller fronthaul rate (a higher compression ratio). More specifically, at 8 dB the MinBits-PEP algorithm saves 1.4 bits and the MinBits-SDR algorithm saves 5 bits. Moving to a higher-SNR regime (from 12 dB in the figure), both adaptive schemes meet the target QoS while significantly improving the compression ratio. Because the No Constraint scheme always uses 12 bits for quantization, its fronthaul rate is 12 bits per sample in the whole SNR range. However, a compression ratio of 350% is observed for both adaptive schemes, which only require 3.4 bits per sample to achieve a BLER less than or equal to 10^{-2}. It is also shown in Fig. 9.5 that MinBits-SDR achieves a performance close to that of Exhaustive Search, which confirms the effectiveness of the SDR optimization. Furthermore, MinBits-SDR obtains a better compression efficiency than MinBits-PEP at low and medium SNR. This result is obvious since MinBits-PEP minimizes the worst PEP while MinBits-SDR targets the average BLER. When the SNR is large, it is preferable to employ MinBits-PEP because it yields a smaller BLER.

Figure 9.6 shows the BLER and compression efficiency of adaptive compressions for different fronthaul bandwidths Q at SNR = 12 dB. As in previous simulations, the target ζ_{tag} is set at 10^{-2}. The results show that all the schemes satisfy the BLER target while significantly reducing the fronthaul rate. The MinBits-SDR does exactly what we require: it satisfies a BLER of 10^{-2}, but no more. In that sense it is the best since it satisfies the constraint but with fewer bits. The MinBits-PEP scheme obtains a similar BLER to that of the No Constraint scheme while MinBits-SDR achieves a slightly worse BLER. However, they all satisfy the target BLER of 10^{-2}. Specifically, the compression efficiency obtained by MinBits-PEP increases from 120% at $Q = 5$ to 240% at $Q = 12$,

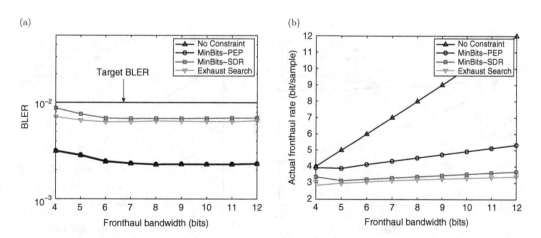

Figure 9.6 (a) Block error rate performance and (b) compression efficiency of adaptive compression with target $\zeta_{\text{tag}} = 10^{-2}$, for different fronthaul bandwidths; SNR = 12 dB.[4]

while MinBits-SDR gains a better compression efficiency, 160% at $Q = 5$ to about 330% at $Q = 12$. This observation is consistent with the result in Fig. 9.3, in which at 12 dB the 4-bit sampling and 12-bit sampling yield approximately similar performances.

The above simulations aim to minimize the actual fronthaul rate given the BLER constraint. On the basis of the analysis of the BER in Section 9.3 a reciprocal problem is how to allocate the fronthaul bandwidth $\{Q_n\}_{n=1}^N$ in order to minimize the BLER for a given a sum $Q_{sum} = \sum_{n=1}^N Q_n$. Because the nth RRH uses all Q_n bits for quantization, we refer to Q_n as the fronthaul rate in this paragraph for simplicity. The optimization problem is described in Section 9.4.2. Figure 9.7(a) shows the BLER of the MinBLER-PEP and two other allocation schemes; uniform rate allocation (Rate allocation 2), e.g., $Q_1 = Q_2 = Q_3$, and non-uniform rate allocation (Rate allocation 1), e.g., $(Q_1, Q_2, Q_3) = (Q - 2, Q, Q + 2)$. The sum-rate $Q_{sum} = 9$. In addition, the performance of Exhaustive Search is also presented. For this scheme in, we checked every combination of the rate allocation $[Q_1, Q_2, Q_3]$ and found the one which gives the minimum BLER based on (9.4) while satisfying the Q_{sum} constraint. As expected, our MinBLER-PEP algorithm achieves almost the same BLER as Exhaustive Search, which yields the best BLER but is limited by its NP-complete complexity. It is observed that the non-uniform rate allocation scheme obtains a worse BLER because there is always one RRH using fewer quantization bits than needed. At low SNR, both uniform and optimal rate allocations give a similar performance because at this SNR the thermal noise is dominant. As the SNR increases, the performance gain provided by the optimal rate allocation is larger.

Figure 9.7b shows the BLER for different values of Q_{sum} at 12 dB. The performance of non-uniform allocation is the worst when Q_{sum} is small but it approaches the optimal BLER when Q_{sum} increases. The optimal allocation outperforms the uniform rate

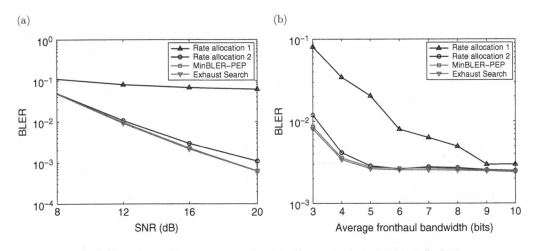

(a) (b)

Figure 9.7 Block error rate comparison of the optimal rate allocation scheme for different Q_{sum}. values. Rate allocation 1, $(Q_1, Q_2, Q_3) = (n, n, n)$; rate allocation 2, $(Q_1, Q_2, Q_3) = (n - 2, n, n + 2)$, where n is the average fronthaul bandwidth. In (a) $Q_{sum} = 9$ bits. In (b) the SNR = 12 dB.

allocation for small Q_{sum} and achieves a similar BLER as Q_{sum} increases. Another observation is that the optimal BLER remains constant for a large range of Q_{sum}. This can be explained from Fig. 9.3, where we see that at 12 dB, 4-bit quantization achieves nearly the same performance as 12-bit quantization. Spending more bits in this case brings only a small improvement in BLER. In conclusion, the MinBLER-PEP-based rate allocation is more effective at high SNR and when the fronthaul bandwidth is small. Otherwise, uniform rate allocation is less complex but still can achieve the best performance.

Remark 9.5   The schemes MinBits-SDR and MinBits-PEP described in this chapter are based on the bound (9.4), which takes into account all PEPs. When the number of users or the modulation order is large, this bound can be too relaxed. In this case, considering the nearest symbols only might be practically preferred owing to its lower complexity (fewer PEP constraints) and better compression gain.

9.6   Conclusions

We have proposed a near-optimal receiver structure for cloud radio access networks which takes into consideration the quantization effects of capacity-limited fronthaul links and exploits the correlation among the remote radio heads. Analytical results were derived for the pairwise error probability. On the basis of the analysed PEP, two adaptive compression schemes were proposed to improve the compression efficiency while satisfying the QoS constraint. The proposed optimization problem comes from practical situations in which most applications tolerate a target QoS, e.g., the BLER. A compression efficiency of 350% can be achieved by the proposed optimization schemes. In addition, an optimal rate allocation, which minimizes the BLER given the compression efficiency, is proposed on the basis of the theoretical PEP. We further considered the counterparts of the adaptive compression schemes for Rayleigh-fading channels and analyzed their asymptotic solutions for extreme cases.

References

[1] ChinaMobile, "C-RAN: the road towards green RAN," 2011, white paper.

[2] Z. Zhu, P. Gupta, Q. Wang, S. Kalyanaraman, Y. Lin, H. Franke, and S. Sarangi, "Virtual base station pool: towards a wireless network cloud for radio access networks," in *Proc. ACM Int. Conf. on Computing Frontiers*, New York, March 2011, pp. 34:1–34:10.

[3] A. Sanderovich, O. Somekh, H. V. Poor, and S. Shamai, "Uplink macro diversity of limited backhaul cellular network," *IEEE Trans. Inf. Theory*, vol. 55, no. 8, pp. 3457–3478, August 2009.

[4] P. Marsch and G. Fettweis, "Uplink CoMP under a constrained backhaul and imperfect channel knowledge," *IEEE Trans. Wireless Commun.*, vol. 10, no. 6, pp. 1730–1742, June 2011.

[5] UMTS, "Mobile traffic forecasts 2010-2020." Technical Report 44, 2011.

[6] Y. Zhou, W. Yu, and D. Toumpakaris, "Uplink multi-cell processing: Approximate sum capacity under a sum backhaul constraint," in *Proc. IEEE Inf. Theory Workshop*, Seville, September 2013, pp. 1–5.

[7] S.-H. Park, O. Simeone, O. Sahin, and S. Shamai, "Joint precoding and multivariate backhaul compression for the downlink of cloud radio access networks," *IEEE Trans. Signal Process.*, vol. 61, no. 22, pp. 5646–5658, August 2013.

[8] P. Patil and W. Yu, "Hybrid compression and message-sharing strategy for the downlink cloud radio-access network," in *Proc. IEEE Information Theory and Applications Workshop*, San Diego, CA, February 2014, pp. 1–6.

[9] J. Kang, O. Simeone, J. Kang, and S. Shamai, "Joint signal and channel state information compression for the backhaul of uplink network MIMO systems," *IEEE Trans. Wireless Commun.*, vol. 13, no. 3, pp. 1555–1567, March 2014.

[10] Y. Wang, H. Wang, and L. Scharf, "Optimum compression of a noisy measurement for transmission over a noisy channel," *IEEE Trans. Signal Process.*, vol. 62, no. 5, pp. 1279–1289, March 2014.

[11] J. Tang, W. P. Tay, and T. Q. S. Quek, "Cross-layer resource allocation with elastic service scaling in cloud radio access network," *IEEE Trans. Wireless Commun.*, vol. 14, no. 9, pp. 5068–5081, September 2015.

[12] A. del Coso and S. Simoens, "Distributed compression for MIMO coordinated networks with a backhaul constraint," *IEEE Trans. Wireless Commun.*, vol. 8, no. 9, pp. 4698–4709, September 2009.

[13] D. Gesbert, S. Hanly, H. Huang, S. Shamai, O. Simeone, and W. Yu, "Multi-cell MIMO cooperative networks: A new look at interference," *IEEE J. Sel. Areas Commun.*, vol. 28, no. 9, pp. 1380–1408, December 2010.

[14] J. Zhao, T. Q. S. Quek, and Z. Lei, "Coordinated multipoint transmission with limited backhaul data transfer constraints," *IEEE Trans. Wireless Commun.*, vol. 12, no. 6, pp. 2762–2775, June 2013.

[15] J. Zhao, T. Q. S. Quek, and Z. Lei, "Heterogeneous cellular networks using wireless backhaul: fast admission control and large system analysis," *IEEE J. Sel. Areas Commun.*, vol. 33, no. 10, pp. 2128–2143, October 2015.

[16] S. H. Park, O. Simeone, O. Sahin, and S. Shamai, "Joint decompression and decoding for cloud radio access networks," *IEEE Signal Process. Lett.*, vol. 20, no. 5, pp. 503–506, March 2013.

[17] M. Peng, Y. Li, J. Jiang, J. Li, and C. Wang, "Heterogeneous cloud radio access networks: a new perspective for enhancing spectral and energy efficiencies," *IEEE Wireless Commun. Mag.*, vol. 21, no. 6, pp. 126–135, December 2014.

[18] T. Q. S. Quek, M. Peng, O. Simeone, and W. Yu, *Cloud Radio Access Networks: Principles, Technologies, and Applications*. Cambridge University Press, 2016.

[19] T. Q. S. Quek, M. Peng, O. Simeone, and W. Yu,"Common public radio intreface (CPRI): interface specification," 2013, CPRI Specification V6.0.

[20] J. Lorca and L. Cucala, "Lossless compression technique for the fronthaul of LTE/LTE-advanced Cloud-RAN architectures," in *Proc. IEEE Int. Symp. and Workshops on World of Wireless, Mobile, and Multimedia Networks*, Madrid, June 2013, pp. 1–9.

[21] A. Nanba, and S. Agata, "A new IQ data compression scheme for front-haul link in centralized RAN," in *Proc. IEEE Int. Symp. on Personal, Indoor, and Mobile Radio Commun.*, London, September 2013, pp. 210–214.

[22] K. F. Nieman and B. L. Evans, "Time-domain compression of complex-baseband LTE signals for cloud radio access networks," in *Proc. IEEE Global Conf. on Signal and Information Processing*, Austin, TX, December 2013, pp. 1198–1201.

[23] K. F. Nieman and B. L. Evans, "Feasibility study for further advancements for E-UTRA (LTE-advanced)," Technical Report ETSI TR 136 912, 2010.

[24] T. X. Vu, H. D. Nguyen, and T. Q. S. Quek, "Adaptive compression and joint detection for fronthaul uplinks in cloud radio access networks," *IEEE Trans. Commun.*, vol. 63, no. 11, pp. 4565–4575, November 2015.

[25] T. X. Vu, H. D. Nguyen, T. Q. S. Quek, and S. Sun, "Cloud radio access networks: compression and optimization," *IEEE Trans. Signal Process.*, to appear.

[26] R. Gray and D. Neuhoff, "Quantization," *IEEE Trans. Inf. Theory*, vol. 44, no. 6, pp. 2325–2383, October 1998.

[27] S. Boyd and L. Vandenberghe, *Convex Optimization*. Cambridge University Press, 2004.

[28] M. Chiani, D. Dardari, and M. K. Simon, "New exponential bounds and approximations for the computation of error probability in fading channels," *IEEE Trans. Commun.*, vol. 2, no. 4, pp. 840–845, July 2003.

[29] Z. Q. Luo, W. K. Ma, A. M. C. So, Y. Ye, and S. Zhang, "Semidefinite relaxation of quadratic optimization problems," *IEEE Signal Process. Mag.*, vol. 27, no. 3, pp. 20–34, March 2010.

[30] P. Li, D. Paul, R. Narasimhan, and J. Cioffi, "On the distribution of SINR for the MMSE MIMO receiver and performance analysis," *IEEE Trans. Inf. Theory*, vol. 52, no. 1, pp. 271–286, January 2006.

Part III

Resource Allocation and Networking in C-RANs

10 Resource Management of Heterogeneous C-RANs

Shao-Yu Lien, Shao-Chou Hung, Chih-Hsiu Zeng, Hsiang Hsu, Qimei Cui, and Kwang-Cheng Chen

10.1 Introduction

Mobile cellular infrastructures, which have been deployed in recent decades, successfully provide seamless and reliable streaming (voice or video) services for billions of mobile users. From GSM/GPRS, UMTS, to LTE/LTE-A, transmission data rates have been enhanced a million-fold. The recent deployment of heterogeneous networks (HetNets) consisting of macro cells, small cells (femtocells, picocells), and/or further relay nodes ubiquitously support basic multimedia and Internet browsing applications. As a result, primitive human-to-human (H2H) communication applications using existing network architectures and technologies seem satisfactory. However, to substantially facilitate human daily activities in addition to basic voice or video and Internet access services, achieving full automation and everything-to-everything (X2X), had been regarded as an ultimate goal not only for the future information communication industry but also for financial transactions, economics, social communities, transportation, agriculture, and energy allocation. Full automation implies a significant enhancement of human beings' sensory and processing capabilities, which embraces unmanned or remotely controlled vehicles, robots, offices, factories, augmented or virtual reality, and sensory human interactions of cyber-physical–social systems. The goal is to employ distributed autonomous control to relieve or simplify network control and evolutive, by which resource utilization can be boosted in dynamic complex networks and be re-optimized after major environmental changes. However, X2X connection implies that diverse entities including human beings and machines are able to form general sense communities other than H2H, such as social networks that are human-to-machine (H2M) or machine-to-machine (M2M), facilitating the ultimate cyber-physical–social systems. Application scenarios include intelligent transportation systems (ITSs), volunteer information networks, the Internet of Things (IoT), smart grids, and much more.

To enable these various applications, boosting transmission data rates is just one of the diverse requirements. The performance in terms of end-to-end transmission latency, energy efficiency, reliability, scalability, cost efficiency as well as stability should also be fundamentally enhanced. As the data traffic from the Internet has gradually been dominating the traffic volume in mobile communication systems, in addition to an improvement in air-interface the migration to more efficient network architecture is definitely a must in technology development. Furthermore, a large portion of the current traffic data is user-generated via social networks such as documents, pictures, videos, and messages. Such data is circulated among users according to their social relationships, which precisely indicates the interplay between mobile communication networks and social networks. More precisely, most data is generated at the edge of networks but stored and sometimes analyzed in clouds. Consequently, the current Internet architecture partitioning networks into layers will not be able to support these heterogeneous applications with affordable cost. All these new technology opportunities suggest the need to evolve state-of-the-art network architecture beyond ultra-efficient air-interface.

This chapter consequently starts from the introduction of C-RANs and heterogeneous C-RANs (H-CRANs), with their unique features, and then reveals particular challenges obstructing the operation of C-RAN and H-CRAN. Centering on resource management in C-RAN and H-CRAN, innovations are demonstrated that meet these unique challenges in the next-generation network architecture.

10.2   Future Network Architectures

10.2.1   C-RAN and H-CRAN

It is foreseeable that wireless data traffic will significantly grow, as reported in [3]. This motivates operators to design new RAN architectures in order to meet the future demand for wireless network. Cloud-RAN was first proposed by IBM [4], and was also introduced by China Mobile Research Institute [5]. In the current cellular radio access network, the baseband unit (BBU) and remote radio head (RRH) are collocated in the base station. However, C-RAN has a geographically separate BBU-RRH structure in which all BBUs, being grouped together and placed in a central room, called a BBU pool, are connected to RRHs via high-speed fronthaul links, say, fiber or microwave. Since the cell site rooms originally occupied by base stations now only have to accommodate antennas and some RF circuits, the energy consumption for air conditioning and other site support equipment can be largely reduced, and the difficulty of base station deployment is also much eased. Widely deploying RRHs, in effect bringing base stations closer to users, can improve network capacity, supporting the ever-increasing data rate service, and consolidating BBUs into BBU pools makes flexible spectrum use, joint signal processing, and scheduling much easier. One of the challenges in C-RAN is the fronthaul link capacity limitation, which constrains the data rate that the RRHs can provide. In general, there are three centralized architecture options [6]: (1) The RRH is equipped only with antennas, and the common public radio interface (CPRI) is used for connecting the RRHs to the BBU pools; (2) the PHY split option has the RRH equipped with some L1 functions, such as FFT/IFFT and cyclic prefix (CP) insertion and framing in addition to the antennas; (3) compared to the PHY split option in which the RRH has partial PHY functions, all PHY functions are shifted to the RRH in the remote PHY option. The more functions the RRH has, the less is the fronthaul link requirement; specifically, the PHY split option needs only 50% of the fronthaul link rate for CPRI and the remote PHY option needs even less, only around 10% of the CPRI rate. Although the PHY split and the remote PHY options alleviate the requirement for the fronthaul link, they have less flexibility in upgrading the system and joint signal processing.

10.2.2 Fog Network

Denser RRHs (which means smaller cells) will incur more inter-cell interference. Besides, that will force mobile users to experience frequent handovers which probably bring more radio link failures. Although the BBUs and RRHs are separate, there exist one-to-one mappings between BBUs and RRHs, i.e., one BBU is assigned to generate (receive) a signal to (from) an RRH. The more RRHs deployed, the more BBUs are needed in the BBU pools, thus there will be a greater consumption of energy and computation resources in the BBU pools. When the traffic load is low in the area covered by multiple small cells, the capacity they can provide may be far beyond what is actually needed at a given moment, making multiple BBUs unnecessary and wasting energy and computation resources in the BBU pool. The FluidNet proposed by Sundaresan et al. [7] adopts a logically reconfigurable fronthaul to use BBU resources more efficiently, taking handover and interference issues into account at the same time. The concept is to dynamically adjust the fronthaul configuration (one-to-one mapping or one-to-many mapping) on the basis of the spatial traffic distribution and demand from users. In one-to-one mapping, the setting supports the maximum amount of traffic, but it does not save computing resources in the BBU. In order to alleviate inter-cell interference, fractional frequency reuse (FFR) is executed. In one-to-many mapping (one BBU to many RRHs), however, the same signal is transmitted simultaneously by multiple RRHs, as in the distributed antenna system (DAS). The BBU-pools resource usage is minimized and the spectrum is under-utilized in this setting, where multiple small cells merge to one bigger cell suitable for mobile (especially vehicular) users.

10.2.3 SoftRAN and V-Cell

The interference and handover issues in a dense environment are considered not only under C-RAN but also in the current LTE architecture. It is argued that the current LTE distributed control plane is not suitable for dense environment, since each base station makes radio resource management decisions on its own without considering the effect on adjacent cells. SoftRAN [8] and V-cell [9] both introduce a logically centralized entity, called the controller, to manage all the radio resources of multiple base stations in a region (which can have the coverage of a macro cell), grouping these base stations as a virtual large base station. The radio resource blocks have three-dimensions time, frequency, and space (base station index). In SoftRAN, the controller periodically receives the local network state from all the base stations and updates the global network state in its database containing the interference map, the flow record, and the network operator preferences. Via a global view of the network state, decisions regarding handover, transmit power, and resource block allocation can be made to balance the loading of cells and reach utility optimization. V-cell has a similar mechanism. However, it focuses more on handover issues.

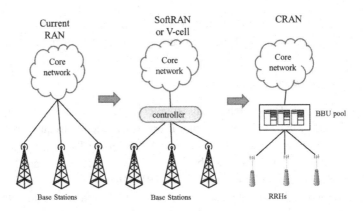

Figure 10.1 The evolution from current RAN to C-RAN.

The development of conventional RAN, SoftRAN, and C-RAN is shown in Fig. 10.1; C-RAN, SoftRAN and V-cell all aim at collaboration between cells. C-RAN has a significant difference in RAN infrastructure, while SoftRAN and V-cell make only a minor change to the current RAN architecture. Centralization is one trend for the future. However, it is not possible for a current distributed RAN suddenly to become a highly centralized C-RAN; there must be a transition period, and the SoftRAN or V-cell would be the intermediate stage in RAN evolution.

10.2.4   Anchor–Booster Architecture

The anchor–booster architecture proposed in LTE-B (beyond LTE-A) [10–12] was introduced mainly to deal with the frequent handover issue in heterogeneous networks formed by the macro cells and various types of small cells. The method splits the C-plane and U-plane, connections, which are supported by the macro base station (MBS) and small base station, respectively. That is, the high-power node (called the anchor eNB) is used for sending the control signal and the low-power node (called the booster eNB) provides a high data rate service. Owing to the wide coverage of the macro cell and the direct C-plane connection between the UE and the anchor eNB, the frequent handovers for mobile users can be largely reduced while the small cells can offer a higher data rate to users since they do not have to waste bandwidth to send the control signals, which include cell-specific primary or secondary synchronization signals (PSS or SSS), cell-specific reference signals, master information block or system information block (MIB or SIB), etc. In this configuration, the booster eNB is no longer the conventional small-cell base station, and the anchor eNB should be able to track the location and the motion of the UE and know the supportable coverage of the booster eNB so that the anchor eNB can inform the appropriate booster eNB to serve an approaching UE via a backhaul link. Thus, the anchor eNB and the under-coverage booster eNBs will appear to the core network as one node.

10.2.5   Heterogeneous C-RAN

The scenario where C-RANs and FluidNet are introduced does not include the hetero-geneous network (HetNet); this was first considered together with C-RAN architecture by Peng *et al.* [13, 14] as H-CRAN. In H-CRAN, the existing MBSs are connected with BBU pools via a backhaul link; some small cells (such as pico-cells and femto-cells) can be formed using RRHs, which can be viewed as highly cognitive-empowered elements due to the BBUs' centralization. Full advantage of C-RAN can be taken for coordination between RRHs, and resource management and cross-tier interference among RRHs and MBSs can be addressed more easily through centralized cloud computing. The article [15] derives downlink performance expressions for a scenario involving the coopera-tion of one MBS and several RRHs, which are both equipped with multiple antennas. Therein three different transmission schemes are analyzed: (1) the best channel among the MBS and the RRHs is selected for transmission; (2) all the RRHs together with the MBS participate in transmission; (3) the minimal number of RRHs needed to reach a predefined data rate is used. In [16] resource sharing in H-CRAN is comprehensively investigated at three levels including spectrum, infrastructure, and network, for each of which the relevant technologies and the pros and cons are discussed.

10.3   Practical Challenges in C-RAN and H-CRAN

10.3.1   Potential Practical Concerns in C-RAN and H-CRAN

In addition to fronthaul limitation and frequent handovers, in practice there are critical challenges in C-RAN and H-CRAN.

1 *Inter-network interference*   The current radio spectrum is very crowded and this leaves only very limited space for future evolution, which results in a compact arrangement of frequency bands between 4G or 5G releases and other wireless sys-tems. For instance, the 2400–2483.5 MHz ISM band is utilized by both WiFi and Bluetooth, while the operating band of LTE-A Band-40 ranges from 2300 to 2400 MHz as an immediate neighbor. Owing to imperfect transceiver components, the compact arrangement of adjacent frequency bands introduces severe inter-network interference, not only among wireless stations but also within a device with mul-tiple radios (i.e., the well known in-device coexistence (IDC) interference), which immensely perplexes system and device designs. From investigations by 3GPP [17], using state-of the-art RF filters alone does not provide sufficient rejection to adjacent channel interference.

2 *Intra-network interference*   Severe interference would occur not only between a small cell and a macro cell but also among small cells under highly dense deploy-ments. There will be different classes of interference in the H-CRAN; among those, the major challenge lies in the interference from a small-cell base station (or RRH) to a macro-cell user located within the coverage of the small cell. To mitigate inter-network interference under this limitation without knowledge on the presence of

other wireless systems, Band-40 may be forbidden to avoid the waste of precious 100 MHz bandwidth.

3 *Latency consideration*   In the LTE or LTE-A, the communication scheme is viewed as closed-loop communication, relying on precise control of the physical layer, which needs considerable signaling overhead such as channel estimation performed 1000 times per second and hybrid automatic repeat request (HARQ) used to make sure of successful frame reception. The signaling overheads in the air interface impose large data-exchange delays in mobile networks. Besides, H-CRAN further induces two new sorts of latency which may be more severe than that in the air interface. The first is latency in resource optimization. Compared with a distributed RAN, the highly centralized H-CRAN should lead to a global optimization over multiple cells in terms of resource allocation, but when the number of RRHs and devices increases, the computational burden becomes heavier and more computational time is consumed, to degrade the latency performance. The second is latency in the routing and paging procedures to forward data to a mobile device in the H-CRAN. In the existing mobile network design, the routing and paging procedures assume that each mobile device may communicate with any other mobile device or server. Therefore, fixed routing and paging information with a hierarchical information-inquiry scheme is adopted. However, this design fully ignores the fact that a mobile device may frequently communicate with the mobile devices or web servers within its social network, while rarely exchanging data with terminals or servers outside its social network. The existing routing and paging procedures may therefore result in a more severe delay than that in the air interface.

10.3.2   Resource Management in C-RAN and H-CRAN

To combat interference in C-RAN or H-CRAN, radio resource management has been shown an effective means [18–21]. Through allocating disjoint radio resources to each user and consequently fully reusing the spectrum in the spatial domain, interference among users can be avoided rather than solely mitigated in layer 1. Centering on radio resource management, this chapter demonstrates the art of leveraging resource management to eliminate potential concerns in practice of C-RAN or H-CRAN.

10.4   Cognitive Radio Resource Management and Software-Defined Design

To solve interference issues in 4G or 5G, it is suggested that the distributive nature of information collection and parameter optimization empowers cognitive radio (CR) technology [22] as a novel design paradigm [23]. However, the favorable part of CR technology for 4G or 5G was not in fact the original idea, which was to develop intelligent devices (with a powerful RF front end, powerful signal processing capability, and powerful computation capability for sophisticated analysis, learning, and decision making) and networks formed by these intelligent devices (i.e., the cognitive radio network (CRN)), owing to the following critical concerns.

1. *Reliability issue*   The major task for a cellular network is to provide reliable services to users while maintaining network stability; specifically, quality of service (QoS) should be guaranteed for the timing of constrained services. In addition, continuous traffic congestion, unrecoverable link failure, and uncontrollable communication behavior should be avoided in a cellular network. These are extremely critical challenges for a CRN with a distributive design. For cellular networks, when a tradeoff between performance and reliability is encountered, reliability should have a higher priority. This is the reason why 4G engages in sophisticated network infrastructures.

2. *Potential system impacts and complexity*   Owing to a conflict in the design philosophy and system architecture between 4G and the conventional CRN, considerable system impacts are introduced in applying the original design paradigm of Cognitive radio (CR) to 4G. This concern has substantially obstructed the development of this direction, at least in this decade.

Since [24], a novel design paradigm based on the CR technology referred to as cognitive radio resource management (CRRM), using radio resources as the core of cross-layer design, opens a practical direction harmonious to the system operations of 4G. With the technical merit of CR technology, the CRRM controls the physical-layer radio operations of communication environment cognition and channel access via the upper-layer resource management. Compatible with the state-of-the-art system architecture of 4G, CRRM smoothly introduces CR technology to 4G. In LTE-A and LTE-B, a variety of communication scenarios will be supported beyond the HetNet, such as those coexisting with systems operating on the ISM band (e.g., WiFi networks), device-to-device (D2D) communications, and heterogeneous coordinated multipoint (Het-CoMP) communications. Targeting at these scenarios, in the literature a variety of CRRM schemes have been proposed recently for different scenarios [25–30]. However, all these schemes suffer from a new dimension of challenge: that the inherent opportunistic channel access results in a severe channel availability variation that significantly harms the QoS guarantees. To combat such an issue, each CRRM scheme needs corresponding optimum control mechanisms. However, such a diverse design perplexes system implementation in 4G or 5G.

10.4.1   Configurations of the Software-Defined Design

Next, we introduce a software-defined design [1] which supports all kinds of communication scenarios, integrating a variety of CRRM schemes to mitigate interference under H-CRAN, D2D communications, and multi-system coexistence (Fig. 10.2).

Scenario 1. Mitigating Interference between a Small Cell and Macro Cells (and WiFi Networks)

As illustrated in Fig. 10.2, all small cells underlying the coverage areas of macro cells (and WiFi networks) may invoke or suffer interference to or from macro cells (and WiFi networks). For such interference mitigation, small cells should be able to identify occupied resource blocks (RBs) in a measurement subframe by measuring the received

By utilizing compressed sensing, the eNB of a picocell can acquire RBs occupation in a subframe on all grids (locations) with limited reports from UE

Game theory can be applied to mitigate interference among small cells

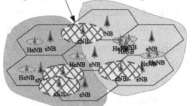

Scenario 1: Mitigating interference between a small cell and macro cells (and WiFi)

Scenario 2: Mitigating interference among small cells

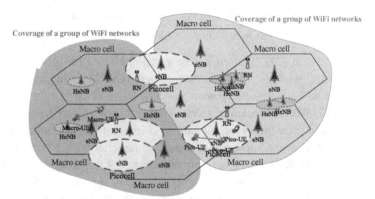

Multi-System Coexistence, HetNet, and D2D Communications

D2D communications underlay macro cells

D2D communications underlay small cells

The traditional CoMP and HetNet multi-hop CoMP can be applied to mitigate interference from WiFi

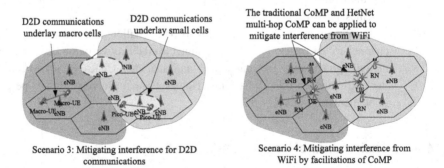

Scenario 3: Mitigating interference for D2D communications

Scenario 4: Mitigating interference from WiFi by facilitations of CoMP

Figure 10.2 Multi-system coexistence, HetNet, and D2D communications in state-of-the-art E-UTRA of 4G or 5G, and a general illustration of the software-defined design.

interference power on all RBs in that measurement subframe. If this quantity exceeds a predetermined threshold on an RB, this RB is identified as occupied by macro cells (or WiFi networks). By utilizing only unoccupied RBs, interference can be consequently mitigated. However, this operation is not generally sufficient for picocells. Since the coverage of a picocell is typically larger than that of a femtocell, in order to completely reflect the interference situation of all UEs distributed over the coverage of the pico-cell, all UEs should measure the corresponding location-based reference-signal received quality (RSRQ) and report the measurement results to the picocell eNB. However, if there is a large number of UEs, the uplink channel for measurement report may suffer from severe congestion. To overcome this huge challenge, compressed sensing technology can be applied [31]. Compressed sensing originated as a signal processing technology which is able to sample an audio or image signal with a sampling rate far lower than the Nyquist rate. The signal can be recovered with an acceptable error rate if certain constraints can be satisfied. Such technical merit makes compressed sensing a powerful technology for picocells. As shown in [1], by utilizing compressed sens-ing, the eNB of a picocell can acquire an RB's occupation profile with limited reports from UEs.

Scenario 2. Mitigating Interference among Small Cells

In a software-defined design, all small cells can autonomously mitigate interference to or from macro cells and WiFi networks. However, the subsequent challenge is interference among small cells deployed close to each other (referred to as collocated small cells). For such collocated small cells, an identical "pool" of unoccupied RBs is identified. To avoid interference among the small cells, each small cell should utilize distinct RBs. Information exchange between small cells formed by RRHs is easy, but this is not the case between these small cells and legacy small cells formed by conventional eNBs. As a result, lacking an appropriate interface for centralized coordination among all small cells, each collocated small cell should randomize the utilization of unoccupied RBs to minimize interference among small cells. However, there is another factor that can significantly affect the performance. Since demands in small cells are typically different from one to another, on the one hand if a small cell decides to utilize as many unoccupied RBs as it needs, the interference may become severe when all collocated small cells have heavy demands. On the other hand, if a small cell decides to utilize unoccupied RBs in a very conservative manner, RBs are under-utilized if all collocated small cells have light demands. Since the traffic demand in a small cell may be unavailable to other small cells, an effective solution to determine the optimum number of unoccupied RBs for utilization lies in game theory [32]. For interference mitigation among femtocells or among picocells, the following enhancement can support software-defined design for collocated small cells.

1. By measuring the number of neighboring small cells (i.e., the number of play-ers in the game) with the layer-1 capability defined in TS 25.967, the number of unoccupied RBs in the measurement subframe, and the number of total RBs in a sub-frame, the utility function can be formulated as the number of available RBs without interference.
2. The optimum number of unoccupied RBs for utilization is determined by the equilibrium solution.
3. After determining the optimum number of unoccupied RBs for utilization, these RBs are utilized by a small cell in a randomized manner.

Scenario 3. Mitigation Interference for a D2D Link

D2D is a new type of communication that will be supported by 3GPP Rel. 12 to reduce the burdens of eNBs by localized communications. There can be two possible cases for D2D communication: an eNB allocates RBs for D2D links; UEs autonomously identify unoccupied RBs for D2D communication. The first case is suitable for D2D links underlaying small cells, as shown in Fig. 10.2. Since a small cell typically handles a smaller number of UEs, scheduling for D2D communication does not introduce significant computational burdens. Furthermore, a small cell typically has a smaller coverage, and thus the impact on other UEs from a small cell can be precisely characterized. The second case is suitable for D2D links underlaying macro cells for similar reasons. For D2D links underlaying small cells, an HeNB or an eNB of a picocell can identify unoccupied RBs from macro cells (and WiFi networks) and neighboring small cells through the configurations for scenario 1 and scenario 2. For D2D links underlaying macro cells, a UE can also utilize the configuration for scenario 1 to identify unoccupied RBs from macro cells (and/or WiFi groups) and neighboring D2D links, respectively. In this case, the measurement period of a D2D link is also adjustable.

Scenario 4. Mitigating Interference between Macro Cells and WiFi

As in the above three scenarios, to mitigate interference to or from WiFi networks, macro cells also adopt measurement subframes for interference estimations. However, unlike the above three scenarios for small cells and D2D communications, macro cells face the huge challenge of handling a large number of UEs. As a result, the measurement period of macro cells and RNs may have to be fixed, as a dynamic measurement period may prevent the system from operating stably to support and coordinate a large number of UEs. In addition, communications in a small cell enjoy a stronger signal strength than that in a macro cell. Therefore, the data rates in a small cell are typically invulnerable to interference from WiFi groups. However, the data rates of longer-distance communications in a macro cell are typically vulnerable to WiFi groups. As a result, most RBs in data subframes may be unavailable for macro cells owing to severe interference from concurrent transmissions of WiFi networks and a macro cell. For IEEE 802.11, the network allocation vector (NAV) for each station can be set to a maximum of around 33 ms for continuous packet transmissions. In this case, communications in a macro cell may be suspended for 33 subframes, which may disable the macro cell from providing QoS guarantees for services with timing constraints. For the traditional link between an eNB and an UE, such interference is never alleviated. To combat this critical issue, a powerful multi-antenna technology of CoMP transmissions [23] can be applied. For traditional CoMP transmissions, multiple eNBs transmit identical packets to a UE via different links. By such transmission diversity, the timing constraint of a service is violated only if none of the links can successfully deliver the packet before the expiration of timing constraints; thus the QoS is enhanced. Although for CoMP transmissions additional radio resources in the spatial domain are required, communication suspension due to interference from WiFi can be alleviated. In future releases of 3GPP, eNBs will no longer be the only transmitter in CoMP transmissions. In addition, relay nodes (RNs) are involved as a major part of CoMP transmissions. Thus, the traditional CoMP evolves to the Het-CoMP, as shown in Fig. 10.2. To support this new feature, the software-defined design can be configured in the following way.

1. All eNBs and RNs involved in CoMP transmissions identify interference from WiFi based on the operation of the software-defined design. The measurement periods of these eNBs and RNs are set to a fixed value.
2. When a packet with a timing constraint is required to be transmitted to a UE, the eNBs involved in CoMP simultaneously transmit identical packets (hence there is duplication of the packet) to the UE, or to an RN if there is one, via different links. The RN transmits the packet to the UE or to the next-hop RN as it receives the packet from an eNB or the previous-hop RN. The eNBs and RNs utilize only unoccupied RBs in each data subframe, and it may take multiple subframes to transmit a packet to the UE. In addition, since eNBs (and RNs) suffer from different levels of interference from WiFi, it also takes distinct numbers of subframes to deliver a packet to a UE from different links (with or without RNs). Since all these eNBs deliver an identical packet, the timing constraint is violated only if none of the links can deliver the packet by the timing constraint expiration.
3. If a packet without timing constraints is required to be transmitted, eNBs are involved in the CoMP transmission of distinct packets to the UE (or to an RN if there is one) via different links to enhance the throughput. The RNs transmit the packet to the UE or the next-hop RN as they receive the packet from an eNB or from the previous-hop RN. The eNBs and RNs utilize only unoccupied RBs in each data subframe.

10.4.2   Optimum Control of The Software Defined Design

To provide QoS guarantees while maximizing resource utilization for intra-network interference mitigation, the measurement period as well as the allocation of unoccupied RBs should be optimized. For this purpose, a novel control theory consisting of a joint queueing theory and information theory, i.e., an effective-bandwidth theory and an effective-capacity theory, can be exploited. The effective-bandwidth theory specifies the minimum service rate for a given arrival process subject to a given QoS requirement. However, the effective-capacity theory specifies the maximum constant-arrival rate that can be supported by the system subject to a given QoS requirement. When the effective bandwidth equals the effective capacity, an equilibrium involving a statistical delay guarantee can be reached (that is, the probability that the packet-delivery delay exceeds the required value is upper bounded by a certain value). To apply this in cellular networks, a HeNB, the eNB of a picocell, or a UE performing D2D communications calculates the effective bandwidth of timing-constrained services [33] and then calculates the effective capacity of the interference configurations in Scenarios 1, 2, or 3 according to the traffic load and RB-allocation correlation of neighboring cells under different measurement periods and numbers of utilized RBs. The measurement period and the number of utilized RBs can consequently be optimized under the condition that the effective bandwidth equals the effective capacity, and the corresponding QoS performance satisfies the requirement. For inter-network interference mitigation, since the measurement period is fixed for macro cells and RNs (and thus the overhead is fixed), the key to providing QoS guarantees while maximizing resource utilization is to determine the optimal packet transmission time for different services in Het-CoMP transmissions, as identical packet transmissions from multiple eNBs may waste spatial-domain radio resources. To achieve this goal, packet transmission delays under Het-CoMP transmissions should be characterized by eNBs for different (voice or video) services on the basis of the different activities of WiFi networks. After analytically characterizing packet transmission delays for different services, eNBs can make the optimum RB allocation as well as admission control such that the timing constraints of all admitted services can be satisfied.

10.4.3   Performance Evaluation

To show the technical merits of the software-defined design of the CRRM, reliability in terms of QoS guarantees is evaluated under different interference scenarios. These simulations are conducted using the system parameters and assumptions defined by 3GPP TR 36.814 for LTE-A with 20 MHz bandwidth. Each RB is composed of 12 subcarriers over seven OFDM symbols. There are seven hexagonal-grid macro cells with wrap-around, and five clusters of small cells deployed over the coverage areas of macro cells. Each cluster is composed of one to five small cells. The transmission powers of a macro cell, a picocell, a femtocell, an RN, and a WiFi network are set to 46, 30, 20, 30, and 3 dBm, respectively. In the software-defined design, a station can autonomously acquire the RB-occupation situations of neighboring cells to proceed to an optimized action according to acquired-RB occupation information. Without the autonomous acquisition of RB-occupation information, a station has no information about RB occupation in a subframe (and the potential trend of such RB occupation) by neighboring cells. Under this constraint, the optimal scheme for interference mitigation lies in a randomized scheme. That is, each eNB, HeNB, or UE in D2D communications randomizes the RB utilization in a subframe to avoid interference on successive RBs. In the following simulations, the performance of the support of real-time voice and video services is in particular evaluated. For intra-network interference mitigation, Table 10.1 shows the simulation results for a small cell on the support of real-time voice and video transmissions. For the voice traffic, a VoIP stream is considered. The arrival process of the VoIP is the well-known ON–OFF fluid model. The holding times in the ON and OFF states are exponentially distributed with means 6.1 s and 8.5 s, respectively. The data rate of the ON state is 32 Kb/s. The delay bound is 20 ms and the delay-bound violation probability is 0.02. For video traffic, a high-quality MPEG4 movie trace is considered. The delay bound of the video traffic is 40 ms and the delay-bound violation probability is 0.02. The results in Table 10.1 show the effectiveness of the support of the statistical delay guarantees by the software-defined design, while the video traffic cannot be supported by this randomized scheme.

For inter-network interference mitigation, Table 10.2 shows simulation results for the timing-constraint-violation probabilities of five VoIPs and video streams. The results are demonstrated in the form of (timing-constraint-violation probability, average activity of WiFi networks on each Het-CoMP link) under different numbers of Het-CoMP links

Table 10.1 Quality of service requirements and simulation results on the delay-bound violation probability for VoIP and high-quality MPEG4 video transmissions in a small cell for intra-network interference mitigation

Traffic	Delay bound	Delay-bound violation prob.	Violation prob. of software-defined design with correlation		Violation prob. of randomization with correlation	
			(low)[a]	(high)[b]	(low)	(high)
Star War	40 ms	≤ 0.02	0.0199	0.0066	0.0464	0.0331
Die hard	40 ms	≤ 0.02	0.0172	0.0058	0.0398	0.0313
Jurassic Park	40 ms	≤ 0.02	0.0181	0.0055	0.0415	0.0299
VoIP	20 ms	≤ 0.02	0.0012	0.0003	0.0013	0.0008

[a] There is a low correlation of RB allocation in macro cells (and WiFi networks) among subframes (that is, if a particular RB is occupied in a subframe then this RB is occupied by macro cells (and WiFi networks) in subsequent subframes with a low probability equal to 0.3).
[b] There is a high correlation of RB allocation in macro cells (and WiFi networks) among subframes (that is, if a particular RB is occupied by macro cells (and WiFi networks) in a subframe then this RB is occupied in subsequent subframes with a high probability equal to 0.8).

(denoted by S). We can observe from Table 10.2 that, for $S = 1$, the QoS of all VoIP and video can be guaranteed when the average activity of the WiFi networks on each Het-CoMP link does not exceed 0.8. For $S = 3$ and $S = 5$, the QoS can be guaranteed when the average activity of the WiFi networks on each Het-CoMP link does not exceed 0.5 or 0.4, respectively. Although Table 10.2 shows that the QoS of all VoIP and video can be guaranteed, it does not reveal the efficiency of utilization of the spatial-domain radio resources in the Het-CoMP; transmitting identical packets via multiple eNBs potentially wastes radio resources. In Fig. 10.3 the numbers of required Het-CoMP (or CoMP) paths of the software-defined design and the state-of-the-art CR transmission scheme are plotted. We can observe from Fig. 10.3 that, to provide QoS guarantees for five VoIP and five video streams, the existing CR scheme requires a larger number of Het-CoMP (or CoMP) paths compared with that of the software-defined design. These results sufficiently demonstrate the effectiveness of the software-defined design at optimum operation.

10.5   Feedbackless Radio Access

Going from the conventional RAN to the H-CRAN is a big change in the infrastructure; this architecture enables cloud computing, joint signal processing, and resource allocation, all of which is about L1, L2, or L3 enabling spectrum efficiency and energy efficiency enhancement. However, if the H-CRAN is were based on the current LTE layered protocols. it would be not able to fully meet future requirements. In the LTE or LTE-A, the communication scheme is viewed as closed-loop communication, relying on precise control of the physical layer, which as mentioned above needs considerable signaling overhead, such as channel estimation performed 1000 times per second and

Table 10.2 Simulation results of QoS provisioning for the scenario of inter-network interference mitigation

VoIP stream	VoIP 1	VoIP 2	VoIP 3	VoIP 4	VoIP 5
$S = 1$	(0.001,0.9), (0.008,0.8)	(0.001,0.9), (0.008,0.8)	(0.001,0.9), (0.008,0.8)	(0.002,0.9), (0.008,0.8)	(0.002,0.9), (0.008,0.8)
$S = 3$	(0,0.9), (0,0.8), (0,0.7), (0.002,0.6), (0.007,0.5)	(0,0.9), (0,0.8), (0,0.7), (0.002,0.6), (0.007,0.5)	(0,0.9), (0,0.8), (0,0.7), (0.002,0.6), (0.007,0.5)	(0,0.9), (0,0.8), (0,0.7), (0.002,0.6), (0.007,0.5)	(0,0.9), (0,0.8), (0,0.7), (0.002,0.6), (0.007,0.5)
$S = 5$	(0,0.9), (0,0.8), (0,0.7), (0,0.6), (0.001,0.5), (0.003,0.4)	(0,0.9), (0,0.8), (0,0.7), (0,0.6), (0.001,0.5), (0.004,0.4)	(0,0.9), (0,0.8), (0,0.7), (0,0.6), (0.001,0.5), (0.004,0.4)	(0,0.9), (0,0.8), (0,0.7), (0,0.6), (0.001,0.5), (0.004,0.4)	(0,0.9), (0,0.8), (0,0.7), (0,0.6), (0.001,0.5), (0.004,0.4)
Video stream	**Video 1**	**Video 2**	**Video 3**	**Video 4**	**Video 5**
$S = 1$	(0,0.9), (0,0.8)	(0,0.9), (0,0.8)	(0,0.9), (0,0.8)	(0,0.9), (0.006,0.8)	(0,0.9), (0.007,0.8)
$S = 3$	(0,0.9), (0,0.8), (0,0.7), (0,0.6), (0.007,0.5)	(0,0.9), (0,0.8), (0,0.7), (0,0.6), (0.007,0.5)	(0,0.9), (0,0.8), (0,0.7), (0,0.6), (0.007,0.5)	(0,0.9), (0,0.8), (0,0.7), (0.001,0.6), (0.005,0.5)	(0,0.9), (0,0.8), (0,0.7), (0.002,0.6), (0.006,0.5)
$S = 5$	(0,0.9), (0,0.8), (0,0.7), (0,0.6), (0,0.5), (0,0.4)	(0,0.9), (0,0.8), (0,0.7), (0,0.6), (0,0.5), (0,0.4)	(0,0.9), (0,0.8), (0,0.7), (0,0.6), (0,0.5), (0,0.4)	(0,0.9), (0,0.8), (0,0.7), (0,0.6), (0.001,0.5), (0.003,0.4)	(0,0.9), (0,0.8), (0,0.7), (0,0.6), (0.001,0.5), (0.003,0.4)

hybrid automatic repeat request (HARQ), used to be sure of successful frame reception. Closed-loop communication will be not suitable in such a high-density environment because it invokes a poor spectrum and energy efficiency in massive uplink transmissions. In order to achieve full automation, it is expected that, in addition to mobile phones and tablets, many other devices will be placed everywhere, including sensors, wearable devices, smart home and smart city devices, etc., and collectively form an Internet of Things (IoT). Even though each device only generates small packets to transmit, in total they create a tremendous traffic in the uplink owing to their huge number. The mechanism of closed-loop communications forces the H-CRAN to acknowledge all these packets in the downlink channel; consequently the downlink channel may become blocked, leading to communication termination. Taking the possibility of the reception of acknowledge messages into account, a link failure is eventually identified when the

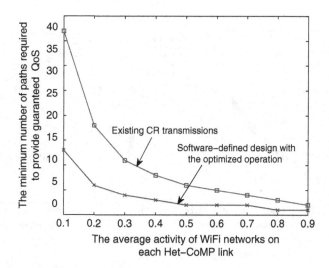

Figure 10.3 The numbers of Het-CoMP paths required to provide QoS guarantees for five VoIP and five video streams.

maximum allowable retransmission number is reached. The latency may not be accept-able for full automation, especially for the case of multi-hop (relay or mesh networks) transmissions, in which the latency at each hop accumulates. In order to tackle this prob-lem, open-loop communication without relying on feedback control to achieve ultra-low latency [2, 34] is proposed. In open-loop communication, a transmitter will not obtain any feedback information (including channel state information, acknowledgment, etc.) from the receiver, no matter whether the transmission is successful or not. Since the transmitter does not need to wait for feedback, much time wastage can be avoided. For reliability concerns, a transmitter can adopt a very conservative modulation and/or a repetitive transmission scheme can be used. The transmitter utilizes duplicates of the radio resources for packet transmission, and the repetition can be in the time domain, the frequency domain, or the spatial domain. The technical merits of open-loop com-munications are revealed in Fig. 10.4. In Fig. 10.4a, seven small cells (i.e., HeNBs or femto-cells) and a number of devices are randomly deployed. Each device attaches to the HeNB with the highest signal strength. However, considering the uplink transmis-sions, as the density of devices increases the SINR for each uplink transmission at the HeNB may decrease owing to increasing interference. To combat interference, in closed-loop communications HARQ and different levels of signaling overheads are needed. In Fig. 10.4a signaling overheads of 10, 20, and 30 percent (compared with the amount of transmitted data) are considered. These overheads include the traffic of acknowledgment messages and channel estimation reports, and the HARQ retransmission number limit is set to 10. In closed-loop communications, when an error occurs in a data transmission, this data transmission will be repeated after the notification of acknowledgment mes-sages. If the density of devices is high (and thus the SINR is low), retransmissions may be performed several times, which severely harms the latency performance. However, in

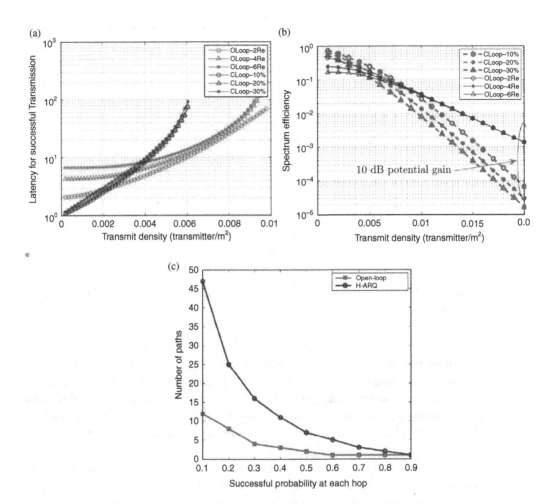

Figure 10.4 (a) Latency performance for successful (error-free) transmission of open-loop and closed-loop communications versus the density of the deployed devices; (b) spectrum efficiency of open-loop and closed-loop communications versus the density of the deployed devices; (c) the number of redundant paths needed to provide QoS guarantees of five VoIP and five MPEG4 streams under different levels of channel fading (in terms of "successful probability") at each hop. In this simulation, the number of hops in each path is randomly selected from the range 0 to 5.

open-loop communications, to combat interference redundant resources (for a conservative modulation and coding scheme and transmission repetitions) are needed, while the amount of these redundant resources is decided in one shot (with two-, four-, and six-fold redundancy compared with the amount of transmitted data). Therefore, we can observe from Fig. 10.4a that, to achieve a successful transmission, the latency performance of open-loop communications must be significantly enhanced compared with that of closed-loop communications, when the density of devices increases. Since redundant resources are required in both closed-loop and open-loop communications, whether the spectrum efficiency of open-loop communications may be worse than that of closed-loop communications needs to be investigated. We can observe from Fig. 10.4b that as the density of devices increases the spectral efficiency of closed-loop communications decays dramatically. However, the spectrum efficiency of open-loop communications is competitive as the density of devices increases. This result demonstrates that open-loop communications are practical for H-CRAN. We also evaluate the latency performance of multi-hop and multi-path transmissions in H-CRAN to support multimedia traffic in Fig. 10.4(c). In multi-hop relay transmissions, if a link failure occurs at a hop then the source node may fail to deliver a packet to the destination node on time. To enhance the latency performance, the source node may replicate (or network-code) packet transmission simultaneously via multiple paths [35]. In this case the latency is unacceptable only if the source node fails to deliver a packet to the destination on time via any path. As a result, there is a tradeoff between latency and the number of redundant paths. In Fig. 10.4(c), 50 disjoint paths were deployed; and each path was composed of a number of links (hops) randomly selected from [36, 37]. In closed-loop communications, HARQ is applied to each hop transmission, and therefore the latency at all hops in a path accumulates. We can observe from Fig. 10.4(c) that closed-loop communication requires more redundant paths than open-loop communication to support five voice (VoIP) and five video (MPEG4) streams. This result confirms the latency-performance improvement via open-loop communication, especially in multi-hop relay transmissions.

In open-loop communication the transmission scheme is autonomously determined by a transmitter. To optimize the transmission scheme without any knowledge at the receiver side, each transmitter thus needs to infer CSI, interference levels, and even the transmission scheme adopted by other transmitters. This operation is very similar to cognitive radio (CR) technology, which can be implemented in different forms for devices with different capabilities. For low-complexity machine-type communication (MTC) devices, the hardware and computing competence are limited. Hence, sophisticated CR functions may not be able to be supported. In this case the CR technology can be realized simply as the listen-before-talk scheme with energy-detection-based clear-channel assessment to detect the presence of interference. If interference is detected, a transmitter is able to transmit data with a large amount of redundancy to protect the data, or to suspend transmissions, to avoid interference. For devices with very large capabilities, CR technology can be realized to be intelligent so as to optimize the overall network performance as well as a user's particular needs. As mentioned earlier, CRRM has been demonstrated to apply CR efficiently to mitigate interference in the HetNet.

The CRRM and subsequent ultra-low-latency network design suggests a novel paradigm of top-down system-design philosophy, by which the radio-resource-control layer of a transmitter jointly optimizes layer-1 and layer-2 resource allocations for channel sensing, data transmissions, and interference avoidance by taking the upper-layer QoS requirements (the establishment of bearings) into account, as illustrated in Fig. 10.5. If interference is detected on certain radio resources, these radio resources are not utilized for data transmission in subsequent subframes (these subframes are

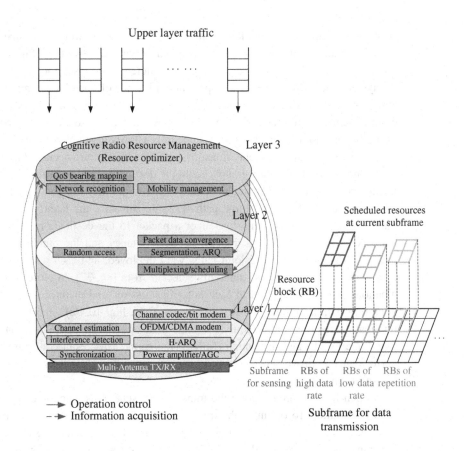

Figure 10.5 The layerless design of CRRM for open-loop communications.

known as data subframes). Frequently allocating measurement subframes can fully capture channel variations as well as interference. However, measurement subframes are a sort of overhead, as data transmissions to detect interference cannot be performed within a measurement subframe. A proper period of allocating measurement subframes is decided on the basis of the latency requirement of the upper-layer traffic. As a result, CRRM converts the existing protocol stacks of mobile networks into a sort of layerless design. Such a design is particularly crucial for small packet transmissions, reducing the latency of packet and protocol conversion between layers, alleviating overheads and processing delays imposed on packets at each layer, and making the radio behavior more flexible. Owing to the coherence of the design goal, CRRM is sufficiently compatible with open-loop radio access for each transmitter to decide autonomously on a transmission scheme that also supports traditional broadband multimedia services.

10.6 Information-Bridled Resource Optimization and Social Data Cache-Based Routing

Another latency issue lies in the computation of high complexity in the cloud. Compared with the distributed RAN, the highly centralized H-CRAN should lead to a global optimization over multiple cells in terms of resource allocation, but when the number of RRHs and devices increases, the computational burden will be heavier; then more time will be consumed in computation, also degrading the latency performance.

To provide reliability in open-loop communications, a transmitter does not need instantaneous CSI. Instead, it needs the long-term statistics of CSI, interference levels, and transmission schemes adopted by other transmitters. Such radio information can be obtained socially through CR technology or provided globally by the network. Given radio information, each transmitter is thus able to optimize its transmission scheme. In this configuration the computing servers in the H-CRAN do not have to conduct centralized resource optimization. Instead, the H-CRAN needs only to tailor the radio information provided to all transmitters. As the optimization procedure performed by each transmitter with given radio information is well known by the H-CRAN, it can control all transmitters through controlling and optimizing the information fed to transmitters. This concept leads to a new architecture of information-bridled resource optimization in the H-CRAN, as illustrated in Fig. 10.6. In this architecture the H-CRAN optimizes only the radio information provided to transmitters; thus the radio accesses of all transmitters are under the control of the H-CRAN, as all transmitters optimize their transmission schemes on the basis of the radio information provided. The principles of information-bridled resource optimization are summarized as follows; we use uplink transmission as an elaboration example.

1. The H-CRAN broadcasts a radio information set $\mathbf{R} = (r_1, r_2, \ldots, r_K)$ regarding radio resources for all mobile devices (transmitters), where K is the total number of radio resources in a scheduling period. By taking the statistics of the CSI, interference levels, and transmission schemes adopted by all mobile devices into account, radio information for the kth radio resource is mapped into a number r_k, which is normalized to $0 \leq r_k \leq 1$. Radio resources can be defined in different domains, such as the time, frequency, eNB, code, or spatial domain. A lower value of r_k reveals that the kth radio resource suffers from severe channel fading, interference, or congestion. This distracts mobile devices from selecting the kth radio resource. However, an r_K of a higher value implies that the kth radio resource enjoys better channel quality, lower interference, or only mild congestion. Such an indication attracts mobile devices to select the kth radio resource.

2. Upon receiving a set of radio information $\mathbf{R} = (r_1, r_2, \ldots, r_K)$, each mobile device autonomously selects a radio resource by taking the resource selection strategies adopted by other mobile devices into consideration. That is, a mobile device should not always select the radio resource with the highest r_k since other mobile devices may also select this radio resource, leading to severe interference and congestion. Game theory can be an effective means for all mobile devices to determine a transmission scheme in this scenario so as to optimize the utility. After selecting a radio resource, each mobile device then transmits data to the H-CRAN via this radio resource in an open-loop fashion.

3. After the transmissions from all mobile devices, the H-CRAN is able to obtain the interference and congestion levels at all the radio resources. In addition, the H-CRAN

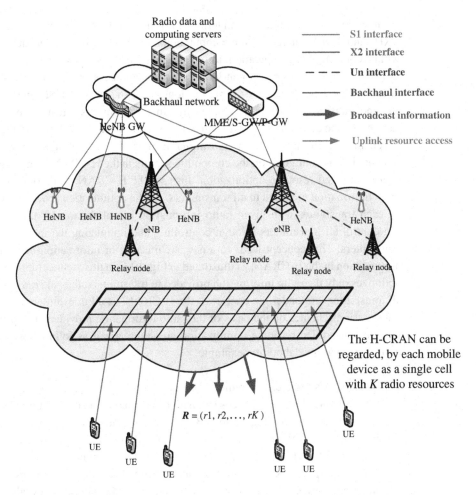

Figure 10.6 Information-bridled resource optimization, in which the H-CRAN optimizes only radio information broadcast to all mobile devices. Given this radio information, each mobile device autonomously optimizes the transmission scheme. Thus, the H-CRAN controls the radio access of all mobile devices implicitly.

needs to estimate the CSI of all radio resources at all locations (known as the spectrum map). The spectrum map can be estimated via the uplink transmissions from all mobile devices [29]. Then the H-CRAN optimizes and updates $\mathbf{R} = (r_1, r_2, \ldots, r_K)$ and broadcasts this radio information.

As mentioned earlier, the complexity of conventional resource optimization becomes subject to the number of available resources K, the number of devices M, and the number of eNBs N by adopting joint resource allocation among all cells. Such a complexity is around the level of $O(KMN)$. However, information-bridled resource optimization optimizes only $\mathbf{R} = (r_1, r_2, \ldots, r_K)$. The complexity is thus $O(KMN)$. Note that information-bridled resource optimization is very different from closed-loop communication in the following respects.

4. In closed-loop communication the receiver feeds back CSI to the corresponding transmitter. However, the H-CRAN broadcasts a common **R** to all transmitters, which does not impose large signaling overheads.

5. The concept of information-bridled resource optimization is infeasible to apply to closed-loop communications. In closed-loop communications the optimization scheme in layers 1 and 2 is fixed to maximize the data rate on the basis of the present SINR. However, H-CRAN controls the SINR via radio resource allocations, and thus, in this framework, H-CRAN cannot further reduce the complexity.

6. In closed-loop communications, layers 1 and 2 of a transmitter need instantaneous CSI. However, exchanging information within the H-CRAN suffers from inevitable latency, which may not support instantaneous CSI exchanges in closed-loop communications.

Social Data Cache-Based Routing and Paging

To further alleviate the third type of latency in routing and paging in the H-CRAN, we should understand the state-of-the-art procedure in existing mobile networks, shown in Fig. 10.7(a). Most wireless services, e.g., user social applications, M2M IoT communications, and remote control, especially for full automation, need an application (app) server to store wireless service data and handle service functions. Suppose that a mobile device (UE-A) wishes to send a message to (or access) another mobile device within the same eNB's coverage (UE-B) or within another eNB's coverage (UE-C); then a six-segment protocol is needed for this procedure.

1st segment   UE-A sends the message to the eNB.
2nd segment   The eNB forwards this message to the S-GW or P-GW.
3rd segment   The S-GW or P-GW routes this message to the APP server.
4th segment   Upon receiving the message, the APP server stores it and then routes it back to the S-GW or P-GW.
5th segment   According to paging information, the S-GW or P-GW forwards the message to the eNB.
6th segment   The eNB sends the message to UE-B (or UE-C).

Such a framework may invoke unacceptable latency in the worst case, which is not allowed for remote control to robots or vehicles, intelligent transportation systems, or immersive sensory experience. Nevertheless, the number of hops in Fig. 10.7a can actually be significantly reduced, as shown in Fig. 10.7b. Then we have the following.

1st segment   UE-A sends the message to the eNB.
2nd segment   If the H-CRAN caches the knowledge about the destination of the message, the eNB is able to forward the message directly to the collocated mobile device (UE-B) or forward to another eNB with the destination mobile device (UE-C). In the meantime, the eNB also routes the message to the S-GW or P-GW.
3rd segment   The eNB forwards the message to UE-C. The S-GW or P-GW forwards the message to the app server.

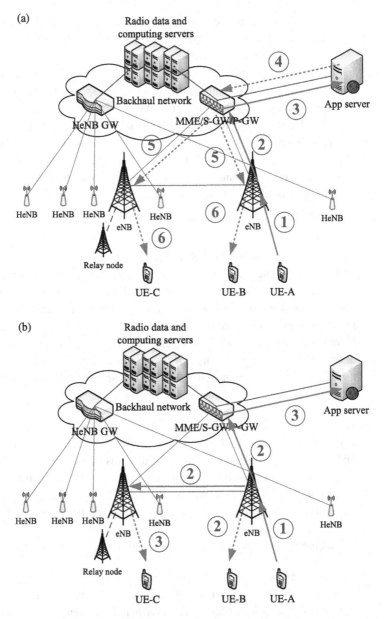

Figure 10.7 (a) In existing mobile networks, six segments are needed to exchange an application message between two mobile devices; (b) by caching the social profiles of mobile devices in the H-CRAN, the number of hops (and thus the latency) can be significantly reduced.

The challenge of applying the state-of-the-art framework in Fig. 10.7(a) to the enhancement in Fig. 10.7(b) is twofold.

1. The existing framework assumes that a mobile device may send messages to every mobile device in the world with equal likelihood. For this purpose, the sophisticated

routing and paging architecture as shown in Fig. 10.7(a) is adopted. However, this assumption is impractical owing to its disregard of the social relationship among mobile devices [38]. The social relationship is the correlation among mobile devices in terms of geographic locations (i.e., collocated with each other), identities (e.g., students of the same university), users' interpersonal connections, or contact platforms (e.g., connection to a common website or email or game server). In other words, the destinations (i.e., mobile devices, servers, and websites) that a mobile device contacts are subject to the social network or social properties of the mobile device user. According to different communication categories, there are three types of social networks: human-to-human, human-to-machine, and machine-to-machine (M2M), as shown in Fig. 10.8(a). In all these three types, each mobile device contacts only a certain set of devices instead of all devices in the network system. Such a social profile of each mobile device does not change rapidly, and this nature is not fully exploited in the existing design of mobile networks [39, 40].

2. The social profile of each mobile device is available in the application layer. However, this information is unavailable to radio layers based on the existing layered architecture of mobile networks. The general concept of a social-network-driven mobile network design first arose in [38], which revealed a number of novel methodologies for future mobile networks. However, to develop a solution compatible to the H-CRAN, the impacts on the existing infrastructures should be minimized. To facilitate this engineering constraint, a promising design lies in the concept of OpenFlow in software-defined networking (SDN) [41, 42]. In OpenFlow the network is able to extract information in packet headers to distinguish data packets and control packets with different QoS requirements, and impose information to packet headers to boost packet routing. However, to enable the performance enhancement in Fig. 10.8(b), a co-design between SDN in the H-CRAN and the app server is needed.

3. The H-CRAN caches the social profiles of all mobile devices provided by app servers, as shown in Fig. 10.8a. From these social profiles, the H-CRAN can construct a connection map for each mobile device in OpenFlow. The connection map specifies the route to possible destinations of a mobile device. Since the number of possible destinations for a mobile device is limited to the cardinality of the mobile device's social network, the size of a connection map is also limited. The construction of a connection map also takes into account the radio resource allocation in the H-CRAN. For example, if two mobile devices are collocated, a D2D link can be configured by the H-CRAN for direct data exchanges among mobile devices.

4. When a mobile device sends a message, the app server allows the H-CRAN to forward the message directly to the destination without passing through the app server. Nevertheless, the H-CRAN could forward the message to the app server whenever necessary.

The performance of the above social data cache-based routing and paging scheme can be evaluated primitively via the routing process using mobile IP. In this experiment, a router (in a home network) is connected by two devices (say, DEV1 and DEV2). Another router (in a foreign network) is connected by the home network router and

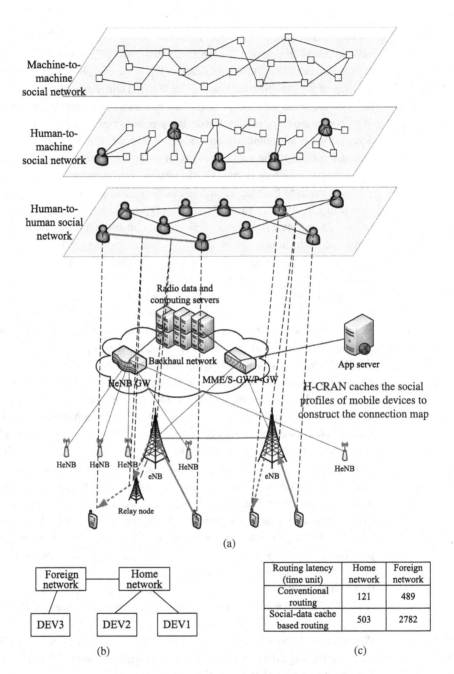

(a)

Foreign network	Home network
DEV3	DEV2 DEV1

(b)

Routing latency (time unit)	Home network	Foreign network
Conventional routing	121	489
Social-data cache based routing	503	2782

(c)

Figure 10.8 (a) The H-CRAN caches social profiles to construct connection maps for all mobile devices; (b) system layout for the performance of the proposed social data cache-based routing and paging scheme, with three devices; (c) performance evaluation results of the conventional scheme and of the proposed scheme.

a device (DEV3), as shown in Fig. 10.8b. In the conventional routing procedure, when DEV1 wishes to communicate with DEV2 the home network router checks whether DEV2 is within its routing domain. If this is so, packets from DEV1 are forwarded to DEV2. However, in the social data cache-based routing scheme, the social profile of DEV1 (in this case represented by DEV2) is available for the home network router. As DEV1 attaches to the home network router, a routing table to DEV2 is ready in the home network router, and therefore the routing latency is significantly reduced. Such performance is demonstrated in Fig. 10.8(c), where the routing latency is shown in time units. A time unit is a logical unit of time used by counters in routing protocols. If DEV1 wishes to communicate with DEV3, in the conventional routing procedure the home network router checks whether DEV3 is within its routing domain. If this is not so, the home network router checks whether DEV3 is within the foreign network router's routing domain. If so, packets from DEV1 can be routed via the home network router and the foreign network router to DEV3. In contrast, in the social data cache-based routing scheme, since both the home network router and the foreign network router have knowledge of the social relationship between DEV1 and DEV3 (i.e., the social profile), when DEV1 attaches to the home network router a routing table to DEV3 is ready for DEV1. The technical merit of the social data cache-based routing scheme can be observed from Fig. 10.8(c).

The concept of the above social data cache-based routing and paging scheme significantly reduces the packet exchanges crossing the H-CRAN and the cyber world. It also precludes destinations outside the social network of a mobile device, which leads to concise routing and paging, eliminating latency in the H-CRAN.

10.7   Conclusion

In order to meet the future wireless communication requirement, RANs should evolve to a brand new type, significantly different from the current architecture. Cloud-RAN, as a candidate for 5G RAN, has been much discussed in the literature. Compared with C-RAN, SoftRAN and V-cell make minor changes to the current RAN to improve its performance and could be intermediate in the RAN evolution. Heterogeneous CRAN is a combination of C-RAN and heterogeneous networks in which the existing macro base stations are connected with BBU pools via backhaul links. Moreover, the C/U plane splitting scheme can be adopted to tackle the frequent handover problems and alleviate the fronthaul link burden. In addition to fronthaul limitation and frequent handover, H-CRAN has other problems such as inter-network interference, intra-network interference, and long latency. As for interference issues, the CRRM could be an effective measure, which reforms the original concept of the CR to a realistic layerless technology. Considering all practical scenarios, a software-defined design with optimum control relieving individual subsystem function should be introduced. Such a design enables the development of the CRRM from a promising concept to successful practical implementation. Regarding the long-latency issue, mainly resulting from the radio access, optimization computation, and routing and paging procedures, new methodologies enabling ultra-low-latency connections in the H-CRAN have also been introduced; these involve a systematic design, with unique open-loop radio access reducing latency in the air interface, information-bridled resource optimization reducing the latency of radio resource optimization, and a social data cache-based routing and paging scheme reducing latency in the backhaul packet forwarding. The methodologies in this chapter resolve several predicaments and open issues in H-CRAN through introducing a new design philosophy to practical H-CRAN. Further harmonization with edge computing and networking may be studied in [43].

References

[1] S.-Y. Lien, K.-C. Chen, Y.-C. Liang, and Y. Lin, "Cognitive radio resource management for future cellular networks," *IEEE Trans. Wireless Commun.*, vol. 21, no. 1, pp. 70–79, February 2014.

[2] S.-Y. Lien, S.-C. Hung, K.-C. Chen, and Y.-C. Liang, "Ultra-low-latency ubiquitous connections in heterogeneous cloud radio access networks," *IEEE Trans. Wireless Commun.*, vol. 22, no. 3, pp. 22–31, June 2015.

[3] Cisco, "Visual networking index: global mobile data traffic forecast update," Cisco White Paper, February 2016.

[4] Y. Lin, L. Shao, Z. Zhu, Q. Wang, and R. Sabhikhi, "Wireless network cloud: architecture and system requirements," *IBM J. Res. Dev.*, vol. 54, no. 1, pp. 4:1–4:12, January 2010.

[5] China Mobile Research Institute, "C-ran. The road towards green ran," Technical Report, October 2011.

[6] H. Niu, C. Li, A. Papathanassiou, and G. Wu, "Ran architecture options and performance for 5g network evolution," in *Proc. 2014 IEEE Wireless Communications and Networking Conf. Workshops*, April 2014, pp. 294–298.

[7] K. Sundaresan, M. Arslan, S. Singh, S. Rangarajan, and S. Krishnamurthy, "Fluidnet: a flexible cloud-based radio access network for small cells," *Netw.*, vol. 24, no. 99, pp. 1–14, 2015.

[8] A. Gudipati, D. Perry, L. E. Li, and S. Katti, "Softran: software defined radio access network," in *Proc. 2nd ACM SIGCOMM Workshop on Hot Topics in Software Defined Networking*, New York, ACM, 2013, pp. 25–30. Online. Available at doi.acm.org/10.1145/2491185.2491207.

[9] R. Riggio, K. Gomez, L. Goratti, R. Fedrizzi, and T. Rasheed, "V-cell: going beyond the cell abstraction in 5g mobile networks," in *Proc. 2014 IEEE Network Operations and Management Symp.*, May 2014, pp. 1–5.

[10] H. Ishii, Y. Kishiyama, and H. Takahashi, "A novel architecture for lte-b: c-plane/u-plane split and phantom cell concept," in *Proc. 2014 IEEE Globecom Workshops*, December 2012, pp. 624–630.

[11] Q. Li, H. Niu, G. Wu, and R. Hu, "Anchor-booster based heterogeneous networks with mmwave capable booster cells," in *Proc. 2014 IEEE Globecom Workshops*, December 2013, pp. 93–98.

[12] A. Mukherjee, "Macro–small cell grouping in dual connectivity lte-b networks with non-ideal backhaul," in *Proc. 2014 IEEE Int. Conf. on Communications*, June 2014, pp. 2520–2525.

[13] M. Peng, Y. Li, J. Jiang, J. Li, and C. Wang, "Heterogeneous cloud radio access networks: a new perspective for enhancing spectral and energy efficiencies," *IEEE Wireless Commun.*, vol. 21, no. 6, pp. 126–135, December 2014.

[14] M. Peng, Y. Li, Z. Zhao, and C. Wang, "System architecture and key technologies for 5g heterogeneous cloud radio access networks," *IEEE Netw.*, vol. 29, no. 2, pp. 6–14, March 2015.

[15] F. Khan, H. He, J. Xue, and T. Ratnarajah, "Performance analysis of cloud radio access networks with distributed multiple antenna remote radio heads," *IEEE Trans. Signal Process.*, vol. 63, no. 18, pp. 4784–4799, September 2015.

[16] M. Marotta, N. Kaminski, I. Gomez-Miguelez, L. Zambenedetti Granville, J. Rochol, L. DaSilva *et al.*, "Resource sharing in heterogeneous cloud radio access networks," *IEEE Trans. Wireless Commun.*, vol. 22, no. 3, pp. 74–82, June 2015.

[17] Z. Hu, R. Susitaival, Z. Chen, I.-K. Fu, P. Dayal, and S. Baghel, "Interference avoidance for in-device coexistence in 3gpp lte-advanced: challenges and solutions," *IEEE Commun. Mag.*, vol. 50, no. 11, pp. 60–67, November 2012.

[18] M. Peng, K. Zhang, J. Jiang, J. Wang, and W. Wang, "Energy-efficient resource assignment and power allocation in heterogeneous cloud radio access networks," *IEEE Trans. Veh. Technol.*, vol. 64, no. 11, pp. 5275–5287, November 2015.

[19] M. Gerasimenko, D. Moltchanov, R. Florea, S. Andreev, Y. Koucheryavy, N. Himayat *et al.*, "Cooperative radio resource management in heterogeneous cloud radio access networks," *IEEE Access*, vol. 3, pp. 397–406, 2015.

[20] M. Marotta, N. Kaminski, I. Gomez-Miguelez, L. Zambenedetti Granville, J. Rochol, L. DaSilva *et al.*, "Resource sharing in heterogeneous cloud radio access networks," *IEEE Trans. Wireless Commun.*, vol. 22, no. 3, pp. 74–82, June 2015.

[21] A. Douik, H. Dahrouj, T. Al-Naffouri, and M.-S. Alouini, "Coordinated scheduling and power control in cloud-radio access networks," *IEEE Trans. Wireless Commun.* vol. 15, no. 4, pp. 2523–2536, April 2015.

[22] Y.-C. Liang, Y. Zeng, E. Peh, and A. T. Hoang, "Sensing–throughput tradeoff for cognitive radio networks," *IEEE Trans. Wireless Commun.*, vol. 7, no. 4, pp. 1326–1337, April 2008.

[23] A. Attar, V. Krishnamurthy, and O. Gharehshiran, "Interference management using cognitive base-stations for umts lte," *IEEE Commun. Mag.*, vol. 49, no. 8, pp. 152–159, August 2011.

[24] S.-Y. Lien, C.-C. Tseng, K.-C. Chen, and C.-W. Su, "Cognitive radio resource management for qos guarantees in autonomous femtocell networks," in *2010 IEEE Int. Conf. on Proc. Communications*, May 2010, pp. 1–6.

[25] S.-Y. Lien, Y.-Y. Lin, and K.-C. Chen, "Cognitive and game-theoretical radio resource management for autonomous femtocells with qos guarantees," *IEEE Trans. Wireless Commun.* vol. 10, no. 7, pp. 2196–2206, July 2011.

[26] S.-Y. Lien and K.-C. Chen, "Statistical traffic control for cognitive radio empowered lte-advanced with network mimo," in *Proc. 2011 IEEE Conf. on Computer Communications Workshops*, April 2011, pp. 80–84.

[27] J. Huang and V. Krishnamurthy, "Cognitive base stations in lte/3gpp femtocells: A correlated equilibrium game-theoretic approach," *IEEE Trans. Commun.*, vol. 59, no. 12, pp. 3485–3493, December 2011.

[28] Y.-Y. Li and E. Sousa, "Cognitive uplink interference management in 4g cellular femtocells," in *Proc. 2010 IEEE 21st Int. Symp. on Personal Indoor and Mobile Radio Communications*, September 2010, pp. 1567–1571.

[29] S.-Y. Lien, S.-M. Cheng, S.-Y. Shih, and K.-C. Chen, "Radio resource management for qos guarantees in cyber-physical systems," *IEEE Trans. Parallel Distrib. Syst.*, vol. 23, no. 9, pp. 1752–1761, September 2012.

[30] Q. Wang, J. Wang, Y. Lin, J. Tang, and Z. Zhu, "Interference management for smart grid communication under cognitive wireless network," in *Proc. 2012 IEEE 3rd Int. Conf. on Smart Grid Communications*, November 2012, pp. 246–251.

[31] D. Donoho, "Compressed sensing," *IEEE Trans. Inf. Theory*, vol. 52, no. 4, pp. 1289–1306, April 2006.

[32] D. Fudenberg and J. Tirole, *Game Theory*. MIT Press, 1991.

[33] C.-S. Chang, *Performance Guarantees in Communication Networks*. Springer, 2000.

[34] S.-Y. Lien, S.-C. Hung, and K.-C. Chen, "Optimal radio access for fully packet-switching 5g networks," in *Proc. 2015 IEEE Int. Conf. on Communications*, June 2015, pp. 3921–3926.

[35] I.-W. Lai, C.-L. Chen, C.-H. Lee, K.-C. Chen, and E. Biglieri, "End-to-end virtual mimo transmission in ad hoc cognitive radio networks," *IEEE Trans. Wireless Commun.*, vol. 13, no. 1, pp. 330–341, January 2014.

[36] D. Lopez-Perez, I. Guvenc, G. de la Roche, M. Kountouris, T. Q. S. Quek, and J. Zhang, "Enhanced intercell interference coordination challenges in heterogeneous networks," *IEEE Trans. Wireless Commun.*, vol. 18, no. 3, pp. 22–30, June 2011.

[37] R. Balakrishnan and B. Canberk, "Traffic-aware qos provisioning and admission control in ofdma hybrid small cells," *IEEE Trans. Veh. Technol.*, vol. 63, no. 2, pp. 802–810, February 2014.

[38] K.-C. Chen, M. Chiang, and H. Poor, "From technological networks to social networks," *IEEE J. Sel. Areas Commun.*, vol. 31, no. 9, pp. 548–572, September 2013.

[39] Y. Yang and T. Q. S. Quek, "Optimal subsidies for shared small cell networks: a social network perspective," *IEEE J. Select. Topics Signal Process.*, vol. 8, no. 4, pp. 690–702, August 2014.

[40] E. Stai, V. Karyotis, and S. Papavassiliou, "Exploiting socio-physical network interactions via a utility-based framework for resource management in mobile social networks," *IEEE Trans. Wireless Commun.*, vol. 21, no. 1, pp. 10–17, February 2014.

[41] H. Kim and N. Feamster, "Improving network management with software defined networking," *IEEE Commun. Mag.*, vol. 51, no. 2, pp. 114–119, February 2013.

[42] S. Yeganeh, A. Tootoonchian, and Y. Ganjali, "On scalability of software-defined networking," *IEEE Commun. Mag.*, vol. 51, no. 2, pp. 136–141, February 2013.

[43] S.-C. Hung, H. Hsu, S.-Y. Lien, , and K.-C. Chen, "Architecture harmonization between cloud radio access networks and fog networks," to appear in *IEEE Access*, 2016.

11 Coordinated Scheduling in C-RANs

Ahmed Douik, Hayssam Dahrouj, Oussama Dhifallah, Tareq Y. Al-Naffouri, and Mohamed-Slim Alouini

11.1 Introduction

In the wireless network literature, scheduling denotes the strategy according to which users are active at each time–frequency resource block. In this classical literature, scheduling is done per base station (BS), i.e., without coordination between the BSs. With the large demand for mobile data services that is straining wireless networks nowadays, coordination among transmitters becomes a real necessity to better manage the high levels of interference.

The C-RAN architecture is a practical platform for the implementation of coordinated multi-point (COMP) systems. By connecting all BSs to a central computing center (i.e., a cloud) via wire or wireless backhaul links, C-RAN allows joint signal processing and coordinated resource allocation at the cloud. Depending on the capacity of the connecting backhaul, the level of coordination among the BSs may vary from a high coordination level (e.g., signal-level coordination as in fiber optic high-capacity backhaul links) to a low coordination level (e.g., resource allocation coordination as in wireless backhaul links). In situations where optical fiber backhaul links are unavailable at the exact location of a BS, or when the extension of optical fiber cables to the base station location is prohibitively expensive, wireless backhauls become a cheaper and easier-to-deploy solution. Intelligently devising schemes for the coordination of resource allocation in a C-RAN setup is, therefore, a real necessity for providing a practical, reliable, and scalable solution to the impending capacity crunch.

From the recent literature, coordinated resource allocation can be classified into three categories:

- coordinated beamforming;
- coordinated power control;
- coordinated scheduling.

While adjustment of the continuous variables (i.e., the beamforming vectors and the power) typically requires high-resolution algorithms and sensitive hardware physical platforms, discrete allocation (i.e., scheduling) is considered a more practical resource-allocation solution. Scheduling problems are, nevertheless, discrete optimization problems which may be often NP-hard problems. This chapter provides a framework for solving coordinated scheduling problems in C-RANs, using graph theory practical techniques that perform close-to-optimal solutions.

In conventional cellular network architecture, scheduling policies are often performed per base station given a pre-known association of users and base stations, e.g., the classical proportionally fair scheduling [1, 2]. One point illustrated in this chapter is that coordinated scheduling not only allows assigning users to resource blocks across the network but also jointly solves for the user-to-BS association problem.

Throughout this chapter, it is assumed that the cloud is solely responsible for the scheduling policy and synchronization of the base stations' transmission, i.e., there is no signal-level coordination. Consider the downlink of a C-RAN, where the BSs are connected to the cloud via low-rate links. Multiple remote users may be served by each BS through time–frequency multiplexing, as each BS transmit frame consists of several time–frequency blocks, each serving one user. The network performance becomes, therefore, a function of the scheduling decisions taken at the cloud. The chapter addresses coordinated scheduling in C-RANs by maximizing a generic network-wide utility subject to connectivity constraints. It presents a framework for solving the problem using techniques from graph theory.

11.1.1   Notation

The following notation is adopted throughout the chapter. Sets are defined using the calligraphic font. For a set $\mathcal{X}$, let $|\mathcal{X}|$ be its cardinality, i.e., the number of elements of $\mathcal{X}$. The Cartesian product of sets $\mathcal{X}$ and $\mathcal{Y}$ is denoted by $\mathcal{X} \times \mathcal{Y}$. The set of all subsets of $\mathcal{X}$, also known as the power set of $\mathcal{X}$, is denoted by $\mathcal{P}(\mathcal{X})$. Finally, the discrete Dirac function is denoted by $\delta(x)$. The function $\delta(x)$ is equal to 1 if and only if its argument is equal to 0; otherwise, $\delta(x)$ is equal to 0.

11.2   Coordinated Scheduling in a Single Cloud-RAN

In this section we consider the coordinated scheduling problem in a single C-RAN. The section first presents a concise overview of both the signal-level and the scheduling-level coordination strategies. For its practical implementation, this section then adopts scheduling-level coordination in which the cloud is responsible only for synchronizing the connected base station and scheduling users to the different radio resource blocks of the base stations. The radio resource blocks, called power zones (PZs) throughout this section, are maintained at fixed transmit power. The section addresses the problem of maximizing a network-wide utility by assigning users to base stations and power zones across the network under the scheduling-level coordination constraint that each user can be associated with at most a single base station, but potentially with various power zones within this base station's transmit frame. We use a graph-theoretical approach to solve the problem at hand by introducing the conflict graph and reformulating the problem as a maximum-weight independent-set problem.[1] Given the NP-hardness of

---

[1] An independent set is a set of vertices in a graph, no two of which are connected. A maximum independent set is the largest possible independent set. The maximum-weight independent, set problem is that of finding the independent set with the highest weight.

the optimal solution, the section suggests a suboptimal low-complexity heuristic. Simulation results reveal that the greedy approach performs near-optimal in low shadowing environments.

11.2.1   Scheduling-Level and Signal-Level Coordination

Consider the downlink of a network comprising several base stations connected to a central unit (the cloud). The transmit frame of each base station contains various time/frequency resource blocks, denoted as power zones. The scheduling problem considered in this chapter is that of assigning users to base stations and their transmit frame power zones, so as to maximize a given network-wide metric.

Traditionally, such scheduling is performed assuming no inter-base-station coordination, i.e., per base-station. A popular approach is the classical proportionally fair scheduling [1, 2] in which scheduling is performed on the basis of a pre-assigned association of users and base stations, in such a way that fairness is guaranteed. With the advent of cloud-enabled networks, the central unit enables synchronization of the transmit frames of the connected base stations. Such base station coordination allows network-wide scheduling, resulting in a more efficient use of the radio resources. Different coordination strategies are possible in C-RANs, namely signal-level and the scheduling-level coordination.

Signal-level coordination [3–5] is achieved by jointly encoding and decoding the users' data. However, such joint resource allocation and signal processing of the data belonging to different users at the cloud necessitates high-capacity low-latency backhaul and fronthaul links so as to share all users' data streams between the connected base stations. Furthermore, given the finite bandwidth of the links, such a substantial amount of backhaul and fronthaul communications introduces quantization noise that is inherent to the compression problem. Nonetheless, the effects of the quantization noise can be intelligently mitigated through the appropriate design of the precoding and the correlation matrices of the quantization noise; see e.g., [5–7], which propose using the majorization–minimization (MM) approach and Wyner–Ziv (WZ) compression to reach a local optimum.

Scheduling-level coordination [8, 9], however, is a simplified resource coordination scheme as it requires a lower level of backhaul communication. In scheduling-level coordination the cloud, besides its role in coordinating the transmit frames of the connected base stations, is responsible only for efficiently allocating the different resources among users. Although such a lower level of coordination may reduce the overall performance of the system as compared with signal-level coordination, it suits well a scenario of low data rate backhaul links. In other words, the coordinated scheduling problem is that of optimally assigning users to base stations and power zones so as to maximize a generic network-wide merit function under the system limitation that each user can be assigned to, at maximum, a single base station; otherwise, signal-level coordination is required. Moreover, to exploit the radio resources efficiently, each power zone serves exactly one user. We focus next on the coordinated scheduling problem. The problem is solved by introducing a conflict graph and showing its equivalence with a well-known graph-theory problem known as the maximum-weight independent set problem.

11.2.2   System Model

Consider the downlink of a cloud radio access network comprising a set $\mathcal{B}$ of B base stations connected to a central processor via low-rate links. Figure 11.1 shows such a C-RAN composed of eight base stations and 12 users. The cloud is responsible for the synchronization of the transmit frames of all base stations and the scheduling of a set $\mathcal{U}$ of U users. The frame structure of every base station consists of a set $\mathcal{Z}$ of Z resource blocks, called power zones (PZs). Hence, the total number of available PZs is $Z_{\text{tot}} = BZ$. In this section we assume that the transmit power of each PZ is maintained at a fixed level. Let $P_{bz} \forall b \in \mathcal{B}$ and $\forall z \in \mathcal{Z}$ be the power level of the zth PZ in the bth BS, as shown in Fig. 11.2, which illustrates the frame structure under consideration. The power levels of each PZ can be optimized for better overall performance; however, this falls outside the scope of this chapter as here we are focusing only on solving the scheduling problem.

Remark 11.1   The scheduling problem considered here is closely related to the concept developed in [8]. The authors in [8] considered the scheduling problem in a soft-frequency reuse setup wherein the number of users U corresponds to the total number of available PZs, i.e., $Z \times B$. Such an assumption allowed the authors to reformulate the scheduling problem as a simple linear assignment problem and to solve it using classical

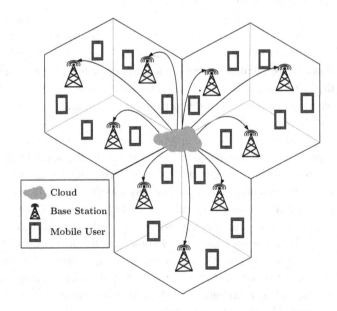

Figure 11.1 Network configuration.

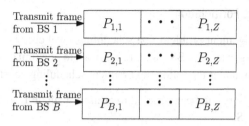

Figure 11.2 Frame structure.

auction methodology [10]. We consider the most general case, without restrictions on the numbers of users, BSs, or PZs.

We employ a general utility function π_{uzb} associating a user $u \in \mathcal{U}$ to PZ $z \in \mathcal{Z}$ in BS $b \in \mathcal{B}$. Such a utility can, *inter alia*, contain fairness terms, achievable rate terms, etc. However, for illustration purposes, we consider maximizing the network throughput in the upcoming discussion, assuming that all BSs and users are equipped with single antennas.

Let R_{uzb} be the rate of the uth user connected to the zth PZ in the bth BS. According to the Shannon formula for capacity, the rate R_{uzb} can be written as follows:

$$R_{uzb} = W \log_2(1 + \text{SINR}_{bz}^u), \tag{11.1}$$

where W is the total bandwidth of the transmission and SINR_{bz}^u is the corresponding signal-to-interference-plus-noise ratio (SINR) of associating user $u \in \mathcal{U}$ with PZ $z \in \mathcal{Z}$ in BS $b \in \mathcal{B}$.

In order to derive an expression for the SINR, define $h_{bz}^u \in \mathbb{C}$, $\forall(u, b, z) \in \mathcal{U} \times \mathcal{B} \times \mathcal{Z}$ as the complex channel gain[2] between the uth user and the zth PZ in the bth BS. The power at the uth user received from the zth PZ in the bth BS is a function of both the transmit power and the channel gains, $P_{bz}|h_{bz}^u|^2$. For the PZs to be orthogonal, interference[3] received at the uth user connected to the zth PZ in the bth BS is seen only from the PZs z in BSs $b' \neq b \in \mathcal{B}$. Hence, the interference term received at the uth user connected to the zth PZ in the bth BS can be written as $\sum_{b' \neq b} P_{b'z}|h_{b'z}^u|^2$. Finally, the SINR of associating user $u \in \mathcal{U}$ with PZ $z \in \mathcal{Z}$ in BS $b \in \mathcal{B}$ is

$$\text{SINR}_{bz}^u = \frac{P_{bz}|h_{bz}^u|^2}{\Gamma(\sigma^2 + \sum_{b' \neq b} P_{b'z}|h_{b'z}^u|^2)}, \tag{11.2}$$

where σ^2 is the Gaussian noise variance and Γ denotes the SINR gap[4] [11] between the achievable capacity and the Shannon limit.

---

[2] This chapter assumes full channel state information at the cloud. In other words, the values of the channel gains are perfectly known.

[3] Note that interference, being a function of only the transmit power of the PZ and the channel between the considered user and PZ, does not depend on other scheduled users.

[4] The SINR gap quantifies the gap in performance between practical, finite-codeword, length modulations and the theoretical achievable rate for an additive white Gaussian noise (AWGN) channel.

11.2.3   Problem Formulation

The assignment problem in a scheduling-level coordinated C-RAN is that of optimally assigning users to BSs and PZs so as to maximize a generic network-wide utility function, under the following practical scheduling constraints.

- C1: Users can be scheduled to at most a single BS but potentially to multiple PZs in that BS.
- C2: A power zone is allocated to exactly one user.

To formally express the scheduling problem, let X_{ubz} be a binary variable indicating, i.e., when $X_{ubz} = 1$, the mapping of the uth user to the zth PZ in the bth BS. It can readily be seen that the system constraint C2 can be written in the form $\sum_{u\in\mathcal{U}} X_{ubz} = 1$, $\forall (b,z) \in \mathcal{B} \times \mathcal{Z}$. To formulate the first scheduling constraint, define Y_{ub} as a binary variable that indicates the mapping of the uth user to any PZ in the bth BS. The variable Y_{ub} is linked to the scheduling variables X_{ubz}, $z \in \mathcal{Z}$, by the rule that $Y_{ub} = 1$ if and only if there exists at least one $z \in \mathcal{Z}$ such that $X_{ubz} = 1$. The variables X_{ubz} are binary, and the relationship can be written as $Y_{ub} = \min(\sum_{z\in\mathcal{Z}} X_{ubz}, 1)$ and further simplified to $Y_{ub} = 1 - \delta(\sum_{z\in\mathcal{Z}} X_{ubz})$. Therefore, the system constraint C1 can be expressed as follows: $\sum_{b\in\mathcal{B}} Y_{ub} \leq 1, \forall u \in \mathcal{U}$.

Finally, the scheduling problem in a cloud-enabled radio access network can be expressed by the following generic network-wide optimization problem:

$$\text{maximize} \sum_{u,b,z} \pi_{ubz} X_{ubz} \tag{11.3a}$$

$$\text{s.t.} \quad Y_{ub} = 1 - \delta\left(\sum_{z\in\mathcal{Z}} X_{ubz}\right), \quad \forall (u,b) \in \mathcal{U} \times \mathcal{B}, \tag{11.3b}$$

$$\sum_{b\in\mathcal{B}} Y_{ub} \leq 1, \quad \forall u \in \mathcal{U}, \tag{11.3c}$$

$$\sum_{u\in\mathcal{U}} X_{ubz} = 1, \quad \forall (b,z) \in \mathcal{B} \times \mathcal{Z}, \tag{11.3d}$$

$$X_{ubz}, Y_{ub} \in \{0,1\}, \quad \forall (u,b,z) \in \mathcal{U} \times \mathcal{B} \times \mathcal{Z}, \tag{11.3e}$$

where X_{ubz} and Y_{ub} are the optimization variables constrained to be binary by (11.3e). The optimization problem (11.3) is a 0–1 mixed integer program that maximizes the sum of the utilities of all the associations (11.3a). The system constraints C1 and C2 are represented by both constraints (11.3b) and (11.3c) and by the constraint (11.3d), respectively.

A brute force approach to solving the discrete optimization problem (11.3) requires searching over all possible user-to-power-zone assignments and thus testing UBZ binary variables, resulting in a complexity of order 2^{UBZ}. Furthermore, the use of a generic 0–1 binary program yields the same order of complexity, which is still prohibitive for any moderately sized network. The rest of this chapter proposes an efficient method for solving a problem with a complexity of α^{UBZ} wherein $1 < \alpha < 2$ is the complexity constant, which depends on the applied algorithm. More specifically, we introduce the conflict graph and reformulate the discrete optimization problem (11.3) as a maximum-weight independent set problem, which can be solved using efficient algorithms, e.g., [12, 13]. For example, using the method proposed in [13] the complexity constant of the solution is $\alpha = 1.21$. Besides, several polynomial-times algorithms, e.g., [14, 15], generally produce satisfactory results.

11.2.4 Coordinated Scheduling in C-RAN

This subsection proposes a graph-theory approach to solving the binary problem illustrated in (11.3). In order to show that the problem is equivalent to a maximum-weight independent set, the first part of our discussion reformulates the scheduling problem in a more tractable form by introducing the set of feasible schedules. Afterward, the conflict graph is constructed using the constraints of the set of feasible schedules. Finally, the proof is concluded by establishing a one-to-one mapping between such set of feasible schedules and the maximum independent sets in the conflict graph.

Problem Reformulation

As mentioned above, the problem reformulation hinges upon the derivation of the set of feasible schedules. According to the problem formulation in equations (11.3), a feasible schedule is a collection of X_{ubz} that satisfy the constraints (11.3b), (11.3c), (11.3d), and (11.3e). It can readily be seen from the constraint (11.3d) that the number of X_{ubz} that are equal to 1 is the total number of PZs, i.e., $Z_{tot} = BZ$. Hence, a feasible schedule can be reduced to BZ elements X_{ubz} that are equal to 1 and that satisfy the constraints. Let these individual elements X_{ubz} be called associations and their collection be called the schedule. More specifically, while an association a is the triplet representing the user, BS, and PZ, a schedule S is a set containing such associations.

The set $\mathcal{A}$ of all possible associations between users, base stations, and power zones can be expressed using the Cartesian product of the sets of users, BSs, and PZs, i.e., $\mathcal{A} = \mathcal{U} \times \mathcal{B} \times \mathcal{Z}$. Furthermore, as the set of all possible schedules, regardless of their feasibility, contains all possible combinations of the elements of $\mathcal{A}$, it can be represented by the power set $\mathcal{P}(\mathcal{A})$ of the set of all possible schedules $\mathcal{A}$.

To clearly establish the link between a schedule $S \in \mathcal{P}(\mathcal{A})$ and the initial scheduling variables X_{ubz}, let φ_u, φ_b, and φ_z be the mapping functions of the set $\mathcal{A}$ to the sets of users $\mathcal{U}$, BSs $\mathcal{B}$, and PZs $\mathcal{Z}$, respectively. In other words, for each association $a \in \mathcal{A}$ represented by the triplet $a = (u, b, z)$, the functions φ_u, φ_b, and φ_z return, respectively, the index of the user $\varphi_u(a) = u$, the index of the BS $\varphi_b(a) = b$, and the index of the PZ $\varphi_z(a) = z$. Exploiting the mapping functions defined above, the schedule $S \in \mathcal{P}(\mathcal{A})$ corresponding to the particular scheduling variables X_{ubz} can be expressed as follows:

$$S = \left\{ a \in \mathcal{A} \mid X_{\varphi_u(a)\varphi_b(a)\varphi_z(a)} = 1 \right\}. \tag{11.4}$$

The following lemma characterizes the feasibility of a schedule as a function of the individual associations it contains.

LEMMA 11.1   *A schedule $S \in P(A)$ is feasible if and only if it verifies $\forall a \neq a' \in S$ the following three properties:*

$$\delta(\varphi_u(a) - \varphi_u(a'))\varphi_b(a) = \varphi_b(a')\delta(\varphi_u(a) - \varphi_u(a')) \tag{11.5a}$$

$$(\varphi_b(a), \varphi_z(a)) \neq (\varphi_b(a'), \varphi_z(a')), \tag{11.5b}$$

$$|S| = Z_{tot}. \tag{11.5c}$$

Proof   Showing the feasibility of a given schedule $S \in P(A)$ is equivalent to showing that the corresponding scheduling variables X_{ubz}, obtained from (11.4), satisfy the constraints (11.3b)–(11.3e). Since the variables are binary by definition, constraint (11.3e) is automatically satisfied. Constraint (11.5a) implies that, given the same users, the associated base stations should be the same as those which correspond to the original constraints (11.3b) and (11.3c). Finally, constraint (11.5b) translates the fact that each PZ is allocated to at most one user. The previous constraint combined with (11.5c) implies that BZ associations are equivalent to the original constraint (11.3d). Therefore, a feasible schedule $S \in P(A)$ satisfies all the constraints appearing in (11.5). A similar approach shows the converse. A complete proof can be found in Appendix A of [9].   □

Let $\mathcal{F}$ be the set of all feasible schedules, i.e., $\mathcal{F}$ contains all $S \in P(A)$ that satisfy the constraints (11.5). The scheduling problem can then be reformulated as follows:

$$\underset{S \in \mathcal{F}}{\text{maximize}} \sum_{a \in S} \pi(a), \tag{11.6}$$

where the objective function $\pi(a)$ is defined in a natural way as $\pi_{\varphi_u(a)\varphi_b(a)\varphi_z(a)}$.

Optimal Scheduling

Given the problem reformulation proposed in (11.6), a conflict graph is constructed such that the set of feasible schedules $\mathcal{F}$ corresponds to the set of maximum independent sets and the weight of such independent sets perfectly matches the objective function to be maximized.

Let $\mathcal{G}(V, \mathcal{E})$ be the conflict graph wherein V represents the set of vertices and $\mathcal{E}$ the set of edges. The set of vertices is constructed by generating a vertex $v \in V$ of each possible association $a \in A$. For an independent set to be a feasible solution, each pair of vertices violating at least one constraint given in (11.5) should be connected. Hence, the set of edges $\mathcal{E}$ is generated by connecting each pair of vertices v and v' in V if one of the following connectivity conditions (CC) is satisfied.

- CC1: $\delta(\varphi_u(v) - \varphi_u(v'))[1 - \delta(\varphi_b(v) - \varphi_b(v'))] = 1$. An edge is generated between any two vertices that represent the same user but are associated with different BSs. In other words, the constraint represents a violation of (11.5a).
- CC2: $(\varphi_b(v), \varphi_z(v)) = (\varphi_b(v'), \varphi_z(v'))$. An edge is generated between any two vertices that represent an association of the same PZ in the same BS. It can readily be seen that such an edge constitutes a violation of (11.5b).

Consider the conflict graph $\mathcal{G}(\mathcal{V}, \mathcal{E})$ generated as explained above and let the weight of the vertices be equal to the utility function of the association represented by that vertex. In other words, for a vertex $v \in \mathcal{V}$ representing the association $a \in \mathcal{A}$, the weight is $w(v) = \pi(a)$. It is worth mentioning that constraint (11.5c), instead of appearing in the connectivity constraints of the conflict graph, is directly expressed as a constraint on the cardinality of the independent set, as shown in the following theorem:

THEOREM 11.2   *The optimal solution to the scheduling problem (11.3) is the maximum-weight independent set in the conflict graph among the sets of size Z_{tot}.*

Proof Given the conflict graph constructed using the connectivity conditions CC1 and CC2, let $\mathcal{C}$ be the set of all independent sets of cardinality Z_{tot}. In virtue of the reformulation proposed in (11.6), showing that the optimization problem (11.3) is equivalent to a maximum-weight independent set problem reduces to showing a one-to-one mapping between the set of feasible schedules $\mathcal{F}$ and the set $\mathcal{C}$. Hence, given a schedule $\mathcal{S} \in \mathcal{F}$, the individual associations $a \in \mathcal{S}$ satisfy the constraints illustrated in (11.5). Therefore, each pair of vertices satisfies neither CC1 nor CC2, resulting in an independent set of vertices of size Z_{tot}. In other words, the following inclusion holds: $\mathcal{F} \subseteq \mathcal{C}$. Similarly, for an independent set of size Z_{tot}, the vertices are not connected and hence they satisfy the conditions of (11.5). Therefore, in virtue of Lemma 11.1, the corresponding schedule $\mathcal{S}$ is feasible, resulting in the second inclusion, $\mathcal{C} \subseteq \mathcal{F}$. Finally, as the total weight of the independent set corresponds to the objective function of (11.3), both problems are equivalent. A complete proof[5] can be found in Appendix B of [9].   □

Maximum-weight independent set problems are NP-hard. However, they can be solved more efficiently than generic binary optimization problems. Indeed, while the optimization problem (11.3) requires a complexity of order 2^{UBZ} to be optimally solved using generic binary program solvers, its reformulation as a maximum-weight independent set requires only α^{UBZ}, where $1 < \alpha < 2$. The scheduling solutions proposed in this section are all centralized by the nature of the maximum-weight independent set problem. The computation is carried by the cloud and the final solution coordinated to the base stations. A decentralized solution that actually exploits the characteristics of the conflict graph, i.e., its number of vertices and distribution of edges, is proposed in the upcoming sections.

Figure 11.3 plots the conflict graph for a C-RAN composed of $U = 2$ users, $B = 2$ BSs, and $Z = 2$ PZs. The vertices are labeled *ubz*, where u, b, and z indicate the indices of users, BSs, and PZs, respectively. The set $\mathcal{C}$ of independent sets of size $Z_{tot} = BZ = 4$ is the following, $\mathcal{C} = \{\{111, 112, 221, 222\}, \{121, 122, 211, 212\}\}$.

By allowing PZs not to serve users, which is also known as the blanking solution, i.e., replacing the strict equality of (11.3d) by an inequality, the solution should produce a higher objective function since the feasible search space is enlarged. Figure 11.4

---

[5] Reference [9] employs a maximum-weight clique formulation in the scheduling graph. A clique, in contrast with an independent set, is a set of vertices all of which are pairwise connected. Therefore, the scheduling graph is defined with the connectivity conditions opposite to those proposed in this chapter. The maximum-weight clique problem in the complementary graph is equivalent to the maximum-weight independent set in that graph.

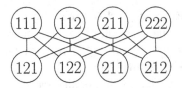

Figure 11.3 Example of conflict graph for two users, two BSs, and two PZs.

π_{1bz}	BS1	BS2
PZ1	1	0.1
PZ2	0.1	0.1
π_{2bz}	BS1	BS2
PZ1	0.1	1
PZ2	0.1	0.1

Figure 11.4 Example of a system with two users, two BSs, and two PZs. The general utility functions π_{1bz} and π_{2bz} associate users 1 and 2 with a BS and a PZ. The table entries give the weights of association. It is more efficient to schedule user 1 to BS1 and PZ1 and user 2 to BS1 and PZ1 than to schedule all the PZs.

illustrates an example for which the total objective function increases a system with a couple of BSs and a couple of users and PZs. It can easily be shown that the solution is equivalent to a maximum-weight independent set in the conflict graph without restriction on the size of the independent set. However, for practical objective functions, such a blanking solution is never observed in the simulations. Therefore, since the proposed solution is more computationally efficient to reach, here we consider the constraint with strict equality.

Low-Complexity Heuristic

As shown previously, the optimal schedule in single C-RAN is equivalent to a maximum-weight independent set problem, i.e., the optimal schedule is NP-hard to discover. This part proposes an heuristic schedule based on approximating the maximum-weight independent set so as to produce a solution in polynomial time for the problem of size UBZ. The fundamental concept of the heuristic is to select sequentially the "feasible" associations that yield the highest utility function, where "feasible" means that the association is combinable with the schedule selected so far.

More specifically, the first step of the heuristic is to generate the conflict graph $\mathcal{G}(\mathcal{V}, \mathcal{E})$ as illustrated previously. Beginning with an empty schedule $\mathcal{S} = \varnothing$, the vertex $v \in \mathcal{V}$ with the highest weight is added to the schedule, i.e., now $\mathcal{S} = \{v\}$. After this, the graph is updated so as to contain only the vertices that are combinable with $\mathcal{S}$. In other words, all vertices $v' \in \mathcal{V}$ that are connected to v are removed from the graph. The process of selecting the vertex with the maximum weight and removing all vertices connected to that vertex is repeated until the graph becomes empty. The steps of the proposed algorithm, summarized in Algorithm 4, ensure that the selected schedule $\mathcal{S}$ is

Table 11.1 System model parameters

Cellular layout	hexagonal
Number of BSs	variable
Number of PZs	variable
Number of users	variable
Cell-to-cell distance	500 meters
Channel model	SUI-3 terrain type B
Channel estimation	perfect
High power	-26.98 dBm/Hz
Background noise power	-168.60 dBm/Hz
SINR gap Γ	0 dB
Bandwidth	10 MHz

Algorithm 4 Heuristic scheduling for a single C-RAN.

Required: $\mathcal{U}, \mathcal{B}, \mathcal{Z}, P_{bz}$, and h_{bz}^u, $\forall (u, b, z) \in \mathcal{U} \times \mathcal{B} \times \mathcal{Z}$

1: Generate conflict graph $\mathcal{G}(\mathcal{V}, \mathcal{E})$.
2: Initialize $\mathcal{S} = \varnothing$.
3: **while** $\mathcal{G} \neq \varnothing$ **do**
4:    Select maximum weight vertex $v^* = \arg\max_{v \in \mathcal{V}} w(v)$.
5:    Set $\mathcal{S} = \mathcal{S} \cup \{v^*\}$
6:    Update $\mathcal{G}$ by removing $v \in \mathcal{V}$ connected to v^*
7: Output schedule $\mathcal{S}$.

---

an independent set in the conflict graph $\mathcal{G}(\mathcal{V}, \mathcal{E})$. Therefore, $\mathcal{S}$ is a feasible[6] solution to the optimization problem (11.3).

11.2.5 Simulation Results

To illustrate the performance of the proposed coordinated scheduling algorithms, simulation results are presented for the downlink of a cloud radio access network where the users are uniformly placed in the system and the cell-to-cell distance is set to 500 meters. The remaining simulation parameters are summarized in Table 11.1. To investigate the performance of the considered scheduling schemes in various scenarios, the number of users and number of power zones per BS frame change in each simulation example.

First, the optimal scheduling solution and the heuristic scheduling solution are compared for a C-RAN consisting of three base stations and four power zones per frame. Figure 11.5 demonstrates the sum-rate for different numbers of users. It can be seen that, for a high-shadowing environment (20 dB log-normal shadowing), the optimal scheduling provides a better performance than the heuristic solution, especially for a

---

[6] The size of the independent set $\mathcal{S}$ selected using Algorithm 4 is not guaranteed to be Z_{tot}. In other words, $\mathcal{S}$ may violate the constraint (11.3d). However, as explained previously, for practical utility functions, a blanking solution is never observed in the simulations.

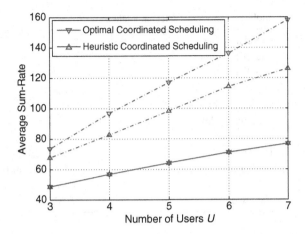

Figure 11.5 Sum-rate in bps/Hz versus number of users U for a network composed of $B = 3$ base stations each containing $Z = 4$ power zones. While the solid line represents a low-shadowing environment, the dashed and dotted lines display a high-shadowing environment.

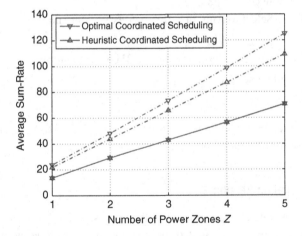

Figure 11.6 Sum-rate in bps/Hz versus the number of power zones Z for a network composed of $B = 3$ base stations and five users. While the solid line represents a low-shadowing environment, the dashed and dotted lines display a high-shadowing environment.

large number of users. This result can be explained by the fact that increasing the number of users leads to higher interference, especially in strong-shadowing environments, and hence coordinated scheduling becomes a crucial technique for mitigating interference. In a low-shadowing environment (8 dB log-normal shadowing), however, the heuristic method provides a similar performance to the optimal solution.

Figure 11.6 illustrates the sum-rate for different numbers of zones per BS frame, for a network consisting of three base stations and five users. Similarly, as the number of zones per frame increases, the optimal scheduling achieves a higher performance than the heuristic solution in a high-shadowing environment. This result is due to the rise in the size of the search space, i.e., in the number of feasible schedules.

11.3 Hybrid Scheduling in a Multicloud-RAN

This section investigates the coordinated scheduling problem in multicloud radio access networks (MC-RANs) wherein the system is composed of multiple clouds each connected to various base stations with many resource blocks. The first part of this section extends the different coordination strategies of the single C-RAN explained in the previous section and proposes a novel hybrid level of coordination. The section addresses the problem of maximizing a network-wide utility by multiplexing users to the different radio resource blocks of the base stations and clouds under the practical constraints of each of the above-mentioned coordination strategies for a fixed transmit power. A particular emphasis is given to hybrid coordination for its complexity–performance tradeoff. The section extends the proposed graph-theoretical approach for single C-RAN, i.e., the conflict graph, to solving the scheduling problem for each coordination strategy in multicloud-enabled networks. The vertex generation and connectivity conditions are derived for each coordination strategy, and the scheduling problems are reformulated as maximum-weight independent set problems. The section, further, suggests a method for reducing the number of vertices so as to decrease the complexity of the proposed solution and sketches an entirely decentralized scheduling process. Through extensive simulations, the proposed hybrid coordination strategy is demonstrated to provide appreciable performance gain, as compared with scheduling-level coordination, for a negligible degradation from the signal-level coordination.

11.3.1 Coordination Levels in Multicloud-RAN

Consider the downlink of a network comprising several clouds each connected to multiple base stations. The transmit frame of each base station contains various time–frequency resource blocks, denoted as earlier by the generic term of power zones. The scheduling problem is that of multiplexing users across the different base stations and their power zones so as to maximize a given network-wide metric.

Traditionally, scheduling in multi-cell networks is performed by synchronizing the connected base stations through extensive signaling, e.g., the authors in [16] and [17] maximize and derive bounds, respectively, on the achievable ergodic capacity of a multi-cell coordinated network on the basis of the complete co-channel gain information. To reduce the signaling, the authors in [18] propose a fully distributed approach based on sharing only the average channel state information. For cloud-enabled networks the clouds, responsible for synchronizing the transmit frames of the different base stations, can coordinate the system with different strategies, e.g., signal-level and scheduling-level coordination. Unlike single C-RAN, the MC-RAN configuration enables a novel

Figure 11.7 Cloud-enabled network composed of three clouds, each connected to three base stations.

coordination level, which we call hybrid coordination. This section extends the coordination strategies illustrated in the previous section to MC-RAN-enabled networks and proposes a hybrid coordination.

As in single C-RAN, the signal-level coordination in MC-RANs allows joint signal processing at the clouds, e.g., joint encoding and decoding of the users' data through joint beamforming. In MC-RANs, such coordination requires the sharing of all users, data streams between all clouds and their connected base stations. Therefore, signal-level coordination necessitates not only high-capacity low-latency backhaul and fronthaul links to connect the base stations to the clouds but also high-rate connections between the different clouds, as shown in Fig. 11.7. Furthermore, the coordination in MC-RAN suffers from the same drawbacks as in single C-RAN, namely the quantization noise inherent in the limited capacity of the links.

At the other extreme, scheduling-level coordination prevents any joint operations. The scheduling problem in MC-RAN becomes that maximizing a given utility function by optimally associating users to base stations and power zones under the constraint that each user can connect to at most one base station but possibly to multiple PZs. Hence, the scheduling problem in scheduling-level coordinated networks can be seen as a problem of associating users with the connected base stations and their resource blocks. Such a coordination eliminates the need for high-capacity cloud-to-BS and cloud-to-cloud links, resulting in much more practical implementation and cheaper backhaul network. However, as it does not involve the clustering of the network into multiple clouds, its performance is severely limited.

In order to see the benefits from both the easy implementation of that goes with scheduling-level coordination and the high performance of signal-level coordination, we consider scheduling from another perspective. The coordination that we consider assumes that, while each cloud is responsible for signal-level coordinating, its connected BSs and the various clouds are scheduling-level coordinated. In other words, the proposed hybrid coordination adopts intra-cell signal-level coordination and inter-cell scheduling-level coordination. The scheduling problem in such coordinated networks becomes one of maximizing a network-wide utility by assigning users to at most a single cloud but possibly to multiple connected base stations and their power zones. Therefore, although high-capacity links are still needed to share the data streams between the connected base stations, only low-rate links are required to connect the clouds.

11.3.2   System Model

Consider the downlink of a multicloud radio access network comprising a set $\mathcal{C}$ of C clouds, each responsible for the signal-level coordination of a set[7] $\mathcal{B}$ of B base stations. The different clouds are responsible for synchronizing the transmit frames of all CB base stations and exchange only the scheduling policy, through low-capacity backhaul or fronthaul links.

The transmit frame of each BS is composed of a set $\mathcal{Z}$ of Z power zones, i.e., time–frequency resource blocks. The total number of PZs in the network is $Z_{\text{tot}} = CBZ$. Let the transmit power of the zth PZ in the bth BS of the cth cloud be fixed at the value $P_{cbz} \forall (c, b, z) \in \mathcal{C} \times \mathcal{B} \times \mathcal{Z}$. Figure 11.8 depicts the transmit frames of the connected BSs. The power levels of each PZ can be further optimized for better overall performance. However, this falls outside the scope of the current section as here we are focussing on solving the scheduling problem.

The network serves a set $\mathcal{U}$ of U users in total. The users and BSs are equipped with single antennas. Figure 11.9 illustrates a multi-C-RAN formed by $U = 21$ users and $C = 3$ clouds, each coordinating $B = 3$ BSs.

As in the case of the single C-RAN setting, this section considers a general utility function π_{cuzb} associating a user $u \in \mathcal{U}$ with PZ $z \in \mathcal{Z}$ in BS $b \in \mathcal{B}$ in cloud $c \in \mathcal{C}$. For illustration purposes, we implicitly assume in the upcoming discussion that the utility is the achievable capacity $R_{cuzb} = W \log_2(1 + \text{SINR}^u_{cbz})$, where W is the total bandwidth of the transmission and SINR^u_{bz} is the corresponding signal-to-interference-plus-noise ratio (SINR) of associating user $u \in \mathcal{U}$ with PZ $z \in \mathcal{Z}$ in BS $b \in \mathcal{B}$ in cloud $c \in \mathcal{C}$.

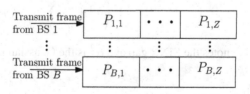

Figure 11.8 Frame structure.

---

[7] For notation simplicity, in this section we assume that the number of base stations in each cloud is the same, i.e., B. However, the results presented in this section are independent of this assumption. For different numbers of BSs, the total number of PZs is $Z_{\text{tot}} = B_{\text{tot}}Z$, where B_{tot} is the total number of BSs in all the clouds.

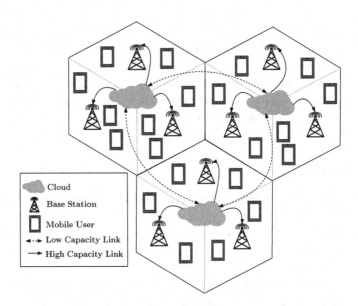

Figure 11.9 Cloud-enabled network composed of three clouds, each connected to three base stations.

Let $h^u_{cbz} \in \mathbb{C}$, $\forall (c, u, b, z) \in \mathcal{C} \times \mathcal{U} \times \mathcal{B} \times \mathcal{Z}$, be the complex channel gain between the uth user and the zth PZ in the bth BS connected to the cth cloud. We assume full channel state information at the cloud, i.e., that the value h^u_{cbz} is perfectly estimated at the cth cloud.

Given that the PZs are orthogonal and that the interference is a function of only the transmit power of a PZ and the channel between the user under consideration and the PZ, the SINR does not depend on the other scheduled users. Hence, the interference experienced by the uth user connected to the zth PZ in the bth BS in the cth cloud is generated by the remaining $CB - 1$ PZs with the label z. In other words, the SINR of the uth user associated with the zth PZ z in the bth BS of the cth cloud can be expressed as:

$$\text{SINR}^u_{cbz} = \frac{P_{cbz}|h^u_{cbz}|^2}{\Gamma\left(\sigma^2 + \displaystyle\sum_{(c',b')\neq(c,b)} P_{c'b'z}|h^u_{c'b'z}|^2\right)}, \tag{11.7}$$

where Γ denotes the SINR gap and σ^2 is the Gaussian noise variance.

11.3.3 Problem Formulation

As mentioned above, hybrid-level coordination allows joint signal processing within the same cloud and only scheduling-level coordination between clouds. Therefore, each user can be mapped with at maximum a single cloud but possibly multiple BSs and PZs within that cloud. However, as users are equipped with a single antenna, a user cannot be served by the same PZ across different BSs. Therefore, the coordinated scheduling problem under investigation in this subsection is that of maximizing a generic utility by optimally assigning users to clouds, BSs, and PZs under the following practical constraints.

- C1: Users can be scheduled to at most a single cloud, but potentially to multiple BSs and PZs in that cloud.
- C2: Power zones are allocated to exactly one user.
- C3: Each user cannot be served by the same PZ across different BSs.

Let X_{cubz} be a binary variable indicating, i.e., for $X_{cubz} = 1$, the mapping of the uth user to the zth PZ of the bth BS in the cth cloud. Similarly, let Y_{uz} be a variable denoting the number of associations of the u th user with any of the z th PZs. It can readily be seen that Y_{uz} can be expressed as $Y_{uz} = \sum_{c,b} X_{cubz}$. Finally, define Z_{cu} as a binary variable that is equal to 1 if and only if the u th user is mapped to the c th cloud.

The first system constraint, implying that a user can be mapped with at most a single cloud, can easily be expressed using the variable Z_{cu} as follows: $\sum_{c \in C} Z_{cu} \leq 1$, $\forall u \in \mathcal{U}$. As for the single C-RAN case, the system constraint C2 can be written as $\sum_{u \in \mathcal{U}} X_{cubz} = 1$, $\forall (c, b, z) \in C \times B \times Z$. Finally, given that a user u can connect, at maximum, to one PZ, i.e., constraint C3, then the variable Y_{uz} is constrained to be binary. Therefore, coordinated scheduling in MC-RAN can be expressed by the following generic network-wide optimization problem:

$$\text{maximize} \quad \sum_{c,u,b,z} \pi_{cubz} X_{cubz} \tag{11.8a}$$

$$\text{s.t.} \quad Z_{cu} = 1 - \delta\left(\sum_{b,z} X_{cubz}\right), \quad \forall (c,u) \in C \times \mathcal{U}, \tag{11.8b}$$

$$\sum_{c \in C} Z_{cu} \leq 1, \quad \forall u \in \mathcal{U}, \tag{11.8c}$$

$$\sum_{u \in \mathcal{U}} X_{cubz} = 1, \quad \forall (c,b,z) \in C \times B \times Z, \tag{11.8d}$$

$$Y_{uz} = \sum_{c,b} X_{cubz} \leq 1, \quad \forall (u,z) \in \mathcal{U} \times Z, \tag{11.8e}$$

$$X_{cubz}, Y_{uz}, Z_{cu} \in \{0, 1\}, \tag{11.8f}$$

where X_{ubz}, Y_{ub}, and Z_{cu} are the optimization variables, constrained to be binary by (11.8f). The optimization problem (11.8) is a 0–1 mixed integer program that maximizes the sum of the utilities of all associations (11.8b). The system constraint C1 is represented by the constraints (11.8c) and (11.8d). The constraints C2 and C3 are represented by equations (11.8d) and (11.8e), respectively.

Solving the problem mentioned above using a brute force approach is even more complex than for the single C-RAN case as the number of variables is $CUBZ$ and thus the complexity is 2^{CUBZ}. Below we suggest the use of a similar graph theory approach to that developed in the previous section. The conflict graph is extended to the MC-RAN and hybrid-coordination configuration, and the problem is reformulated as a maximum-weight independent set problem that can be solved using efficient algorithms, e.g., [12–15]. Furthermore, we show that the proposed solution applies to other coordination techniques, e.g., signal-level and scheduling-level coordination.

11.3.4   Hybrid Multicloud-RAN Association

This subsection proposes an efficient method for solving the optimization problem (11.8) using a graph theory approach similar to that developed in Section 11.2. In the first part we derive the connectivity conditions of the conflict graph. Afterwards, the optimization problem is reformulated as a maximum-weight independent set problem. Finally, we suggest a method for removing vertices from the conflict graph so as to reduce the complexity of the proposed solution.

Optimal Hybrid Scheduling

To derive the connectivity conditions of the conflict graph, the problem is first reformulated in a more tractable form involving the set of feasible schedules $\mathcal{F}$. As in the previous section, any schedule S can be written as a function of the individual association $a \in \mathcal{A}$ with $\mathcal{A} = \mathcal{C} \times \mathcal{U} \times \mathcal{B} \times \mathcal{Z}$ the set of all associations between clouds, users, BSs, and PZs. Hence, the set of all possible schedules, regardless of their feasibility, coincides with the power set of $\mathcal{A}$, i.e., $S \in \mathcal{P}(\mathcal{A})$.

Finally, to link a schedule $S \in \mathcal{P}(\mathcal{A})$ to the initial optimization variables X_{cubz}, let φ_c, φ_u, φ_b, and φ_z be mapping functions from the set of associations $\mathcal{A}$ to the set of clouds $\mathcal{C}$, the set of users $\mathcal{U}$, the set of BSs $\mathcal{B}$, and the set of PZs $\mathcal{Z}$, respectively. In other words, for an association $a = (c, u, b, z) \in \mathcal{A}$, the mapping functions return the following values: $\varphi_c(a) = c$, $\varphi_u(a) = u$, $\varphi_b(a) = b$, and $\varphi_z(a) = z$. Let the definition of the utility function be extended to the association set, so that $\pi(a)$ represents the utility of the association indicated by element $a \in \mathcal{A}$.

Let $\mathcal{G}(\mathcal{V}, \mathcal{E})$ be the conflict graph of a multi-cloud system wherein $\mathcal{V}$ represents the set of vertices and $\mathcal{E}$ the set of edges. As in the single C-RAN setting, a vertex $v \in \mathcal{V}$ is generated for each association $a \in \mathcal{A}$. An edge between vertices represents a conflict between the two corresponding associations. Therefore, vertices v and v' are adjacent if and only if one of the following connectivity conditions (CC) is satisfied:

- CC1: $\delta(\varphi_u(v) - \varphi_u(v'))[1 - \delta(\varphi_c(v) - \varphi_c(v'))] = 1$.
- CC2: $(\varphi_c(v), \varphi_b(v), \varphi_z(v)) = (\varphi_c(v'), \varphi_b(v'), \varphi_z(v'))$.
- CC3: $\delta(\varphi_u(v) - \varphi_u(v'))\delta(\varphi_z(v) - \varphi_z(v')) = 1$.

The connectivity constraint CC1 states that the same user cannot connect to more than one cloud that reflects the system constraint C1 (see Section 11.3.3). The second connectivity constraint implies that each PZ should be associated with at most one user. Such a constraint, along with the limitation on the size of the independent set of the proposed solution, translates the system constraint C2. Finally, the system constraint C3 and the connectivity constraint CC3 are equivalent. The weight of each vertex represents the benefit of the corresponding association. In other words, for a vertex $v \in \mathcal{V}$ corresponding to the association $a \in \mathcal{A}$, the weight $w(v) = \pi(a)$.

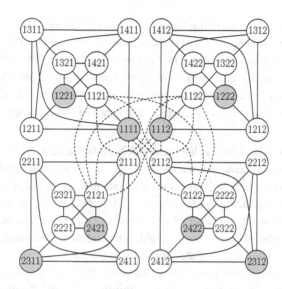

Figure 11.10 The conflict graph for a network composed of two clouds (the upper and lower parts of the figure), two BSs per cloud, two PZs per BS and a total of four users. Intra-cloud connection are plotted in solid lines. Inter-cloud connections are illustrated for user 1 only in dashed lines.

Figure 11.10 shows an example of the conflict graph in an MC-RAN composed of $C = 2$ clouds, $B = 2$ BSs per cloud, $Z = 2$ PZs per BS, and $U = 4$ users. The vertices are labeled *cubz*, where c, u, b and z represent the indices of cloud, user, BS, and PZ, respectively. Figure 11.10 shows that the independent set of size $Z_{tot} = CBZ = 8$ can be written in the following form:

1. $\{1a11, 1a12, 1b21, 1b22, 2c11, 2c12, 2d21, 2d22\}$,
2. $\{1a11, 1a12, 1b21, 1b22, 2c11, 2d12, 2d21, 2c22\}$,
3. $\{1a11, 1b12, 1b21, 1a22, 2c11, 2c12, 2d21, 2d22\}$,
4. $\{1a11, 1b12, 1b21, 1a22, 2c11, 2d12, 2d21, 2c22\}$,

where $a, b, c,$ and $d \in \{1, 2, 3, 4\}$ with $a \neq b \neq c \neq d$. For example, substituting (a, b, c, d) in $\{1a11, 1a12, 1b21, 1b22, 2c11, 2c12, 2d21, 2d22\}$ by $(1, 2, 3, 4)$ gives the independent set shown in gray in Figure 11.10.

The following theorem characterizes the optimal coordinated scheduling proposed in problem (11.8):

THEOREM 11.3 *The optimal solution to the scheduling problem (11.8) for a multi-cloud system is the maximum-weight independent set in the conflict graph among the sets of size $Z_{tot} = CBZ$.*

Proof To demonstrate the theorem, the optimization problem (11.8) is reformulated using the expression for the feasible schedules $\mathcal{F}$. As for the single C-RAN system, such a set of feasible schedules allows one to derive the connectivity conditions of the conflict graph. Assume a conflict graph constructed using the connectivity conditions CC1, CC2, and CC3. Let $\mathcal{C}$ be the set of all maximal independent sets of cardinality Z_{tot}.

A one-to-one mapping between the set of maximal independent sets C and the set of feasible schedules $\mathcal{F}$ can be established. If the weight of the independent set is identical to the utility represented by the corresponding feasible schedule, the two optimization problems are equivalent. Therefore, the optimal solution to the discrete optimization problem (11.8) is the maximum-weight independent set in the conflict graph among sets of size $Z_{\text{tot}} = CBZ$. A complete proof can be found in Appendix A of [19].   □

Heuristics For Multicloud-RAN Association

As shown above, the maximum-weight independent set problem is an NP-hard. Instead of proposing a polynomial-time solution to approximate the exact solution, here we suggest reducing the total number of vertices. In other words, the proposed heuristic proposes utilizing a subset of the entire conflict graph so as to minimise the complexity of solving the maximum-weight independent-set problem. Such an approach is motivated by the rationale that, for a vast multi-cloud system, the benefit of some users from base stations and power zones located in far away clouds may be low and does not contribute much to the network-wide utility. Therefore, the fundamental idea is to consider only a fraction $0 < p \leq 1$, resulting in $\lfloor pCUBZ \rfloor$, of the total available vertices having the highest benefits to the system, where $\lfloor \ldots \rfloor$ is the floor function. Then, the maximum-weight independent set algorithm is performed using the newly generated smaller-size graph, resulting in a complexity of α^{pCUBZ} instead of α^{CUBZ}.

11.3.5   Multicloud Scheduling-Level and Signal-Level Coordination

This subsection extends the results, discussed earlier, of the scheduling problem in hybrid-level coordination to signal-level and scheduling-level coordination. Signal-level coordination, corresponding to a fully coordinated network, allows joint data processing at all clouds and BSs and hence requires the sharing of all data streams, resulting in massive backhaul communication. At the other extreme, scheduling-level coordination prohibits joint processing and focuses only on the optimal assignment of users to base stations. Such a constraint, although it sacrifices optimality, is more practical to implement as it requires low-rate BS–cloud and cloud–cloud backhaul links. For each coordination technique the connectivity conditions of the conflict graph are derived and the problems reformulated as maximum-weight independent set problems.

Signal-Level Coordination

In signal-level-coordinated systems, each user can be associated with multiple clouds, BSs, and PZs. However, as explained earlier, the same user cannot be served by the same PZ across the various BSs as it has a single antenna. Therefore, the scheduling problem in such a coordinated network is the one of assigning users to clouds, BSs, and PZs under the following practical constraints.[8]

---

[8] Note that the first and second constraints are similar to the constraints C2 and C3 of hybrid-level coordination, respectively. This is due to the fact that the only limitation of hybrid scheduling is that it prevents users from being associated with multiple clouds.

- A power zones is allocated to exactly one user.
- Each user cannot be served by the same PZ across different BSs.

Given the constraints above, the scheduling problem can be expressed as:

$$\text{maximize} \quad \sum_{c,u,b,z} \pi_{cubz} X_{cubz} \tag{11.9a}$$

$$\text{s.t.} \quad \sum_{u \in \mathcal{U}} X_{cubz} = 1, \quad \forall (c,b,z) \in \mathcal{C} \times \mathcal{B} \times \mathcal{Z}, \tag{11.9b}$$

$$Y_{uz} = \sum_{c,b} X_{cubz} \leq 1, \quad \forall (u,z) \in \mathcal{U} \times \mathcal{Z}, \tag{11.9c}$$

$$X_{cubz}, Y_{uz} \in \{0,1\}, \tag{11.9d}$$

where X_{ubz} and Y_{cu} are the binary optimization variables.

Using an approach similar to that proposed for hybrid-level scheduling, it can easily be shown that the conflict graph $\mathcal{G}'(\mathcal{V}', \mathcal{E}')$ can be generated using only the connectivity constraints CC2 and CC3. The following corollary characterizes the solution to the scheduling problem (11.9) for signal-level coordinated networks:

COROLLARY *The optimization problem (11.9) is equivalent to a maximum-weight independent set problem among the independent sets of size CBZ in the conflict graph $\mathcal{G}'(\mathcal{V}', \mathcal{E}')$.*

Scheduling-Level Coordination

In scheduling-level coordination, each user can be associated with, at most, a single BS. Therefore, the scheduling problem in such a coordinated network is that of assigning users to clouds, BSs, and PZs under the following practical constraints.[9]

- Users can be scheduled to at most a single BS, but potentially to multiple PZs in that BS.
- Power zones are allocated to exactly one user.

Therefore, the scheduling problem in scheduling-level coordinated MC-RAN can be expressed as follows:

$$\text{maximize} \quad \sum_{c,u,b,z} \pi_{cubz} X_{cubz} \tag{11.10a}$$

$$\text{s.t.} \quad Y_{cub} = \min \left(\sum_z X_{cubz}, 1 \right), \quad \forall (c,u,b), \tag{11.10b}$$

$$\sum_{c,b} Y_{cub} \leq 1, \quad \forall u \in \mathcal{U}, \tag{11.10c}$$

$$\sum_u X_{cubz} = 1, \quad \forall (c,b,z) \in \mathcal{C} \times \mathcal{B} \times \mathcal{Z}, \tag{11.10d}$$

$$X_{cubz}, Y_{cub} \in \{0,1\}, \quad \forall (c,u,b,z), \tag{11.10e}$$

[9] Note that the constraint that each user cannot be served by the same PZ across different BSs is readily satisfied by preventing users from connecting to multiple BSs.

Table 11.2 System model parameters

Cellular layout	hexagonal
Number of clouds	variable
Number of BSs	variable
Number of PZs	variable
Number of users	variable
Cell-to-cell distance	500 meters
Channel model	SUI-3 terrain type B
Channel estimation	perfect
High power	-26.98 dBm/Hz
Background noise power	-168.60 dBm/Hz
SINR Gap Γ	0 dB
Bandwidth	10 MHz

where X_{ubz} and Y_{cu} are the binary optimization variables.

Extending the solution proposed in the previous section for single C-RAN, a conflict graph $\mathcal{G}''(\mathcal{V}'', \mathcal{E}'')$ is obtained by generating a vertex $v \in \mathcal{V}''$ for each association $a \in \mathcal{A}$. Vertices v and v' are conflicting, and thus are connected by an edge in $\mathcal{E}''$,[10] if one of the following connectivity conditions is true:

- $\delta(\varphi_u(v) - \varphi_u(v'))[1 - \delta(\varphi_c(v) - \varphi_c(v'))] = 1$.
- $(\varphi_c(v), \varphi_b(v), \varphi_z(v)) = (\varphi_c(v'), \varphi_b(v'), \varphi_z(v'))$.
- $\delta(\varphi_u(v) - \varphi_u(v'))[1 - \delta(\varphi_b(v) - \varphi_b(v'))] = 1$.

The following corollary extends the result of Theorem 11.2 to the MC-RAN setting:

COROLLARY   *The optimal solution to the scheduling problem (11.10) is the maximum-weight independent set in the conflict graph among the sets of size $Z_{tot} = CBZ$.*

11.3.6   Simulation Results

In this subsection, the performances of the proposed scheduling schemes are investigated in the downlink of a cloud radio access network where the users are uniformly placed in the system, and the cell-to-cell distance is set to 500 meters. The remaining simulation parameters are summarized in Table 11.2. To assess the performance of the three scheduling schemes in different scenarios, the number of clouds, base-stations per cloud, and users were varied in each simulation example.

First, consider a network consisting of $C = 3$ clouds, $B = 3$ BSs per cloud, and $Z = 5$ power zones per BSs transmit frame. Figure 11.11 illustrates the sum-rate in bps/Hz for different numbers of users U. The plot demonstrates that for a small number of users the proposed hybrid coordination policy significantly outperforms the scheduling-level

---

[10] The edges are generated between vertices using the same method as that proposed in the connectivity conditions CC1 and CC2 given in Section 11.2. However, as the number of base stations is CB instead of B, a third condition is necessary to ensure that a given user does not connect to the same BS across different clouds.

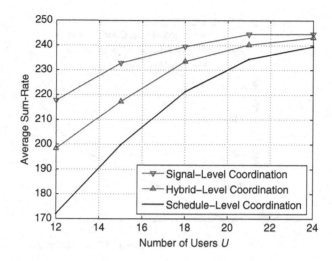

Figure 11.11 Sum-rate in bps/Hz versus the number of users U for a network composed of $C = 3$ clouds each having $B = 3$ base stations, each containing $Z = 5$ power zones.

coordinated system. Since increasing the number of users in the network provides more scheduling opportunities, the performance gap between the three different scheduling policies becomes smaller, which is clearly illustrated in Figure 11.11.

Figure 11.12 illustrates the sum-rate in bps/Hz for different number of base stations B per cloud for a network consisting of $C = 3$ clouds, $Z = 5$ power zones per BS transmit frame, and $U = 24$ users. The figure in particular shows that the proposed policies provide the same performance when the number of BSs is small. Besides, increasing the number of BSs leads to a higher level of coordination, which improves the performance by increasing the scheduling opportunities. It is also observed that the proposed hybrid coordination scheme leads to a 15% performance increase as compared with the scheduling-level-coordinated network and around 5% performance decrease as compared with signal-level coordination.

Figure 11.13 illustrates the sum-rate in bps/Hz for different numbers of clouds C for a network formed by $B = 3$ base stations per cloud, $Z = 5$ power zones per BS transmit frame, and $U = 8$ users per cloud. It can clearly be seen that the proposed hybrid coordination provides an appreciable performance gain as compared with scheduling-level coordination and provides only a slight performance loss as compared with the signal-level coordinated system.

Finally, Figure 11.14 illustrates the sum-rate for different fractions of associations p in a network composed of $C = 3$ clouds, each containing $B = 2$ base stations and 4 power zones per frame, and a total number of $U = 15$ users. The figure in particular illustrates the performance of the lower-complexity heuristic scheduling algorithm DAS compared with the optimal and full-association heuristic. For a suitable choice of the fraction of associations p, it can be seen that both heuristics provides a similar performance. Figure 11.14 especially demonstrates that the proposed algorithms provide the same performance in a low shadowing environment.

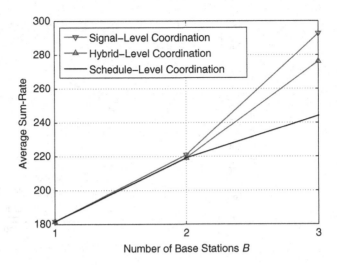

Figure 11.12 Sum-rate in bps/Hz versus number of base stations B per cloud each containing $Z = 5$ power zones. The network is composed of $C = 3$ clouds and a total number $U = 24$ of users.

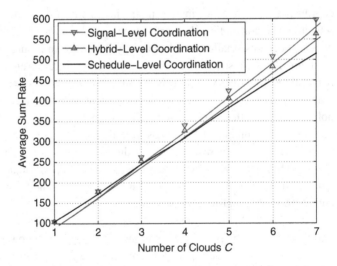

Figure 11.13 Sum-rate in bps/Hz versus number of clouds C for a network composed of $B = 3$ base stations per cloud, each containing $Z = 5$ power zones and a total number $U = 8$ users.

11.4   General Framework and Future Applications

The coordinated scheduling strategies considered in this chapter constitute a framework for solving scheduling problems of similar nature, i.e., the problems of maximizing network-wide utilities subject to association constraints. In other terms, as discussed above, on the basis of the connectivity constraints, a scheduling graph is first constructed in which each vertex represents an association of users, BSs, and time–frequency blocks.

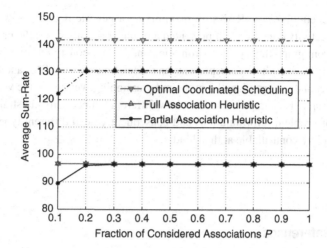

Figure 11.14 Sum-rate in bps/Hz versus the fraction of considered associations p for a network composed of $C = 3$ clouds each having $B = 2$ base stations, each containing $Z = 4$ power zones and a total number $U = 15$ of users. While the solid lines represent a low-shadowing environment, the dashed lines display a high-shadowing environment.

Second, the problem is reformulated by assigning weights to the vertices of the scheduling graph. The above sections show how different connectivity constraints give different types of problem reformulation and consequently need different types of solvers that are solely controllable at the cloud.

Furthermore, such a generic approach can in fact be adopted as a framework for the discrete mapping problem. It can be used in several resource allocation problems in C-RANs (e.g., joint coordinated scheduling and power control, user-to-cloud association in MC-RANs, channel assignment, antenna selection, etc.). Joint coordinated scheduling and power control, especially, can be seen as a higher-dimensional application of the approach proposed in this chapter. Such an application requires the construction of a higher-dimensional scheduling and power control graph consisting of several clusters, where each cluster is formed by a set of vertices representing the possible associations of users, BSs, and power level for one specific PZ. More details of such an approach can be found in reference [20].

11.5 Conclusion

This chapter presented a network utility maximization framework for coordinated scheduling in C-RANs. A downlink scenario was considered, where BSs are connected to the cloud via low-rate links. Multiple remote users may be served by each BS through time–frequency multiplexing, as each BS transmit frame consists of several power zones, each serving one user. The cloud is solely responsible for the scheduling policy and synchronization of base stations. The chapter addressed the coordinated scheduling in C-RANs by maximizing a generic network-wide utility subject to connectivity constraints. It presented a framework for solving the problem using techniques from graph theory. First, from the connectivity constraints construct a scheduling graph in which each vertex represents an association of users, BSs, and time–frequency blocks. Second, reformulate the problem by assigning weights to the vertices of the scheduling graph. The chapter discusses how different connectivity constraints give different types of problem reformulations and consequently need different types of solvers, which are solely controllable at the cloud.

References

[1] W. Yu, T. Kwon, and C. Shin, "Multicell coordination via joint scheduling, beamforming, and power spectrum adaptation," *IEEE Trans. Wireless Commun.*, vol. 12, no. 7, pp. 1–14, July 2013.

[2] A. Stolyar and H. Viswanathan, "Self-organizing dynamic fractional frequency reuse for best-effort traffic through distributed inter-cell coordination," in *Proc. 28th IEEE Conf. on Computer Communications*, Rio de Janeiro, April 2009, pp. 1287–1295.

[3] B. Dai and W. Yu, "Sparse beamforming for limited-backhaul network mimo system via reweighted power minimization," in *Proc. IEEE Global Telecommunications Conf.*, Atlanta, GA, December 2013, pp. 1962–1967.

[4] Y. Shi, J. Zhang, and K. Letaief, "Group sparse beamforming for green cloud-ran," *IEEE Trans. Wireless Commun.*, vol. 13, no. 5, pp. 2809–2823, May 2014.

[5] S.-H. Park, O. Simeone, O. Sahin, and S. Shamai, "Joint precoding and multivariate backhaul compression for the downlink of cloud radio access networks," *IEEE Trans. Signal Process.*, vol. 61, no. 22, pp. 5646–5658, November 2013.

[6] S.-H. Park, O. Simeone, O. Sahin, and S. Shamai, "Inter-cluster design of precoding and fronthaul compression for cloud radio access networks," *IEEE Wireless Commun. Lett.*, vol. 3, no. 4, pp. 369–372, Auguest 2014.

[7] Y. Zhou and W. Yu, "Optimized backhaul compression for uplink cloud radio access network," *IEEE J. Sel. Areas Commun.*, vol. 32, no. 6, pp. 1295–1307, June 2014.

[8] H. Dahrouj, W. Yu, T. Tang, J. Chow, and R. Selea, "Coordinated scheduling for wireless backhaul networks with soft frequency reuse," in *Proc. 21st European Signal Processing Conf.*, Marrakesh, September 2013, pp. 1–5.

[9] A. Douik, H. Dahrouj, T. Y. Al-Naffouri, and M.-S. Alouini, "Coordinated scheduling for the downlink of cloud radio-access networks," in *Proc. IEEE Int. Conf. Communications*, London, 2015.

[10] D. P. Bertsekas, "The auction algorithm: a distributed relaxation method for the assignment problem," *Ann. Oper. Res.*, vol. 14, pp. 105–123, 1988.

[11] C.-H. Fung, W. Yu, and T. J. Lim, "Precoding for the multiantenna downlink: multiuser snr gap and optimal user ordering," *IEEE Trans. Commun.*, vol. 55, no. 1, pp. 188–197, January 2007.

[12] F. V. Fomin, F. Grandoni, and D. Kratsch, "A measure and conquer approach for the analysis of exact algorithms," *J. ACM*, vol. 56, no. 5, pp. 25:1–25:32, August 2009.

[13] N. Bourgeois, B. Escoffier, V. T. Paschos, and J. M. M. van Rooij, "A bottom-up method and fast algorithms for max independent set," in *Proc. 12th Scandinavian Conf. Algorithm Theory*, Bergen, 2010.

[14] P. jun Wan, X. Jia, G. Dai, H. Du, and O. Frieder, "Fast and simple approximation algorithms for maximum weighted independent set of links," in *Proc. 33th IEEE Conf. Computer Communications*, Toronto, April 2014, pp. 1653–1661.

[15] N. Esfahani, P. Mazrooei, K. Mahdaviani, and B. Omoomi, "A note on the p-time algorithms for solving the maximum independent set problem," in *Proc. 2nd Conf. Data Mining and Optimization*, Bandar Baru Bangi, Malaysia, October 2009, pp. 65–70.

[16] S. Kiani and D. Gesbert, "Optimal and distributed scheduling for multicell capacity maximization," *IEEE Trans. Wireless Commun.*, vol. 7, no. 1, pp. 288–297, January 2008.

[17] W. Choi and J. Andrews, "The capacity gain from intercell scheduling in multi-antenna systems," *IEEE Trans. Wireless Commun.*, vol. 7, no. 2, pp. 714–725, February 2008.

[18] R. Bendlin, Y.-F. Huang, M. Ivrlac, and J. Nossek, "Fast distributed multi-cell scheduling with delayed limited-capacity backhaul links," in *Proc. IEEE Int. Conf. on Communications*, Dresden, June 2009, pp. 1–5.

[19] A. Douik, H. Dahrouj, T. Y. Al-Naffouri, and M.-S. Alouini, "Hybrid scheduling/signal-level coordination in the downlink of multi-cloud radio-access networks," in *Proc. IEEE Global Telecommunications Conf. (GLOBECOM 2015)*, San Diego, CA. Available at Arxiv e-prints, vol. abs/1504.01552, 2015.

[20] A. Douik, H. Dahrouj, T. Y. Al-Naffouri, and M.-S. Alouini, "Coordinated scheduling and power control in cloud-radio access networks," *IEEE Trans. Wireless Commun.*, vol. 15, no. 4, pp. 2523–2536, April 2016.

12 Delay-Aware Radio Resource Allocation Optimization in Heterogeneous C-RANs

Mugen Peng, Jian Li, Hongyu Xiang, and Yuling Yu

12.1 Introduction

12.1.1 Challenges

Owing to the increased level of frequency sharing, node density, interference, and network congestion in heterogeneous C-RANs, obtaining signal processing techniques and dynamic radio resource allocation optimization algorithms are the most important tasks [2]. Multi-user interference, which is a major performance-limiting factor, should be astutely manipulated through advanced signal processing techniques in the physical (PHY) layers. In addition, to a satisfy the quality-of-service (QoS) requirement, it is crucial to study radio resource allocation optimization in heterogeneous C-RANs, which is usually more challenging than that in a traditional cellular network, considering practical issues such as fronthaul capacity limitations, channel state information (CSI) overhead, and the parallel implementation of algorithms [3]. In heterogeneous C-RANs radio resource optimization algorithms should support the bursty mobile traffic data, which is usually delay-sensitive. Most traditional methods are based on heuristics and there is lack of theoretical understanding on how to design delay-aware radio resource allocation optimization algorithms in a time-varying system. Therefore, it is very important to consider random bursty arrivals and delay performance metrics, in addition to the conventional PHY-layer performance metrics, in cross-layer radio resource optimization, which may embrace the PHY, medium access control (MAC), and network layers [4]. A combined framework taking into account both queueing delay and PHY-layer performance is not trivial as it involves both queueing theory (to model the queue dynamics) and information theory (to model the PHY-layer dynamics). The system state involves both the CSI and the queue state information (QSI), and a delay-aware cross-layer radio resource optimization policy should be adaptive to both the CSI and the QSI of heterogeneous C-RANs. Furthermore, radio resource allocation optimization algorithms have to be scalable with respect to network size, while traditional algorithms become infeasible due to the in huge computational complexity as well as signaling latency involved [5]. The situation is even worse for heterogeneous C-RANs because there are more thin RRHs connected to the BBU pool via fronthaul links. Unlike conventional radio resource allocation optimization, which is designed to optimize the resource of a single base station, that for heterogeneous C-RANs involves radio resources from many RRHs or traditional macro base stations (MBSs), and thus the scalability in terms of computation and signaling is also a key obstacle.

12.1.2   Contributions

In this chapter, the contributions present a comprehensive survey on major systematic approaches and outline some recent advances in dealing with delay-aware radio resource allocation optimization problems in heterogeneous C-RANs. The two major approaches are the Markov decision process (MDP) approach and the Lyapunov optimization approach, which have their own implementation challenges to deal with. A brief overview of the general assumption, theorems and design methodology for the two approaches is given first, followed by the introduction of various delay-aware radio-resource-allocation optimization algorithms. The proposed algorithms involve optimization across different layers: advanced cooperative beamforming and power allocation in the PHY layer for interference control; effective MAC layer algorithms for rate allocation, RRH on/off control, and resource block allocation; network layer solutions to control individual flows and avoid traffic congestion. Specifically, we illustrate how approximate MDP, stochastic learning, and continuous-time MDP could help to obtain low-complexity solutions. We also describe how to integrate the equivalent weighted minimum square error (WMMSE) method in Lyapunov optimization in order to derive algorithms that can be easily implemented in a parallel fashion. The main content of this chapter is based on published papers [14, 17, 26, 30]. These surveyed references are by no means exhaustive. Nonetheless, they still help to present the main roadmap of ongoing attempts to design delay-aware radio-resource-allocation-optimization algorithms for future heterogeneous C-RANs. While we have highlighted a few recent developments, some future research directions are also identified.

12.1.3   Organization of this Chapter

The chapter is organized as follows. In Section 12.2, we elaborate on the general model and methodology to deal with the delay-aware radio resource allocation optimization problem. In Section 12.3, we introduce various delay-aware radio resource allocation algorithms to illustrate how how to overcome the complexity as well as the parallel implementation requirement by utilizing the special structures of the problems we formulate. Finally, future research directions are identified in Section 12.4.

12.2   General Model and Methodology

For simplicity, the discussion on delay-aware radio resource allocation optimization is restricted to downlink heterogeneous C-RANs, where the traffic flows from the BBU pool to the UEs through the wired or wireless fronthaul link and the radio access network. The networks are assumed to operate in slotted time, with the time dimension partitioned into decision slots indexed by $t \in \{0, 1, 2, \ldots\}$ with slot duration τ. Let $\mathcal{K} = \{1, 2, \ldots, K\}$ and $\mathcal{I} \in \{1, 2, \ldots, I\}$ denote the set of RRHs and the set of UEs, respectively. The instantaneous transmit rate achieved by each UE is a function of a collection of system states and system parameters defined across different layers. The system states include the CSI, QSI, and some performance state information if possible. The system parameters include the beamforming precoders, power allocations, resource block allocations, RRH clustering, and so on. Let $\mathbf{Q}(t) = \{Q_1(t), \ldots, Q_I(t)\} \in \mathcal{Q}$ denote the QSI for the I UEs at the beginning of slot t, where $\mathcal{Q}$ is is the system QSI state space. Let $\mathbf{H}(t) = \{\mathbf{H}_{ki}(t) : k \in \mathcal{K}, i \in \mathcal{I}\} \in \mathcal{H}$ denote the CSI in slot t, where $\mathcal{H}$ is the system CSI state space. We have the following assumptions on the CSI.

Assumption 12.1 (Assumption on the CSI)   Each element in $\mathbf{H}$ takes a value from the discrete state space $\mathcal{H}$ (which may be arbitrarily large), and the system CSI $\mathbf{H}(t)$ is a Markov process, i.e.

$$P[\mathbf{H}(t)|\mathbf{H}(t-1), \mathbf{H}(t-2), \ldots, \mathbf{H}(0)] = P[\mathbf{H}(t)|\mathbf{H}(t-1)]. \tag{12.1}$$

The BBU pool maintains I traffic queues for random traffic arrivals at the I UEs. Let $\mathbf{A}(t) = \{A_1(t), \ldots, A_I(t)\}$ be the amount of stochastic traffic data arrivals (bits) at the end of slot t. We have the following assumptions on the traffic arrivals.

Assumption 12.2 (Assumption on the traffic arrival process)   The traffic arrival $A_i(t)$ at UE i is independent of i and i.i.d. over the slots, according to a general distribution with mean $\mathbb{E}[A_i(t)] = \lambda_i$.

The queue dynamic for UE i is given by

$$Q_i(t+1) = \max[Q_i(t) - R_i(t), 0] + A_i(t). \tag{12.2}$$

Let $\chi(t) = \mathbf{H}(t), \mathbf{Q}(t) \in \chi$, be the system state which can be obtained by the BBU at slot t, where $\chi = \mathcal{H} \times \mathcal{Q}$ is the full system state space. A delay-aware radio-resource-allocation policy has control actions that are adaptive to the full system state, i.e., the CSI and the QSI. The control policy $\Omega(\chi)$ is a mapping from the full system state space to the action space. Under a control policy Ω, the average queue length for UE i is given by

$$\bar{Q}_i = \limsup_{T \to \infty} \frac{1}{T} \sum_{t=1}^{T} \mathbb{E}^{\Omega}\{Q_i(t)\} \tag{12.3}$$

where $\mathbb{E}^{\Omega}\{\cdot\}$ means the expectation operation taken w.r.t. the measure induced by the given policy Ω. The average throughput and the average power consumption can be given in a similar way. Delay-aware radio resource allocation optimization problems for heterogeneous C-RANs can be solved using various criteria of the system performance, for example, the average weighted sum system throughput, the average weighted sum delay, the average weighed sum power consumption, or a combination of these. In the following subsections, after a brief survey of the two major systematic approaches in dealing with delay-aware radio resource allocation optimization problems, various practical problems will then be discussed on the basis of those approaches.

12.2.1   Overview of the Markov Decision Process

The MDP is a systematic approach to dealing with delay-aware control problems. The system states of wireless systems are characterized by the aggregation of the CSI and the QSI. In fact, under Assumptions 12.1 and 12.2, the system state dynamics evolve as a controlled Markov chain and delay-aware radio-resource-allocation optimization can be modeled as an infinite-horizon average cost MDP [6]. In general, an MDP can be characterized by four elements, namely the state space, the action space, the state transition probability, and the system cost, which are defined as follows:

- $\chi = \{\chi^1, \chi^2, \ldots\}$: the finite state space;
- $\mathcal{A} = \{a^1, a^2, \ldots\}$: the action space;
- $P[\chi'|\chi, a]$: the transition probability from state χ to state χ' under action a;
- $g(\chi, a)$: the system cost in state χ under action a.

Therefore, an MDP is a 4-tuple $\chi, \mathcal{A}, P[\chi'|\chi, a], g(\chi, a)$. A stationary and deterministic control policy $\Omega : \chi \rightarrow \mathcal{A}$ is a mapping from the state space χ to the action space $\mathcal{A}$, which determines the specific action taken when the system is in state χ. Given a policy Ω, the corresponding random process of the system state and the per-stage cost, $\chi, g(\chi, a)$, evolve as a Markov chain with probability measure induced by the transition kernel $P[\chi'|\chi, a]$. The goal of the infinite-horizon average cost problem is to find an optimal policy Ω such that the long-term average cost is minimized among all feasible policies, i.e.

$$\underset{\Omega}{\text{minimize}} \limsup_{T \to \infty} \frac{1}{T} \sum_{t=1}^{T} \mathbb{E}^{\Omega}\{g(\chi(t), \Omega(\chi(t)))\}. \tag{12.4}$$

If the feasible policies are unichain policies, then the optimization problem can be written as

$$\underset{\Omega}{\text{minimize}} \limsup_{T \to \infty} \frac{1}{T} \sum_{t=1}^{T} \mathbb{E}^{\Omega}\{g(\chi(t), \Omega(\chi(t)))\} = \min_{\Omega} \mathbb{E}^{\pi(\Omega)}\{g(\chi(t), \Omega(\chi(t)))\}, \tag{12.5}$$

where $\pi(\Omega)$ is the unique steady state distribution, given policy Ω.

Assume that the buffer size is finite and that the maximum buffer size is denoted as N_Q. The system queue dynamics $\mathbf{Q}(t)$ evolves according to (12.2) with projection onto $[0, N_Q]$. The traffic arrival departure, and the CSI processes are Markovian under Assumptions 12.1 and 12.2. Hence, the system state $\chi(t)$ is a finite-state controlled Markov chain with the following transition kernel:

$$P[\chi(t+1)|\chi(t), \Omega(\chi(t))] = P[\mathbf{Q}(t+1)|\chi(t), \Omega(\chi(t))]P[\mathbf{H}(t+1)|\mathbf{H}(t)]. \tag{12.6}$$

Under the unichain-policy-space assumption, the delay-aware control policy of (12.5) is given by the solution of the Bellman equation in the following theorem [6].

THEOREM 12.1 (Bellman equation)   *If there exists a scalar θ and a vector $V = [V(\chi^1), V(\chi^2), \ldots]$ that satisfy the following Bellman equation for the delay-aware radio resource allocation optimization in (12.5),*

$$\theta + V(\chi) = \min_{\Omega} \left\{ g(\chi, \Omega(\chi)) + \sum_{\chi'} P[\chi'|\chi, \Omega(\chi)] V(\chi') \right\}, \quad \forall \chi \in \chi, \qquad (12.7)$$

then θ is the optimal average cost per stage and the policy Ω that attains the minimum in (12.5) for any $\chi \in \chi$ is the optimal control policy.

It can be seen from Theorem 12.1 that the value functions $V(\chi)$ are critical to the solution, which is usually too complicated to compute owing to the extremely high complexity. Besides, the extension to the C-RAN scenario will incur the *curse of dimensionality*, where the number of system queueing states as well as the corresponding number of value functions will grow exponentially with the number of traffic queues maintained by the centralized BBU. From an implementation perspective, it is desirable to obtain low-complexity and parallel solutions. To this end, the techniques of approximate MDP in [7] and stochastic learning in [8] could be used to overcome the complexity as well as the distributed implementation requirement in delay-aware resource control. The details of approximate MDP and stochastic learning will be discussed in the following subsections according to the specific problem structure of the model under consideration.

12.2.2   Overview of Lyapunov Optimization

The second approach to deal with delay-aware radio-resource-allocation optimization is the Lyapunov optimization approach, which can stabilize the queues of networks while additionally optimizing some performance metrics or satisfying additional constraints across the layers.

To model the impacts of radio resource allocation on the average queue delay and other system-level performance metrics, the definitions of queue stability, stability region, and throughput optimal policy are first given formally as follows [9].

DEFINITION 12.2 (Queue stability)   A discrete-time queue $Q(t)$ is strongly stable if

$$\limsup_{T \to \infty} \frac{1}{T} \sum_{t=1}^{T} \mathbb{E}\{|Q(t)|\} < \infty. \qquad (12.8)$$

Furthermore, a network of queues is stable if all individual queues of the network are stable.

DEFINITION 12.3 (Stability region and throughput-optimal policy)   The stability region $\mathcal{C}$ is the closure of the set of all the arrival rate vectors $\{\lambda_i : i \in \mathcal{I}\}$ that can be stabilized. A throughput-optimal radio resource allocation policy is a policy that stabilizes all the arrival rate vectors $\{\lambda_i : i \in \mathcal{I}\}$ within the stability region $\mathcal{C}$.

In order to show the stability property of the queueing systems, the well-developed stability theory in Markov chains using the Lyapunov drift will be utilized. To represent a scalar metric of queue congestion in a C-RAN, we define the quadratic Lyapunov function as $L(\mathbf{Q}(t)) = \frac{1}{2} \sum_{i \in \mathcal{I}} Q_i(t)^2$. To keep the system stable by persistent pushing of the

Lyapunov function towards a lower congestion state, a one-step conditional Lyapunov drift is difined. It is given by

$$\Delta(\mathbf{Q}(t)) = \mathbb{E}\{L(\mathbf{Q}(t+1)) - L(\mathbf{Q}(t))|\mathbf{Q}(t)\}. \tag{12.9}$$

Let $g(t)$ be a concave function representing the utility given by a radio resource allocation action. Lyapunov optimization aims to stabilize the $\mathbf{Q}(t)$ process while maximizing the time average utility, which suggests that a good control strategy is to greedily minimize the following drift-minus-utility metric at every time slot,

$$\Delta(\mathbf{Q}(t)) - V\mathbb{E}\{g(t)\,|\mathbf{Q}(t)\}. \tag{12.10}$$

Let g^* represent a desired target utility value; then the relationship between the Lyapunov drift-minus-utility and queue stability is established in the following theorem [10].

THEOREM 12.4 (Lyapunov optimization) *Suppose there are positive constants B, ϵ, and V such that, for all slots $t \in \{0, 1, 2, \ldots\}$ and all possible values of $\mathbf{Q}(t)$, the Lyapunov drift-minus-utility function satisfies*

$$\Delta(\mathbf{Q}(t)) - V\mathbb{E}\{g(t)|\mathbf{Q}(t)\} \le B + Vp^* - \epsilon \sum_{i=1}^{I} Q_i(t); \tag{12.11}$$

then all queues $Q_i(t)$ are strongly stable. The average queue length satisfies

$$\limsup_{T\to\infty} \frac{1}{T}\sum_{t=0}^{T-1}\sum_{i\in\mathcal{I}} \mathbb{E}^{\Omega}\{Q_i(t)\} \le \frac{B + V(\bar{g} - g^*)}{\varepsilon}, \tag{12.12}$$

and the average achieved utility satisfies

$$\limsup_{T\to\infty} \frac{1}{T}\sum_{t=0}^{T-1} \mathbb{E}^{\Omega}\{g(t)\} \ge g^* - \frac{B}{V}. \tag{12.13}$$

The above theorem indicates that the delay–utility tradeoff can be achieved by adjusting the positive parameter V when minimizing the drift-minus-utility. With the queue dynamics, we have the following theorem about the upper bound of the Lyapunov drift-minus-utility [10].

LEMMA 12.5 (Upper bound of Lyapunov drift-minus-utility) *Under any control policy Ω, the drift-minus-utility has the following upper bound for all t, all possible values of $\mathbf{Q}(t)$, and all parameters $V > 0$:*

$$\Delta(\mathbf{Q}(t)) - V\mathbb{E}\{g(t)|\mathbf{Q}(t)\} \le B - V\mathbb{E}\{g(t)|\mathbf{Q}(t)\} + \sum_{i\in\mathcal{I}} Q_i(t)\mathbb{E}\{A_i(t) - R_i(t)|\mathbf{Q}(t)\}, \tag{12.14}$$

where $B \ge \frac{1}{2}\sum_{i\in\mathcal{I}}\mathbb{E}\{(A_i(t))^2 + (R_i(t))^2|\mathbf{Q}(t)\}$.

The above lemma indicates that the control policy can also be obtained by greedily minimizing the derived upper bound over all possible policies instead of directly minimizing the drift-plus-penalty at each slot. The Lyapunov optimization approach provides a simple framework to deal with delay-aware cross-layer radio-resource-allocation optimization problems. In a summary, the procedure can be summarized as follows [11]:

1. If needed, transform the average performance constraints into queue stability problems using the technique of virtual cost queues;
2. Choose a Lyapunov function and calculate the Lyapunov drift-minus-utility $\Delta(\mathbf{Q}(t)) - V\mathbb{E}\{g(t)|\mathbf{Q}(t)\}$, where $g(t)$ is the utility to be maximized;
3. On the basis of the system state observations, minimize the upper bound of $\Delta(\mathbf{Q}(t)) - V\mathbb{E}\{g(t)|\mathbf{Q}(t)\}$ over all polices at each time slot.

12.3   Delay-Aware Radio-Resource-Optimization Algorithms

12.3.1   Delay-Aware Joint Power and Rate Allocation with Hybrid-CoMP Scheme

As the RRHs in C-RANs are connected to the BBU pool via high-speed fronthaul links, coordinated multi-point (CoMP) transmission and reception is shown to be very effective in improving the overall spectrum efficiency. Generally, CoMP techniques can be characterized into two classes: joint processing (JP) and coordinated beamforming (CB) [12]. For JP, the traffic payload is shared and transmitted jointly by all RRHs within the CoMP cluster, which means multiple delivery of the same traffic payload from the centralized BBU pool to each cooperative RRH through capacity-limited fronthaul links. For CB, the traffic payload is transmitted only by the serving RRH, but the corresponding beamformer is jointly calculated at the centralized BBU pool so as to coordinate the interference to all other UEs within the CoMP cluster. Obviously, JP achieves a higher average spectrum efficiency than CB does, at the expense of more fronthaul consumption, while CB requires more antennas at each RRH to achieve full intra-cluster interference coordination. However, the practical non-ideal fronthaul with limited capacity restricts the overall performance of CoMP in C-RANs.

An important design issue in this context is how to make a flexible tradeoff between cooperation gain and average fronthaul consumption [13]. In this context, the hybrid CoMP scheme is proposed for delay-sensitive traffic in C-RANs. Consider the transmission of K delay-sensitive traffic payloads within a CoMP cluster of K RRHs in downlink C-RANs. For each RRH, there is a served UE. Let $\mathcal{K} = \{1, 2, \ldots, K\}$ and $\mathcal{I} = \{1, 2, \ldots, K\}$ denote the set of RRHs and the set of UEs within the CoMP cluster. Each RRH and UE are equipped with M and N antennas respectively, where $KN \geq M > (K-1)N$. With the perfect CSIT, at most $L_{K,M,N} = M - (K-1)N$ private streams can be zero-forced at the serving RRH of UE i (denoted as RRH i) in order to eliminate interference to UE $j \neq i$; then we have $L_{(i,p)} \leq L_{K,M,N}$. Furthermore, to recover fully the $L_{(i,p)}$ private streams and $L_{(i,s)}$ shared streams at UE i, the constraint $L_{(i,p)} + L_{(i,s)} \leq N$ should be satisfied. Although the shared streams and private streams are superimposed in the downlink transmission of C-RANs, it is possible to transmit the shared streams and private streams simultaneously by constructing the private streams and shared streams and designing appropriate precoders at the BBU.

The reader should refer to [14] for the design details. The CSIT obtained by BBU is usually imperfect; thus there will be uncertain residual interference of the recovered streams. The amount of traffic data successfully transmitted to UE i is given by $R_i = R_{(i,s)}\tau\mathbf{1}(R_{(i,s)} \leq C_{(i,s)}) + R_{(i,p)}\tau\mathbf{1}(R_{(i,p)} \leq C_{(i,p)})$, where $C_{(i,s)} = \sum_{a=1}^{L_{(i,s)}} C_{(i,s)}^a$ and $C_{(i,p)} = \sum_{a=1}^{L_{(i,p)}} C_{(i,p)}^a$ are the mutual information for the shared streams and private streams, respectively; $R_{(i,s)}$ and $R_{(i,p)}$ are the allocated data rates for the shared streams and private streams of UE i, respectively. The total transmit power consumed by RRH i is given by $P_i = \sum_{a=1}^{L_{(i,p)}} P_{(i,p)}^a + \sum_{j\in\mathcal{I}} \sum_{a=1}^{L_{(j,s)}} P_{(j,s)}^a \rho_{(j,i)}^a$, where $P_{(i,p)}^a$ and $P_{(i,s)}^a$ are the power for ath private and shared streams, respectively. Considering the energy-efficient fronthaul-constrained transmission of delay-sensitive traffic, the delay-aware rate and power allocation problem can be formulated as

$$\underset{\Omega}{\text{minimize}}\ D(\boldsymbol{\beta},\Omega) = \lim_{T\to\infty} \sup \frac{1}{T} \sum_{t=1}^{T} \mathbb{E}^{\Omega}\left\{\sum_{i\in\mathcal{I}} \beta_i Q_i/\lambda_i\right\}$$

$$\text{s.t.}\qquad P_i(\Omega) = \lim_{T\to\infty} \sup \frac{1}{T} \sum_{t=1}^{T} \mathbb{E}^{\Omega}\{P_i(t)\} \leq P_i^{\max},$$

$$R_i^f(\Omega) = \lim_{T\to\infty} \sup \frac{1}{T} \sum_{t=1}^{T} \mathbb{E}^{\Omega}\{R_{(i,p)}(t) + \sum_{j\in\mathcal{I}} R_{(j,s)}(t)\} \leq R_i^{\max}, \qquad (12.15)$$

where β_i is the weight for each traffic queue and Q_i/λ_i is the average traffic delay cost for UE i. A feasible stationary resource allocation policy $\Omega(\hat{\chi})$ is a mapping from the global observed system states $\hat{\chi} = \{\mathbf{Q},\hat{\mathbf{H}}\}$, instead of the global system states $\chi = \{\mathbf{Q},\mathbf{H},\hat{\mathbf{H}}\}$, to the resource allocation actions, where $\mathbf{Q}$, $\mathbf{H}$, and $\hat{\mathbf{H}}$ are the global QSI, global CSI, and global CSIT, respectively. Therefore, the problem is a constrained partially observed MDP (POMDP). Using Lagrange duality theory, the corresponding Lagrange dual function is given by

$$J(\gamma) = \min_{\Omega} L(\beta,\gamma,\Omega(\hat{\chi})) = \lim_{T\to\infty} \sup \frac{1}{T} \sum_{t=1}^{T} \mathbb{E}^{\Omega}[g(\boldsymbol{\beta},\gamma,\Omega(\hat{\chi}))], \qquad (12.16)$$

where

$$g(\boldsymbol{\beta},\gamma,\Omega(\hat{\chi})) = \sum_{i\in\mathcal{I}} (\beta_i Q_i/\lambda_i) + \gamma_{(i,P)}(P_i - P_i^{\max}) + \gamma_{(i,R)}(R_i^f - R_i^{\max})$$

is the per-stage system cost and $\gamma_{(i,P)}$ and $\gamma_{(i,R)}$ are the nonnegative Lagrange multipliers (LMs) w.r.t the power consumption constraints and fronthaul consumption constraints.

Although (12.16) is an unconstrained POMDP, the solution is generally nontrivial. To substantially reduce the global observed system state space and derive an equivalent Bellman equation, partitioned actions with the i.i.d. property of the CSIT are defined as follows.

DEFINITION 12.6 (Partitioned actions)   Given the stationary resource allocation policy Ω, $\Omega(\mathbf{Q}) = \{\Omega(\hat{\chi}) : \forall\ \hat{\mathbf{H}}\}$ is defined as the collection of power and rate allocation actions for all possible CSITs $\hat{\mathbf{H}}$ on a given QSI $\mathbf{Q}$; therefore Ω is equal to the union of all partitioned actions, i.e., $\Omega = \cup_Q \Omega(\mathbf{Q})$.

As the distribution of the traffic arrival process is unknown to the BBU, the post-decision state potential function, instead of the potential function, will be introduced and we have the following theorem about the equivalent Bellman equation.

THEOREM 12.7 (Equivalent Bellman equation with reduced state space)   *(a) Given the LMs, the unconstrained POMDP problem can be solved by the equivalent Bellman equation as follows:*

$$U(\tilde{\mathbf{Q}}) + \theta = \sum_A P(A) \min_{\Omega(\mathbf{Q})} g(\boldsymbol{\beta}, \boldsymbol{\gamma}, \mathbf{Q}, \Omega(\mathbf{Q})) + \sum_{\tilde{\mathbf{Q}}'} P[\tilde{\mathbf{Q}}' | \mathbf{Q}, \Omega(\mathbf{Q})] U(\tilde{\mathbf{Q}}'), \quad (12.17)$$

where $g(\boldsymbol{\beta}, \boldsymbol{\gamma}, \mathbf{Q}, \Omega(\mathbf{Q})) = \mathbb{E}\{g(\boldsymbol{\beta}, \boldsymbol{\gamma}, \Omega(\hat{\mathbf{S}})) | \mathbf{Q}\}$ is the conditional per-stage cost and $P\{\tilde{\mathbf{Q}}' | \mathbf{Q}, \Omega(\mathbf{Q})\} = \mathbb{E}\{P[\mathbf{Q}' | \mathbf{H}, \mathbf{Q}, \Omega(\hat{\mathbf{S}})] | \mathbf{Q}\}$ is the conditional average transition kernel, and $U(\tilde{\mathbf{Q}})$ is the post-decision value function; $\tilde{\mathbf{Q}}$ is the post-decision state and $\tilde{\mathbf{Q}}' = (\mathbf{Q} - \mathbf{R})^+$ is the next post-decision state, where $\mathbf{Q} = \min\{\tilde{\mathbf{Q}} + \mathbf{A}, N_Q\}$ and $\mathbf{R} = \{R_i : i \in \mathcal{I}\}$.
   (b) If there exists a unique $(\theta, \{U(\tilde{\mathbf{Q}})\})$ that satisfies (12.17) then
$\theta = \min_{\Omega(\mathbf{Q})} \mathbb{E}\{g(\boldsymbol{\beta}, \boldsymbol{\gamma}, \mathbf{Q}, \Omega(\mathbf{Q}))\}$ *is the optimal average per-stage cost for the unconstrained POMDP and the optimal resource allocation policy Ω is obtained by minimizing the right-hand side of (12.17).*

The post-decision value functions are critical to the derivation of a queue-aware resource allocation policy; they can be obtained by solving N_Q fixed point nonlinear Bellman equations with N_Q+1 variables. The offline calculation requires explicit knowledge of the conditional average transition kernel, which is infeasible. To substantially reduce the enormous computing complexity of estimating post-decision value functions in a centralized BBU pool, a linear approximation for the post-decision value functions is adopted as $U(\tilde{\mathbf{Q}}) \approx \sum_{i \in \mathcal{M}} U_i(\tilde{Q}_i)$ [7], where $U_i(\tilde{Q}_i)$ is the per-queue post-decision value function which satisfies the following per-queue fixed point Bellman equation:

$$U_i(\tilde{Q}_i) + \theta_i = \sum_{A_i} P(A_i) \min_{\Omega_i(Q_i)} \left[g_i(\beta_i, \boldsymbol{\gamma}_i, Q_i, \Omega_i) + \sum_{\tilde{Q}_i'} P[\tilde{Q}_i' | Q_i, \Omega_i] U(\tilde{Q}_i') \right],$$
$$(12.18)$$

where

$$g_i(\beta_i, \boldsymbol{\gamma}_i, Q_i, \Omega_i) = \mathbb{E} \left\{ \beta_i Q_i / \lambda_i + \gamma_{(i,P)} \left(\sum_{a=1}^{L_{(i,s)}} P_{(i,s)}^a \rho_{(i,i)}^a + \sum_{a=1}^{L_{(i,p)}} P_{(i,p)}^a - P_i^{\max} \right) \right.$$
$$+ \sum_{j \in \mathcal{M}, j \neq i} \gamma_{(j,P)} \sum_{a=1}^{L_{(i,s)}} P_{(i,s)}^a \beta_{(i,j)}^a$$
$$\left. + \gamma_{(i,R)} \left(R_{(i,p)}(t) + \sum_{j \in \mathcal{M}} R_{(j,s)}(t) - R_i^{\max} \right) | Q_i \right\}$$

is the per-queue per-stage cost function; $Q_i = \min\{\tilde{Q}_i + A_i, N_Q\}$ is the pre-decision state and $\tilde{Q}_i' = (Q_i - R_i)^{\dagger}$ is the next post-decision state. The linear approximation is optimal when the CSIT is perfect, which means that interference is completely eliminated with H-CoMP scheme, and the queue dynamics of the M UEs are decoupled. Motivated by [8], the online learning of per-queue post-decision value functions on the basis of equation (12.18) is proposed. At the end of each slot, the per-queue post-decision value function $U_i(\tilde{Q}_i)$ is online-learned at the BBU pool as follows, from observation of the post-decision QSI $\tilde{Q}_i$ and the pre-decision QSI Q_i:

$$U_i^{t+1}(\tilde{Q}_i) = U_i^t(\tilde{Q}_i) + \zeta_u(t)[g_i(\gamma_i^t, \hat{S}_i, P_i, R_i + U_i^t(Q_i - U_i) - U_i^t(\tilde{Q}_i^0) - U_i^t(\tilde{Q}_i)]. \quad (12.19)$$

With observation of the current system states and the estimated post-decision value functions, the per-stage optimization is given by

$$\Omega^*(\hat{\chi}) = \{\Omega_P^*(\hat{\chi}), \Omega_R^*(\hat{\chi})\} = \arg \min_{\Omega_P(\hat{\chi}), \Omega_R(\hat{\chi})} B(\hat{\chi}, \Omega_P(\hat{\chi}), \Omega_R(\hat{\chi})), \quad (12.20)$$

where $B(\hat{\chi}, \Omega_P(\hat{\chi}), \Omega_R(\hat{\chi}))$ is the per-stage objective simplified from the right-hand side of the Bellman equation (12.17). The delay-aware power and rate allocations for each UE can be obtained by solving the per-stage optimization problem with the following stochastic gradient-based iteration [15]:

$$e_i^t(\hat{\chi}_i) = [e_i^{t-1}(\hat{\chi}_i) - \gamma_e(t-1)d(e_i^{t-1}(\hat{\chi}_i))]^+, \quad (12.21)$$

where $\gamma_e(t)$ is the step size satisfying $\gamma_e(t) > 0, \sum_t \gamma_e(t) = \infty, \sum_t (\gamma_e(t))^2 < \infty$ and $d(e_i^t(\hat{\chi}_i))$ is the stochastic gradient of (12.20) w.r.t power and rate allocation [14]. With the obtained power and rate allocations, the LMs are updated as follows:

$$\gamma_{(i,P)}^{t+1} = [\gamma_{(i,P)}^t + \zeta_\gamma(t)(P_i - P_i^{\max})]^+, \quad (12.22)$$

$$\gamma_{(i,R)}^{t+1} = [\gamma_{(i,R)}^t + \zeta_\gamma(t)(R_{(i,p)}(t) + \sum_{j \in \mathcal{M}} R_{(j,s)}(t) - R_i^{\max})]^+. \quad (12.23)$$

The proposed stochastic gradient-based algorithm gives the asymptotically local optimal solution at small CSIT errors, which means that the explicit knowledge of imperfect CSIT is unnecessary, and it is robust against the uncertainties caused by imperfect CSIT.

Simulations were conducted to compare the performances of the proposed delay-aware resource allocation with the hybrid-CoMP transmission scheme (labeled as QSI-adaptive Hybrid) with three baselines in C-RANs: the CSI-adaptive CB scheme, the CSI-adaptive JP scheme, and CSI-adaptive Hybrid scheme. These three baselines all make rate and power allocations to maximize the average system throughput with the same fronthaul capacity and average power consumption constraint as the proposed QSI-adaptive Hybrid scheme. Figure 12.1 shows the average packet delay versus the maximum transmit power when the packet arrival rate is $\lambda_i = 2.5$ packets/s and the maximum fronthaul consumption is $R_i^{\max} = 30$ Mbit/s. The figure corresponds to the medium fronthaul-consumption regime, in which the CSI-adaptive JP scheme outperforms the CSI-adaptive CB scheme owing to its higher spectrum efficiency. The CSI-adaptive hybrid scheme outperforms both the CSI-adaptive CB scheme and CSI-adaptive JP scheme while its outperformance is not so obvious with relatively enough fronthaul capacity. There is a significant performance gain of the proposed QSI-adaptive Hybrid scheme compared with all the baselines across a wide range of maximum power consumption, owing to the delay-aware resource allocation.

Figure 12.2 shows the average packet delay versus the maximum fronthaul consumption when the packet arrival rate is $\lambda_i = 2.5$ packets/s and the maximum transmit

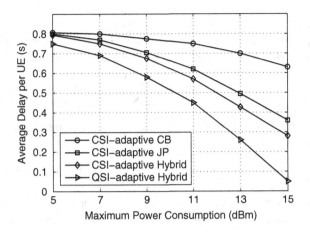

Figure 12.1 Average packet delay versus maximum transmit power.

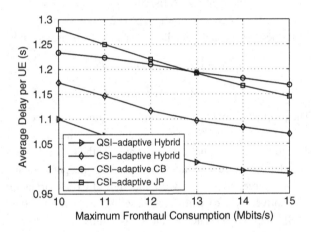

Figure 12.2 Average packet delay versus maximum fronthaul consumption.

power is $P_i^{\mathrm{max}} = 10$ dBm. The figure corresponds to the small fronthaul-consumption regime, in which the CSI-adaptive hybrid scheme clearly outperforms both the CSI-adaptive CB scheme and the CSI-adaptive JP scheme; this is done to flexible adjustment of the cooperation level when the fronthaul capacity is limited. Note that the CSI-adaptive JP scheme has a worse delay performance than the CSI-adaptive CB scheme owing to limited fronthaul capacity at first but eventually the performance improves with increasing fronthaul capacity. Similarly, owing to the delay-aware power and rate allocation, the QSI-adaptive hybrid scheme substantially outperforms the three baselines.

12.3.2 Delay-Aware Asymmetrical Cooperative Beamforming with Limited Fronthaul Transfer

Full cooperation among all RRHs usually requires a significant overhead across the fronthaul links. In practice, it is preferabke to perform asymmetrical cooperation beamforming within an RRH cluster for each UE [16]. Consider the downlink cooperative transmission of K delay-sensitive traffic flows by K RRHs. Let $\mathcal{K} = \{1, 2, \ldots, K\}$ and $\mathcal{I} = \{1, 2, \ldots, K\}$ denote the sets of RRHs and UEs, respectively. Each RRH is equipped with M antennas and each UE is equipped with one antenna. Let $Q_i(t)$ denote the QSI of UE i at the beginning of slot t. Let $\mathbf{h}_i$ denote the CSI vector from all RRHs to UE i. Let $\mathbf{w}_i = [\mathbf{w}_{1i}^T\ \mathbf{w}_{2i}^T \cdots \mathbf{w}_{Ki}^T]^T$ denote the joint beamforming vector for UE i, where $\mathbf{w}_{ki}$ is the beamforming vector at RRH k for UE i. Then the average system cost, consisting of the average delay cost and the average transmit power cost, is given by

$$\bar{g}(\Omega) = \lim_{T \to \infty} \sup \frac{1}{T} \sum_{t=0}^{T-1} \mathbb{E}^{\Omega} \left\{ \beta_i \frac{Q_i(t)}{\lambda_i(t)} + \sum_{k \in \mathcal{K}} \sum_{i \in \mathcal{I}} ||\mathbf{w}_{ki}||^2 \right\} \tag{12.24}$$

where β_i is the positive delay cost. Suppose that the set of cooperative RRHs for each UE has been properly scheduled and we consider only the delay-aware asymmetrical beamforming optimization problem for delay-sensitive traffic. Specifically, for UE i, we define $z_i = [z_{1i}\ z_{2i} \cdots z_{Ki}] \in R^{M \times 1}$, where $z_{ki} = 1$ means that the information signal for UE i is not available to RRH k for cooperative transmission and the beamforming vector $\mathbf{w}_{ki}$ is forced to zero, while $z_{ki} = 0$ means the opposite. Let $\hat{\mathbf{z}}_i = z_i \otimes \mathbf{1}_{1 \times M}$ and $\mathbf{Z}_i = \text{diag}(\hat{\mathbf{z}}_i)$, where $\otimes$ denotes the Kronecker product and $\mathbf{1}_{1 \times M}$ denotes a vector with every element equal to 1. To force a specified $\mathbf{w}_{ki}$ to be zero according to the cooperative RRH sets and so to alleviate the fronthaul consumption, the delay-aware beamforming control problem with limited fronthaul transfer is formulated as

$$\underset{\Omega}{\text{minimize}}\ \bar{g}(\Omega)$$

$$\text{s.t.}\ \mathbf{Z}_i \mathbf{w}_i = 0. \tag{12.25}$$

The problem is an infinite-horizon average-cost discrete-time MDP. The optimal beamforming control policy can be obtained by solving a similar Bellman equation to that in the previous subsection and the important value function is estimated by a similar online-learning algorithm. However, iteration of the online-learning algorithm still suffers from the *challenge of slow convergence* and lack of insight. To address this challenge, we will consider a continuous-time MDP instead of a discrete-time MDP. Let $\mathbf{H}(t)$ and $\mathbf{W}(t)$ denote the system CSI and the beamforming action at slot t. When τ is asymptotically close to 0, according to the discrete-time queue dynamics, the trajectory of the ith traffic queue can be described by the following differential equation in the continuous-time system:

$$\frac{dQ_i(t)}{dt} = \lambda_i - \mathbb{E}\{R_i(\mathbf{H}(t), \mathbf{W}(t)) | \mathbf{Q}(t)\} \tag{12.26}$$

and the first-order Taylor expansion of $V(\mathbf{Q})$ is given by

$$V(\mathbf{Q}') = V(\mathbf{Q}) + \sum_{i \in \mathcal{I}} \frac{\partial V(\mathbf{Q})}{\partial Q_i} (\lambda_i - \mathbb{E}\{R_i(\mathbf{H}, \mathbf{W}) | \mathbf{Q}(t)\}) \tau + o(\tau). \tag{12.27}$$

The equivalence between the discrete time Bellman equation and the continuous time Bellman equation is established in the following theorem [6].

THEOREM 12.8 (Equivalent continuous-time Bellman equation) *For sufficiently small* τ, *if there exist* $\tilde{\theta}$ *and* $J(\mathbf{Q})$ *that solve the following continuous time Bellman equation,*

$$\tilde{\theta} = \min_{\Omega(\mathbf{Q})} \left\{ g(\mathbf{Q}, \Omega(\mathbf{Q})) + \sum_{i \in \mathcal{I}} \frac{\partial J(\mathbf{Q})}{\partial Q_i} (\lambda_i - \mathbb{E}[R_i(\mathbf{H}, \mathbf{W}) | \mathbf{Q}(t)]) \right\}, \tag{12.28}$$

then we have

$$\tilde{\theta} = \theta + o(1), \quad J(\mathbf{Q}) = V(\mathbf{Q}) + o(1), \tag{12.29}$$

and the error is negligible as $\tau \to 0$.

According to Theorem 12.8, we can solve the continuous-time Bellman equation, with an equation instead of the discrete-time version, and the solution is asymptotically accurate as $\tau \to 0$. The calculation of $J(\mathbf{Q})$ involves solving an N-dimensional non-linear partial differential equation. To facilitate the calculation of $J(\mathbf{Q})$, the value function can be linearly approximated as $J(\mathbf{Q}) \approx \sum_{i \in \mathcal{I}} J_i(Q_i)$, which will result in some loss of strict optimality but can be easily calculated, making substantial use of well-established calculus theory to reduce the complexity of this calculation.

Let $\mathbf{H}_i = \mathbf{h}_i \mathbf{h}_i^H$ and let $\mathbf{W}_i = \mathbf{w}_i \mathbf{w}_i^H$. It is obvious that $\mathbf{W}_i \succeq 0$ and $\text{rank}(\mathbf{W}_i) = 1$. By introducing the slack variables η_i and using the semidefinite relaxation technique, we can remove the non-convex rank constraint $\text{rank}(\mathbf{W}_i) = 1$ and we obtain the following conservation formulation of (12.28):

$$\underset{\{\mathbf{W}_i\}, \{\eta_i\}}{\text{minimize}} \quad \sum_{i \in \mathcal{I}} \left[\text{Tr}(\mathbf{W}_i) - \frac{\partial J(\mathbf{Q})}{\partial Q_i} B \log_2 \eta_i \right]$$

$$\text{s.t.} \quad \text{Tr}(\mathbf{H}_i \mathbf{W}_i) - \eta_i \sum_{j \neq i} \text{Tr}(\mathbf{H}_i \mathbf{W}_j) - \eta_i \sigma_i^2 \geq 0,$$

$$\text{Tr}(\mathbf{Z}_i \mathbf{W}_i) = 0,$$

$$\mathbf{W}_i \succeq 0, \tag{12.30}$$

which proves to be a convex optimization problem. Furthermore, when $\{\mathbf{W}_i\}$ is fixed, (12.30) is a convex problem w.r.t. $\{\eta_i\}$ and when $\{\eta_i\}$ is fixed (12.30) is a semidefinite programming problem (SDP) w.r.t. $\{\mathbf{W}_i\}$. Therefore, the convex problem can be solved effectively by the following alternating optimization algorithm [18].

With the alternating optimization algorithm, the optimal $\{\mathbf{W}_i^*\}$ is obtained. If $\text{rank}(\mathbf{W}_i^*) = 1$, it can be be expressed as $\mathbf{W}_i^* = \mathbf{w}_i^*(\mathbf{w}_i^*)^H$; then $\mathbf{w}_i^* = [(\mathbf{w}_{1i}^*)^T (\mathbf{w}_{2i}^*)^T \cdots (\mathbf{w}_{Ki}^*)^T]^T$ is the optimal solution. Otherwise, the standard rank-reduction techniques can be applied to obtain an approximate rank-1 matrix from $\{\mathbf{W}_i^*\}$. Noted that the objective value after every iteration is nonincreasing and is lower

---

Algorithm 5 Alternating optimization algorithm

1: **Initialize** Set $n = 0$ and choose $\{\eta_i(n) > 0\}$.
2: **repeat**
3:     For fixed $\{\eta_i(n)\}$, update $\{\mathbf{W}_i(n)\}$ by solving the following semidefinite problem (SDP) using the interior-point method:

$$\underset{\{\mathbf{W}_i\}}{\text{minimize}} \sum_{i \in \mathcal{I}} \text{Tr}(\mathbf{W}_i)$$

$$\text{s.t.} \qquad \text{Tr}(\mathbf{H}_i \mathbf{W}_i) - \eta_i \sum_{j \neq i} \text{Tr}(\mathbf{H}_i \mathbf{W}_j) - \eta_i \sigma_i^2 \geq 0,$$

$$\text{Tr}(\mathbf{Z}_i \mathbf{W}_i) = 0,$$

$$\mathbf{W}_i \succeq 0.$$

4:     For fixed $\{\mathbf{W}_i(n)\}$, update $\{\eta_i(n)\}$ by solving the following convex problem using the cutting-plane method:

$$\underset{\{\eta_i\}}{\text{minimize}} \sum_{i \in \mathcal{I}} -\frac{\partial J(\mathbf{Q})}{\partial Q_i} B \log_2 \eta_i$$

$$\text{s.t.} \qquad \text{Tr}(\mathbf{H}_i \mathbf{W}_i) - \eta_i \sum_{j \neq i} \text{Tr}(\mathbf{H}_i \mathbf{W}_j) - \eta_i \sigma_i^2 \geq 0$$

5: **until** the termination condition is satisfied.

---

bounded. Since problem (12.30) is convex, the algorithm will converge to the global optimal point. Furthermore the computation complexity of the proposed solution is low. It mainly comes from the calculation of $J_i'(Q_i)$ and the alternating optimization of (12.30), which has polynomial complexity w.r.t. K.

Simulations were carried out to compare the performances of the proposed delay-aware asymmetric cooperation algorithm (labeled as QSI-adaptive AC) with three baselines: CSI-only full cooperation (CSI-only FC), CSI-only asymmetric cooperation (CSI-only AC), and QSI-adaptive full cooperation (QSI-adaptive FC); see [17] for the scenario and parameter settings. Figure 12.3 gives the average delay performance of each scheme versus the average transmit power per traffic when the average traffic arrival rate is 1.25 packets/slot. The average delay of all schemes decreases with increasing average transmit power per traffic. The proposed QSI-adaptive AC scheme outperforms the CSI-only FC and CSI-only AC schemes, which is due to its adaptiveness to QSI. The QSI-adaptive FC scheme slightly outperforms the proposed QSI-adaptive AC scheme at the expense of excessive fronthaul consumption.

Figure 12.4 gives the average transmit power per traffic versus average traffic arrival rate when the average delay requirement is 0.8 second. It can be seen that the proposed QSI-adaptive AC scheme achieves significant power performance gain compared with the CSI-only AC and CSI-only FC schemes across a wide range of average packet

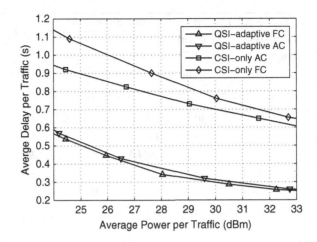

Figure 12.3 Average delay versus average transmit power per traffic.

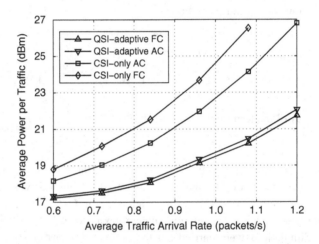

Figure 12.4 Average transmit power per traffic versus average traffic arrival rate.

arrival rates. The observation is the same as above, that the performance gain of the QSI-adaptive FC scheme compared with the proposed QSI-adaptive AC scheme is achieved by excessive fronthaul consumption, which is far from desired.

12.3.3   Delay-Aware Beamforming with Fronthaul and Interference Constraints

In a heterogeneous C-RANs with both traditional high-power MBSs and low-power RRHs, the inter-tier interference between MBSs and RRHs is severe when the two tiers share the same frequency bandwidth. Such characteristics in heterogeneous C-RANs bring challenges for optimization of the overall SE or energy efficiency (EE), because too many factors and challenges must be jointly considered, such as collaborative signal processing to suppress intra-tier and inter-tier interference in the PHY layer, and cooperative radio resource allocation and delay-aware packet scheduling in the MAC and upper layers. In addition interference problems result in frequent information exchanges, which make it necessary to consider capacity constraints on fronthaul and backhaul links as well.

Consider a downlink heterogeneous C-RAN system operateed in the slotted time mode with the unit time slot $t \in \{0, 1, 2, \ldots\}$. There are in total one MBS and K RRHs, where the RRHs, with M_R antennas, are deployed within the same coverage of the single MBS with M_M antennas in an underlaying manner. The RRHs and MBS are connected to a BBU pool with fronthaul and backhaul links, respectively. Define the set of MBS and RRHs as $\{0, 1, 2, \ldots, K\}$, where the index 0 refers to the MBS, which serves I_M sets of single-antenna MBS user equipment (MUE), and $\mathcal{K} = \{1, 2, \ldots, K\}$ denotes the set of RRHs, which cooperatively serve I_R single-antenna RRH user equipments (RUEs) with user-centric clustering. The set of RUEs is denoted as $\mathcal{I}_R = \{1, 2, \ldots, I_R\}$, and the set of MUEs as $\mathcal{I}_M = \{1, 2, \ldots, I_M\}$. Under the assumption that the BBU pool centrally processes all the RUEs' signals and distributes each RUE's data to an individually selected cluster of RRHs through fronthaul links, each RUE is cooperatively served by its serving cluster through the joint beamforming technique and receives an independent data stream from the RRH at the time slot t. It is assumed that the scalar-valued data stream $s_i(t)$ is temporally white, with zero mean and unit variance.

The RRHs transmit a precoding signal $\mathbf{v}_i(t)s_i(t)$ to RUE i, where $\mathbf{v}_i(t) = [\mathbf{v}_{1,i}^T(t) \cdots \mathbf{v}_{K,i}^T(t)]^T$ and $s_i(t) \sim \mathcal{CN}(0, \sigma^2)$ are respectively the network-wide beamforming vector and the scalar-valued data stream for RUE i; $\mathbf{v}_{k,i}(t) \in \mathbb{C}^{M_R \times 1}$ is the transmit beamformer from RRH k to RUE i at the time slot t. Then we have

$$\mathbf{v}_{k,i}(t) = \mathbf{D}_k \mathbf{v}_i(t),$$

$$\mathbf{D}_k = \left\{ \underbrace{\mathbf{0}_{M_R}, \ldots, \mathbf{0}_{M_R}}_{k-1}, \mathbf{I}_{M_R}, \ldots, \mathbf{0}_{M_R} \right\} \in \mathbb{C}^{M_R \times K M_R} \quad (n > 0). \tag{12.31}$$

In particular, the transmit beamformer from the MBS to MUE i is denoted by $\mathbf{v}_{0,i}(t) \in \mathbb{C}^{M_M \times 1}$. Though all RRHs can potentially serve each scheduled RUE, in fact each RUE is mainly contributed to by only a small number of adjacent RRHs and thus the network-wide beamforming vector is often group-sparse [19].

With the linear transmit beamforming scheme at the RRHs, the received signal at the RUE i is

$$y_i(t) = \mathbf{h}_i(t)\mathbf{v}_i(t)s_i(t) + \sum_{j=1, j \neq i}^{I_R} \mathbf{h}_i(t)\mathbf{v}_j(t)s_j(t) + \sum_{j=1}^{I_M} \mathbf{g}_{0,i}(t)\mathbf{v}_{0,j}(t)s_j(t) + n_i(t), \tag{12.32}$$

where $\mathbf{h}_i(t) \in \mathbb{C}^{1 \times K M_R}$ denotes the CSI matrix from all RRHs' transmit antennas to RUE i, and $\mathbf{g}_{0,i}(t) \in \mathbb{C}^{1 \times M_M}$ denotes the CSI matrix from the MBS's transmit antennas to RUE i; $n_i(t)$ is the received noise at RUE i with distribution $\mathcal{CN}(0, \sigma^2)$, where σ^2 is the noise variance at RUEs. Equation (12.32) suggests that both $\mathbf{v}_i(t)$ and $\mathbf{v}_{0,j}(t)$ should be carefully designed to suppress intra-tier and inter-tier interference, respectively.

Suppose that the queues maintained for RUEs in heterogeneous C-RANs are represented by $\mathbf{Q}(t) = \{Q_i(t) | i = 1, \ldots, I_R\}$, where $Q_i(t)$ denotes the queue backlog for RUE i at time slot t. The random traffic arrival for RUE i at time slot t is denoted by $A_i(t)$, which is assumed to be independent and identically distributed (i.i.d.) over the time slots with peak arrival rate $A_i^{\max}$. Define the set of $A_i(t)$ as $\mathbf{A}(t) = \{A_i(t) | i = 1, \ldots, I_R\}$ and the arrival rates at queues are $\lambda = \mathbb{E}\{\mathbf{A}(t)\}$.

At each time slot, the arrival and departure rates of RUE i are $A_i(t)$ and $R_i(t)$, respectively. Therefore, $Q_i(t)$ evolves according to

$$Q_i(t+1) = \{Q_i(t) - R_i(t)\}^+ + A_i(t). \tag{12.33}$$

To optimize the EE performance of H-CRANs, the transmission rate and power consumption performance metrics should be jointly considered. The EE performance can be characterized by the following expression:

$$\eta_{\text{EE}}(t) = f(R_i(t), P_k(t)) = \frac{\alpha}{I_R} \sum_{i=1}^{I_R} \omega_i R_i(t) - \frac{1-\alpha}{K} \sum_{k=1}^{K} \mu_k P_k(t), \tag{12.34}$$

where $\alpha \in [0, 1]$ is a weighting factor representing the ratio of the transmit rate and the power consumption, and $\omega_i \geq 0$ (channels/bits) and $\mu_k \geq 0$ (W^{-1}) represent the transmit weight of user i and the power consumption weight of the kth RRH, respectively. The solution to the problem of maximizing $\eta_{\text{EE}}(t)$ is a Pareto-optimal solution to the problem of maximizing $\tilde{\eta}_{\text{EE}}(t)$ [20]. The maximization of the averaged weighted EE-utility objective function for RRHs in H-CRANs can be formulated as the following stochastic optimization problem:

$$\underset{\{\mathbf{v}_i(t)\}}{\text{maximize}} \quad \overline{\eta_{\text{EE}}} = \lim_{t \to \infty} \frac{1}{t} \sum_{\tau=0}^{t-1} \mathbb{E}\{\eta_{\text{EE}}(\tau)\}$$

$$\text{s.t.} \quad \text{C1:} \quad \overline{P}_k \leq P_k^{\text{avg}}, \quad k \in \mathcal{K},$$

$$\text{C2:} \quad \lim_{t \to \infty} \frac{\mathbb{E}\{|Q_i(t)|\}}{t} = 0, \quad i \in \mathcal{I}_R,$$

$$\text{C3:} \quad P_k(t) \leq P_k^{\max}, \quad k \in \mathcal{K},$$

$$\text{C4:} \quad \sum_{i=1}^{I_R} \mathbf{v}_i^H(t) \mathbf{g}_j^H(t) \mathbf{g}_j(t) \mathbf{v}_i(t) \leq \varphi_j, \quad j \in \mathcal{I}_M, \quad \forall t,$$

$$\text{C5:} \quad \sum_{j=1}^{I_R} \mathbb{1}\{\mathbf{v}_j^H(t) \mathbf{D}_k^H \mathbf{D}_k \mathbf{v}_j(t)\} R_j(t) \leq C_k, \quad k \in \mathcal{K}, \tag{12.35}$$

where the constraint C1 ensures the long-term energy consumption of the kth RRH under the predefined level where P_k^{avg} denotes the average power consumption threshold; C2 is the network stability constraint guaranteeing a finite queue length for each queue; C3 is the energy-saving constraint for the kth RRH, where $P_k^{\max}$ denotes the maximum transmit power of the kth RRH; C4 is the constraint on interference from the RRHs to the MUEs, where $\mathbf{g}_j \in \mathbb{C}^{1 \times NL_R}$ denotes the CSI matrix from all RRHs' transmit antennas to MUE j, and C5 is the constraint on the fronthaul consumption for the kth RRH.

As earlier, we can transform the constraint C1 in (12.35) into a queue-stable problem by constructing a virtual queue $H_k(t)$ for each RRH k:

$$H_k(t+1) = \left\{ H_k(t) - P_k^{\text{avg}} + P_k(t) \right\}^+. \tag{12.36}$$

Denoting $\Theta(t) = [\mathbf{Q}(t) \ \mathbf{H}(t)]$ as the combined matrix of all the actual and virtual queues, where $\mathbf{H}(t) = \{ H_k(t) | k = 1, \ldots, K \}$, we have a Lyapunov function as a scalar metric of the queue congestion:

$$L(\Theta(t)) \triangleq \frac{1}{2} \left\{ \sum_{i=1}^{I_R} Q_i^2(t) + \sum_{k=1}^{K} H_k^2(t) \right\}, \tag{12.37}$$

with the Lyapunov drift $\Delta(\Theta(t))$ defined as in (12.9).

In terms of Lyapunov optimization, the underlying objective of optimal network-wide beamformer design is to minimize an infimum bound on the following drift-plus-penalty expression in each time slot:

$$\Delta(\Theta(t)) - V\mathbb{E} \left\{ \eta_{\text{EE}}(t) | \Theta(t) \right\}. \tag{12.38}$$

According to Lemma 12.5, a proper network-wide beamformer $\mathbf{v}_i(t)$ should be chosen to minimize the corresponding upper bound on the basis of the observed QSI and CSI at each time slot t. Using the concept of opportunistically minimizing an expectation, this is accomplished by greedily minimizing as follows:

$$\underset{\{\mathbf{v}_i(t)\}}{\text{minimize}} \left(\sum_{K=1}^{K} X_k(t) P_k(t) - \sum_{i=1}^{I_R} Y_i(t) R_i(t) \right)$$

$$\text{s.t.} \qquad \text{C3, C4, C5 hold}, \tag{12.39}$$

where $X_k(t) = H_k(t) + V(1 - \alpha)\mu_k/K$ and $Y_i(t) = Q_i(t) + V\alpha\omega_i/I_R$. However, the above problem is non-convex and the solution is difficult to obtain. To deal with this challenge effectively, an approximation and generalized WMMSE equivalence is utilized as follows.

THEOREM 12.9 *The problem (12.39) has the same optimal solution as the following WMMSE minimization problem* [21]:

$$\underset{\{w_i(t), u_i(t), \mathbf{v}_i(t)\}}{\text{minimize}} \sum_{i=1}^{I_R} Y_i(t) \left[w_i(t) e_i(t) - \log w_i(t) \right]$$

$$+ \sum_{k=1}^{K} X_k(t) \sum_{i=1}^{I_R} \mathbf{v}_i^H(t) \mathbf{D}_k^H \mathbf{D}_k \mathbf{v}_i(t),$$

$$\text{s.t.} \qquad \text{C3, C4, C5 hold}, \tag{12.40}$$

where $w_i(t)$ denotes the mean-square-error (MSE) weight for user i at time slot t and $e_i(t)$ is the corresponding MSE, defined as

$$
e_i(t) \triangleq \mathbb{E}\left\{(u_i(t)y_i(t) - s_i(t))^2\right\}
$$

$$
= u_i^H(t)\left(\sum_{j=1}^{I_R} \mathbf{v}_j^H(t)\mathbf{h}_i^H(t)\mathbf{h}_i(t)\mathbf{v}_j(t)\right)u_i(t)
$$

$$
+ u_i^H(t)\left(\sum_{j=1}^{I_M} \mathbf{v}_{0,j}^H(t)\mathbf{g}_{0,i}^H(t)\mathbf{g}_{0,i}(t)\mathbf{v}_{0,j}(t)\right)u_i(t) - 2\,\Re\{u_i(t)\mathbf{h}_i(t)\mathbf{v}_i(t)\}
$$

$$
+ \sigma^2\Re\{u_i^2(t)\} + 1,
$$

(12.41)

under the receiver $u_i(t) \in \mathbb{C}$.

Thus, based on the equivalent WMMSE problem (12.40), which is convex with respect to each individual optimization variable, the averaged weighted EE utility objective maximization problem (12.35) can be solved. This crucial observation allows the problem (12.35) to be solved efficiently through the block-coordinate descent method by iterating with respect to $\mathbf{v}_i(t), u_i(t)$, and $w_i(t)$. For more details of the solution, see [22]. Algorithm 6 summarizes the procedure.

Algorithm 6 Averaged weighted EE maximization with power and interference constraints

Require: Initial network-wide beamforming vector $\mathbf{v}_i(t)$;
Ensure: Calculate the optimal $\mathbf{v}_i^*(t)$ and corresponding $\eta_{EE}^*(t)$.
 1: **repeat**
 2:     With $\mathbf{v}_i(t)$ fixed, compute the MMSE receiver $u_i(t)$ and the corresponding MSE $e_i(t)$;
 3:     Update the MSE weight $w_i(t) = 1/e_i(t)$;
 4:     Find the optimal transmit network-wide beamformer $\mathbf{v}_i(t)$ by solving the problem (12.40), which is convex when $u_i(t)$ and $w_i(t)$ are fixed;
 5:     Compute the EE function $\eta_{EE}(t)$;
 6:     Update $\mathbf{v}_i^*(t) = \mathbf{v}_i(t)$, $\beta_k^i(t)$, $\tilde{R}_i(t)$ and $\eta_{EE}^*(t) = \eta_{EE}(t)$;
 7: **until** convergence.

As shown in Fig. 12.5, the average queue backlog achieved by the proposed dynamic network-wide beamformer-design algorithm grows linearly in $O(V)$ under a given traffic arrival rate λ, which satisfies (12.12). In Fig. 12.6, the average weighted EE performance increases with V for any given arrival rate λ, which supports (12.13). This can be intuitively understood from the fact that greater emphasis is placed on the average weighted EE when V increases. The lower the traffic arrival rate λ is, the higher the average weighted EE performance under a given control parameter V will be. This happens because both the transmit rate and power consumption decrease with decreasing λ and the logarithmic rate–power function has the characteristic of a diminishing slope.

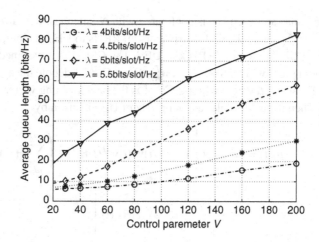

Figure 12.5 Average queue length versus V.

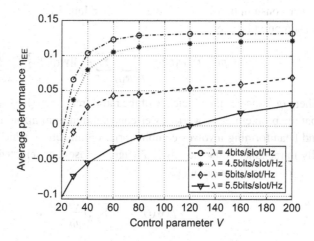

Figure 12.6 Average weighted EE performance versus V.

12.3.4 Delay-Aware Energy-Efficient Joint RRH Activation and Beamforming

With a dense deployment of RRHs, the too-close proximity of many active RRHs incurs increased interference and hence increased power consumption to meet a given QoS requirement. Moreover, the power consumption of fronthaul links to support high-capacity connection between active RRHs and the BBU pool becomes comparable to the transmission power consumption of the RRHs. Energy savings can be achieved by reducing the transmit power consumption of the RRHs and the number of active RRHs together with their corresponding fronthaul links [23]. Therefore, it is necessary to incorporate RRH activation into the beamforming models and devise a more energy-efficient joint RRH activation and beamforming algorithm.

Consider the downlink of a C-RAN, where the set of RRHs and UEs is denoted as $\mathcal{K} = \{1, 2, \ldots, K\}$ and $\mathcal{I} = \{1, 2, \ldots, I\}$, respectively. Each RRH is configured with M

antennas and each UE is configured with N antennas. Let $\mathbf{H}_i(t) = [\mathbf{H}_{1i}(t) \cdots \mathbf{H}_{Ki}(t)] \in \mathbb{C}^{N \times MK}$ denote the CSI matrix from all RRHs to UE i at slot t, where $\mathbf{H}_{ki}(t) \in \mathbb{C}^{N \times M}$ is the CSI matrix from RRH k to UE i. The CSI $\mathbf{H}(t)$ is assumed to be quasi-static block fading. Let $\mathbf{w}_i(t) = [\mathbf{w}_{1i}^T(t) \ldots \mathbf{w}_{Ki}^T(t)]^T \in \mathbb{C}^{MK \times 1}$ denote the aggregated beamformer for UE i at slot t, where $\mathbf{w}_{ki}(t) \in \mathbb{C}^{M \times 1}$ is the beamforming vector at RRH k for UE i at slot t. Then the achievable rate in bps/Hz of UE i is given by

$$
\mu_i(t) = \log_2 \det \left[\mathbf{I} + \mathbf{H}_i(t)\mathbf{w}_i(t)\mathbf{w}_i^H(t)\mathbf{H}_i^H(t) \left(\sum_{j \neq i} \mathbf{H}_i(t)\mathbf{w}_j(t) + \sigma^2 \mathbf{I} \right)^{-1} \right.
$$

$$
\left. \times \; \mathbf{w}_j^H(t)\mathbf{H}_i^H(t) \right]. \tag{12.42}
$$

Let $\mathcal{R}(t)$ denote the set of active RRHs at slot t. As there is a fixed amount of fronthaul power consumption P_k^c when the corresponding RRH k is active, the network power consumption is given by

$$
p(\mathcal{R}(t), \mathbf{w}(t)) = \sum_{k \in \mathcal{R}(t)} \sum_{i \in \mathcal{I}} \frac{1}{\eta_k} ||\mathbf{w}_{ki}(t)||_2^2 + \sum_{k \in \mathcal{R}(t)} P_k^c, \tag{12.43}
$$

where η_k indicates the efficiency of the radio frequency (RF) power amplifier. It is found that the problem of network power-consumption minimization by joint RRH activation and beamforming involves combinatorial optimization, which is usually computationally intractable. To tackle this challenge, we use the weighted mixed ℓ_1/ℓ_2-norm of $\mathbf{w}$ as a convex estimation of (12.43):

$$
\hat{p}(\mathbf{w}(t)) = 2 \sum_{k \in \mathcal{K}} \sqrt{\frac{P_k^c}{\eta_k}} ||\tilde{\mathbf{w}}_k(t)||_2,
$$

where $\tilde{\mathbf{w}}_k(t) = [\mathbf{w}_{k1}^T(t) \cdots \mathbf{w}_{kI}^T(t)]^T \in \mathbb{C}^{MI \times 1}$ is the aggregated beamformer used by RRH k at slot t. The mixed ℓ_1/ℓ_2-norm is an effective way to enforce desired group-sparsity of the beamformer [24]. Intuitively, an RRH with a higher fronthaul link power consumption and a lower RF power-amplifier efficiency will have a high probability of being forced to switch off.

Let $\mathbf{Q}(t) = [Q_1(t) \cdots Q_I(t)]$ denote the global QSI at the beginning of slot t. To maintain the network queue stability and minimize the network power consumption simultaneously, by joint RRH activation and beamforming, the stochastic optimization problem can be formulated as follows:

$$
\underset{\mathbf{w}}{\text{minimize}} \; \bar{p} = \lim_{T \to \infty} \frac{1}{T} \sum_{t=1}^{T} \mathbb{E}\{\hat{p}(\mathbf{w}(t))\}
$$

$$
\text{s.t.} \quad \text{C1, Queue } Q_i(t) \text{ is strongly stable, } \forall i, \tag{12.44}
$$

$$
\text{C2,} \; ||\tilde{\mathbf{w}}_k||_2^2 \leq P_k,
$$

where C2 comprises the instantaneous per-RRH power constraints. To solve this stochastic optimization problem efficiently using Lyapunov optimization, we use the Lyapunov drift-plus-penalty, given by

$$\Delta(\mathbf{Q}(t)) + V\mathbb{E}\left\{\hat{p}(\mathbf{w}(t))|\mathbf{Q}(t)\right\}, \tag{12.45}$$

where $\Delta(\mathbf{Q}(t)) = \mathbb{E}\{L(\mathbf{Q}(t+1)) - L(\mathbf{Q}(t))|\mathbf{Q}(t)\}$ is the Lyapunov drift and $L(\mathbf{Q}(t)) = \frac{1}{2}\sum_{i\in\mathcal{I}} Q_i(t)^2$ is the Lyapunov function. According to Lemma 12.5 in Section 12.2.2, the optimization problem can be simplified to

$$\underset{\mathbf{w}}{\text{maximize}} \sum_{i\in\mathcal{I}} Q_i(t)\mu_i(t) - V\hat{p}(\mathbf{w}(t)), \tag{12.46}$$

which is a penalized weighted sum-rate (WSR) maximization problem and can be proven to be NP-hard for a C-RAN with interference. By utilizing the equivalence, the penalized WSR maximization problem is equivalent to the following penalized WMMSE problem:

$$\underset{\alpha,\mathbf{u},\mathbf{w}}{\text{minimize}} \sum_{i\in\mathcal{I}} Q_i(\alpha_i e_i - \log\alpha_i) + \sum_{k\in\mathcal{K}} \beta_k\|\tilde{\mathbf{w}}_k\|_2$$
$$\text{s.t.} \quad \|\tilde{\mathbf{w}}_k\|_2^2 \leq P_k, \tag{12.47}$$

where $\alpha = \{\alpha_i|i \in \mathcal{I}\}$ is the set of nonnegative mean-square-error (MSE) weights and $\mathbf{u} = \{\mathbf{u}_i \in \mathbb{C}^{N\times 1}|i \in \mathcal{I}\}$ is the set of receivers for all UEs. Furthermore, $e_i = \mathbf{u}_i^H(\sum_{j\in\mathcal{I}} \mathbf{H}_i\mathbf{w}_j\mathbf{w}_j^H\mathbf{H}_i^H + \sigma^2\mathbf{I})\mathbf{u}_i - 2\Re\{\mathbf{u}_i^H\mathbf{H}_i\mathbf{w}_i\} + 1$ is the MSE for estimating s_i and

$$\beta_k = \frac{V}{W\tau\log_2^e}\sqrt{\frac{P_k^c}{\eta_k}}$$

is the parameter influencing the number of active RRHs.

Since (12.47) is convex with respect to each of the individual optimization variables, it can be solved efficiently through the block coordinate descent (BCD) method by iteratively optimizing over α, $\mathbf{u}$, and $\mathbf{w}$. Under fixed $\mathbf{w}$ and α, minimizing the weighted-sum MSE leads to the well-known MMSE receiver: $\mathbf{u}_i = (\sum_{j\in\mathcal{I}} \mathbf{H}_i\mathbf{w}_j\mathbf{w}_j^H\mathbf{H}_i^H + \sigma^2\mathbf{I})^{-1}\mathbf{H}_i\mathbf{w}_i$. Under fixed $\mathbf{w}$ and $\mathbf{u}$, a closed-form α is given by $\alpha_i = e_i^{-1}$ according to the first-order optimality conditions. Under fixed $\mathbf{u}$ and α, the optimal $\mathbf{w}$ can be obtained by solving the following convex problem:

$$\underset{\mathbf{w}}{\text{minimize}} \sum_{i\in\mathcal{I}}\mathbf{w}_i^H\mathbf{C}\mathbf{w}_i - 2\sum_{i\in\mathcal{I}}\Re\{\mathbf{d}_i^H\mathbf{w}_i\} + \sum_{k\in\mathcal{K}}\beta_k\|\tilde{\mathbf{w}}_k\|_2$$
$$\text{s.t.} \quad \|\tilde{\mathbf{w}}_k\|_2^2 \leq P_k, \tag{12.48}$$

where $\mathbf{C} = \sum_{j\in\mathcal{I}} Q_j\alpha_j\mathbf{H}_j^H\mathbf{u}_j\mathbf{u}_j^H\mathbf{H}_j$ and $\mathbf{d}_i = Q_i\alpha_i\mathbf{H}_i^H\mathbf{u}_i$. The solution is given in a summary, by Algorithm 7.

For a large-sized C-RAN, the computational complexity of solving the convex problem (12.48) using the traditional convex optimization method is usually intensive. The alternating direction method of multipliers (ADMM) algorithm can be utilized to reduce the complexity and facilitate parallel implementation [25]. Specifically, introduce a copy

---

Algorithm 7 Equivalent penalized WMMSE algorithm

1: For each slot t, observe the current QSI $\mathbf{Q}(t)$ and CSI $\mathbf{H}(t)$, then make queue-aware joint RRH activation and beamforming according to the following steps:
2: **Initialize w, u**, and $\boldsymbol{\alpha}$;
3: **repeat**
4:     Fix **w**, compute the MMSE receiver **u** and update the MSE weight $\boldsymbol{\alpha}$;
5:     Calculate the optimal beamformer **w** under fixed **u** and $\boldsymbol{\alpha}$ by solving convex problem(12.48);
6: **until** the stopping criteria on $(f' - f)/|f'| \leq \xi$ is met;
7: **Update** the traffic queue $Q_i(t)$ according to (12.2).

$\tilde{\mathbf{v}}_k$ of the original beamformer $\tilde{\mathbf{w}}_k$, and define $\mathbf{v} = [\tilde{\mathbf{v}}_1^T \cdots \tilde{\mathbf{v}}_K^T]^T \in \mathbb{C}^{MKI \times 1}$. Then the above problem can be equivalently expressed as

$$
\underset{\mathbf{w},\mathbf{v}}{\text{minimize}} \sum_{i \in \mathcal{I}} \mathbf{w}_i^H \mathbf{C} \mathbf{w}_i - 2 \sum_{i \in \mathcal{I}} \Re\{\mathbf{d}_i^H \mathbf{w}_i\} + \sum_{k \in \mathcal{K}} \beta_k \|\tilde{\mathbf{v}}_k\|_2
$$

$$
\text{s.t.} \qquad \|\tilde{\mathbf{v}}_k\|_2^2 \leq P_k,
$$

$$
\tilde{\mathbf{v}}_k = \tilde{\mathbf{w}}_k. \tag{12.49}
$$

The partially augmented Lagrangian function for the above problem is given by

$$
L(\mathbf{w}, \mathbf{v}, \mathbf{y}) = \min_{\mathbf{w},\mathbf{v}} \sum_{i \in \mathcal{I}} \mathbf{w}_i^H \mathbf{C} \mathbf{w}_i - 2 \sum_{i \in \mathcal{I}} \Re\{\mathbf{d}_i^H \mathbf{w}_i\} + \sum_{k \in \mathcal{K}} \beta_k \|\tilde{\mathbf{v}}_k\|_2
$$

$$
+ \sum_{k \in \mathcal{K}} \Re\{\tilde{\mathbf{y}}_k^H (\tilde{\mathbf{v}}_k - \tilde{\mathbf{w}}_k)\} + \frac{\rho}{2} \sum_{k \in \mathcal{K}} \|\tilde{\mathbf{v}}_k - \tilde{\mathbf{w}}_k\|_2^2, \tag{12.50}
$$

where $\mathbf{y} = \{\tilde{\mathbf{y}}_k | k \in \mathcal{K}\}$, with $\tilde{\mathbf{y}}_k = [\mathbf{y}_{k1}^T \cdots \mathbf{y}_{kI}^T]^T \in \mathbb{C}^{MI \times 1}$, is the set of Lagrangian dual variables for equality constraints in (12.49), and $\rho > 0$ is some constant. Then the main steps of the ADMM algorithm are summarized in Algorithm 8.

As can been seen from Algorithm 8, the step to obtain **v** is decomposed into K subproblems, each of which is associated with an RRH and has the following closed-form solution;

$$
\tilde{\mathbf{v}}_k^* = \begin{cases} \mathbf{0}, & \|\mathbf{b}_k\|_2 \leq \dfrac{\beta_k}{\rho}, \\[2ex] \dfrac{(\rho\|\mathbf{b}_k\|_2 - \beta_k)\mathbf{b}_k}{\rho\|\mathbf{b}_k\|_2}, & \dfrac{\beta_k}{\rho} < \|\mathbf{b}_k\|_2 < \dfrac{\beta_k}{\rho} + \sqrt{P_k}, \\[2ex] \dfrac{\mathbf{b}_k \sqrt{P_k}}{\|\mathbf{b}_k\|_2}, & \text{otherwise}, \end{cases} \tag{12.51}
$$

where $\mathbf{b}_k = \tilde{\mathbf{w}}_k - \tilde{\mathbf{y}}_k/\rho$. Similarly, the step to obtain **w** is decomposed into I subproblems, each of which is associated with a UE and has the following closed-form solution:

$$
\mathbf{w}_i^* = (2\mathbf{C} + \rho\mathbf{I})^{-1}(2\mathbf{d}_i + \rho\mathbf{v}_i + \mathbf{y}_i), \tag{12.52}
$$

where $\mathbf{y}_i = [\mathbf{y}_{1i}^T \cdots \mathbf{y}_{Ki}^T]^T \in \mathbb{C}^{MK \times 1}$, with $\mathbf{y}_{ki} \in \mathbb{C}^{M \times 1}$ the ith block of $\tilde{\mathbf{y}}_k$. Each step of the algorithm is closed form and can be carried out in a parallel manner, which makes the algorithm computationally efficient.

Algorithm 8 ADMM Algorithm for (12.48)

1: **Initialize** all primal variables $\mathbf{w}^{(0)}$, $\mathbf{v}^{(0)}$ and all dual variables $\mathbf{y}^{(0)}$.

2: **repeat**

3:　　Solve the following problem and obtain $\mathbf{v}^{(n+1)}$,

$$\underset{\tilde{\mathbf{v}}_k}{\text{minimize}}\ \beta_k||\tilde{\mathbf{v}}_k||_2 + \left(\frac{\rho}{2}||\tilde{\mathbf{v}}_k - \tilde{\mathbf{w}}_k + \frac{\tilde{\mathbf{y}}_k}{\rho}||_2^2\right)$$
$$\text{s.t.}\quad ||\tilde{\mathbf{v}}_k||_2^2 \le P_k;$$

4:　　Solve the following problem and obtain $\mathbf{w}^{(n+1)}$,

$$\underset{\mathbf{w}_i}{\text{minimize}}\ \mathbf{w}_i^H \mathbf{C} \mathbf{w}_i \left(-2\,\mathfrak{R}\{\mathbf{d}_i^H \mathbf{w}_i\} + \frac{\rho}{2}\sum_{k \in \mathcal{K}} ||\tilde{\mathbf{w}}_k - \tilde{\mathbf{v}}_k - \tilde{\mathbf{y}}_k/\rho||_2^2\right);$$

5:　　Update the multipliers $\mathbf{y}^{(n+1)}$ by

$$\tilde{\mathbf{y}}_k^{(n+1)} = \tilde{\mathbf{y}}_k^{(n)} + \rho(\tilde{\mathbf{v}}_k^{(n+1)} - \tilde{\mathbf{w}}_k^{(n+1)});$$

6: **until** stopping criterion $||\mathbf{v}^{(n+1)} - \mathbf{w}^{(n+1)}||_2 \le \delta$ is met.

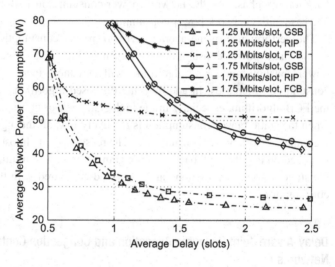

Figure 12.7 Quantitative power–delay tradeoff.

The resulting power–delay performance tradeoffs in the proposed algorithms may be compared in Fig. 12.7 with the relaxed integer programming (RIP) based algorithm and the full cooperative beamforming (FCB) algorithm, by varying the control parameter V in the formulated problems; see [26] for more details. It can be observed that the GSB-based WMMSE algorithm provides a significantly better power–delay performance tradeoff than the FCB algorithm and is very close to the RIP-based algorithm, which, however, has a higher computational complexity. Figure 12.7 also demonstrates that the larger V is, the less the network power consumption and the larger the delay. This is so because a system with a larger V will place less emphasis on the delay performance

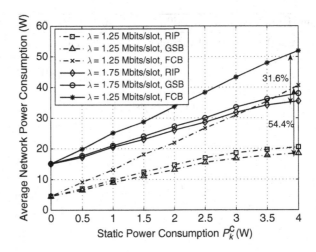

Figure 12.8 Average network power consumption versus static power consumption.

but more emphasis on the network power-consumption performance; this, leads to a decrease in the network power consumption.

The performance of the average network power consumption with different values of the static power consumption is shown in Fig. 12.8. It can be observed that the power consumption for all algorithms will inevitably increase when the static power consumption increases. In addition, the proposed algorithms significantly outperform the FCB algorithm, especially in the high-static-power-consumption regime. However, when the static power consumption is relatively low, all the algorithms produce almost the same network power consumption. The reason is that almost all the RRHs need to be switched on to get a high beamforming gain on order to minimize the traffic delay; then the static power consumption can be ignored compared with the RRH transmit power consumption.

12.3.5   Delay-Aware Joint Resource Allocation and Congestion Control in OFDMA-based Networks

Orthogonal frequency-division multiple access (OFDMA) is a promising multi-access technique for exploiting channel variations in both the frequency and time domains to provide high-data-rate transmission in 4G and beyond cellular networks [27]. To achieve favorable system performances and maintain satisfactory QoS for UEs, the optimal resource block (RB) and the power allocation with the optimal UE association should be investigated. Furthermore, since heterogeneous C-RANs maintain transmission for UEs with various levels of QoS requirement and the traffic requests from UEs arrive in an unpredictable and bursty fashion, congestion control should be taken into account to guarantee fairness among UEs and the efficient utilization of radio resources.

Consider the downlink transmission in an OFDMA-based heterogeneous C-RAN with several low-power RRHs and one traditional MBS. Since the MBS is mainly used to deliver the control signaling and to guarantee basic coverage, UEs with low traffic arrival rates are more likely to be served by the MBS and are labeled as MUEs. The UEs with high traffic arrival rates will be served by the RRHs and are labeled as RUEs. Let $\mathcal{K}$, $\mathcal{I}_M$, and $\mathcal{I}_R$ denote the set of RRHs, MUEs, and RUEs, respectively, which are each equipped with a single antenna. To avoid severe inter-tier interference, orthogonal frequency resources $\mathcal{J}_R$ and $\mathcal{J}_M$ are assigned to the RRHs and the MBS, respectively. Any UE associated with an RRH tier receives signals simultaneously from multiple cooperative RRHs on allocated RBs, and the RBs allocated to different UEs are orthogonal; thus inter-RRH signals are interference-free among UEs. The network is assumed to operate in slotted time with slot duration τ and index t.

Some MUEs tend to associate with the RRH tier in order to get more transmission opportunities when the traffic load of the MBS becomes heavier, which is usually what happens in practice. Let binary variable $s_m(t)$ indicate the UE association of MUE m. Let $g_{kij}(t)$, and $g_{kmj}(t)$ represent the channel conditions on RB j from RRH k to RUE i and MUE m, respectively. Let $g_{ml}(t)$ represent the channel condition on RB l from MBS to MUE m at slot t. Let $p_{kij}(t)$ denote the allocated power for RUE i on RB j from RRH k, and let $p_{kmj}(t)$ denote the allocated power for MUE m on RB j from RRH k if MUE m is associated with the RRH tier. Let $p_{ml}(t)$ denote the allocated power for MUE m on RB l from MBS if MUE m is associated with the MBS. Moreover, let $a_{ij}(t)$, $a_{mj}(t)$, and $b_{ml}(t)$ denote the corresponding RB allocations. Then non-reuse constraints for the RRH tier, $c_k^R(t) = \sum_{i \in \mathcal{I}_R} a_{ij}(t) + \sum_{m \in \mathcal{I}_H} s_m(t) a_{mj}(t) \leq 1$, should be satisfied and that for the MBS tier, $c_l^M(t)$, is similar. For the UEs that are served by RRHs, maximum-ratio combining (MRC) is adopted at the receivers; thus the transmit rate of RUE i is given by

$$\mu_i(t) = \sum_{j \in \mathcal{J}_R} a_{ij}(t) W_0 \log_2 \left(1 + \sum_{k \in \mathcal{K}} p_{kij}(t) g_{kij}(t) \right), \tag{12.53}$$

and that of MUE m is given similarly. The transmit power of RRH i is given by

$$p_k(t) = \sum_{i \in \mathcal{I}_R} \sum_{j \in \mathcal{J}_R} a_{ij}(t) p_{kij}(t) + \sum_{m \in \mathcal{I}_M} \sum_{j \in \mathcal{J}_R} s_m(t) a_{mj}(t) p_{kmj}(t), \tag{12.54}$$

and again that of the MBS is similar. By comparison with [29], the definition of the EE is given as follows.

DEFINITION 12.10   The EE of a heterogeneous C-RAN is defined as the ratio of the long-term time-averaged total transmit rate and the corresponding long-term time-averaged total power consumption, with units of bits/Hz per joule, and is given by

$$\eta_{EE} = \frac{\lim_{T \to \infty} T^{-1} \sum_{t=0}^{T-1} \mathbb{E}\{\mu_{sum}(t)\}}{W \lim_{T \to \infty} T^{-1} \sum_{t=0}^{T-1} \mathbb{E}\{p_{sum}(t)\}} = \frac{\bar{\mu}_{sum}}{W \bar{p}_{sum}}, \tag{12.55}$$

where $\mu_{sum}(t) = \sum_{m \in \mathcal{I}_M} \mu_m(t) + \sum_{i \in \mathcal{I}_R} \mu_i(t)$ is the total transmit rate, $p_{sum}(t) = \sum_{k \in \mathcal{K}} \varphi_{eff}^R p_k(t) + p_c^R + \varphi_{eff}^M p_M(t) + p_c^M$ is the total power consumption, φ_{eff}^R and φ_{eff}^M

are the drain efficiency of the RRHs and the MBS, respectively, and p_c^R and p_c^M are the static power consumptions of the RRHs and the MBS, respectively.

Let $Q_m(t)$ and $Q_i(t)$ denote the length of the buffering queues maintained for MUE $m \in \mathcal{I}_M$ and RUE $i \in \mathcal{I}_R$, respectively. Let $A_m(t)$ and $A_i(t)$ denote the amount of random traffic arrivals at slot t destined for MUE $m \in \mathcal{I}_M$ and RUE $i \in \mathcal{I}_R$, respectively. Since the statistics of $A_m(t)$ and $A_i(t)$ are usually unknown to the network, and the achievable capacity region is usually difficult to estimate [28], a situation for which the exogenous arrival rates are outside the network capacity region may occur. In this situation, the traffic queues cannot be stabilized without a transport-layer flow-control mechanism to determine the amount of admitted traffic data. Let $R_m(t)$ and $R_i(t)$ denote the amount of admitted traffic data; then the traffic buffering queues for RUE i evolve as

$$Q_i(t+1) = \max\{Q_i(t) - \mu_i(t)\tau, 0\} + R_i(t), \tag{12.56}$$

and that for MUE m evolves similarly. To maximize the throughput utility of networks and at the same time ensure the strong stability of traffic queues by joint congestion control and resource optimization, the stochastic optimization problem can be formulated as follows:

$$\begin{aligned}
& \underset{\{r,s,p,a\}}{\text{maximize }} U(\bar{r}) \\
& \text{s.t.} \quad \text{C1}, c_j^R(t) \leq 1, \\
& \qquad \text{C2}, c_l^M(t) \leq 1, \\
& \qquad \text{C3}, p_k(t) \leq p_k^{\max}, \\
& \qquad \text{C4}, p_M(t) \leq p_M^{\max}, \\
& \qquad \text{C5}, \eta_{EE} \geq \eta_{EE}^{\text{req}}, \\
& \qquad \text{C6}, Q_m(t) \text{ and } Q_i(t) \text{ are strongly stable}, \\
& \qquad \text{C7}, R_m(t) \leq A_m(t), R_i(t) \leq A_i(t), \\
& \qquad \text{C8}, a_{ij}(t), a_{mj}(t), b_{ml}(t), s_m(t) \in \{0,1\}, \tag{12.57}
\end{aligned}$$

where $U(\bar{r}) = \alpha \sum_{i \in \mathcal{I}_R} \lim_{T \to \infty} T^1 \sum_{t=0}^{T-1} R_i(t) + \beta \sum_{m \in \mathcal{I}_M} \lim_{T \to \infty} T^1 \sum_{t=0}^{T-1} R_m(t)$ is the utility function of the average throughput and α and β are positive utility prices, which indicate the relative importance of the corresponding utility functions.

To ensure that the EE performance constraint C5 is satisfied, a virtual queue $Z(t)$ with initial value $Z(0) = 0$ is also introduced, and it evolves as

$$Z(t+1) = \max\{Z(t) - \mu_{\text{sum}}(t), 0\} + W\eta_{EE}^{\text{req}} p_{\text{sum}}(t), \tag{12.58}$$

which can be satisfied only when the virtual queue $Z(t)$ is stable. To stabilize the traffic queues, while additionally optimizing the system throughput utility, the Lyapunov conditional drift-minus-utility function is defined as $\Delta(\chi(t)) = \mathbb{E}\{L(\chi(t+1)) - L(\chi(t)) - VU(\mathbf{r}(t))|\chi(t)\}$, where $L(\chi(t)) = \frac{1}{2}\left(\sum_{m \in \mathcal{I}_M} Q_m^2(t) + \sum_{i \in \mathcal{I}_R} Q_i^2(t) + Z^2(t)\right)$ is the quadratic Lyapunov function. According to the theory of Lyapunov optimization, instead of minimizing the drift-minus-utility expression directly, a good joint congestion control and resource optimization strategy can be obtained by minimizing the upper bound of the drift-minus-utility at each slot. This can be decoupled into the two following independent subproblems, which can be solved concurrently with the real-time online observation of traffic queues and virtual queues at each slot [30].

(1) *Optimal traffic admission control*   The optimal traffic admission control can be decoupled, to be computed for each UE separately as

$$\underset{R_m}{\text{maximize}} \quad [V - Q_m(t)]R_m(t)$$

$$\text{s.t.} \qquad R_m(t) \leq A_m(t), \tag{12.59}$$

$$\underset{R_i}{\text{maximize}} \quad [V - Q_i(t)]R_i(t)$$

$$\text{s.t.} \qquad R_i(t) \leq A_i(t), \tag{12.60}$$

which are linear problems with simple threshold-based admission control strategy.

(2) *Optimal radio resource allocation*   The optimal radio resource allocation can be obtained by minimizing the remaining items of the right-hand side of the drift-minus-utility inequality: which is given by

$$\underset{\text{s,p,a}}{\text{minimize}} - \sum_{m \in \mathcal{I}_M} \left(B_m(t)\mu_m(t) - \sum_{j \in \mathcal{I}_R} B_i(t)\mu_i(t) + Y_R(t) \sum_{k \in \mathcal{K}} p_k(t) + Y_M(t)p_M(t) \right)$$

$$\text{s.t. C1,C2,C3,C4,C8 hold,} \tag{12.61}$$

where $B_m(t) = Q_m(t)\tau + Z(t)$, $B_i(t) = Q_i(t)\tau + Z(t)$, $Y_R(t) = W\eta_{EE}^{req}\varphi_{eff}^R Z(t)$, and $Y_M(t) = W\eta_{EE}^{req}\varphi_{eff}^M Z(t)$. However, since the transmission rates $\mu_m(t)$, $\mu_i(t)$ and the transmit power consumptions $p_i(t)$ and $p_M(t)$ are functions of UE association, power, and RB allocations, this subproblem is a mixed-integer non-convex problem and is usually prohibitively difficult to solve. By relaxing the multiplicative binary variables in (12.53) and (12.54) to take continuous values in [0,1], and introducing auxiliary variables for power allocation, a convex optimization problem can be finally obtained, which can be solved by Lagrange dual decomposition efficiently. Further details can be found in [30]. The overall procedure of joint congestion control and resource optimization is summarized in Algorithm 9.

Simulations were carried out to evaluate the performances of the proposed joint-congestion-control-and-resource-optimization (JCCRO) scheme in heterogeneous C-RANs; see [30] for more details. Figure 12.9 plots the achieved EE versus the control parameter V; it shows that the achieved EE is always larger than or equal to η_{EE}^{req}. The EE archived by the case without an EE requirement is $\eta_{EE}^{thr} = 1.12$ in the simulations. When the required EE is below η_{EE}^{thr}, the actually achieved EE and the delay–throughput tradeoff is almost the same as for the situation without an EE requirement. Once the required EE is above the threshold η_{EE}^{thr}, to guarantee the required EE the network will decrease the transmit power, which further results in a decrease in transmit rate followed by a decrease in achieved throughput and an increase in the average delay.

Algorithm 9 Joint congestion control and resource optimization algorithm

1: For each slot, record the traffic queues $Q_m(t)$, $Q_i(t)$ and the virtual queue $Z(t)$;
2: Determine the optimal amount of admitted traffic $R_m(t)$ and $R_i(t)$;
3: **repeat**
4:     Obtain the optimal power allocation p_{kmj} and p_{kij} of the RRH tier;
5:     Obtain the optimal power allocation p_{ml} of the MBS;
6:     Obtain the optimal RB allocations a_{ij} and a_{mj} of the RRH tier and derive the optimal UE association s_m;
7:     Obtain the optimal RB allocation b_{ml} of the MBS;
8:     Update the vector of Lagrangian dual variables $\boldsymbol{\theta}$;
9: **until** a certain stopping criterion is met;
10: Update the traffic queues $Q_m(t)$, $Q_i(t)$ and the virtual queue $Z(t)$ according to the corresponding queue dynamics.

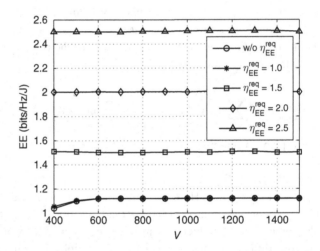

Figure 12.9 Achieved EE versus control parameter V. The plots for $\eta_{EE}^{req} = 1.0$ and for the situation without an EE requirement nearly coincide.

In Fig. 12.10, the total average power consumption of the proposed JCCRO scheme may be compared with the CSI-only maximum sum-rate (CSI-only MSR) scheme. We can see that the power consumption of the CSI-only MSR scheme remains unchanged as the traffic arrival rate varies and is much greate than that of the JCCRO scheme for the relatively light traffic states. This is so because the CSI-only MSR scheme delivers data under the full buffer assumption and fails to adapt to traffic arrival, which thus leads to a waste of energy despite achieving the same EE performance. We can conclude that, for relatively light traffic states, more energy can be saved with adaptive resource optimization, and, for relative heavy traffic states, traffic queues can be stabilized with the traffic admission control.

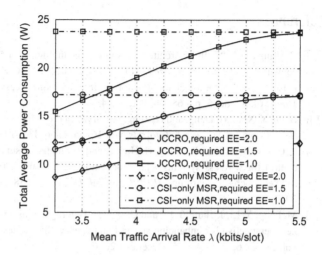

Figure 12.10 Total average power consumption versus mean traffic arrival rate λ.

12.4 Concluding Remarks

An important feature of heterogeneous C-RANs is the large number of RRHs needed, each connected to the BBU via a fronthaul network to serve the users' traffic requests. How best to manage such networks to meet the users' high-rate low-delay requirement is a major challenge. The delay-aware radio resource allocation optimization algorithms described in this chapter address many challenging issues, mainly using two major systematic approaches, namely, the Lyapunov optimization optimization approach and the approximate Markov decision process (MDP) approach. We have shown that these algorithms can guarantee a desirable delay performance and can be implemented in a parallel manner using the efficient algorithmic framework of the two approaches.

While we have highlighted a few recent developments in delay-aware radio resource allocation optimization algorithms in heterogeneous C-RANs, much remains to be done. An important direction of work involves addressing the tradeoffs among the SE, EE, and delay performances of delay-aware radio resource optimization algorithms. Recent work [31] unveils this intertwined tradeoff in interference-free wireless networks, but whether the same conclusions will still hold in heterogeneous C-RANs with interference remains an open question. Another important direction is to design algorithms that can optimize system parameters over different time scales. In particular, the signalings and controls are usually enforced at the frame level in the PHY or lower MAC layers, while this is usually done at longer time scales in the upper MAC and network layers. Therefore, the time scale of delay-aware radio resource optimization algorithms across the different layers in heterogeneous C-RANs should be separated, and the entire stochastic optimization problem should be decomposed into a number of lower-dimension subproblems by exploiting its structural properties.

References

[1] C. Wang et al., "Cellular architecture and key technologies for 5G wireless communication networks," *IEEE Commun. Mag.*, vol. 52, no. 2, pp. 122–130, February 2014.

[2] P. Rost et al., "Cloud technologies for flexible 5G radio access networks, *IEEE Commun. Mag.*, vol. 52, no. 5, pp. 68–76, May 2014.

[3] B. Fu, Y. Xiao, H. Deng, and H. Zeng, "A survey of cross-layer design in wireless networks," *IEEE Commun. Surveys & Tutorials*, vol. 16, no. 1, pp. 110–126, first quarter, 2014.

[4] Y. Cui, V. K. N. Lau, R. Wang, H. Huang, and S. Zhang, "A survey on delay-aware resource control for wireless systems – large deviation theory, stochastic Lyapunov drift, and distributed stochastic learning," *IEEE Trans. Inf. Theory*, vol. 58, no. 3, pp. 1677–1701, March 2012.

[5] M. Peng, Y. Li, J. Jiang, J. Li, and C. Wang, "Heterogeneous cloud radio access networks: a new perspective for enhancing spectral and energy efficiencies," *IEEE Wireless Commun.*, vol. 21, no. 6, pp. 126–135, December 2014.

[6] D. P. Bertsekas, *Dynamic Programming and Optimal Control. Third Edition*. Athena Scientific, 2007.

[7] W. B. Powell, *Approximate Dynamic Programming: Solving the Curses of Dimensionality*. Wiley-Interscience, 2007.

[8] X. Cao, *Stochastic Learning and Optimization: A Sensitivity-Based Approach*. Springer, 2008.

[9] S. Meyn and R. Tweedie, *Markov Chains and Stochastic Stability*. Springer-Verlag, 1993.

[10] L. Georgiadis, M. J. Neely, and L. Tassiulas, "Resource allocation and cross-layer control in wireless networks," *Found. Trends Netw.*, vol. 1, no. 1, pp. 1–44, 2006.

[11] M. Neely, *Stochastic Network Optimization with Application to Communication and Queueing Systems*. Morgan & Claypool, 2010.

[12] R. Irmer et al., "Coordinated multipoint: concepts, performance, and field trial results," *IEEE Commun. Mag.*, vol. 49, no. 2, pp. 102–111, February 2011.

[13] R. Zakhour and D. Gesbert, "Optimized data sharing in multicell MIMO with finite backhaul capacity," *IEEE Trans. Signal Process.*, vol. 59, no. 12, pp. 6102–6111, December 2011.

[14] J. Li, M. Peng, A. Cheng, Y. Yu, and C. Wang, "Resource allocation optimization for delay-sensitive traffic in fronthaul constrained cloud radio access networks," *IEEE Syst. J.*, vol. pp, no. 99, pp. 1–12, November 2014.

[15] S. Boyd and A. Mutapcic, *Stochastic Subgradient Methods*. Notes for EE364b, Stanford University, 2008.

[16] F. Zhuang and V. K. N. Lau, "Backhaul limited asymmetric cooperation for MIMO cellular networks via semidefinite relaxation," *IEEE Trans. Signal Process.*, vol. 62, no. 3, pp. 684–693, February 2014.

[17] J. Li, M. Peng, A. Cheng, and Y. Yu, "Delay-aware cooperative multipoint transmission with backhaul limitation in cloud-RAN," *in Proc. IEEE ICC*, Sydney, June 2014, pp. 665–670.

[18] S. Boyd and L. Vandenberghe, *Convex Optimization*. Cambridge University Press, 2004.

[19] B. Dai and W. Yu, "Sparse beamforming and user-centric clustering for downlink cloud radio access network," *IEEE Access*, vol. 2, pp. 1326–1339, December 2014.

[20] C. He, B. Sheng, P. Zhu, X. You, and G. Y. Li, "Energy- and spectral-efficiency tradeoff for distributed antenna systems with proportional fairness," *IEEE J. Sel. Areas Commun.*, vol. 31, no. 5, pp. 894–902, May 2013.

[21] Q. Shi, M. Razaviyayn, Z. Luo, and C. He, "An iteratively weighted MMSE approach to distributed sum-utility maximization for a MIMO interfering broadcast channel," *IEEE Trans. Signal Process.*, vol. 59, no. 9, pp. 4331–4340, September 2011.

[22] H. Xiang, Y. Yu, Z. Zhao, Y. Li, and M. Peng, "Energy-efficient resource allocation optimization for delay-aware heterogeneous cloud radio access networks", *in Proc. IEEE ICC*, London, June 2015, pp. 1–6.

[23] G. Auer *et al.*, "How much energy is needed to run a wireless network?" *IEEE Wireless Commun.*, vol. 18, pp. 40–49, October 2011.

[24] Y. C. Eldar and G. Kutyniok, *Compressed Sensing: Theory and Applications*. Cambridge University Press, 2012.

[25] S. Boyd *et al.*, "Distributed optimization and statistical learning via the alternating direction method of multipliers," *Found. Trends Mach. Learn.*, vol. 3, no. 1, pp. 1–122, 2011.

[26] J. Li, J. Wu, M. Peng, W. Wang, and V. K. N. Lau, "Queue-aware joint remote radio head activation and beamforming for green cloud radio access networks," in *Proc. IEEE Globecom*, San Diego, December 2015.

[27] M. Peng, K. Zhang, J. Jiang, J. Wang, and W. Wang, "Energy-efficient resource assignment and power allocation in heterogeneous cloud radio access networks," *IEEE Trans. Veh. Tech.*, vol. 64, no. 11, pp. 5275–5287, November 2015.

[28] H. Ju, B. Liang, J. Li, and X. Yang, "Dynamic joint resource optimization for LTE-Advanced relay networks," *IEEE Trans. Wireless Commun.*, vol. 12, no. 11, pp. 5668–5678, November 2013.

[29] Y. Li, M. Sheng, Y. Zhang, X. Wang, and J. Wen, "Energy-efficient antenna selection and power allocation in downlink distributed antenna systems: a stochastic optimization approach," in *Proc. IEEE ICC*, Sydney, June 2014, pp. 4963–4968.

[30] J. Li, M. Peng, Y. Yu, and A. Cheng, "Dynamic resource optimization with congestion control in heterogeneous cloud radio access networks," in *Proc. IEEE Globecom*, Austin, December 2014, pp. 906–911.

[31] Y. Li, M. Sheng, C. X. Wang, X. Wang, Y. Shi, and J. Ji, "Throughput–delay tradeoff in interference-free wireless networks with guaranteed energy efficiency," *IEEE Trans. Wireless. Commun.*, vol. 14, no. 13, pp. 1608–1621, March 2015.

13 C-RAN Using Wireless Fronthaul: Fast Admission Control and Large System Analysis

Jian Zhao, Tony Q. S. Quek, and Zhongding Lei

13.1 Introduction

The fifth generation (5G) mobile communication systems are expected to provide ultra-high data rate services and seamless user experiences across the whole network [1, 2]. The 5G system capacity is expected to be 1000 times greater than current fourth generation (4G) mobile systems. In order to meet such demanding requirements, as well as to reduce the capital investment and operational cost, C-RAN has been proposed as a promising network architecture for future mobile communication systems. In C-RAN, most signal processing functions are performed at the centralized baseband unit (BBU) pool, while data transmission to the users is provided by remote radio heads (RRHs), which are usually low-power nodes serving local area users. It is known that the hyper-dense deployment of RRHs will be a key factor in serving large numbers of users and achieving tremendous capacity enhancement in future mobile systems [3]. The transportation of user data and control signals between the BBU pool and the RRHs is carried out in the fronthaul [4].

The fronthaul is a major constraint for the practical implementation of C-RANs. In order to provide high quality-of-experience services to users in the network, fast and reliable fronthaul connections between the BBU pool and the RRHs must be established [5, 6]. Wired fronthaul, using optical fiber cables, can provide high-rate data links between fixed stations. However, the cost of providing wired fronthaul to all RRHs may be prohibitive when the number of RRHs is large. Moreover, certain locations that are difficult to reach by wired access may restrict the universal deployment of wired fronthaul. Wireless fronthaul, which can overcome many drawbacks of wired fronthaul, offers a cost-effective alternative [7–9]. Compared with wired fronthaul, the management of wireless fronthaul resources, e.g., their power and spectrum, is more complicated owing to finite power and radio spectrum constraints.

Recent works have proposed analysis and design methods for fronthaul and backhaul technologies from many aspects. A linear programming framework for determining the optimum routing and scheduling of data flows in wireless mesh backhaul networks was proposed in [10]. Zhao et al. [11] considered the problem of minimizing backhaul user data transfer in multi-cell joint processing networks, where algorithms involving the joint design of transmit beamformers and user data allocation at base stations (BSs) were proposed to effectively reduce backhaul user data transfer. Zhou et al. [12] presented an information-theoretical study of an uplink multi-cell joint processing network in which the BSs are connected to a centralized processing server via rate-limited digital backhaul links employing the compress-and-forward technique. A similar scenario for the downlink of a C-RAN setup was investigated in [13], where the multivariate compression of different RRHs' signals was exploited to combat additive quantization noise. The spectral efficiency and energy efficiency tradeoff in a homogeneous cellular network was investigated in [14], where the backhaul power consumption was taken into consideration. Considering the overall network power consumption, including that of the fronthaul, Shi et al. [15] proposed schemes to improve the energy efficiency in C-RAN cellular networks. For C-RAN fronthauling, not only the routing of data but also the additional cost of baseband processing in the cloud infrastructure was investigated in [13, 15]. Wireless fronthaul and backhaul technologies were discussed in [7–9]. Lee et al. [7] provided several admission control schemes for multi-hop wireless backhaul networks under rate and delay requirements. Flexible high-capacity hybrid wireless and optical mobile backhauling for small cells was investigated in [8]. The energy efficiencies of wireless fronthaul and backhaul networks for different system architectures and frequency bands were compared in [9].

A vital task in wireless communications is to design schemes to meet quality-of-service (QoS) requirements subject to a given amount of resources. The resource management of wireless systems with stringent QoS requirements has been discussed in [16–20]. A decentralized method to minimize the sum transmit power of BSs under given QoS requirements was proposed in [16] for a multi-cell network, relying on limited backhaul information exchange between BSs. Iterative algorithms to maximize the minimum QoS measure of users in multi-cell joint processing networks under per-BS power constraints were proposed in [17]. When it is not possible to meet the QoS requirements for all wireless stations using the given resources, only a subset of transmission links can be selected to be active. Zhai et al. [18] investigated the link activation problem in cognitive radio networks with single-antenna primary and secondary BSs and users. A price-driven spectrum-access algorithm was proposed and the energy–infeasibility tradeoff was analyzed. The uplink user admission control and user clustering under given QoS and power constraints were considered in [19], where algorithms relying on the ℓ_1-norm relaxation were proposed. Transmission schemes using semidefinite relaxation and Gaussian randomization to select active antenna ports were proposed in [20] to maximize the minimum user rate in multi-cell distributed antenna systems.

In order to achieve an enormous enhancement in spectral efficiency, the technique of "massive multiple-input multiple-output (MIMO)" is envisioned to be an important ingredient in 5G communication systems [21, 22]. In massive MIMO, the number of antennas equipped at BSs is much larger than that of active users in the same time–frequency channel. Hoydis et al. [23] analyzed the number of required antennas in massive MIMO systems using different linear beamformers. They showed that more sophisticated beamformers may reduce the number of antennas in massive MIMO systems required to achieve the same performance. Fernandes *et al.* [24] provided an asymptotic performance analysis of both the downlink and the uplink for a cellular network as the number of BS antennas tends to infinity. Huang *et al.* [25] studied joint beamformer design and power allocation in multi-cell massive MIMO networks, where efficient algorithms to maximize the minimum weighted QoS measure of all the users were proposed.

In this chapter, we consider a C-RAN with densely deployed RRHs. The user data at the BBU pool is transferred to the RRHs using wireless fronthaul. A wireless fronthaul hub (WBH) with multiple antennas is deployed at the BBU pool to transmit wireless fronthaul signals. Minimum data rate requirements must be satisfied when the RRHs receive data from the WBH, so that their users can be served at the required data rates. In order to allow more RRHs and their corresponding users to be served, it is desirable for the WBH to support as many RRHs as possible. Given QoS constraints at RRHs and power constraints at the WBH, we propose wireless fronthaul transmission schemes aiming to admit the maximum number of RRHs into the network. Such transmission schemes involve the joint design of the beamformers and the power control at the WBH, as well as selection of the subset of RRHs to be supported. We tackle this difficult problem by applying ℓ_1-norm relaxation to the original non-convex non-smooth problem and utilize uplink–downlink duality to transform the transmit beamforming problem into an equivalent receive beamforming problem. On the basis of the optimality conditions, we propose an iterative algorithm that jointly updates the values of the primal and dual variables. Such an algorithm is fast, and we prove that it can converge locally to the optimum solution of the ℓ_1-relaxed problem. Its convergence property is validated by numerical simulations and we observe that its range of convergence is large. Using the solution of the ℓ_1-relaxed problem, we then iteratively remove the RRH that corresponds to the largest QoS gap until all the remaining RRHs can be supported. Furthermore, we provide a large-system analysis of the above RRH admission control problem. As the system dimensions become large, we show that certain system parameters may be approximated as deterministic quantities irrespective of the actual channel realization. Random matrix theory is leveraged to transform our proposed finite-system iterative algorithm to large systems, so that only large-scale statistical information about the channel is required. As long as the large-scale channel coefficients and the QoS requirements remain unchanged, the selected RRHs will satisfy the QoS and power constraints almost surely. Simulations are carried out to verify the proposed algorithms in a simplified C-RAN setup. The proposed algorithms demonstrate fast convergence and low computational complexity. For the finite-system iterative algorithm, the average number of admitted RRHs is very close to the optimum result. For the large-system iterative algorithm, the results for selected RRHs accurately match those obtained by Monte Carlo simulations.

The remainder of the chapter is organized as follows. In Section 13.2, we introduce the system model and formulate the RRH admission control problem. An iterative algorithm for finite systems and its convergence property are presented in Section 13.3. The large-system analysis and its corresponding iterative algorithm are presented in Section 13.4.

Section 13.5 shows the convergence behavior and the simulation results in a C-RAN scenario. Finally, conclusions are drawn in Section 13.6.

Notation As before, bold uppercase letters are used to denote matrices and bold lowercase letters to denote vectors; $\mathcal{CN}(0, \sigma^2)$ denotes a circularly symmetric complex normal zero-mean random variable with variance σ^2; $(\cdot)^T$ and $(\cdot)^H$ stand for the transpose and conjugate transpose, respectively; $\mathbb{C}^M$ and $\mathbb{R}_+^M$ denote M-dimensional complex vectors and nonnegative real vectors, respectively; $\mathbb{E}\{\cdot\}$ stands for the mathematical expectation; $\{\mathbf{u}_i\}$ denotes the set containing the $\mathbf{u}_i$, $\forall i$. $||\mathbf{x}||$, $||\mathbf{x}_0||$, and $||\mathbf{x}_1||$ stand for the Euclidean norm, the ℓ_0-norm, and the ℓ_1-norm of the vector $\mathbf{x}$, respectively; $\xrightarrow{a.s.}$ denotes almost sure convergence; $\rho(\mathbf{A})$ stands for the spectral radius of matrix $\mathbf{A}$; $\mathbf{a} \circ \mathbf{b}$ denotes the Hadamard product of $\mathbf{a}$ and $\mathbf{b}$. For a vector $\mathbf{q}$, $\mathbf{q}^{-1}$ stands for the element-wise inverse of $\mathbf{q}$.

13.2 System Model and Problem Formulation

We consider a C-RAN with N single-antenna RRHs as in Fig. 13.1. The RRHs obtain their user data from the BBU pool via wireless fronthaul. A WBH with M antennas is responsible for transmitting wireless fronthaul signals from the BBU pool to the RRHs. The wireless fronthaul spectrum is out-of-band, and so does not interfere with users in the network. In order to meet the data rate requirements for serving users, the received data at the RRHs must satisfy certain minimum rate requirements. In order to achieve high-capacity enhancement in future mobile communication systems, the number of RRHs in a C-RAN is expected to be large. Under given power and spectrum constraints at the WBH, it may not be possible to serve all RRHs. The users that cannot be served by their RRHs will have to be served by the macro cell base station (MBS) directly [26]. In order to allow more users to be served by their corresponding RRHs, the WBH should support as many RRHs as possible. Under given QoS requirements at RRHs and the transmit power constraint at the WBH, we propose schemes aiming to admit the maximum subset of RRHs that can be simultaneously supported by the wireless fronthaul.

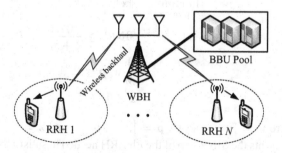

Figure 13.1 A C-RAN with wireless fronthaul.

We denote the wireless fronthaul channel from the WBH to the ith RRH as $\mathbf{h}_i^H$, where $\mathbf{h}_i \in \mathbb{C}^M$, $\forall i = 1, \ldots, N$. Linear transmit processing is applied at the WBH to deliver user data to the RRHs using wireless fronthaul links. The transmit beamformer at the WBH for the ith RRH is $\mathbf{u}_i \in \mathbb{C}^M$ such that $\|\mathbf{u}_i\| = 1$, $\forall i$. We denote the normalized data symbol for the ith RRH's users that is transmitted via wireless fronthaul as d_i, where $E\{|d_i|^2\} = 1$. The received signal at the ith RRH can be expressed as

$$y_i = \mathbf{h}_i^H \mathbf{u}_i \sqrt{\frac{p_i}{M}} d_i + \sum_{j=1, j \neq i}^{N} \mathbf{h}_i^H \mathbf{u}_j \sqrt{\frac{p_j}{M}} d_j + \eta_i \tag{13.1}$$

where η_i is the additive white Gaussian noise (AWGN) at the ith RRH, such that $\eta_i \sim \mathcal{CN}(0, n_i)$ and n_i denotes the noise variance. The transmit power for the data of the ith RRH is p_i/M. We denote the power constraint for fronthaul transmission at the WBH as P. The fronthaul transmit power must satisfy $\sum_{i=1}^{N} w_i p_i / M \leq P$, where $w_i > 0$ is the weight for the ith transmit power. The signal-to-interference-plus-noise ratio (SINR) at the ith RRH is given by

$$\mathrm{SINR}_i^D = \frac{p_i M^{-1} \left|\mathbf{h}_i^H \mathbf{u}_i\right|^2}{\sum_{j=1, j \neq i}^{N} p_j M^{-1} \left|\mathbf{h}_i^H \mathbf{u}_j\right|^2 + n_i}. \tag{13.2}$$

In order to meet the QoS requirements of its corresponding users, the receive SINR at each RRH must satisfy a minimum SINR requirement. The SINR requirements at RRHs can be determined from the QoS requirements of their served users and they can be easily fed back to the WBH.[1] We denote the SINR requirement for the ith RRH as γ_i, $\forall i$.

In a C-RAN with densely deployed RRHs, the WBH may not be able to support all the RRHs simultaneously for given SINR and power constraints. The users that cannot be served by RRHs need to be served by the MBS. In order to let as many RRHs and their corresponding users be served by the C-RAN as possible, it is desirable to select the subset of RRHs with the maximum cardinality that can be simultaneously supported by the wireless fronthaul. Since it is cheaper to build wireless fronthaul than wired fronthaul, we also minimize the total cost of building fronthaul for C-RAN in this way. Such a problem of RRH admission control can be formulated as

$$\begin{aligned}
\underset{\mathbf{p}, \{\mathbf{u}_i\}, \mathbf{x}}{\text{minimize}} \quad & \|\mathbf{x}\|_0 \\
\text{s.t.} \quad & \mathrm{SINR}_i^D \geq \frac{\gamma_i}{1 + x_i}, \quad \forall i = 1, \ldots, N, \\
& \frac{1}{M} \mathbf{w}^T \mathbf{p} \leq P, \\
& \mathbf{x} \geq \mathbf{0},
\end{aligned} \tag{13.3}$$

where $\mathbf{w} = [w_1 \cdots w_N]^T$, $\mathbf{p} = [p_1 \cdots p_N]^T$, and $\mathbf{x} = [x_1 \cdots x_N]^T \in \mathbb{R}_+^N$. Here $x_i \geq 0$ represents the SINR gap of the ith RRH needed to satisfy its SINR requirement. If $x_i = 0$ in the solution, it shows that the SINR requirement γ_i can be satisfied by the ith RRH.

---

[1] Denote the users of the ith RRH as $\mathbb{J}$. Assume the data rate requirement of the ith user to be r_i, $\forall i \in \mathbb{J}$. Then the SINR requirement for the ith RRH must be chosen such that $\log_2(1 + \gamma_i) \geq \sum_{i \in \mathbb{J}} r_i$.

The objective function $\|\mathbf{x}\|_0$ denotes the ℓ_0-norm of $\mathbf{x}$, which is the number of non-zero elements in $\mathbf{x}$.

13.3 Analysis and Algorithm Design for Finite Systems

The problem (13.3) is combinatorial and NP-hard due to the non-convex ℓ_0-norm in the objective function [27]. Approximate solutions of non-convex optimization problems can be obtained by applying convex relaxation [11, 19] and replacing the ℓ_0-norm in (13.3) with its convex envelop, i.e., the ℓ_1-norm. The ℓ_1-relaxed problem can then be expressed as

$$\begin{aligned} \underset{\mathbf{p},\{\mathbf{u}_i\},\mathbf{x}}{\text{minimize}} \quad & \|\mathbf{x}\|_1 \\ \text{s.t.} \quad & \text{the constraints in (13.3) hold.} \end{aligned} \tag{13.4}$$

The problem (13.4) is still difficult to solve owing to the need to jointly optimize the transmit power $\mathbf{p}$, the transmit beamformers $\{\mathbf{u}_i\}$, and the SINR gap $\mathbf{x}$. The following lemma shows that the downlink transmit optimization problem (13.4) can be converted to an uplink receive optimization problem.

LEMMA 13.1 *The downlink transmit optimization problem (13.4) is equivalent to the following dual uplink receive optimization problem:*

$$\begin{aligned} \underset{\mathbf{q},\{\mathbf{u}_i\},\mathbf{x}}{\text{minimize}} \quad & \sum_i x_i \\ \text{s.t.} \quad & SINR_i^U \geq \frac{\gamma_i}{1 + x_i}, \quad \forall i = 1, \ldots, N, \\ & \frac{1}{M} \mathbf{n}^T \mathbf{q} \leq P, \\ & \mathbf{x} \geq \mathbf{0}, \end{aligned} \tag{13.5}$$

where $\mathbf{n} = [n_1 \cdots n_N]^T$ *and* $\mathbf{q} = [q_1 \cdots q_N]^T \in \mathbb{R}_+^N$. *Here we define*

$$SINR_i^U = \frac{q_i M^{-1} |\mathbf{u}_i^H \mathbf{h}_i|^2}{\sum_{j=1, j \neq i}^N q_j M^{-1} |\mathbf{u}_i^H \mathbf{h}_j|^2 + w_i}. \tag{13.6}$$

Furthermore, the optimum solutions of $\{\mathbf{u}_i\}$ *and* $\mathbf{x}$ *in (13.4) are equal to those in (13.5). The optimum solution of* $\mathbf{p}$ *in (13.4) has a one-to-one correspondence with the optimum solution of* $\mathbf{q}$ *in (13.5). The problem (13.5) is a receive optimization problem in the dual uplink with the same SINR constraints, where* $\mathbf{h}_i$ *is the dual uplink channel from the ith RRH to the WBH and* w_i *becomes the uplink noise variance at the ith RRH. Here* q_i/M *represents the dual uplink transmit power of the ith RRH, and* $\mathbf{u}_i^H$ *is the uplink receive beamformer for the ith RRH transmission.*

Proof For any set of given beamformers $\{\mathbf{u}_i\}$, according to [28, Proposition 27.2], both the uplink and downlink have the same SINR feasible region under the power constraint P with the uplink and downlink SINR definitions in (13.6) and (13.2), respectively. The target SINRs $\gamma_i/(1 + x_i)$, $\forall i$, are feasible in the downlink if and only if the same targets are feasible in the uplink. Therefore, the same set of $\mathbf{x}$ is feasible for (13.4) and (13.5) under the power constraint P. Since this is true for any set of beamformers $\{\mathbf{u}_i\}$, the optimum solutions of $\{\mathbf{u}_i\}$ and $\mathbf{x}$ in (13.4) and (13.5) are the same. For given $\{\mathbf{u}_i\}$ and $\mathbf{x}$, the mapping between the uplink power $\mathbf{q}$ and downlink power $\mathbf{p}$ that achieves the same SINRs can be obtained by uplink–downlink power mapping [29]. $\qquad\square$

We denote the optimum solution of (13.4) as $\mathbf{p}^\star$, $\{\mathbf{u}_i^\star\}$, $\mathbf{x}^\star$. According to Lemma 13.1 the optimum solution of (13.5) can be denoted as $\mathbf{q}^\star$, $\{\mathbf{u}_i^\star\}$, $\mathbf{x}^\star$. The following lemma shows that the optimum beamformer $\{\mathbf{u}_i^\star\}$ can be determined from the optimum uplink power $\mathbf{q}^\star$.

LEMMA 13.2   *The optimum receive beamformer $\{\mathbf{u}_i^\star\}$ of (13.5) is the minimum-mean-square-error (MMSE) receiver, which can be obtained from the optimum uplink power $\mathbf{q}^\star$ as*

$$\mathbf{u}_i^\star = \mathbf{u}_i^{MMSE}(\mathbf{q}^\star) = \frac{1}{\xi_i}\left(\sum_{j=1,j\neq i}^N \frac{q_j^\star}{M}\mathbf{h}_j\mathbf{h}_j^H + w_i\mathbf{I}\right)^{-1}\mathbf{h}_i, \quad \forall i = 1,\ldots,N, \qquad (13.7)$$

where ξ_i is a normalization factor such that $\|\mathbf{u}_i^\star\| = 1$. The corresponding uplink SINR in (13.6) is

$$SINR_i^U\left(\mathbf{q}^\star, \{\mathbf{u}_i^\star\}\right) = \frac{q_i^\star}{M}\mathbf{h}_i^H\left(\sum_{j=1,j\neq i}^N \frac{q_j^\star}{M}\mathbf{h}_j\mathbf{h}_j^H + w_i\mathbf{I}\right)^{-1}\mathbf{h}_i, \quad \forall i = 1,\ldots,N. \qquad (13.8)$$

Proof   Considering the uplink power allocation $\mathbf{q}^\star$, it is known [30] that the uplink SINR (13.6) is maximized by the MMSE receiver, i.e., $SINR_i^U\left(\mathbf{q}^\star, \{\mathbf{u}_i^{MMSE}\}\right) \geq SINR_i^U\left(\mathbf{q}^\star, \{\mathbf{u}_i\}\right)$. If $\mathbf{u}_i \neq \mathbf{u}_i^{MMSE}$ in the solution of (13.5), substituting $\mathbf{u}_i$ by $\mathbf{u}_i^{MMSE}$ will also satisfy (13.5). Therefore, $\{\mathbf{u}_i^{MMSE}\}$ must be the solution for optimum receive beamformers in (13.5). The corresponding SINR can be obtained by substituting (13.7) into (13.6). $\qquad\square$

For notational brevity, we define an equivalent channel matrix $\mathbf{G}$, where

$$G_{ij} = \left|\mathbf{h}_i^H\mathbf{u}_j\right|^2, \quad \forall i,j = 1,\ldots,N; \qquad (13.9)$$

we have omitted the dependence of G_{ij} on $\mathbf{u}_j$. After a change of variables $\tilde{q}_i = \log(q_i/M)$, $\forall i$, the problem (13.5) can be equivalently expressed as

$$\begin{aligned}
\underset{\tilde{\mathbf{q}},\{\mathbf{u}_i\},\mathbf{x}}{\text{minimize}} \quad & \sum_i x_i \\
\text{s.t.} \quad & \log\frac{\gamma_i\left(\sum_{j=1,j\neq i}^N G_{ji}e^{\tilde{q}_j} + w_i\right)}{G_{ii}e^{\tilde{q}_i}} \leq \log(1 + x_i), \quad \forall i, \\
& \sum_{i=1}^N n_i e^{\tilde{q}_i} \leq P, \\
& \mathbf{x} \geq \mathbf{0}. \qquad (13.10)
\end{aligned}$$

For fixed $\{\mathbf{u}_i\}$, the problem (13.10) is a geometric programming (GP) problem with variables $\tilde{\mathbf{q}}$ and $\mathbf{x}$. The standard way of solving a GP problem involves using interior-point algorithms, which are employed in software such as cvx [31]. Considering $\{\mathbf{u}_i\}$ also as optimization variables, one method to obtain solutions of (13.10) is alternately optimizing $\{\tilde{\mathbf{q}}, \mathbf{x}\}$ and $\{\mathbf{u}_i\}$ by solving (13.10) with fixed $\{\mathbf{u}_i\}$ and then updating the MMSE receiver $\{\mathbf{u}_i^{\text{MMSE}}\}$ using the obtained $\tilde{\mathbf{q}}$. However, such an alternative optimization needs to solve (13.10) using standard convex optimization software [31] in each iteration, which makes it relatively slow in practice. In the following, we provide a low-complexity iterative algorithm.

We associate the ith SINR constraint in (13.10) with the Lagrange dual variable v_i, the power constraint with μ, and the nonnegativity constraint on x_i with α_i. The GP problem (13.10) satisfies Slater's condition. Hence, the Karush–Kuhn–Tucker (KKT) conditions are necessary and sufficient for the optimality of (13.10). The following lemma provides these optimality conditions for (13.10), which are key to our iterative algorithm for solving (13.4).

LEMMA 13.3 *The optimum primal and dual solutions of (13.10) satisfy the following conditions:*

$$x_i^* = \max\left(v_i^* - 1, 0\right), \quad \forall i = 1, \ldots, N, \tag{13.11}$$

$$\frac{Mv_i^*}{q_i^*} = \sum_{j=1, j\neq i}^{N} \frac{MG_{ij}\gamma_j v_j^*}{(1+x_j^*)G_{jj}q_j^*} + \mu^* n_i, \quad \forall i, \tag{13.12}$$

$$\frac{(1+x_i^*)G_{ii}q_i^*}{M\gamma_i} = \sum_{j=1, j\neq i}^{N} \frac{G_{ji}q_j^*}{M} + w_i, \quad \forall i, \tag{13.13}$$

$$\sum_{i=1}^{N} \frac{n_i q_i^*}{M} = P, \tag{13.14}$$

$$\mu^* > 0, \quad v_i^* > 0, \forall i. \tag{13.15}$$

The optimum downlink power $\mathbf{p}^*$ of (13.4) can be obtained directly from the optimum primal and dual solutions of (13.10) as

$$\frac{p_i^*}{M} = \frac{M\gamma_i v_i^*}{(1+x_i^*)G_{ii}q_i^*\mu^*}, \quad \forall i. \tag{13.16}$$

Furthermore, we have

$$\frac{(1+x_i^*)G_{ii}p_i^*}{M\gamma_i} = \sum_{j=1, j\neq i}^{N} \frac{G_{ij}p_j^*}{M} + n_i, \quad \forall i, \tag{13.17}$$

$$\sum_{i=1}^{N} \frac{w_i p_i^*}{M} = P. \tag{13.18}$$

Proof The Lagrangian of (13.10) can be expressed as

$$\mathcal{L}(\tilde{\mathbf{q}}, \{\mathbf{u}_i\}, \mathbf{x}) = \sum_i (1-\alpha_i)x_i + \mu\left(\sum_i n_i e^{\tilde{q}_i} - P\right)$$

$$+ \sum_i v_i\left(\log\frac{\gamma_i\left(\sum_{j\neq i} G_{ji}e^{\tilde{q}_j} + w_i\right)}{G_{ii}e^{\tilde{q}_i}} - \log(1+x_i)\right). \tag{13.19}$$

According to the KKT conditions, we have

$$\frac{\partial \mathcal{L}}{\partial x_i} = 1 - \alpha_i - \frac{v_i}{1 + x_i} = 0, \quad \forall i = 1, \ldots, N, \tag{13.20}$$

$$\frac{\partial \mathcal{L}}{\partial \tilde{q}_i} = -v_i + \sum_{j \neq i} v_j \frac{G_{ij} e^{\tilde{q}_i}}{\sum_{k \neq j} G_{kj} e^{\tilde{q}_k} + w_j} + \mu n_i e^{\tilde{q}_i} = 0, \quad \forall i. \tag{13.21}$$

Therefore, we have

$$x_i = \max (v_i - 1, 0), \quad \forall i, \tag{13.22}$$

$$v_i = \left(\sum_{j \neq i} \frac{v_j G_{ij}}{\sum_{k \neq j} G_{kj} e^{\tilde{q}_k} + w_j} + \mu n_i \right) e^{\tilde{q}_i}, \quad \forall i. \tag{13.23}$$

Because $e^{\tilde{q}_i} > 0$, any $v_i = 0$ requires $\mu = 0$ and $v_j = 0$, $\forall j = 1, \ldots, N$, simultaneously. Abandoning this trivial solution, we have $\mu > 0$ and $v_i > 0$, $\forall i$. According to the complementary slackness conditions [32] and substituting $e^{\tilde{q}_i} = q_i / M$, we have

$$\frac{(1 + x_i) G_{ii} q_i}{M \gamma_i} = \sum_{j \neq i} \frac{G_{ji} q_j}{M} + w_i, \quad \forall i, \tag{13.24}$$

$$\sum_i \frac{n_i q_i}{M} = P. \tag{13.25}$$

Substituting (13.24) into (13.23), we have

$$\frac{M v_i}{q_i} = \sum_{j \neq i} \frac{M G_{ij} \gamma_j v_j}{(1 + x_j) G_{jj} q_j} + \mu n_i, \quad \forall i. \tag{13.26}$$

Therefore

$$\frac{(1 + x_i)}{\gamma_i} G_{ii} \frac{M \gamma_i v_i}{(1 + x_i) G_{ii} q_i \mu} = \sum_{j \neq i} G_{ij} \frac{M \gamma_j v_j}{(1 + x_j) G_{jj} q_j \mu} + n_i, \quad \forall i. \tag{13.27}$$

We define

$$\frac{p_i}{M} \triangleq \frac{M \gamma_i v_i}{(1 + x_i) G_{ii} q_i \mu}, \quad \forall i \tag{13.28}$$

and we have

$$\frac{(1 + x_i) G_{ii} p_i}{M \gamma_i} = \sum_{j \neq i} \frac{G_{ij} p_j}{M} + n_i, \quad \forall i. \tag{13.29}$$

Multiply both sides of (13.24) by p_i and sum them for all i. We also multiply both sides of (13.29) by q_i and sum them for all i. Because $\sum_i \sum_{j \neq i} G_{ji} q_j p_i = \sum_i \sum_{j \neq i} G_{ij} q_i p_j$, we have

$$\sum_i \frac{w_i p_i}{M} = \sum_i \frac{n_i q_i}{M} = P. \tag{13.30}$$

Equation (13.29) shows that the power allocation $\mathbf{p}$ defined by (13.28) achieves the same SINR as the uplink $\mathbf{q}$. Equation (13.30) shows that $\mathbf{p}$ satisfies the same power constraint as $\mathbf{q}$. When the variables on the right-hand side of (13.28) are the optimum primal and dual solutions of (13.5), the power $\mathbf{p}$ defined by (13.28) corresponds to the optimum downlink power that solves (13.4), according to Lemma 13.1. □

COROLLARY *The optimum downlink transmit power $\mathbf{p}^\star$ of (13.4) and the optimum primal and dual solutions of (13.10) satisfy*

$$v_i^\star = \left(\sum_{j=1,j\neq i}^{N} \frac{G_{ij}p_j^\star}{M} + n_i \right) \frac{\mu^\star q_i^\star}{M}, \quad \forall i = 1,\ldots,N, \tag{13.31}$$

$$\mu^\star = \sum_{i=1}^{N} \frac{M\gamma_i v_i^\star w_i}{(1+x_i^\star)G_{ii}q_i^\star P}; \tag{13.32}$$

$$\frac{\gamma_i}{(1+x_i^\star)} = SINR_i^U \left(\mathbf{q}^\star, \{\mathbf{u}_i^\star\} \right)$$

$$= \frac{q_i^\star}{M}\mathbf{h}_i^H \left(\sum_{j=1,j\neq i}^{N} \frac{q_j^\star}{M}\mathbf{h}_j\mathbf{h}_j^H + w_i\mathbf{I} \right)^{-1} \mathbf{h}_i, \quad \forall i. \tag{13.33}$$

Here (13.31) was obtained by substituting (13.16) into (13.12). Equation (13.32) was obtained by substituting (13.16) into (13.18), and (13.33) was obtained from (13.13) and Lemma 13.2.

On the basis of Lemma 13.3 and the above Corollary, we propose Algorithm 10, which iteratively updates the values of the primal and dual variables to obtain the optimum $\mathbf{p}^\star$, $\{\mathbf{u}_i^\star\}$, and $\mathbf{x}^\star$ in (13.4). In Algorithm 10, we first store the old values of $\mathbf{q}$ in (13.34). The fixed-point iteration (13.35) and normalization (13.36) are obtained according to (13.33) and (13.14), respectively. Here (13.37) ensures contraction mapping for the algorithm, which will be made clear in the proof of Theorem 13.4. After obtaining the uplink power $\mathbf{q}$, we calculate the corresponding MMSE receiver and the equivalent channel, in (13.38) and (13.39), according to (13.7) and (13.9), respectively. The value of μ in (13.40) is obtained according to (13.32). The corresponding downlink power (13.41) is calculated according to (13.16) after obtaining the value of μ, and the value of v_i in (13.42) is updated according to (13.31). In this algorithm, the optimum solution for $\mathbf{p}^\star$ is obtained directly in (13.41). There is no need to perform uplink–downlink power mapping [29]. The convergence property of Algorithm 10 is shown in the following theorem.

THEOREM 13.4 *Starting from an initial point that is sufficiently close to the optimum solution of (13.5), Algorithm 10 converges to the optimum solution of (13.5) that satisfies the KKT conditions.*

Proof We give a proof only for $v_i^\star \geq 1$, $\forall i$, near the optimum solution of (13.5). The proof for some $v_i^\star < 1$ can be obtained likewise. Since we are considering points that are sufficiently close to the optimum solution, we can assume the beamformers $\{\mathbf{u}_i\}$ to be sufficiently close to the optimum beamformers $\{\mathbf{u}_i^\star\}$. Then the equivalent channel is

---

Algorithm 10 Finite-system iterative algorithm to solve (13.4)

1: Initialization: $v_i \geq 1$ and $q_i > 0$, $\forall i$, such that $\sum_{i=1}^{N} n_i q_i = MP$.

2: **repeat**

3:　　Store old values of **q**

$$\tilde{\mathbf{q}} \mathbf{q} = \mathbf{q}. \tag{13.34}$$

4:　　Update

$$\bar{q}_i = \frac{M\gamma_i}{\max(v_i, 1) \left[\mathbf{h}_i^H \left(\sum_{j=1, j\neq i}^{N} \frac{q_j}{M} \mathbf{h}_j \mathbf{h}_j^H + w_i \mathbf{I} \right)^{-1} \mathbf{h}_i \right]}, \quad \forall i = 1, \ldots, N. \tag{13.35}$$

5:　　Normalize

$$\bar{\mathbf{q}} = \frac{MP}{\mathbf{n}^T \bar{\mathbf{q}}} \bar{\mathbf{q}}. \tag{13.36}$$

6:　　Set

$$\mathbf{q} = \frac{1}{2}(\bar{\mathbf{q}} + \tilde{\mathbf{q}}). \tag{13.37}$$

7:　　Calculate $\mathbf{u}_i$

$$\mathbf{u}_i = \frac{1}{\xi_i} \left(\sum_{j=1, j\neq i}^{N} \frac{q_j^\star}{M} \mathbf{h}_j \mathbf{h}_j^H + w_i \mathbf{I} \right)^{-1} \mathbf{h}_i, \quad \forall i. \tag{13.38}$$

8:　　Calculate the equivalent channel

$$G_{ij} = \left| \mathbf{h}_i^H \mathbf{u}_j \right|^2, \quad \forall i, j = 1, \ldots, N. \tag{13.39}$$

9:　　Calculate μ

$$\mu = \sum_{i=1}^{N} \frac{M\gamma_i v_i w_i}{\max(v_i, 1) G_{ii} q_i P}. \tag{13.40}$$

10:　　Calculate the downlink power

$$\frac{p_i}{M} = \frac{M\gamma_i v_i}{\max(v_i, 1) G_{ii} q_i \mu}, \quad \forall i. \tag{13.41}$$

11:　　Update v_i

$$v_i = \left(\sum_{j=1, j\neq i}^{N} \frac{G_{ij} p_j}{M} + n_i \right) \frac{\mu q_i}{M}, \quad \forall i. \tag{13.42}$$

12: **until** $|q_i - \tilde{q}_i| \leq \epsilon$, $\forall i$.

13: Set $x_i = \max(v_i - 1, 0)$, $\forall i$. **return x, p,** and $\{\mathbf{u}_i\}$.

$G_{ij} = \left| \mathbf{h}_i^H \mathbf{u}_j^\star \right|^2$, $\forall i, j$, and it can be assumed to be fixed. The updates in Algorithm 10 reduce to updates of q_i and v_i, $\forall i$. We define a vector $\boldsymbol{\omega} \in \mathbb{R}_+^N$ and a matrix $\mathbf{F} \in \mathbb{R}_+^{N \times N}$, where $\omega_i = Mw_i\gamma_i/G_{ii}$, $\forall i = 1, \ldots, N$, and

$$F_{ij} = \begin{cases} \dfrac{G_{ij}\gamma_j}{G_{jj}}, & \text{if } i \neq j, \\ 0, & \text{if } i = j. \end{cases} \tag{13.43}$$

Furthermore, we introduce a vector $\mathbf{y} = [y_1 \cdots y_N]^T$, where $y_i = v_i/q_i$, $\forall i$. Therefore, $\mathbf{v} = \mathbf{y} \circ \mathbf{q}$.

The updating of $\mathbf{q}$ in (13.35) is actually obtained from the KKT condition (13.13) using MMSE receivers $\{\mathbf{u}_i\}$. Since we are assuming $v^\star \geq 1$ and $\mathbf{u}_i \approx \mathbf{u}_i^\star$ here, the updates of $\mathbf{q}$ can be expressed as

$$\mathbf{q}^{(m+1)} = \text{diag}\left(\mathbf{v}^{(m)}\right)^{-1}\left(\mathbf{F}^T\mathbf{q}^{(m)} + \omega\right) \tag{13.44}$$

$$= \text{diag}\left(\mathbf{q}^{(m)} \circ \mathbf{y}^{(m)}\right)^{-1}\left(\mathbf{F}^T\mathbf{q}^{(m)} + \omega\right). \tag{13.45}$$

By substituting (13.40) and (13.41) into (13.42), the updates of $\mathbf{y}$, which are obtained by $\mathbf{y} = \mathbf{v} \circ \mathbf{q}^{-1}$, can be expressed as

$$\mathbf{y}^{(m+1)} = \mathbf{F}\left(\mathbf{q}^{(m)}\right)^{-1} + MP^{-1}\mathbf{n}\omega^T\left(\mathbf{q}^{(m)}\right)^{-1} \tag{13.46}$$

$$= \left(\mathbf{F} + MP^{-1}\mathbf{n}\omega^T\right)\left(\mathbf{q}^{(m)}\right)^{-1}. \tag{13.47}$$

By dropping the time indices and letting $\mathbf{z} = [\mathbf{q}^T \ \mathbf{y}^T]^T$, the fixed-point updates of $\mathbf{z}$ can be expressed as

$$T(\mathbf{z}) = \begin{bmatrix} f_1(\mathbf{q}, \mathbf{y}) \\ f_2(\mathbf{q}, \mathbf{y}) \end{bmatrix} = \begin{bmatrix} \text{diag}(\mathbf{q} \circ \mathbf{y})^{-1}(\mathbf{F}^T\mathbf{q} + \omega) \\ \left(\mathbf{F} + MP^{-1}\mathbf{n}\omega^T\right)\mathbf{q}^{-1} \end{bmatrix}. \tag{13.48}$$

Its Jacobian matrix can be written as

$$\mathbf{J} = \begin{bmatrix} \partial f_1/\partial \mathbf{q}^T & \partial f_1/\partial \mathbf{y}^T \\ \partial f_2/\partial \mathbf{q}^T & \partial f_2/\partial \mathbf{y}^T \end{bmatrix} \tag{13.49}$$

$$= \begin{bmatrix} \text{diag}(\mathbf{q} \circ \mathbf{y})^{-1}\mathbf{F}^T & 0 \\ 0 & 0 \end{bmatrix} - \mathbf{E}\begin{bmatrix} \text{diag}(\mathbf{q} \circ \mathbf{y})^{-2} & 0 \\ 0 & \text{diag}(\mathbf{q} \circ \mathbf{y})^{-2} \end{bmatrix} \tag{13.50}$$

where

$$\mathbf{E} = \begin{bmatrix} \text{diag}(\mathbf{F}^T\mathbf{q} + \omega)\,\text{diag}(\mathbf{y}) & \text{diag}(\mathbf{F}^T\mathbf{q} + \omega)\,\text{diag}(\mathbf{q}) \\ \left(\mathbf{F} + MP^{-1}\mathbf{n}\omega^T\right)\text{diag}(\mathbf{y})^2 & 0 \end{bmatrix}. \tag{13.51}$$

At the optimum solution we have

$$\mathbf{q}^\star = \text{diag}(\mathbf{q}^\star \circ \mathbf{y}^\star)^{-1}(\mathbf{F}^T\mathbf{q} + \omega), \tag{13.52}$$

$$\mathbf{y}^\star = \left(\mathbf{F} + MP^{-1}\mathbf{n}\omega^T\right)(\mathbf{q}^\star)^{-1}. \tag{13.53}$$

Substitute (13.52) back into (13.51). Let $\mathbf{J}^\star = \mathbf{J}(\mathbf{q} = \mathbf{q}^\star, \mathbf{y} = \mathbf{y}^\star)$ and $\mathbf{A} = \mathbf{J}^\star + \mathbf{I}$; then we have

$$\mathbf{A} = \begin{bmatrix} \text{diag}(\mathbf{q}^\star \circ \mathbf{y}^\star)^{-1}\mathbf{F}^T & \text{diag}(\mathbf{q}^\star)\,\text{diag}(\mathbf{y}^\star)^{-1} \\ \left(\mathbf{F} + MP^{-1}\mathbf{n}\omega^T\right)\text{diag}(\mathbf{q}^\star)^{-2} & \mathbf{I} \end{bmatrix}. \tag{13.54}$$

The matrix $\mathbf{A}$ is nonnegative and irreducible. Furthermore, we have

$$\mathbf{A} \begin{bmatrix} \mathbf{q}^\star \\ \mathbf{y}^\star \end{bmatrix} + \begin{bmatrix} \text{diag}(\mathbf{q}^\star \circ \mathbf{y})^{-1}\omega \\ \mathbf{0} \end{bmatrix} = 2 \begin{bmatrix} \mathbf{q}^\star \\ \mathbf{y}^\star \end{bmatrix}, \tag{13.55}$$

according to (13.52) and (13.53). Therefore,

$$\mathbf{A} \begin{bmatrix} \mathbf{q}^\star \\ \mathbf{y}^\star \end{bmatrix} \lneqq 2 \begin{bmatrix} \mathbf{q}^\star \\ \mathbf{y}^\star \end{bmatrix} \tag{13.56}$$

and $\left[(\mathbf{q}^\star)^T \ (\mathbf{y}^\star)^T \right]^T \gneqq \mathbf{0}$. Because $\mathbf{A}$ is nonnegative and irreducible, (13.56) implies that its spectral radius is $\rho(\mathbf{A}) < 2$, according to [33, Theorem 1.11].

To ensure the contraction mapping of the algorithm, the step (13.37) must be invoked. Consider the update

$$\mathbf{z}^{(m+1)} = \tfrac{1}{2}\mathbf{z}^{(m)} + \tfrac{1}{2}T(\mathbf{z}^{(m)}). \tag{13.57}$$

Let $\mathbf{z}^{(m)} = \mathbf{z}^\star - \boldsymbol{\varepsilon}^{(m)}$ and $\mathbf{z}^{(m+1)} = \mathbf{z}^\star - \boldsymbol{\varepsilon}^{(m+1)}$, where $\mathbf{z}^\star$ is the optimum solution. Then

$$\mathbf{z}^\star - \boldsymbol{\varepsilon}^{(m+1)} = \tfrac{1}{2}(\mathbf{z}^\star - \boldsymbol{\varepsilon}^{(m)}) + \tfrac{1}{2}T(\mathbf{z}^\star - \boldsymbol{\varepsilon}^{(m)}) \tag{13.58}$$

$$\approx \tfrac{1}{2}(\mathbf{z}^\star - \boldsymbol{\varepsilon}^{(m)}) + \tfrac{1}{2}\left(T(\mathbf{z}^\star) - \mathbf{J}^\star \boldsymbol{\varepsilon}^{(m)} \right) \tag{13.59}$$

$$= \mathbf{z}^\star - \left(\tfrac{1}{2}\mathbf{I} + \tfrac{1}{2}\mathbf{J}^\star \right) \boldsymbol{\varepsilon}^{(m)}. \tag{13.60}$$

Here we have used $\mathbf{z}^\star = T(\mathbf{z}^\star)$. Therefore,

$$\boldsymbol{\varepsilon}^{(m+1)} \approx \left(\tfrac{1}{2}\mathbf{I} + \tfrac{1}{2}\mathbf{J}^\star \right) \boldsymbol{\varepsilon}^{(m)}. \tag{13.61}$$

Because $\rho(\mathbf{A}) < 2$, we have $\rho(\tfrac{1}{2}\mathbf{I} + \tfrac{1}{2}\mathbf{J}^\star) < 1$. This shows that there exists a neighborhood of the optimum solution, which ensures that the mapping in Algorithm 10 is a contraction mapping which satisfies the Lipschitz condition [34]. Therefore, Algorithm 10 converges to the optimum solution if the starting point is within this neighborhood. □

Theorem 13.4 shows that Algorithm 10 converges locally to the optimum solution of (13.4). However, its range of convergence cannot be obtained from Theorem 13.4, and the result may depend on the initialization point. Whether this algorithm can converge globally is still an open problem and remains for future work. In our simulations we observed that its range of convergence is large and even random initialization will converge to the correct results.

13.3.1   Connections with Max–Min SINR

If all the RRHs can be supported, $\mathbf{x}^\star = \mathbf{0}$ in the solution of (13.10). The values of v_i, $\forall i$, decrease monotonically towards zero in Algorithm 10. In this case the updates of power in (13.35), (13.36) become the power iteration steps in the max–min SINR algorithm of [35]. Algorithm 10 will still converge. The outputs $\mathbf{p}$ and $\{\mathbf{u}_i\}$ of Algorithm 10 are the power allocation and beamformers that maximize the minimum SINR of all RRHs in the system.

13.3.2   Iterative RRH Removal

Owing to the convex relaxation, the solution of (13.4) may not be always optimal for the ℓ_0-norm optimization problem (13.3). Therefore, (13.4) cannot be simply used as a substitution of (13.3) and we need to refine the selection of RRHs on the basis of the solution of (13.4). Since the solution of x_i in (13.4) represents the gap between the ith RRH's SINR and its SINR requirement, it is natural for us to select those RRHs with small x_i values. We propose to iteratively remove RRHs with decreasing values of x_i until the remaining set of RRHs becomes feasible to satisfy the SINR and power constraints, i.e., until $x_i = 0$, $\forall i$. In this way, we obtain the final results for RRH admission control.

13.3.3   Admission Control between RRHs and Users

The data rate requirements at the admitted RRHs can be guaranteed by the wireless fronthaul with the proposed RRH admission control algorithm. After the admitted RRHs receive data from the WBH, they need to transmit those data to their corresponding users. Existing coordinated-multi-point (CoMP) transmission schemes can be applied [16, 17]. However, owing to inter-RRH interference, there exists the possibility that the users of the admitted RRHs cannot be simultaneously supported with their given SINR requirements. If this happens, we need to perform user admission control within the admitted RRHs in order to allows as many users as possible to be served by their corresponding RRHs. We briefly discuss user admission control, as follows. For simplicity of discussion we assume that each RRH serves one user.

We denote the set of admitted RRHs as $\mathcal{S}$ and the power constraint for the ith admitted RRH as P_i, where $i \in \mathcal{S}$. The channel gain between the jth RRH and the ith user is denoted as g_{ij}, where $i,j \in \mathcal{S}$. Within the set of admitted RRHs the user admission control problem can be formulated similarly to (13.3). After ℓ_1-relaxation, we need to solve the following GP problem:

$$
\begin{aligned}
\underset{\mathbf{p},\mathbf{x}}{\text{minimize}} \quad & \sum_i x_i \\
\text{s.t.} \quad & \frac{g_{ii}p_i}{\sum_{j\in\mathcal{S},j\neq i} g_{ij}p_j + n_i} \geq \frac{\gamma_i}{1+x_i}, \quad \forall i \in \mathcal{S}, \\
& p_i \leq P_i, \quad \forall i \in \mathcal{S}, \\
& \mathbf{x} \geq \mathbf{0},
\end{aligned}
\tag{13.62}
$$

where p_i and x_i denote the transmit power from the ith RRH and the SINR gap for the ith user, respectively. The problem (13.62) is similar to (13.10), the only difference being in the power constraints. The problem (13.62) with per-RRH power constraints can be solved by solving a series of weighted-sum power-constrained problems employing Algorithm 10. Following a similar discussion as in [36, 37], it can be shown that the optimum values of (13.62) are equal to the optimum values of the problem $\max_{\{w_i \geq 0\}} f(\{w_i\})$, where $f(\{w_i\})$ with $\{w_i\}$ as the parameter denotes the optimum objective value of the following weighted-sum power-constrained problem:

$$\underset{\mathbf{p},\mathbf{x}}{\text{minimize}} \quad \sum_i x_i = f(\{w_i\})$$

$$\text{s.t.} \quad \frac{g_{ii}p_i}{\sum_{j=1, j\neq i}^{N} g_{ij}p_j + n_i} \geq \frac{\gamma_i}{1 + x_i}, \quad \forall i \in \mathcal{S},$$

$$\sum_{i \in \mathcal{S}} w_i p_i \leq \sum_{i \in \mathcal{S}} w_i P_i,$$

$$\mathbf{x} \geq \mathbf{0}, \tag{13.63}$$

where $w_i \geq 0$, $\forall i \in \mathcal{S}$, are the weights of the powers and also the parameters of $f(\{w_i\})$. The problem (13.63) can be solved by the proposed Algorithm 10. Furthermore, it can be shown that $P_i - p_i^*$ is a subgradient for w_i, where p_i^* is the solution of (13.63) for the current iteration $\{w_i\}$. Therefore, the solution of (13.62) can be obtained by the projected subgradient method: for a set of given weights $\left\{w_i^{(n)}\right\}$, we can obtain $f\left(\left\{w_i^{(n)}\right\}\right)$ and the corresponding $\{p_i^*\}$ by Algorithm 10 in the nth iteration; after that, the weights can be updated as

$$w_i^{(n+1)} = \max\left(w_i^{(n)} + t_n(P_i - p_i^*), 0\right), \quad \forall i \in \mathcal{S}. \tag{13.64}$$

By iteratively updating the weights $\{w_i\}$ and solving (13.63), we can obtain the solution of (13.62). Finally, the iterative removal of users with their corresponding RRHs, as discussed in Section 13.3.2, can be applied according to the solution of (13.62) until all the remaining users can be admitted.

13.4   Asymptotic Analysis and Algorithm Design for Large Systems

The algorithm proposed in Section 13.3 requires instantaneous channel state information. However, this may change rapidly with time, which can incur frequent resumptions of Algorithm 10 to determine the RRH admission. However, 5G wireless networks are envisioned to be characterized by dense deployments of RRHs and thus large numbers of antennas at the WBH. Under those circumstances, certain system parameters tend to become deterministic quantities that depend on only large-scale channel statistical information and the QoS requirements at RRHs. The large-scale channel statistical information includes path loss, shadowing, and antenna gain, which do not change rapidly with time. The optimum power allocation $\mathbf{p}^*$ in (13.4) and the final selection of RRHs based on $\mathbf{x}^*$ will also tend to be deterministic irrespective of the actual channel changes. In the following, we use random matrix theory to provide an iterative algorithm for the RRH admission control problem of large systems. Such an iterative algorithm can produce an asymptotically optimum solution for (13.4). As long as the large-scale channel coefficients and the QoS requirements remain unchanged, the selected RRHs using large-system analysis will almost surely be the same as those in the results obtained by the method of Section 13.3, based on instantaneous channel state information.

We assume that the number of transmit antennas M at the WBH and the number of RRHs N go to infinity while the ratio N/M remains bounded, i.e., let $M, N \to \infty$ while $0 < \liminf N/M \le \limsup N/M < \infty$. Such an assumption is abbreviated as $M \to \infty$. We use $\xrightarrow{\text{a.s.}}$ to denote almost sure convergence, where $f(\mathbf{x}) \xrightarrow{\text{a.s.}} a$ means that a is the deterministic equivalent of $f(\mathbf{x})$ as $M \to \infty$. The following fading-channel model is used for large-system analysis:

$$\mathbf{h}_i = \sqrt{d_i}\, \tilde{\mathbf{h}}_i, \quad \forall i = 1, \dots, N, \tag{13.65}$$

where d_i represents the large-scale fading coefficient between the WBH and the ith RRH. Here $\tilde{\mathbf{h}}_i$ denotes a normalized channel vector whose elements are independent and identically distributed (i.i.d.) $\mathcal{CN}(0, 1)$ random variables.

Under the channel model (13.65), the uplink SINR (13.8) using MMSE receive beamformers can be expressed as

$$\text{SINR}_i^{\text{U}}\left(\mathbf{q}, \left\{\mathbf{u}_i^{\text{MMSE}}\right\}\right) = \frac{q_i}{M}\mathbf{h}_i^H \left(\sum_{j=1, j\neq i}^{N} \frac{q_j}{M}\mathbf{h}_j\mathbf{h}_j^H + w_i\mathbf{I}\right)^{-1}\mathbf{h}_i \tag{13.66}$$

$$= \frac{q_i d_i}{M}\tilde{\mathbf{h}}_i^H \left(\sum_{j=1, j\neq i}^{N} \frac{q_j d_j}{M}\tilde{\mathbf{h}}_j\tilde{\mathbf{h}}_j^H + w_i\mathbf{I}\right)^{-1}\tilde{\mathbf{h}}_i \quad \forall i. \tag{13.67}$$

Calculating the uplink SINR (13.67) involves a matrix inversion of dimension M, which becomes increasingly complex as $M \to \infty$. However, the following lemma shows that the uplink SINR (13.67) will tend to deterministic quantities asymptotically. Such deterministic quantities depend only on the large-scale fading coefficients $\{d_i\}$, irrespective of the instantaneous channel state information $\{\mathbf{h}_i\}$.

LEMMA 13.5 *As $M \to \infty$, the uplink SINRs (13.67) using MMSE receive beamformers approach deterministic quantities for a given $\mathbf{q}$, i.e.,*

$$SINR_i^{\text{U}}\left(\mathbf{q}, \left\{\mathbf{u}_i^{\text{MMSE}}\right\}\right) \xrightarrow{\text{a.s.}} q_i d_i \varphi_i(\mathbf{q}), \quad \text{as } M \to \infty, \tag{13.68}$$

where

$$\varphi_i(\mathbf{q}) = \left(w_i + \frac{1}{M}\sum_{j=1, j\neq i}^{N} \frac{q_j d_j}{1 + q_j d_j \varphi_i(\mathbf{q})}\right)^{-1}, \quad \forall i = 1, \dots, N. \tag{13.69}$$

Proof According to Lemma 13.8 in Appendix section 13.7, we have

$$\frac{1}{M}\tilde{\mathbf{h}}_i^H \left(\sum_{j\neq i} \frac{q_j d_j}{M}\tilde{\mathbf{h}}_j\tilde{\mathbf{h}}_j^H + w_i\mathbf{I}\right)^{-1}\tilde{\mathbf{h}}_i$$

$$- \frac{1}{M}\text{Tr}\left(\sum_{j\neq i} \frac{q_j d_j}{M}\tilde{\mathbf{h}}_j\tilde{\mathbf{h}}_j^H + w_i\mathbf{I}\right)^{-1} \xrightarrow{\text{a.s.}} 0. \tag{13.70}$$

Furthermore, by applying Lemma 13.10 in Appendix section 13.7 and substituting $z = -w_i$, $t_j = q_j d_j K/M$, $\forall j \neq i$, we have

$$\frac{1}{M} \text{Tr} \left(\sum_{j \neq i} \frac{q_j d_j}{M} \tilde{\mathbf{h}}_j \tilde{\mathbf{h}}_j^H + w_i \mathbf{I} \right)^{-1} - \varphi_i(\mathbf{q}) \xrightarrow{a.s.} 0, \qquad (13.71)$$

where

$$\varphi_i(\mathbf{q}) = \left(w_i + \frac{1}{M} \sum_{j \neq i} \frac{q_j d_j}{1 + q_j d_j \varphi_i(\mathbf{q})} \right)^{-1}, \quad \forall i = 1, \ldots, N. \qquad (13.72)$$

Therefore, we have $\text{SINR}_i^U \left(\mathbf{q}, \{\mathbf{u}_i^{\text{MMSE}}\} \right) \xrightarrow{a.s.} q_i d_i \varphi_i(\mathbf{q})$. $\qquad \square$

The value of $\varphi_i(\mathbf{q})$ is defined implicitly by the fixed-point equation (13.69) and it is deterministic for given $\mathbf{q}$.

As for the uplink SINR using MMSE receive beamformers, the equivalent channels of the system using MMSE beamformers also approach deterministic quantities asymptotically. Using the channel model (13.65), we have

$$\frac{1}{M} \left| \mathbf{h}_i^H \mathbf{u}_j^{\text{MMSE}} \right|^2 = \frac{d_i}{M} \left| \tilde{\mathbf{h}}_i^H \mathbf{u}_j^{\text{MMSE}} \right|^2, \quad \forall i, j = 1, \ldots, N. \qquad (13.73)$$

Depending on whether $i = j$, the equivalent channels asymptotically approach different deterministic values. The following lemma shows the asymptotically equivalent channels when $i = j$.

LEMMA 13.6   *As $M \to \infty$, the equivalent channels $M^{-1} \left| \mathbf{h}_i^H \mathbf{u}_i^{\text{MMSE}} \right|^2$ using MMSE beamformers approach deterministic quantities, i.e.,*

$$\frac{1}{M} \left| \mathbf{h}_i^H \mathbf{u}_i^{\text{MMSE}} \right|^2 \xrightarrow{a.s.} \frac{d_i \varphi_i^2(\mathbf{q})}{-\varphi_i'(\mathbf{q})} \qquad (13.74)$$

where

$$\varphi_i'(\mathbf{q}) = -\varphi_i(\mathbf{q}) \left(w_i + \frac{1}{M} \sum_{j=1, j \neq i}^{N} \frac{q_j d_j}{[1 + q_j d_j \varphi_i(\mathbf{q})]^2} \right)^{-1}. \qquad (13.75)$$

Proof   Substituting the expression (13.7) for $\mathbf{u}_i^{\text{MMSE}}$ into (13.73), we have

$$\frac{1}{M} \left| \tilde{\mathbf{h}}_i^H \mathbf{u}_i^{\text{MMSE}} \right|^2 = \frac{\left(M^{-1} \tilde{\mathbf{h}}_i \left(\sum_{j \neq i} q_j d_j M^{-1} \tilde{\mathbf{h}}_j \tilde{\mathbf{h}}_j^H + w_i \mathbf{I} \right)^{-1} \tilde{\mathbf{h}}_i^H \right)^2}{M^{-1} \tilde{\mathbf{h}}_i \left(\sum_{j \neq i} q_j d_j M^{-1} \tilde{\mathbf{h}}_j \tilde{\mathbf{h}}_j^H + w_i \mathbf{I} \right)^{-2} \tilde{\mathbf{h}}_i^H}. \qquad (13.76)$$

According to Lemma 13.5, the numerator of (13.76) converges almost surely to $\varphi_i^2(\mathbf{q})$.

A deterministic equivalent for the denominator can be obtained from Lemma 13.10. Under the assumptions of Lemma 13.10 and taking derivative of the Stieltjes transform in (13.95) below, we have

$$\frac{1}{M} \operatorname{Tr}\left(\mathbf{B} - z\mathbf{I}\right)^{-2} \xrightarrow{a.s.} -\psi'(z), \quad \text{for } z < 0, \tag{13.77}$$

where

$$\psi'(z) = \frac{d\psi(z)}{dz}. \tag{13.78}$$

By making use of the rules for implicit differentiation [38], the value of $\psi'(z)$ can be expressed as

$$\psi'(z) = -\psi(z)\left(\frac{1}{K}\sum_{k=1}^{K}\frac{t_k}{[1 + ct_k\psi(z)]^2} - z\right)^{-1}. \tag{13.79}$$

By substituting $z = -w_i$, $t_j = q_j d_j K/M$, $\forall j \neq i$, into (13.79) we obtain

$$\frac{1}{M}\tilde{\mathbf{h}}_i\left(\sum_{j\neq i}\frac{q_j d_j}{M}\tilde{\mathbf{h}}_j\tilde{\mathbf{h}}_j^H + w_i\mathbf{I}\right)^{-2}\tilde{\mathbf{h}}_i^H \xrightarrow{a.s.} -\varphi_i'(\mathbf{q}), \tag{13.80}$$

where

$$\varphi_i'(\mathbf{q}) = -\varphi_i(\mathbf{q})\left(w_i + \frac{1}{M}\sum_{j\neq i}\frac{q_j d_j}{[1 + q_j d_j\varphi_i(\mathbf{q})]^2}\right)^{-1}. \qquad \square$$

The asymptotic equivalent channel when $i \neq j$ can be obtained by the following lemma.

LEMMA 13.7   *As $M \to \infty$, the equivalent channels $\left|\mathbf{h}_i^H\mathbf{u}_j^{\text{MMSE}}\right|^2$ using MMSE beamformers approach deterministic quantities, i.e.,*

$$\left|\mathbf{h}_i^H\mathbf{u}_j^{\text{MMSE}}\right|^2 \xrightarrow{a.s.} \frac{d_i}{[1 + q_i d_i\varphi_j(\mathbf{q})]^2}, \quad \text{when } i \neq j. \tag{13.81}$$

Proof   The expression for $\left|\tilde{\mathbf{h}}_i^H\mathbf{u}_j^{\text{MMSE}}\right|^2$ can be further expanded as follows:

$$\left|\tilde{\mathbf{h}}_i^H\mathbf{u}_j^{\text{MMSE}}\right|^2 = \frac{1}{M}\tilde{\mathbf{h}}_i^H\left(\sum_{k\neq j}\frac{q_k d_k}{M}\tilde{\mathbf{h}}_k\tilde{\mathbf{h}}_k^H + w_j\mathbf{I}\right)^{-1}$$

$$\times \tilde{\mathbf{h}}_i\tilde{\mathbf{h}}_i^H\left(\sum_{k\neq j}\frac{q_k d_k}{M}\tilde{\mathbf{h}}_k\tilde{\mathbf{h}}_k^H + w_j\mathbf{I}\right)^{-1}\tilde{\mathbf{h}}_j$$

$$\times \left[\frac{1}{M}\tilde{\mathbf{h}}_j^H\left(\sum_{k\neq j}\frac{q_k d_k}{M}\tilde{\mathbf{h}}_k\tilde{\mathbf{h}}_k^H + w_j\mathbf{I}\right)^{-2}\tilde{\mathbf{h}}_j\right]^{-1}. \tag{13.82}$$

The last factor of (13.82) converges almost surely to $-\varphi'_j(\mathbf{q})$ according to the proof of Lemma 13.6. The other two factors of (13.82), since $\tilde{\mathbf{h}}_i$ and

$$\sum_{k \neq j} \frac{q_k d_k}{M} \tilde{\mathbf{h}}_k \tilde{\mathbf{h}}_k^H$$

are not independent, can be transformed into the following equivalent form using the matrix inversion lemma:

$$\frac{1}{M} \tilde{\mathbf{h}}_j^H \left(\sum_{k \neq j,i} \frac{q_k d_k}{M} \tilde{\mathbf{h}}_k \tilde{\mathbf{h}}_k^H + w_j \mathbf{I} \right)^{-1}$$

$$\times \tilde{\mathbf{h}}_i \tilde{\mathbf{h}}_i^H \left(\sum_{k \neq j,i} \frac{q_k d_k}{M} \tilde{\mathbf{h}}_k \tilde{\mathbf{h}}_k^H + w_j \mathbf{I} \right)^{-1} \tilde{\mathbf{h}}_j$$

$$\times \left(1 + \frac{q_i d_i}{M} \tilde{\mathbf{h}}_i^H \left(\sum_{k \neq j,i} \frac{q_k d_k}{M} \tilde{\mathbf{h}}_k \tilde{\mathbf{h}}_k^H + w_j \mathbf{I} \right)^{-1} \tilde{\mathbf{h}}_i \right)^{-2}. \tag{13.83}$$

According to the proof of Lemma 13.6 and employing the rank-1 perturbation lemma, the numerator of (13.83) converges almost surely to $-\varphi'_j(\mathbf{q})$ and the denominator of (13.83) converges almost surely to $[1 + q_i d_i \varphi_j(\mathbf{q})]^2$.   □

As $M \to \infty$ we can substitute the deterministic equivalents of the system parameters into Algorithm 10 and obtain an iterative algorithm for large systems that solves (13.4) asymptotically, as shown in Algorithm 11. Lemma 13.5 is applied to (13.35) when updating $\mathbf{q}$. Lemma 13.6 is applied to (13.40) and (13.41) when updating μ and $\mathbf{p}$. Lemma 13.7 is applied to (13.42) when updating $\{v_i\}$. Algorithm 11 can produce asymptotically optimum solutions for $\mathbf{p}^\star$, $\mathbf{q}^\star$, and $\mathbf{x}^\star$ of (13.4) using MMSE beamformers when $M \to \infty$. Note their the beamformers $\{\mathbf{u}_i^\star\}$ do not approach deterministic quantities because their values depend on the actual channel realization. In real system implementations, the outputs of Algorithm 11 can be calculated and stored. Depending on the actual channel state information $\{\mathbf{h}_i\}$, the MMSE beamformers $\{\mathbf{u}_i^\star\}$ can be directly obtained using (13.7). Unlike Algorithm 10, there is no need to resume the algorithm for different channel realizations. Furthermore, in reality the large-scale fading coefficients of RRHs change slowly. Algorithm 11 can be re-invoked as soon as the QoS requirements of RRHs change. Therefore, it can save much computation in large systems. After we obtain the output of Algorithm 11, iterative RRH removal can be applied based on the result of $\mathbf{x}$ achieved by Algorithm 11.

13.4.1   Complexity Analysis

For the finite system analysis, the iterative Algorithm 10 needs to perform iterative updates of the primal and dual variables. The most intensive computationally steps in the iteration are the steps involving matrix inversion and matrix–vector multiplication in (13.35) and (13.38), where each step has the complexity of $O(NM^2 + M^3)$ for each user. Considering that there are N users, the complexity of each iteration is $O(N^2 M^2 + NM^3)$. By applying the convergence results of fixed-point algorithms [39], we obtain that the complexity of Algorithm 10 is $O((N^2 M^2 + NM^3) \log c_1^{-1})$, where c_1

Algorithm 11 Large-system iterative algorithm to solve (13.4)

1: Initialization: $\varphi_i(\mathbf{q}) > 0$, $v_i > 0$, and $q_i > 0$, $\forall i$, such that $\sum_{i=1}^{N} n_i q_i = MP$.
2: **repeat**
3:　　Store old values of $\mathbf{q}$

$$\tilde{\mathbf{q}} = \mathbf{q}.$$

4:　　Update

$$\bar{q}_i = \frac{\gamma_i}{\max(v_i, 1)\,\varphi_i(\mathbf{q})d_i}, \quad \forall i = 1, \ldots, N. \tag{13.84}$$

5:　　Normalize

$$\bar{\mathbf{q}} = \frac{MP}{\mathbf{n}^T \bar{\mathbf{q}}}\bar{\mathbf{q}}. \tag{13.85}$$

6:　　Set

$$\mathbf{q} = \frac{1}{2}(\bar{\mathbf{q}} + \tilde{\mathbf{q}}). \tag{13.86}$$

7:　　Update $\varphi_i(\mathbf{q})$

$$\varphi_i(\mathbf{q}) = \left(w_i + \frac{1}{M}\sum_{j=1, j \neq i}^{N} \frac{q_j d_j}{1 + q_j d_j \varphi_i(\mathbf{q})}\right)^{-1}, \quad \forall i. \tag{13.87}$$

8:　　Calculate the corresponding $\varphi_i'(\mathbf{q})$

$$\varphi_i'(\mathbf{q}) = -\varphi_i(\mathbf{q})\left(w_i + \frac{1}{M}\sum_{j=1, j \neq i}^{N} \frac{q_j d_j}{\left(1 + q_j d_j \varphi_i(\mathbf{q})\right)^2}\right)^{-1}, \quad \forall i. \tag{13.88}$$

9:　　Calculate μ

$$\mu = \sum_{i=1}^{N} \frac{-M v_i \gamma_i \varphi_i'(\mathbf{q})}{\max(v_i, 1)q_i d_i \varphi_i^2(\mathbf{q})}. \tag{13.89}$$

10:　　Calculate the downlink power

$$p_i = \frac{-M v_i \gamma_i \varphi_i'(\mathbf{q})}{\max(v_i, 1)q_i d_i \varphi_i^2(\mathbf{q})\mu}, \quad \forall i. \tag{13.90}$$

11:　　Update v_i

$$v_i = \left(n_i + \frac{1}{M}\sum_{j=1, j \neq i}^{N} \frac{p_j d_i}{[1 + q_i d_i \varphi_j(\mathbf{q})]^2}\right)\frac{\mu q_i}{M}, \quad \forall i. \tag{13.91}$$

12: **until** $|q_i - \tilde{q}_i| \leq \epsilon$, $\forall i$.
13: Set $x_i = \max(v_i - 1, 0)$, $\forall i$. **return x, p**, and **q**.

---

is a constant that determines the convergence speed of the iterative operations in Algorithm 10. Following similar considerations, we obtain that the computational complexity of the large-system iterative Algorithm 10 is $O(N^2 \log c_2^{-1})$, where c_2 is a constant that determines the convergence speed of the iterative operations in the Algorithm. If we use the exhaustive search method then the computational complexity will be $O(2^N(N^2M^2 + NM^3)N_{max})$, where N_{max} denotes the maximum number of iterations made using the max–min SINR algorithm [35] to check the feasibility of the selected RRHs. Therefore, the proposed algorithms enjoy polynomial complexity with respect to N, which is much lower than the exponential complexity required by the exhaustive search method.

13.5   Simulation Results

13.5.1   Convergence Behavior of Finite System Iterative Algorithm

To verify the iterative algorithm for finite systems, we assume that the number of antennas at the WBH is $M = 3$ and the number of RRHs is $N = 4$. Each RRH has a single receive antenna for the wireless fronthaul. The SINR requirement at each RRH for the wireless fronthaul is set at $\gamma_i = 3.01$ dB, $\forall i = 1, \ldots, N$. The noise variance at each RRH is normalized to $n_i = 1$ W, $\forall i$. The transmit power constraint for wireless fronthaul at the WBH is $P = 10$ W. In all the whole simulations, the weights for the transmit power are set to $w_i = 1$, $\forall i$. A randomly generated channel is used to show the convergence result. We compare the results obtained by two methods: Algorithm 10 and 'alternate updates using cvx' (AUC). In AUC we solve (13.4) by alternately solving (13.10) with fixed $\{u_i\}$ using cvx [31] to obtain the uplink power $\mathbf{q}$ and calculating the MMSE receiver $\{u_i^{MMSE}\}$ using the obtained $\mathbf{q}$. Finally, we map the obtained uplink power $\mathbf{q}$ into the corresponding downlink power $\mathbf{p}$ that achieves the same SINR using uplink–downlink power mapping [40].

The convergence behavior of Algorithm 10 is shown in Fig. 13.2. Algorithm 10 converges to the same result obtained by AUC in less than 8 iterations. Fig. 13.2(a) shows the convergence results of $\{p_i\}$, and we initialize all p_i with the same value. Fig. 13.2(b) shows the convergence results of $\{v_i\}$. After 8 iterations, the difference between Algorithm 10 and the final result of AUC is negligible.

In order to compare the computational complexities of Algorithm 10 and AUC, we performed simulations on a PC with a Core 2 Duo CPU with internal clock 3.00 GHz and 4 GB RAM. Twenty channel realizations were performed for each algorithm. The number of RRHs varied from $N = 5$ to 20. The SINR requirements at the RRHs were set to $\gamma_i = 7.78$ dB, $\forall i$. The stopping criteria for both algorithms are such that the maximum uplink power difference between neighboring iterations should be smaller than $\epsilon = 10^{-5}$. The average simulation times for each channel realization using Algorithm 10 and AUC are shown in Fig. 13.3. We observe that the average simulation time using AUC is 5000 to 7000 times longer than that of Algorithm 10. This is so because solving the GP problem (13.10) in each iteration requires a significant amount of time using standard interior-point algorithms, in contrast with the iterative Algorithm 10. Moreover, the ratio of the simulation times increases with the number of RRHs N. This shows that the proposed iterative algorithm greatly reduces the amount of computation needed to solve (13.4), without invoking standard convex optimization software.

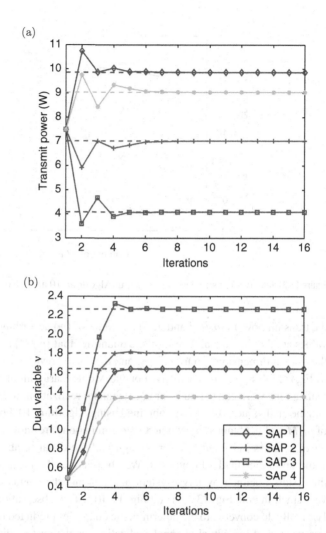

Figure 13.2 Convergence results for finite system iterative algorithms. (a) Convergence result for the power p_i, $\forall i$; (b) Convergence result for dual variable v_i, $\forall i$. The solid lines show the convergence of Algorithm 10, and the dashed lines show the final results obtained by AUC.

13.5.2 Convergence Behavior of the Large-System Iterative Algorithm

Numerical simulations were carried out to verify the iterative algorithm for large systems. In the simulation setup the WBH has $M = 64$ antennas and there are $N = 32$ single-antenna RRHs. The weights of the power are set to $w_i = 1$, $\forall i$. The power constraint at the WBH is $P = 20$ W and the noise variance at each RRH is normalized to $n_i = 1$ W, $\forall i$. The SINR requirement at each RRH is set to $\gamma_i = 6.02$ dB, $\forall i$. The large-scale fading coefficients $\{d_i\}$ between the WBH and the RRHs are assigned some randomly generated positive values, and they are fixed in the simulation.

From Fig. 13.4 we can compare the results obtained by large-system analysis with those obtained by Monte Carlo simulations. For the large-system analysis, we calculate

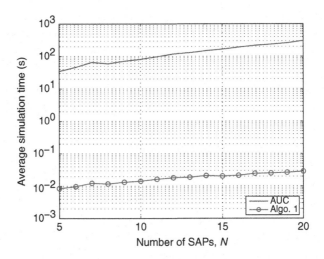

Figure 13.3 The average simulation time using Algorithm 10 and AUC.

the transmit powers $\{p_i/M\}$ and the dual values of $\{v_i\}$ according to Algorithm 11. Those values are the theoretical deterministic quantities that $\{p_i/M\}$ and $\{v_i\}$ converge to when the system dimensions go to infinity, and they are shown in vertical bars for each RRH in Figs. 13.4(a), (b), respectively. For the Monte Carlo simulations, 100 i.i.d. channel realizations were carried out. For each channel realization, the transmit powers $\{p_i/M\}$ and the dual values of $\{v_i\}$ were obtained using Algorithm 10. For each RRH i, the values of p_i/M and v_i obtained by Monte Carlo simulations are shown by error bars, where the mean value is indicated by a cross sign and the distance above and below the mean value gives the standard deviation. We observe that the mean values accurately match the results obtained by large-system analysis and the standard deviations are usually very small in both Fig. 13.4(a) and Fig. 13.4(b). This observation shows that $\{p_i/M\}$ and $\{v_i\}$ really do converge to the deterministic quantities predicted by large-system analysis, irrespective of the actual channel realization, as the system dimensions become large. Instead of instantaneous channel information, the large-system-analysis Algorithm 11 utilizes the large-scale fading coefficients $\{d_i\}$ as channel inputs and requires infrequent updates.

13.5.3   C-RAN Simulations

We performed numerical simulations for RRH on user admission control using a simplified C-RAN setup with multiple RRHs. The WBH was located at the center of the cell with multiple antennas. The cell radius was 1 km. The transmit antenna gain at the WBH was 5 dB. We used a pathloss model from the WBH to the RRHs given by

$$L(\text{dB}) = 128 + 37.6 \log_{10} D, \tag{13.92}$$

where D represents the distance between the WBH and the RRH in km. The lognormal shadowing parameter was 10 dB; the bandwidth of the wireless fronthaul was 10 MHz; the WBH transmit power constraint was 30 dBm; the noise variance at each

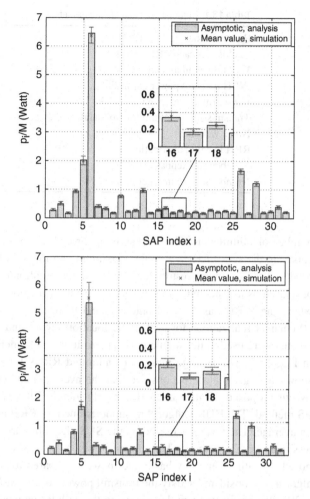

Figure 13.4 Final results for **p** and **v** using large-system analysis and Monte Carlo simulations. Monte Carlo simulations were performed in 100 i.i.d. channel realizations, and their results are represented using error bars. (a) Transmit powers p_i/M, $\forall i$; (b) values of v_i, $\forall i$.

RRH was -93.98 dBm; The cell radius served by that each RRH was 30 m. We assumed that RRHs are randomly and uniformly distributed within the cell. The values of the channel path loss, shadowing parameters, transmit power constraints, and antenna gains were based on [41]. Table 13.1 shows a summary of the above simulation parameters.

In the numerical simulations for finite-system RRH admission, we compared the proposed RRH admission control method, which employs Algorithm 10 and iterative RRH removal, with other two methods commonly reported in the literature: the Lagrange-duality-based *heuristic removal user admission* (HRUA) algorithm [42] and the *semi-orthogonal user selection* (SUS) algorithm [43]. We also compared the results with those of the exhaustive search (ES) method. The ES method obtains the maximum

Table 13.1 Summary of simulation parameters

Radius of the whole C-RAN	1 km
Radius that each RRH serves	30 m
WBH transmit antenna gain	5 dB
WBH transmit power constraint	30 dBm
Log-normal shadowing	10 dB
Transmission spectrum for fronthaul	10 MHz
Noise variance	−93.98 dBm
RRH location distribution	uniform
Number of antennas per RRH	1
Number of antennas per user	1

number of admitted RRHs by searching through all possible choices of RRHs and choosing the set of RRHs with the maximum cardinality. Even though the ES method produces the optimum results for (13.3), its computational load is very high and it is only used as a benchmark for comparing different methods. We considered a network where the WBH has $M = 4$ antennas and there are $N = 12$ RRHs. We generated 60 channel realizations for each RRH location layout and 20 different RRH location layouts were used in the simulations. The results for the methods considered are shown in Fig. 13.5. The average number of admitted RRHs for each channel realization is shown in Fig. 13.5(a). We observe that the average number of admitted RRHs found by our proposed iterative method is nearly identical to that obtained by the optimum ES method. The HURA algorithm supports one less RRH compared to the ES method on average. The results achieved by the SUS algorithm are the worst; they are 2 to 1.2 less than the ES method. The transmit power $\sum_i p_i$ is shown in Fig. 13.5(b). Our proposed algorithm generally has higher power compared to HURA and SUS, but all the algorithms considered satisfy the transmit power constraint P.

We also performed simulations using the cellular network parameters in Table 13.1 for the case when the QoS requirements at different RRHs are unequal and change with time. In the simulation, each RRH is responsible for serving one user. We generated 60 different RRH location layouts. For each layout, 20 different user locations in each RRH were considered. In each time slot, the SINR requirement of each user was drawn randomly and uniformly from 4.3 dB to 18.7 dB. The simulation results for the different RRH admission schemes are shown in Table 13.2. After the admission of RRHs, we used the max–min SINR algorithm [17] to verify whether the users in the admitted set of RRHs could be simultaneously supported for their SINR requirements. If not, we applied the user admission control discussed in Section 13.3.3 to select the maximum set of users whose QoS requirements could be satisfied within the admitted set of RRHs. The user rates are obtained from the finally admitted users. For the ES method, there may be more than one set of RRHs that have the maximum cardinality. In that case, we randomly chose one set of RRHs from them, owing to complexity issues. We observe from Table 13.2 that the average number of admitted RRHs and users for the proposed scheme is very close to the optimum results obtained by the ES method, which

Table 13.2 RRH admission control schemes with unequal SINR requirements

	Ex. Search	Proposed	HURA	SUS
Ave. admitted RRHs/users	3.39/3.31	3.20/3.10	1.85/1.82	1.78/1.75
Ave. WBH power (dBm)	28.75	27.34	26.07	24.66
RRH/user sum-rate (b/s/Hz)	11.34/10.99	11.42/10.98	7.50/7.36	6.71/6.60

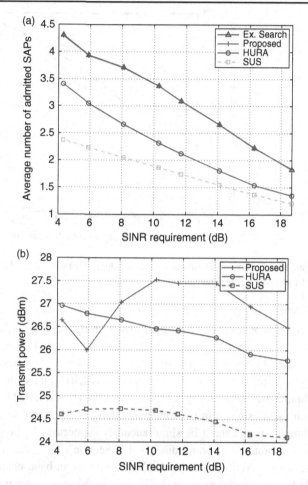

Figure 13.5 Cellular network simulations for finite system RRH admission. (a) Average number of admitted RRHs; (b) transmit power $\sum_i p_i$.

far outperforms the HURA and SUS methods. The average user sum-rate achieved by the proposed scheme is 49.2% and 66.4% higher than the HURA and SUS methods, respectively. The empirical distributions of individual user rates for all the users in the network are shown in Fig. 13.6, where the rates for the unadmitted users are set to 0. Compared with the HURA and SUS methods, the proposed method and the ES method reduce the percentage of users that cannot be served from about 85% to 74% and 72%, respectively. This show that the proposed method, as well as the ES method, can improve the fairness for users in the system.

Table 13.3 Comparison of user throughput with and without RRHs

	With RRHs	Without RRHs
Ave. admitted (RRHs)/users	3.31/3.21	0.43
Ave. (RRH)/user sum-rate (b/s/Hz)	11.17/10.75	1.30

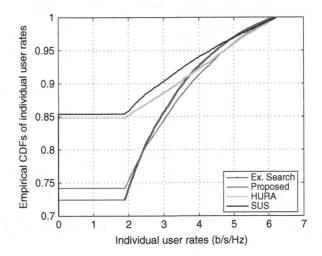

Figure 13.6 Empirical distributions of individual user rates for all the users. The key relates to the left-hand part of the figure.

Simulation results comparing the user throughput with and without RRHs in a cellular network are shown in Table 13.3. We used the simulation parameters in Table 13.1 and the simulation setup is similar to that described for Table 13.2. We compared two cases. In the first case the WBH performs RRH admission and transmits data to the admitted RRHs using wireless fronthaul. Then the admitted RRHs transmit those data to their users using an in-band channel taking into account inter-RRH interference. If the admitted users cannot be simultaneously supported for the given SINR requirements, the user admission control discussed in Section 13.3.3 is applied. In the second case the users are served directly by the MBS using an in-band channel. We assumed that the MBS is co-located with the WBH and has the same number of antennas. In both cases the transmit power of the in-band channel is the same. From Table 13.3, we observe that the user throughput with RRH admission is about 8.3 times that without RRHs. This is so mainly because the deployment of RRHs significantly reduces the distance between the transmitters and users. The signals to the users will experience higher attenuation and become much weaker if transmitted directly from the MBS. This shows that the deployment of RRHs can boost the system throughput considerably.

Simulations for large systems were carried out employing the parameters in Table 13.1. In the simulations, the WBH was equipped with $M = 64$ antennas and there were $N = 32$ single-antenna RRHs. The power weights were set to $w_i = 1$, $\forall i$. The power constraints were set to $P = 30$ dBm, 27.0 dBm, or 24.8 dBm. The

(a)

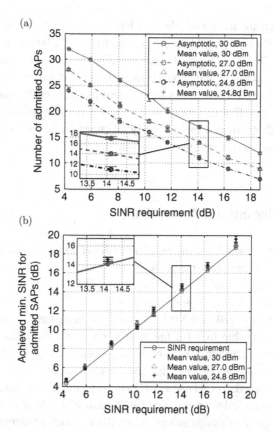

(b)

Figure 13.7 Results using large-system analysis and using Monte Carlo simulations. The Monte Carlo simulations were performed in 100 i.i.d. channel realizations, and their results are represented using error bars. (a) Average number of admitted RRHs; (b) minimum SINR for RRHs admitted using large-system analysis.

simulation results obtained using large-system analysis and those obtained using Monte Carlo simulations are given Figure 13.7. Figure 13.7(a) shows how the number of admitted RRHs depends on the given power and SINR constraints. The asymptotic results are shown by circles; they were obtained by the RRH admission control method employing Algorithm 11 and iterative RRH removal using large-scale fading coefficients. Those asymptotic results may be compared with the Monte Carlo simulation results, where 100 i.i.d. channel realizations were carried out. For each channel realization, the Monte Carlo simulation performs RRH admission employing Algorithm 10 and iterative RRH removal. Those Monte Carlo results are shown by error bars. We observe from Fig. 13.7(a) that the standard deviations of the Monte Carlo results are very small, which shows that the Monte Carlo results tend to be deterministic. Such deterministic values are accurately predicted by the asymptotic results using large-system analysis. In Fig. 13.7(b), we show the achieved minimum SINR of the RRHs that have been selected by employing large-system analysis. For a given SINR constraint and large-scale fading coefficients, the set of admitted RRHs was selected using the RRH admission control method employing Algorithm 11 and iterative RRH removal. The max–min SINR algorithm [35] was utilized for each channel realization to maximize the minimum SINR for that set of admitted RRHs over 100 i.i.d. channel realizations. The error bars show the achieved SINR values for different channel realizations, which are very close to or above the SINR requirements. Figure 13.7(b) shows that the admitted RRHs can satisfy the SINR requirements in nearly all channel realizations even though the RRHs were selected using large system analysis.

13.6   Conclusions and Future Work

We have considered the problem of RRH admission control in a C-RAN using wireless fronthaul. In order to serve large numbers of users, as well as to minimize the total cost of building fronthaul in the C-RAN, the WBH needs to simultaneously serve as many RRHs as possible under given power and SINR constraints. Such a problem is combinatorial and NP-hard. We applied ℓ_1-norm relaxation and proposed an iterative algorithm to solve the relaxed problem. The local convergence property of the iterative algorithm was proved. Its convergence has been validated by numerical simulations and we observe that its range of convergence is large. Using the solution of the ℓ_1-relaxed problem, the RRHs were iteratively removed until all the remaining RRHs satisfied the power and SINR constraints. We also proposed a large-system iterative algorithm using random matrix theory. Such an algorithm requires only large-scale channel coefficients to perform RRH admission control for large systems irrespective of the instantaneous channel information. Simulations showed that the finite system iterative algorithm achieved nearp-optimum results in a simplified C-RAN setup and the large-system iterative algorithm predicted the Monte Carlo simulation results accurately.

In our future work, system-level evaluations of the proposed algorithms will be carried out, especially in ultra-dense 5G network scenarios. In such scenarios inter-RRH interference will be very strong, which may impact the results of the proposed algorithms. Whether the proposed algorithms can still provide near optimum results for admission control needs to be evaluated. Furthermore, as the network becomes denser, the required overhead of fronthaul-channel knowledge feedback and the processing delay induced by the admission control algorithms need to be considered in a practical setup. Distributed admission control algorithms could be investigated.

Another important problem that will be investigated in our future work is the joint optimization of RRH admission control in the fronthaul and user admission. This is a very difficult problem because the selection of RRHs in the wireless fronthaul and the final admission control of users are interconnected. Solving this joint optimization problem is challenging. Our discussion in Section 13.3 considered the RRH and the user admission control separately, and therefore it provides only a suboptimal solution to the joint optimization problem. One research direction is to take into account the possible interference that each RRH could generate to other RRH users if it is admitted and to give less priority to RRHs with high interference when RRH admission control is performed.

13.7   Appendix

13.7.1   Useful Results from Random Matrix Theory

We summarize some useful results obtained from random matrix theory.

LEMMA 13.8   (Trace lemma [44, 45]) *Let* $\mathbf{A} \in \mathbb{C}^{M \times M}$ *and let* $\mathbf{x} \in \mathbb{C}^M$ *be a random vector of i.i.d. entries, independent of* $\mathbf{A}$. *Assume* $\mathbb{E}\{x_i\} = 0$, $\mathbb{E}\{|x_i|^2\} = 1$, $\mathbb{E}\{|x_i|^8\} < \infty$, *and* $\lim \sup_M \|\mathbf{A}\| < \infty$. *Then*

$$\frac{1}{M} \mathbf{x}^H \mathbf{A} \mathbf{x} - \frac{1}{M} \operatorname{Tr}(\mathbf{A}) \xrightarrow{a.s.} 0. \tag{13.93}$$

LEMMA 13.9   (Rank-1 perturbation lemma [46]) *Let* $z < 0$, $\mathbf{A} \in \mathbb{C}^{M \times M}$, *and* $\mathbf{B} \in \mathbb{C}^{M \times M}$, *with* $\mathbf{B}$ *Hermitian nonnegative definite, and* $\mathbf{x} \in \mathbb{C}^M$. *Assume that* $\lim \sup_M \|\mathbf{A}\| < \infty$. *Then*

$$\left| \frac{1}{M} \operatorname{Tr} \left((\mathbf{B} - z\mathbf{I})^{-1} - (\mathbf{B} + \mathbf{x}\mathbf{x}^H - z\mathbf{I})^{-1} \right) \mathbf{A} \right| \le \frac{\|\mathbf{A}\|}{M |z|} \to 0, \quad \text{as } M \to \infty. \tag{13.94}$$

This shows that finite rank perturbations of $\mathbf{B}$ are asymptotically negligible.

LEMMA 13.10   (Right-sided correlation model [47, 48]) *Let* $\mathbf{X} \in \mathbb{C}^{M \times K}$, $[\mathbf{X}]_{ij} \sim \mathcal{CN}(0, \frac{1}{K})$ *i.i.d., and let* $\mathbf{T} = \operatorname{diag}\{t_1, \ldots, t_K\} \in \mathbb{R}_+^{K \times K}$ *be deterministic, satisfying* $\max_k t_k < \infty$. *Denote* $\mathbf{B} = \mathbf{X}\mathbf{T}\mathbf{X}^H$, $c = M/K$, *and let* $M, K \to \infty$ *while* $0 < \lim \inf c \le \lim \sup c < \infty$. *Then*

$$\frac{1}{M} \operatorname{Tr} (\mathbf{B} - z\mathbf{I})^{-1} - \psi(z) \xrightarrow{a.s.} 0, \quad \text{for } z < 0 \tag{13.95}$$

where $\psi(z)$ *is the unique Stieltjes transform [49], which satisfies*

$$\psi(z) = \left(\frac{1}{K} \sum_{k=1}^{K} \frac{t_k}{1 + ct_k \psi(z)} - z \right)^{-1}. \tag{13.96}$$

LEMMA 13.11   (Matrix inversion lemma [47]) *Let* $\mathbf{A} \in \mathbb{C}^{M \times M}$ *be invertible. Then, for any vector* $\mathbf{x} \in \mathbb{C}^M$ *and any scalar* $c \in \mathbb{C}$ *such that* $\mathbf{A} + c\mathbf{x}\mathbf{x}^H$ *is invertible,*

$$\mathbf{x}^H \left(\mathbf{A} + c\mathbf{x}\mathbf{x}^H \right)^{-1} = \frac{\mathbf{x}^H \mathbf{A}^{-1}}{1 + c\mathbf{x}^H \mathbf{A}^{-1} \mathbf{x}}. \tag{13.97}$$

References

[1] J. G. Andrews, S. Buzzi, W. Choi, S. Hanly, A. Lozano, A. C. K. Soong *et al.*, "What will 5G be?" *IEEE J. Sel. Areas Commun.*, vol. 32, no. 6, pp. 1065–1082, June 2014.

[2] 3GPP TR 36.913, "Requirements for further advancements for E-UTRA (LTE Advanced)," v11.0.0, November 2012.

[3] T. Q. S. Quek, G. de la Roche, I. Guvenc, and M. Kountouris, *Small Cell Networks: Deployment, PHY Techniques, and Resource Management*. Cambridge University Press, 2013.

[4] D. Chen, T. Q. S. Quek, and M. Kountouris, "Backhauling in heterogeneous cellular networks – modeling and tradeoffs," *IEEE Trans. Wireless Commun.*, vol. 14, no. 6, pp. 3194–3206, June 2015.

[5] S. Chia, M. Gasparroni, and P. Brick, "The next challenge for cellular networks: Backhaul," *IEEE Microw. Mag.*, vol. 10, no. 5, pp. 54–66, August 2009.

[6] Y. Yang, T. Q. S. Quek, and L. Duan, "Backhaul-constrained small cell networks: refunding and QoS provisioning," *IEEE Trans. Wireless Commun.*, vol. 13, no. 9, pp. 5148–5161, September 2014.

[7] S. Lee, G. Narlikar, M. Pal, G. Wilfong, and L. Zhang, "Admission control for multihop wireless backhaul networks with QoS support," in *Proc. IEEE Wireless Communications Networking Conf.*, Las Vegas, NV, April 2006.

[8] D. Bojic, E. Sasaki, N. Cvijetic, T. Wang, J. Kuno, J. Lessmann *et al.*, "Advanced wireless and optical technologies for small-cell mobile backhaul with dynamic software-defined management," *IEEE Commun. Mag.*, vol. 51, no. 9, pp. 86–93, September 2013.

[9] X. Ge, H. Cheng, M. Guizani, and T. Han, "5G wireless backhaul networks: challenges and research advances," *IEEE Netw.*, vol. 28, no. 6, pp. 6–11, November 2014.

[10] H. Viswanathan and S. Mukherjee, "Throughput–range tradeoff of wireless mesh backhaul networks," *IEEE J. Sel. Areas Commun.*, vol. 24, no. 3, pp. 593–602, March 2006.

[11] J. Zhao, T. Q. S. Quek, and Z. Lei, "Coordinated multipoint transmission with limited backhaul data transfer," *IEEE Trans. Wireless Commun.*, vol. 12, no. 6, pp. 2762–2775, June 2013.

[12] L. Zhou and W. Yu, "Uplink multicell processing with limited backhaul via per-base-station successive interference cancellation," *IEEE J. Sel. Areas Commun.*, vol. 31, no. 10, pp. 1981–1993, October 2013.

[13] S.-H. Park, O. Simeone, O. Sahin, and S. Shamai, "Joint precoding and multivariate backhaul compression for the downlink of cloud radio access networks," *IEEE Trans. Signal Process.*, vol. 61, no. 22, pp. 5646–5658, November 2013.

[14] J. Rao and A. Fapojuwo, "On the tradeoff between spectral efficiency and energy efficiency of homogeneous cellular networks with outage constraint," *IEEE Trans. Veh. Technol.*, vol. 62, no. 4, pp. 1801–1814, May 2013.

[15] Y. Shi, J. Zhang, and K. B. Letaief, "Group sparse beamforming for green Cloud-RAN," *IEEE Trans. Wireless Commun.*, vol. 13, no. 5, pp. 2809–2823, May 2014.

[16] A. Tölli, H. Pennanen, and P. Komulainen, "Decentralized minimum power multi-cell beamforming with limited backhaul signaling," *IEEE Trans. Wireless Commun.*, vol. 10, no. 2, pp. 570–580, February 2011.

[17] D. W. H. Cai, T. Q. S. Quek, C. W. Tan, and S. H. Low, "Max–min SINR coordinated multipoint downlink transmission – duality and algorithms," *IEEE Trans. Signal Process.*, vol. 60, no. 10, pp. 5384–5395, October 2012.

[18] X. Zhai, L. Zheng, and C. W. Tan, "Energy–infeasibility tradeoff in cognitive radio networks: price-driven spectrum access algorithms," *IEEE J. Sel. Areas Commun.*, vol. 32, no. 3, pp. 528–538, March 2014.

[19] J. Zhao, T. Q. S. Quek, and Z. Lei, "User admission and clustering for uplink multiuser wireless systems," *IEEE Trans. Veh. Technol.*, vol. 64, no. 2, pp. 636–651, Feburary 2015.

[20] T. Ahmad, R. Gohary, H. Yanikomeroglu, S. Al-Ahmadi, and G. Boudreau, "Coordinated port selection and beam steering optimization in a multi-cell distributed antenna system using semidefinite relaxation," *IEEE Trans. Wireless Commun.*, vol. 11, no. 5, pp. 1861–1871, May 2012.

[21] T. L. Marzetta, "Noncooperative cellular wireless with unlimited numbers of base station antennas," *IEEE Trans. Wireless Commun.*, vol. 9, no. 11, pp. 3590–3600, September 2010.

[22] F. Rusek, D. Persson, B. K. Lau, E. G. Larsson, T. L. Marzetta, O. Edfors, *et al.*, "Scaling up MIMO: opportunities and challenges with very large arrays," *IEEE Signal Process. Mag.*, vol. 30, no. 1, pp. 40–60, January 2013.

[23] J. Hoydis, S. Ten Brink, and M. Debbah, "Massive MIMO in the UL/DL of cellular networks: How many antennas do we need?" *IEEE J. Sel. Areas Commun.*, vol. 31, no. 2, pp. 160–171, February 2013.

[24] F. Fernandes, A. E. Ashikhmin, and T. L. Marzetta, "Inter-cell interference in noncooperative TDD large scale antenna systems," *IEEE J. Sel. Areas Commun.*, vol. 31, no. 2, pp. 192–201, February 2013.

[25] Y. Huang, C. W. Tan, and B. Rao, "Joint beamforming and power control in coordinated multicell: max–min duality, effective network and large system transition," *IEEE Trans. Wireless Commun.*, vol. 12, no. 6, pp. 2730–2742, June 2013.

[26] M. Peng, Y. Li, J. Jiang, J. Li, and C. Wang, "Heterogeneous cloud radio access networks: a new perspective for enhancing spectral and energy efficiencies," *IEEE Wireless Commun. Mag.*, vol. 21, no. 6, pp. 126–135, December 2014.

[27] E. Matskani, N. Sidiropoulos, Z. Luo, and L. Tassiulas, "Convex approximation techniques for joint multiuser downlink beamforming and admission control," *IEEE Trans. Wireless Commun.*, vol. 7, no. 7, pp. 2682–2693, July 2008.

[28] H. Boche and M. Schubert, "Duality theory for uplink downlink multiuser beamforming," in *Smart Antennas – State of the Art*, Hindawi Publishing Corporation, 2005, pp. 545–576.

[29] M. Bengtsson and B. Ottersten, "Optimal and suboptimal transmit beamforming," in *Handbook of Antennas in Wireless Communications*, CRC Press, 2001.

[30] H. Van Trees, *Optimum Array Processing*. John Wiley & Sons, 2002.

[31] M. Grant and S. Boyd, "CVX: Matlab software for disciplined convex programming, version 2.1." Available at cvxr.com/cvx, March 2014.

[32] S. Boyd and L. Vandenberghe, *Convex Optimization*. Cambridge University Press, 2004.

[33] A. Berman and R. J. Plemmons, *Nonnegative Matrices in the Mathematical Sciences*. Academic Press, 1979.

[34] A. Granas and J. Dugundji, *Fixed Point Theory*. Springer, 2003.

[35] D. W. H. Cai, T. Q. S. Quek, and C. W. Tan, "A unified analysis of max–min weighted SINR for MIMO downlink system," *IEEE Trans. Signal Process.*, vol. 59, no. 8, pp. 3850–3862, August 2011.

[36] L. Zhang, R. Zhang, Y.-C. Liang, Y. Xin, and H. V. Poor, "On Gaussian MIMO BC-MAC duality with multiple transmit covariance constraints," *IEEE Trans. Inf. Theory*, vol. 58, no. 4, pp. 2064–2078, April 2012.

[37] H. Dahrouj and W. Yu, "Coordinated beamforming for the multicell multi-antenna wireless system," *IEEE Trans. Wireless Commun.*, vol. 9, no. 5, pp. 1748–1759, May 2010.

[38] W. Rudin, *Principles of Mathematical Analysis*. McGraw-Hill, 1976.

[39] H. R. Feyzmahdavian, M. Johansson, and T. Charalambous, "Contractive interference functions and rates of convergence of distributed power control laws," *IEEE Trans. Wireless Commun.*, vol. 11, no. 12, pp. 4494–4502, December 2012.

[40] M. Schubert and H. Boche, "Solution of the multiuser downlink beamforming problem with individual SINR constraints," *IEEE Trans. Veh. Technol.*, vol. 53, no. 1, pp. 18–28, March 2004.

[41] 3GPP TS 36.814, "Evolved universal terrestrial radio access (E-UTRA); further advancements for E-UTRA physical layer aspects." v9.0.0, March 2010.

[42] R. Stridh, M. Bengtsson, and B. Ottersten, "System evaluation of optimal downlink beamforming with congestion control in wireless communication," *IEEE Trans. Wireless Commun.*, vol. 5, no. 4, pp. 743–751, April 2006.

[43] T. Yoo and A. Goldsmith, "On the optimality of multiantenna broadcast scheduling using zero-forcing beamforming," *IEEE J. Sel. Areas Commun.*, vol. 24, no. 3, pp. 528–541, March 2006.

[44] Z. D. Bai and J. W. Silverstein, *Spectral Analysis of Large Dimensional Random Matrices.* Second edition, Springer, 2009.

[45] R. Couillet and M. Debbah, *Random Matrix Methods for Wireless Communications.* Cambridge University Press, 2011.

[46] Z. D. Bai and J. W. Silverstein, "On the signal-to-interference ratio of CDMA systems in wireless communications," *Ann. Appl. Prob.*, vol. 17, no. 1, pp. 81–101, 2007.

[47] J. W. Silverstein and Z. D. Bai, "On the empirical distribution of eigenvalues of a class of large dimensional random matrices," *J. Multivariate Analysis*, vol. 54, no. 2, pp. 175–192, 1995.

[48] R. Couillet, M. Debbah, and J. W. Silverstein, "A deterministic equivalent for the analysis of correlated MIMO multiple access channels," *IEEE Trans. Inf. Theory*, vol. 57, no. 6, pp. 3493–3514, June 2011.

[49] A. M. Tulino and S. Verdú, "Random matrix theory and wireless communications," *Found. Trends Commun. Inf. Theory*, vol. 1, no. 1, pp. 1–182, 2004.

14 Toward Green Deployment and Operation for C-RANs

Sheng Zhou, Jingchu Liu, Tao Zhao, and Zhisheng Niu

14.1 Introduction

The boost in the number of mobile devices such as smart phones and tablets, together with the diverse applications enabled by mobile Internet, has triggered the exponential growth of mobile data traffic [1]. It is estimated that the next generation (5G) cellular networks will need to support a 1000-fold increase in traffic capacity [2]. With limited spectrum resources, it is challenging to accommodate the huge volume of traffic demand with conventional radio access network (RAN) architecture, in which the processing functionalities are packed into stand-alone base stations (BSs) and the cooperation between BSs is limited. In addition, 5G is expected to support massive connections including not only human-to-human connections but also machine-to-machine connections. Some demand a high data rate, while others have a low capacity requirement but require a real-time response and high reliability. As a result, the cellular network must be flexible enough to adapt to the various characteristics of different types of connections. Besides, under the influence of innovative applications from IT companies, the average revenue per user of network operators tends to increase slowly or even decrease in some cases, while expenditure increases rapidly [3]. Such trend imposes a great challenge to the sustainability of the cellular network.

Therefore, it is crucial to renovate cellular network architectures to meet the requirements of 5G systems in terms of high efficiency, flexibility, and sustainability. One of the promising architecture evolution trends is integrating cloud computing technology into cellular networks, and accordingly cloud-RAN (C-RAN) [3] is proposed to move base band units (BBUs) of BSs to a centralized cloud computing platform, and only leaving remote radio heads (RRHs) in the front end. A similar idea is also proposed under the name wireless network cloud (WNC) [4]. In cloud-based cellular network architectures, software-defined BS functionalities are implemented on general purpose platforms (GPP) with virtualization technologies, making them *virtual base stations* (VBSs) [5–7]. Compared with conventional BSs, C-RAN is more flexible in terms of the implementation of new functionalities and the management of computational resource. The pooling of BBU processing brings *statistical multiplexing gain* not only for radio resources, via cooperative signal processing, but also for computational resources via BS function consolidation, thus potentially reducing operational cost [8]. In addition, centralized processing can also be combined with dynamic fronthaul switching to address the mobility and energy efficiency issues of small cells [9, 10]. The deployment of C-RAN also demonstrates its capability to reduce deployment cost and improve radio access performance [3]. In the C-RAN context, we have also proposed a novel architecture for the convergence of cloud and cellular systems (CONCERT) [11], with features including heterogeneous physical resources, logically centralized resource virtualization, and software-defined services.

Despite its evident advantages, C-RAN still faces many deployment and operation challenges. Its unique feature of BBU centralization requires fronthaul with a massive bandwidth: transmitting the baseband sample of a single 20 MHz LTE antenna-carrier (AxC) requires around 1 Gbps link bandwidth [12]. Therefore, large-scale centralization will potentially incur enormous fronthaul expenditure. The most straightforward option for the fronthaul physical layer is dark fiber, but since one fiber core can only carry one fronthaul link, enormous fiber resources are needed. Another option is to multiplex fronthaul links into a single fiber core using wave-division multiplexing (WDM), but WDM modules are much more expensive than black and white modules [13]. Thanks to virtualization technology, when operating C-RAN computational resources such as the number of CPU cores and the CPU speed are pooled and can be dynamically allocated to each VBS so as to adapt to the dynamics of the traffic demand over time and space, which brings possible energy savings. However, computational resource dynamics are not captured in the existing BS energy consumption models [14, 15], making it hard to evaluate the energy saving gain brought by C-RAN and to optimize the green resource management schemes to realize that gain.

To this end we aim to study cost-efficient deployment and energy-efficient operation for C-RANs in this chapter, in order to reveal their advantages to green wireless access. To help save the huge deployment cost led by the fronthaul, rather than optimizing the fronthaul itself with cost-efficient design [13], which is not the target of this chapter in Section 14.2 we try to find the best size of VBS pool for C-RAN deployment while maintaining sufficient performance gain of the C-RAN. This is motivated by the fact that a substantial statistical multiplexing gain can be achieved with even small-scale centralization [6]. On the basis of a novel session-level VBS model, we provide a quantitative analysis of the statistical multiplexing gain from the pooling of VBSs with respect to the pool size; this analysis indicates that that VBS pools can readily obtain a considerable pooling-gain at medium size. To inspire green operations in C-RAN, in Section 14.3 we investigate the energy consumption of C-RANs with dynamic BSs on or off, exploiting the flexibility of virtualization. The energy consumption from both computing and radio transmission are considered. We analyze the energy–delay tradeoff in C-RAN, and provide an efficient algorithm to jointly optimize the computational and wireless resources. This minimizes the energy consumption while satisfying the given delay requirement. In Section 14.4 we conclude this chapter and look at some directions to advance green deployment and operation for C-RANs.

14.2 On the Size of VBS Pools in C-RANs

From the deployment perspective, we focus on the cost introduced by the fronthaul. The key is to characterize the tradeoff between the operation performance brought by statistical multiplexing and the fronthauling cost due to the excessive amount of fiber needed for a large BBU pool. We propose a realistic session-level model for heterogeneous video bitstream (VBS) pools with both radio and computational resource constraints in Section 14.2.1, enabling semidynamic resource management. This model can describe heterogeneous VBS pools, in which there are multiple classes of VBSs with different session arrivals, resource configurations, and service strategies. On the basis of model, in Section 14.2.2 we give a recursive method to compute the session blocking probability and thus in Section 14.2.3 we are able to analyze quantitatively the statistical multiplexing gain, defined as the reduction in the computational resources needed, compared with the conventional cellular architecture under identical blocking-probability constraints. Applications of our model and derivations on both real-time and delay-tolerant traffic are demonstrated in Section 14.2.4. By evaluating the gains from centralization the expenditure on fronthaul can be justified, as discussed in Section 14.2.5.

14.2.1 Session-Level Model of VBS Pools

This subsection is devoted to characterizing the computation and radio resource utilization dynamics in VBSs, in order to assist further analysis on the statistical multiplexing gain of VBS pools in C-RANs. Previously, a session-level VBS pool model was proposed in [16], under the assumption of unconstrained radio resources and dynamic resource management, where a user session refers to the time period in which the user occupies a unit of computational resource in the pool. Nevertheless, this model does not consider the influence of radio resources, which, rather than the computation resources, are often the main performance bottleneck for real networks. To this end we introduce a Markovian model for VBS pools, which captures the session-level dynamics in a VBS pool, including the dynamics of both the computational resources and the radio resources. To be general, we assume V different classes of VBSs. The total number of VBSs in class v ($v = 1, 2, \ldots, V$) is M_v. To each VBS in class v is assigned K_v units of radio resource. To perform baseband signal processing on user sessions in the BBU, all VBSs share a total of N units of computational resource. The overall setting is illustrated in Fig. 14.1. We assume homogeneous resource demands: every active session is assumed to occupy one unit of radio resource and one unit of computational resource. For simplicity, hereafter we denote the radio and computational resources as r-servers and c-servers, respectively.

Session Arrival, Service Discipline, and Admission Control
User session arrivals follow independent Poisson processes in the coverage area of their serving VBSs. In fact, the overall session arrival rate in a VBS is proportional to its coverage area. We allow VBSs in different classes to have different sizes of coverage

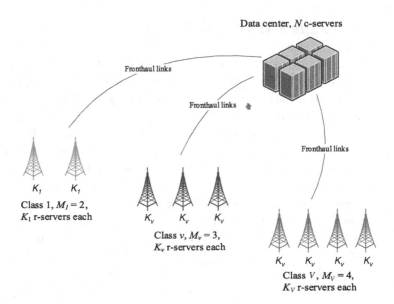

Figure 14.1 An example of a heterogeneous VBS pool.

areas and, consequently, different aggregated session arrival rates. We denote the arrival rate of class-v VBSs by λ_v.

Each user session demands an exponentially distributed amount of service capacity, defined as the time duration for voice call sessions or the amount of information bits for cellular data sessions. We assume that a VBS pool scheduler manages the service capacity so that the service capacity assigned to a class-v VBS is a function of the total number of sessions in this VBS and the assigned capacity is equally divided among these sessions for fairness. Denote the number of sessions in the mth VBS of class v at time t as $U_{v,m}(t)$. Then the above service strategy can be translated into a session departure rate $f_v(U_{v,m}(t))$ for sessions in the mth VBS of class v at time t.

To guarantee that active sessions always have enough r-servers and c-servers, the VBS pool has to enforce admission control on arriving sessions: for class-v sessions, the new session is accepted only if the number of r-servers in its serving VBS is less than K_v and the number of c-servers in the pool is less than N, otherwise the session is blocked.

State Transitions

We describe the session dynamics in the VBS pool by a continuous-time stochastic process

$$U(t) = [U_{1,1}(t) \cdots U_{1,M_1} \cdots U_{V,1} \cdots U_{V,M_V}(t)]^T.$$

Given the Markovian property of the arrival and service of processes, it is obvious that $U(t)$ is a Markov chain. Taking the admission control policy into account, we obtain the set of possible system states:

$$U(t) \in \mathcal{U} = \left\{ u0 \le u_{v,m} \le K_v, 0 \le \sum_{v=1}^{V} \sum_{m=1}^{M_v} u_{v,m} \le N, u_{v,m} \in \mathbb{N} \right\}, \qquad (14.1)$$

where $u = [u_{1,1} \cdots u_{1,M_1} \cdots u_{V,1} \cdots u_{V,M_V}]^T$ is the state vector. Because the session arrivals and departures are Markovian, $U(t)$ is a multi-dimensional birth-and-death process. The transition rate of $U(t)$ from an arbitrary state $u^{(i)}$ to state $u^{(j)}$ is:

$$q_{u^{(i)}u^{(j)}} = \begin{cases} \lambda_v, & \text{if } u^{(j)} - u^{(i)} = e_{v,m}, \\ f_v(u_{v,m}^{(i)}), & \text{if } u^{(j)} - u^{(i)} = -e_{v,m}, \\ 0, & \text{otherwise}, \end{cases} \qquad (14.2)$$

where $u_{v,m}^{(i)}$ is the $(\sum_{w=1}^{v-1} M_w + m)$th entry of $u^{(i)}$, and

$$e_{v,m} = [0 \cdots 1\ 0 \cdots 0]^T$$

is a column vector, of length $\sum_{v=1}^{V} M_v$, of which the $\left(\sum_{w=1}^{v-1} M_w + m \right)$th element is 1.

We illustrate the state transition graph for a simple example with only one class of VBSs, $V = 1$, and set $M_1 = 2$, $K_1 = 3$, $N = 4$ in Fig. 14.2. Here the VBSs serve a voice traffic of exponentially distributed call durations, with mean call duration $1/\mu$, and thus we have the session departure rate $f_1(n) = n\mu$, where n is the number of active sessions. In the figure, each state (in indicated by a shaded ellipse) indicates the number of active sessions in these two VBSs, respectively, and the states in the hatched region are prohibited because of the computational and radio resource constraints.

Our model is mathematically equivalent to the stochastic knapsacks problem [17] under *coordinate convex* admission control policies [18]. While the existing analysis

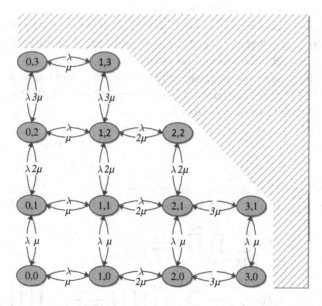

Figure 14.2 Transition graph of a VBS pool with two VBSs.

was limited to problems with small dimensionality, we aim at evaluating the blocking probability under high-dimensional problems, which corresponds to the generally large size of the VBS pools in C-RANs.

Stationary Distribution

The model that we formulated guarantees the reversibility of $U(t)$ [19], which in turn results in a product-form expression for the stationary distribution of $U(t)$ as

$$P\{u\} = P_0 \prod_{v=1}^{V} \prod_{m=1}^{M_v} \frac{\lambda_v^{u_{v,m}}}{\prod_{i=1}^{u_{v,m}} f_v(i)}, \tag{14.3}$$

in which

$$P_0 = P\{0,\ldots,0,\ldots,0\} = \left(\sum_{u \in \mathcal{U}} \prod_{v=1}^{V} \prod_{m=1}^{M_v} \frac{\lambda_v^{u_{v,m}}}{\prod_{i=1}^{u_{v,m}} f_v(i)} \right)^{-1}. \tag{14.4}$$

It is also proved in [19] that the product-form distribution is valid for any service-time distributions with rational Laplace transforms, including the exponentially distributed service demand considered in this chapter.

14.2.2   Session Blocking Probability

The admission control policy enforced on the VBS pool will cause two kinds of session blocking: radio blocking (denoted by B_r) and computational blocking (denoted by B_c). Radio blocking is *solely* due to insufficient r-servers, i.e. $U_{v,m}(t^-) = K$ and $\sum_{v=1}^{V} \sum_{m=1}^{M_v} U_{v,m}(t^-) < N$, while computational blocking is due to insufficient c-servers *regardless of* r-servers, i.e. $\sum_{v=1}^{V} \sum_{m=1}^{M_v} U_{v,m}(t^-) = N$. Here t^- means the epoch right before a session arrival. Note that because we define radio blocking as an event that is solely due to insufficient r-servers, blocking events due to simultaneously insufficient r-servers and c-servers are explicitly classified as computational blocking. Therefore, radio and computational blocking events are mutually exclusive, i.e. $B_r \cap B_c = \varnothing$ and thus the overall blocking $B = B_r \cup B_c$.

Next we derive expressions for the probabilities of radio and computational blocking. Since "Poisson arrivals see time-averages" (PASTA) [20], the radio blocking probability for sessions in class-v VBSs is

$$P_v^{\text{br}} = \sum_{m=1}^{M_v} \frac{1}{M_v} \sum_{u \in \mathcal{U}_{\text{br}}^{v,m}} P\{u\}, \tag{14.5}$$

where $\mathcal{U}_{\text{br}}^{v,m} = \{u \mid u_{v,m} = K_v, u_{1,1} + \cdots + u_{1,M_1} + \cdots + u_{V,1} + \cdots + u_{V,M_V} < N\}$. From expressions (14.3), (14.4) for the stationary distribution, we further have

$$P_v^{\text{br}} = P_0 \sum_{u \in \mathcal{U}_{\text{br}}^{v,1}} \prod_{w=1}^{V} \prod_{m=1}^{M_w} \frac{\lambda_w^{u_{w,m}}}{\prod_{i=1}^{u_{w,m}} f_w(i)} \tag{14.6}$$

$$= P_0 \frac{\lambda_v^{K_v}}{\prod_{i=1}^{K_v} f_v(i)} \sum_{u \in \mathcal{U}_{\text{br}}^{v,1}} \left[\left(\prod_{w \neq v} \prod_{m=1}^{M_w} \frac{\lambda_w^{u_{w,m}}}{\prod_{i=1}^{u_{w,m}} f_w(i)} \right) \left(\prod_{m=2}^{M_v} \frac{\lambda_v^{u_{v,m}}}{\prod_{i=1}^{u_{v,m}} f_v(i)} \right) \right], \tag{14.7}$$

Equation (14.6) holds because (14.3) is symmetric for entries with the same index v, i.e.

$$P\left\{\ldots, u_{v,i}, \ldots, u_{v,j}, \ldots\right\} = P\left\{\ldots, u_{v,j}, \ldots, u_{v,i}, \ldots\right\}. \tag{14.8}$$

Similarly, the computational blocking probability is:

$$P^{bc} = \sum_{u \in \mathcal{U}_{bc}^N} P\{u\} = P_0 \sum_{u \in \mathcal{U}_{bc}^N} \prod_{v=1}^{V} \prod_{m=1}^{M_v} \frac{\lambda_v^{u_{v,m}}}{\prod_{i=1}^{u_{v,m}} f_v(i)}, \tag{14.9}$$

where $\mathcal{U}_{bc}^N = \{u \mid u_{1,1} + \cdots + u_{1,M_1} + \cdots + u_{V,1} + \cdots + u_{V,M_V} = N, u_{v,m} \leq K_v\}$. The overall blocking probability for class-v VBSs can then be evaluated by summing the radio and computational blocking probabilities:

$$P_v^b = P\{B\} = P_v^{br} + P^{bc}. \tag{14.10}$$

Recursive Evaluation

The calculation complexity of (14.7) and (14.9) increases exponentially with the pool size. To reduce the complexity, we provide a recursive method to calculate the blocking probability. We will first introduce two auxiliary functions and reexpress the blocking probability with these functions. Then we will establish a recursive relationship to evaluate those two auxiliary functions.

These two auxiliary functions are defined as follows

$$C(N, M) = \sum_{u \in \mathcal{U}_{bc}^N} \prod_{w=1}^{V} \prod_{m=1}^{M_w} \frac{\lambda_w^{u_{w,m}}}{\prod_{i=1}^{u_{w,m}} f_w(i)}, \tag{14.11}$$

$$R(N, M) = \sum_{u \in (\mathcal{U}_{bc}^N)^C} \prod_{w=1}^{V} \prod_{m=1}^{M_w} \frac{\lambda_w^{u_{w,m}}}{\prod_{i=1}^{u_{w,m}} f_w(i)}, \tag{14.12}$$

where $M = [M_1 \cdots M_v \cdots M_V]^T$ and $\left(\mathcal{U}_{bc}^N\right)^C = \{u \mid u_{1,1} + \cdots + u_{1,M_1} + \cdots + u_{V,1} + \cdots + u_{V,M_V} < N, u_{v,m} \leq K_v\}$ is the complement set of set $\mathcal{U}_{bc}^N$ in set $\mathcal{U}$. Clearly, $C(N, M)$ and $R(N, M)$ are proportional to the sum of probability terms over $\mathcal{U}_{bc}^N$ and $\left(\mathcal{U}_{bc}^N\right)^C$, respectively. Therefore, the blocking probabilities (14.7) and (14.9) can be reexpressed as

$$P_v^{br} = P_0 \frac{\lambda_v^{K_v}}{\prod_{i=1}^{K_v} f_v(i)} R(N - K_v, M - \hat{e}_v),$$

$$P^{bc} = P_0 C(N, M), \tag{14.13}$$

$$P_0 = R^{-1}(N + 1, M),$$

where $\hat{e}_v = [0 \cdots 0\, 1\, 0 \cdots 0]^T$ is a column vector of length V, whose vth element is 1.

From the definition of $C(N, M)$ and $R(N, M)$, the following recursive relationships exist:

$$C(N, M) = \begin{cases} \dfrac{\lambda_v^{N_2(v)}}{\prod_{i=1}^{N_2(v)} f_v(i)}, & M = \hat{e}_v, \\[4ex] \displaystyle\sum_{n=N_1(v)}^{N_2(v)} \dfrac{\lambda_v^n}{\prod_{i=1}^{n} f_v(i)} C(N - n, M - \hat{e}_v), & M > \hat{e}_v, \end{cases} \tag{14.14}$$

$$R(N, M) = \begin{cases} 0, & N = 1, \\[1ex] R(N + 1, M) - C(N, M), & 1 < N < M^T K + 1, \\[2ex] \displaystyle\prod_{w=1}^{V} \left(\sum_{n=1}^{K_w} \dfrac{\lambda_w^n}{\prod_{i=1}^{n} f_w(i)} \right)^{M_w}, & N = M^T K + 1, \end{cases} \tag{14.15}$$

where

$$N_1(v) = \max\left(0, N - \sum_{w \neq v} M_w K_w - (M_v - 1) K_v \right),$$

$$N_2(v) = \min(K_v, N),$$

and $M^T K = \sum_{w=1}^{V} M_w K_w$. The recursion in (14.14) can start from any $\hat{e}_v$, and the recursion in (14.15) can start from either $N = 0$ or $N = M^T K + 1$. Note that the comparison between M and $\hat{e}$ in (14.14) is element-wise, and, accordingly, M is always greater or equal to $\hat{e}$ in non-empty pools.

Large-Pool Approximation

To enable further analysis, we present a closed-form approximation for the blocking probability for large VBS pools. Although the above recursive expression cannot lead us to a direct approximation, the product-form stationary distribution of U does provide an implicit approximation.

First, we define some auxiliary variables. Let $\tilde{U}_{w,m}$ be the number of active sessions in the mth class-w VBS when $N \geq M^T K$; then its mean and variance for class-w VBSs are defined as

$$\xi_w = \mathbb{E}\left\{ \tilde{U}_{w,m} \right\}, \quad \sigma_w^2 = \text{Var}\left\{ \tilde{U}_{w,m} \right\}.^1$$

Also, let $\tilde{S}_M = |M|^{-1} \sum_{w=1}^{V} \sum_{m=1}^{M_w} \tilde{U}_{w,m}$ and $\tilde{S}_{M_w} = M_w^{-1} \sum_{m=1}^{M_w} \tilde{U}_{w,m}$, indicating the average number of active sessions per VBS at a given time instance for all VBSs and for class-w VBSs, respectively. With this notation, the large-pool approximation is stated in the following theorem.

---

[1] It should be obvious that $\tilde{U}_{w,m}$ are i.i.d random variables for $m = 1, 2, \ldots, M_w$.

THEOREM 14.1 (Large-pool blocking probability) *For* $N > |M|\xi$, *the session blocking probability for class-v VBSs is*

$$\lim_{|M|\to\infty} P_v^b = \frac{1}{\sqrt{2\pi|M|\sigma^2}}\frac{1}{e^{\alpha^2/2}-1} + \tilde{P}_v^{br}, \tag{14.16}$$

where $\xi = \sum_{w=1}^{V}\beta_w\xi_w$, $\sigma^2 = \sum_{w=1}^{V}\sigma_w^2$, *provided that* $\beta_w = \lim_{|M|\to\infty}M_w/|M|$; $\alpha = (N - |M|\xi)/\sqrt{|M|}\sigma$ *is the normalized number of c-servers;* $\tilde{P}_v^{br}$ *is the overall blocking probability in class-v VBSs when* $N > M^T K$.

Proof From the definition $\tilde{S}_M = |M|^{-1}\sum_{w=1}^{V}\sum_{m=1}^{M_w}\tilde{U}_{w,m}$, we have

$$\lim_{|M|\to\infty}\tilde{S}_M = \lim_{|M|\to\infty}\frac{1}{|M|}\sum_{w=1}^{V}\sum_{m=1}^{M_w}\tilde{U}_{w,m} = \lim_{|M|\to\infty}\sum_{w=1}^{V}\frac{M_w}{|M|}\frac{\sum_{m=1}^{M_w}\tilde{U}_{w,m}}{M_w}$$

$$= \lim_{|M|\to\infty}\sum_{w=1}^{V}\beta_w\tilde{S}_{M_w}. \tag{14.17}$$

According to the central limit theorem, when $|M|\to\infty$, $\tilde{S}_{M_w}$ converges in distribution to a normal random variable:

$$\lim_{M_v\to\infty}\tilde{S}_{M_w} \sim \mathcal{N}\left(\xi_w, \frac{\sigma_w^2}{M_w}\right). \tag{14.18}$$

Since the $\tilde{U}_{v,m}$ are independent random variables for all v and m, the $\tilde{S}_{M_v}$ are also independent. Therefore $\tilde{S}_M$ will also converge to a normal random variable:

$$\lim_{|M|\to\infty}\tilde{S}_M = \lim_{|M|\to\infty}\sum_{w=1}^{V}\beta_w\tilde{S}_w \sim \mathcal{N}\left(\sum_{w=1}^{V}\beta_w\xi_w, \sum_{w=1}^{V}\beta_w\frac{\sigma_w^2}{M_w}\right)$$

$$\sim \mathcal{N}\left(\xi, \frac{\sigma^2}{|M|}\right). \tag{14.19}$$

To express the blocking probability in terms of this normal distribution, we next establish a relationship between the stationary distributions of U and $\tilde{U}$. Let $\tilde{P}_0$ be the probability of the zero state of $\tilde{U}$; then the product-form stationary distributions of U and $\tilde{U}$ result in the following scaling relationship between the stationary distributions of U and $\tilde{U}$:

$$\frac{P\{U = u\}}{P_0} = \frac{P\{\tilde{U} = u\}}{\tilde{P}_0}. \tag{14.20}$$

From the definitions of P_0 and $\tilde{P}_0$, the following relationship exists:

$$\frac{P_0}{\tilde{P}_0} = P\left\{\sum_{w=1}^{V}\sum_{m=1}^{M_w}\tilde{U}_{w,m} \le N\right\}^{-1} = P\left\{\frac{1}{|M|}\sum_{w=1}^{V}\sum_{m=1}^{M_w}\tilde{U}_{w,m} \le \frac{N}{|M|}\right\}^{-1}$$

$$= P\left\{\tilde{S}_M \le \frac{N}{|M|}\right\}^{-1}. \tag{14.21}$$

Notice that, in the above relationship, $P_0/\tilde{P}_0$ is determined by the probability distribution of $\tilde{S}_M$. Therefore we can use the large-pool limit of $\tilde{S}_M$ to get the following approximation:

$$\lim_{|M|\to\infty} \frac{P_0}{\tilde{P}_0} = \left[1 - Q\left(\frac{N}{|M|}\right)\right]^{-1}, \tag{14.22}$$

where $q(x)$ and $Q(x)$ are respectively the probability density function (PDF) and cumulative tail distribution of $\mathcal{N}(\xi, \sigma^2/|M|)$. We can then approximate the blocking probabilities:

$$\begin{aligned}
\lim_{|M|\to\infty} P^{bc} &= \lim_{|M|\to\infty} P\left\{\sum_{w=1}^{V}\sum_{m=1}^{M_w} U_{w,m} = N\right\} \\
&= \lim_{|M|\to\infty} \frac{P_0}{\tilde{P}_0} P\left\{\sum_{w=1}^{V}\sum_{m=1}^{M_w} \tilde{U}_{w,m} = N\right\} \\
&= \lim_{|M|\to\infty} \frac{P_0}{\tilde{P}_0} P\left\{\tilde{S}_M = \frac{N}{|M|}\right\} \\
&= \frac{P_0}{\tilde{P}_0} \frac{1}{|M|} q\left(\frac{N}{|M|}\right)
\end{aligned} \tag{14.23}$$

and

$$\begin{aligned}
\lim_{|M|\to\infty} P_v^{br} &= \lim_{|M|\to\infty} P\left\{U_{v,1} = K_v, \sum_{w=1}^{V}\sum_{m=1}^{M_w} U_{w,m} < N\right\} \\
&= \lim_{|M|\to\infty} \frac{P_0}{\tilde{P}_0} P\left\{\tilde{U}_{v,1} = K_v, \sum_{w=1}^{V}\sum_{m=1}^{M_w} \tilde{U}_{w,m} < N\right\} \\
&= \lim_{|M|\to\infty} \frac{P_0}{\tilde{P}_0} P\{\tilde{U}_{v,1} = K_v\} P\left\{\sum_{m=2}^{M_v} \tilde{U}_{v,m} + \sum_{w\neq v}\sum_{m=1}^{M_w} \tilde{U}_{w,m} < N - K_v\right\} \\
&= \lim_{|M|\to\infty} \frac{P_0}{\tilde{P}_0} \tilde{P}_v^{br} \left\{\tilde{S}_M - \frac{\tilde{U}_{v,1}}{|M|-1} < \frac{N-K_v}{|M|-1}\right\} \\
&= \frac{P_0}{\tilde{P}_0} \tilde{P}_v^{br} \left\{1 - Q\left(\frac{N}{|M|}\right)\right\}.
\end{aligned} \tag{14.24}$$

Notice that the last equality of (14.24) holds because, since $|M| \to \infty$, N should also approach infinity as $N > |M|\xi$. Hence

$$\lim_{|M|\to\infty} Q\left(\frac{N-K_v}{|M|}\right) = Q\left(\frac{N}{|M|}\right). \tag{14.25}$$

Also, $\lim_{|M|\to\infty} Q(N/|M|) = e^{-\alpha^2/2}$. Therefore, the approximation for the overall session blocking probability of class-v VBSs is

$$\lim_{|M|\to\infty} P_v^b = \lim_{|M|\to\infty} (P^{bc} + P_v^{br})$$

$$= \left[1 - Q\left(\frac{N}{|M|}\right)\right]^{-1} \left\{\frac{1}{|M|}q\left(\frac{N}{|M|}\right) + \tilde{P}_v^{br}\left[1 - Q\left(\frac{N}{|M|}\right)\right]\right\} \qquad (14.26)$$

$$= \frac{1}{\sqrt{2\pi|M|\sigma^2}} \frac{1}{e^{\alpha^2/2} - 1} + \tilde{P}_v^{br}. \qquad \square$$

Remark 14.1   For ease of understanding, we will explain the physical meaning of the notation in Theorem 14.1. The meaning of setting $\beta_w = \lim_{|M|\to\infty} M_w/|M|$ is that we keep the proportion of class-v VBSs constant whenever the total number of VBSs $|M|$ changes. From on the definition of β_w, we define the weighted mean of the number of sessions in the pool as $\xi = \sum_{w=1}^{V} \beta_w\xi_w$ and the corresponding variance as $\sigma^2 = \sum_{w=1}^{V} \sigma_w^2$. In addition, in (14.19) shows that the average number of active sessions per VBS is a normal distributed random variable, i.e., $\lim_{|M|\to\infty} \tilde{S}_M \sim \mathcal{N}(\xi, \sigma^2/|M|)$, then

$$\alpha = \frac{N - |M|\xi}{\sqrt{|M|}\sigma}$$

is the number of c-servers N, normalized with respect to the mean and squared variance of this normal distribution.

Remark 14.2   In the large-pool regime, the blocking probability (14.16) is decomposed into two terms. The first term,

$$\frac{1}{\sqrt{2\pi|M|\sigma^2}} \frac{1}{e^{\alpha^2/2} - 1},$$

reflects the portion of blocking due solely to insufficient computational resources, while the second term $\tilde{P}_v^{br}$ reflects the portion due merely to insufficient radio resources. This structure reveals the decoupling feature between radio and computational blocking in large VBS pools.

Remark 14.3   Although we assumed that $K_v < \infty$ in our derivation, (14.16) is still true when $K \to \infty$. In this case, the approximation used in (14.25) may not hold any more. But this will not cause any problem since the radio blocking probability $\tilde{P}_v^b$ will be 0 when there are infinite radio resources. This will force the second term of (14.16) to be zero, canceling out any inconsistency due to approximation.

14.2.3   Statistical Multiplexing Gain of Centralization in C-RAN

Since user sessions from different VBSs are consolidated, one can expect a reduction in the required amount of computational resource compared with non-pooling schemes, owing to statistical multiplexing. Next we provide a theoretical analysis for the statistical multiplexing gain. Notice that, if $N > M^T K$, then $N - M^T K$ c-servers are definitely wasted and thus, from now on, we consider only the case $N \le M^T K$.

We first derive the asymptotic utilization ratio for the computational resources in large VBS pools.

THEOREM 14.2 (Large-pool utilization)   *When $|M| \to \infty$, given sufficient c-servers $(N = M^T K)$, the utilization ratio of the computational resources converges almost surely to a constant that is smaller than 1:*

$$\lim_{|M| \to \infty} \eta \triangleq \frac{\sum_{w=1}^{V} \sum_{m=1}^{M_w} \tilde{U}_{w,m}}{M^T K} \xrightarrow{a.s.} \frac{|M|\xi}{M^T K} < 1. \tag{14.27}$$

Proof   The first part of the proof is straightforward using (14.19). Since $\tilde{S}_M$ will also converge to a normal distributed random variable $\mathcal{N}(\xi, \sigma^2/|M|)$ as $|M| \to \infty$, according to the strong law of large numbers we have

$$P\left\{ \lim_{|M| \to \infty} \eta = \frac{|M|\xi}{M^T K} \right\} = P\left\{ \lim_{|M| \to \infty} \frac{\sum_{w=1}^{V} \sum_{m=1}^{M_w} \tilde{U}_{w,m}}{M^T K} = \frac{|M|\xi}{M^T K} \right\}$$

$$= P\left\{ \lim_{|M| \to \infty} \tilde{S}_M = \xi \right\} = 1. \tag{14.28}$$

Hence

$$\eta \xrightarrow{a.s.} \frac{|M|\xi}{M^T K}.$$

Also, it is easy to see that $U_{v,m} \le K_v$ and $P\left\{ U_{v,m} < K_v \right\} > 0$. Therefore $\xi_v < K_v$ and

$$\xi = \sum_{w=1}^{V} \xi_w \beta_w < \sum_{w=1}^{V} K_w \beta_w = \frac{\sum_{w=1}^{V} K_w M_w}{|M|} = \frac{M^T K}{|M|}. \tag{14.29}$$

Thus

$$\frac{|M|\xi}{M^T K} < 1.$$

$\square$

This theorem implies that a fraction $1 - \eta$ portion of the computational resources is redundant when the VBS pool is large enough. This limit can also be regarded as the maximum portion of c-servers that one can turn down in order to save computational resources. In what follows, the potential to further turn down c-servers, given the total number of c-servers N, is defined as the difference between the current utilization ratio of c-servers $N/(M^T K)$ and the large-pool limit η.

DEFINITION 14.3 (Residual pooling gain)   Given the total number of c-servers $N \in [\eta M^T K, M^T K]$, the residual pooling gain of a VBS pool is

$$g_r \triangleq \frac{N}{M^T K} - \eta. \tag{14.30}$$

Although some c-servers can be turned down owing to the statistical multiplexing effect, the negative effect is that the overall blocking probability P^b will increase, as can be seen from (14.16). Hence we have to trade QoS for a statistical multiplexing gain, which can still be favorable as long as the QoS degradation is not very significant. Via Theorem 14.1, we can directly derive the following corollary to quantify such "significance" and approximate the gain of VBS pools.

COROLLARY (Knee point) *When $|M| \to \infty$, the minimum normalized computational resource α^* required to keep the overall blocking probability $P_v^b \leq \tilde{P}_v^{br} + \delta$ ($\delta \approx 0$) for all v is*

$$\alpha^* = \sqrt{2 \ln \left(\frac{1}{\sqrt{2\pi |M| \sigma^2 \delta^2}} + 1 \right)}. \tag{14.31}$$

This point is independent of the VBS class index v. From Fig. 14.3 it can be seen that this "knee point" is essentially the point where the blocking probability start to increase at a significantly higher speed as we reduce N or, in other words, the blocking probability hardly decreases if we increase N.

The residual pooling gain at this critical tradeoff point is bounded as follows:

$$g_r^* = \frac{N - |M|\xi}{M^T K} = \sigma \frac{\alpha^* \sqrt{|M|}}{M^T K} \in \frac{\alpha^*}{\sqrt{|M|}} \left[\frac{\sigma}{\max_v K_v}, \frac{\sigma}{\min_v K_v} \right]. \tag{14.32}$$

Thus g_r^* is roughly proportional to $\alpha^* / \sqrt{|M|}$. Note that α^* is a function of the pool size $|M|$, so g_r^* is not necessarily proportional to $1/\sqrt{|M|}$. Investigating two extreme cases will help to reveal the true relationship between g_r^* and the pool size $|M|$.

Extreme case 1 If $|M|$ is not very large, so that $\sqrt{2\pi |M| \sigma^2 \delta^2} \ll 1$ $\sqrt{|M|} \ll 1/\delta^2$, then α^* is approximately a constant because

$$\alpha^* \approx \sqrt{2 \ln \left(\frac{1}{\sqrt{2\pi |M| \sigma^2 \delta^2}} \right)} = \sqrt{\ln \left(\frac{1}{2\pi \sigma^2} \right) + \ln \left(\frac{1}{\delta^2} \right) + \ln \left(\frac{1}{|M|} \right)}$$

$$\approx \sqrt{\ln \left(\frac{1}{2\pi \sigma^2 \delta^2} \right)}. \tag{14.33}$$

Therefore in this case $g_r^* \propto |M|^{-1/2}$, which decreases slowly with $|M|$. Even so, regarding the fact that the residual pooling gain is at most 1, we can still get a considerable pooling gain with a small value of $|M|$.

Extreme case 2 If $|M|$ is very large, so that $\sqrt{2\pi |M| \sigma^2 \delta^2} \gg 1$, since $\lim_{x \to 0} \ln(1 + x) \approx x$ we have

$$\alpha^* \approx \sqrt{2 \frac{1}{\sqrt{2\pi |M| \sigma^2 \delta^2}}} \propto |M|^{-1/4}. \tag{14.34}$$

In this case $g_r^* \propto |M|^{-3/4}$, which means that the decrease in the residual pooling gain will speed up as $|M|$ grows large.

14.2.4 Application Scenarios of the Session-Level VBS Model

In this section we will apply the derived analysis to two specific traffic models: real-time traffic and delay-tolerant traffic. For each scenario, we will first explain how they can be mapped to our model and then we will illustrate the corresponding results.

Real-Time Traffic

For real-time traffic such as voice calls, active sessions will bring in a signal processing workload constantly. Therefore, dedicated r-servers and c-servers need to be provisioned upon admission to guarantee the QoS of active sessions. As a result, the departure rate function can be simplified to $f_v(i) = i\mu_v$, where i is the number of active sessions in class-v VBSs and $1/\mu_v$ is the mean call duration. Note that here we consider an exponentially distributed call duration. The QoS target in this case is to keep the overall session blocking probability for class-v VBSs under a small threshold $P_v^{\text{bth}} \approx 0$. Obviously, the session dynamics in different VBSs are mutually independent when the computational resources are sufficiently provisioned ($N = M^T K$). Therefore the radio blocking probability $\tilde{P}_v^{\text{br}}$ is

$$\tilde{P}_v^{\text{br}} = \frac{a_v^{K_v}}{K_v!} \left(\sum_{i=0}^{K_v} \frac{a_v^i}{i!} \right)^{-1} \leq P_v^{\text{bth}} \approx 0, \tag{14.35}$$

where $a_v = \lambda_v/\mu_v$. Then the first-order and second-order statistics of $\tilde{U}_{v,m}$ can be approximated as follows:

$$\xi_v = \mathbb{E}\left\{ \tilde{U}_{v,m} \right\} = \frac{\sum_{i=0}^{K_v} i a_v^i/i!}{\sum_{i=0}^{K_v} a_v^i/i!} = a_v \left[1 - \frac{a_v^{K_v}}{K_v!} \left(\sum_{i=0}^{K_v} \frac{a_v^i}{i!} \right)^{-1} \right] \approx a_v, \tag{14.36}$$

$$\mathbb{E}\left\{ \tilde{U}_{v,m}^2 \right\} = \frac{\sum_{i=0}^{K_v} i^2 a_v^i/i!}{\sum_{i=0}^{K_v} a_v^i/i!} = \frac{a_v \left(\sum_{i=0}^{K_v-1} a_v^i/i! + a_v \sum_{i=0}^{K_v-2} a_v^i/i! \right)}{\sum_{i=0}^{K_v} a_v^i/i!} \approx a_v + a_v^2. \tag{14.37}$$

Thus we have

$$\sigma_v^2 = a_v. \tag{14.38}$$

Delay-Tolerant traffic

For delay-tolerant traffic such as packet data, the pool scheduler can opportunistically divide the total service capacity among active sessions. Here we will assume a constant service capacity rate $f_v(i) = \mu_v$ for class-v VBSs and that the capacity is equally allocated to active sessions with round-robin-type scheduling schemes. This corresponds to a processor-sharing service model; it is essentially equivalent to a Markovian queueing model with the same λ_v and μ_v.

To derive the statistics in this scenario, first let $a_v = \lambda_v/\mu_v$ be the traffic load of the class-v VBSs and define the following auxiliary function $A(a, K)$:

$$A(a, K) = \sum_{i=0}^{K} a^i = \frac{1 - a^{K+1}}{1 - a}. \tag{14.39}$$

Accordingly,

$$A'_a(a, K) = \left(\sum_{i=0}^{K} a^i\right)'_a = \sum_{i=1}^{K} ia^{i-1} = \frac{1 - (K+1)a^K + Ka^{K+1}}{(1-a)^2},$$

$$A''_a(a, K) = \left(\sum_{i=0}^{K} a^i\right)''_a = \sum_{i=2}^{K} i(i-1)a^{i-2}.$$

(14.40)

With these definitions, the average and covariance of $\tilde{U}_{v,m}$ can be expressed as

$$\mathbb{E}\left\{\tilde{U}_{v,m}\right\} = \frac{\sum_{i=1}^{K_v} ia_v^i}{\sum_{i=0}^{K_v} a_v^i} = \frac{a_v A'_a(a_v, K_v)}{A(a_v, K_v)},$$

(14.41)

$$\mathbb{E}\left\{\tilde{U}_{v,m}^2\right\} = \frac{\sum_{i=1}^{K_v} i^2 a_v^i}{\sum_{i=0}^{K_v} a_v^i} = \frac{a_v A'_a(a_v, K_v) + a_v^2 A''_a(a_v, K_v)}{A(a_v, K_v)}.$$

(14.42)

Again, when the computational resources are sufficiently provisioned ($N = M^T K$), we have

$$\tilde{P}_v^{\text{br}} = \frac{a_v^k}{\sum_{i=0}^{K_v} a_v^i} = \frac{a_v^k}{A(a_v, K_v)}.$$

(14.43)

Although these formulas are already enough for us to evaluate the performance of a VBS pool, the evaluation is nevertheless quite cumbersome. To simplify the formulas we further assume that K_v for all v is large enough that $K_v^2 a_v^{K_v} \to 0$, which is realistic because $a_v^{K_v}$ will diminish exponentially when $a_v < 1$. In this regime,

$$A(a, K) \approx \frac{1}{1-a}, \quad A'_a(a, K) \approx \frac{1}{(1-a)^2}, \quad A''_a(a, K) \approx \frac{2}{(1-a)^3}.$$

(14.44)

Using (14.44), (14.41) and (14.42) can be simplified as follows:

$$\xi_v = \mathbb{E}\left\{\tilde{U}_{v,m}\right\} \approx \frac{a_v}{1-a_v}, \quad \mathbb{E}\left\{\tilde{U}_{v,m}^2\right\} \approx \frac{a_v}{1-a_v} + \frac{2a_v^2}{(1-a_v)^2}.$$

(14.45)

Thus

$$\sigma_v^2 \approx \frac{a_v}{1-a_v} + \frac{a_v^2}{(1-a_v)^2}.$$

(14.46)

14.2.5   Numerical Results and Deployment Insights

In this section, we will use the recursive method to numerically evaluate the blocking probability and compare the results with those for the large-pool approximation, in order to evaluate our model and derivations. Moreover, on the basis of observations, we will justify the relationship between fronthaul expenditure and system performance, from which we will provide some hints for C-RAN deployment.

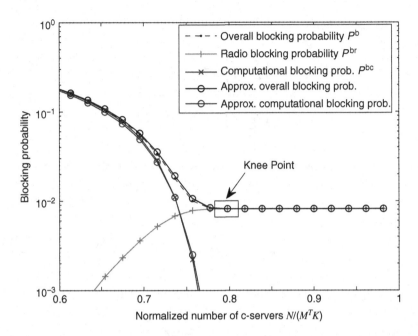

Figure 14.3 Blocking probability for a homogeneous VBS pool as a function of the normalized number of c-servers $(N/M^T K)$ under real-time traffic. Parameters: $M_1 = 50$, $a_1 = 32$, $P_1^{bth} = 10^{-2}$, $K_1 = 44$.

Basic Characteristics

Figure 14.3 shows the exact blocking probability and its large-pool approximation under real-time traffic and different numbers of c-servers, and Fig. 14.4 shows the same metrics for a VBS pool under delay-tolerant traffic. For both figures we varies the number of c-servers N, and this number is normalized by $M^T K$ and plotted along the x-axis. The sufficient deployment of c-servers corresponds to $N = M^T K$. Note that in conventional cellular systems where no computing multiplexing is enabled, the computational resources are statistically matched to the radio resources in BSs, which in our model corresponds to $N = M^T K$ as well. Therefore $N/M^T K$ can also be regarded as the ratio of the required computational resource and that of the conventional cellular system. As can be seen, the trends in both figures are similar. This coincides with our large-pool approximation: that the blocking probabilities are affected only by the first- and second-order statistics of the number of sessions in the VBS pool. Therefore, we will present results for real-time traffic only and the conclusions should apply to the delay-tolerant case as well.

First, Figs. 14.3 and 14.4 prove that the results large-pool approximation using are coherent with the exact values. Moreover, we have the following key observations from these two figures. (1) When the number of c-servers is sufficient, the computational blocking probability P^{bc} is very small and below the scale of this figure; the overall blocking probability is dominated by the radio blocking P^{br}, which is around the desired threshold P^{bth}. (2) As the number of c-servers decreases from

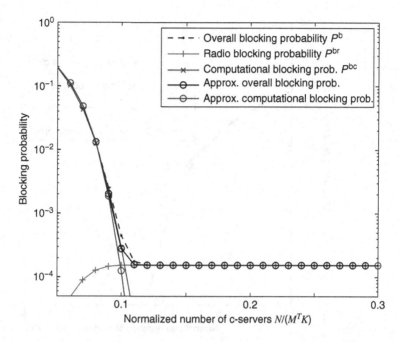

Figure 14.4 Blocking probability for a homogeneous VBS pool as a function of the normalized number of c-servers ($N/M^T K$) under delay-tolerant traffic. Parameters: $M_1 = 100$, $a_1 = 0.3$, $P_1^{bth} = 5 \times 10^{-4}$, $K_1 = 7$.

$M^T K$, the computational blocking probability increases while the radio blocking probability starts to decrease slightly; the joint effect of these two trends is a plateau to the right of the "knee point" and a significant increase to the left. (3) If the number of c-servers decreases further, the overall blocking probability will be dominated by the computational blocking, while the influence of radio blocking on the overall blocking probability will diminish. From these observations, *it is intuitive to deploy the number of c-servers corresponding to the "knee point"*, in order to achieve the optimal tradeoff between computational resource saving and the blocking probability.

Heterogeneous VBSs

In Fig. 14.5, we show the blocking probability of a VBS pool with two classes of VBSs. The two classes have the same number of VBSs and the same traffic load, but the QoS requirements and the corresponding numbers of provisioned r-servers are different. A similar trend in the blocking probability as in the single-class case is observed. Also, we can see that the overall blocking probabilities for the two classes are different: they are respectively close to their threshold blocking probabilities when the c-servers are sufficient, since the overall blocking probabilities are dominated by radio blocking, but converge to the same curve when the c-servers become insufficient and the computational blocking probability begins to be overwhelming.

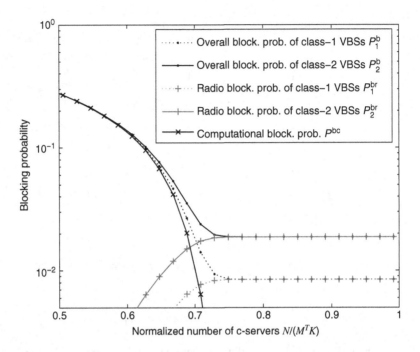

Figure 14.5 Blocking probability of a heterogeneous VBS pool as a function of the normalized number of c-servers ($N/M^T K$) under real-time traffic. Parameters: $M = [23\ 23]^T$, $a = [20\ 20]^T$, $P^{\mathrm{bth}} = [1\ 2]^T \times 10^{-2}$, $K = [30\ 28]^T$.

Statistical Multiplexing Gain and Deployment Principles

One of the most important features of our model is that it is able to quantify the statistical multiplexing gain of the simulated VBS pool. We find that, as the pool size increases, the blocking probability curve (and therefore the "knee point") is pushed to the left. But, the closer the "knee point" is to the large-pool limit, the more slowly does the distance to the large-pool limit decrease with the pool size $|M|$. This indicates a decreasing marginal statistical multiplexing gain. Equivalently, increasing the pool size $|M|$ significantly enhances neither the saving in computational resources nor the system performance in terms of blocking probability.

To further investigate this influence, we show the knee-point position versus varying pool size in Fig. 14.6. In this figure, the dotted line represents the large-pool approximation whereas the solid line is obtained directly from the numerical results; the dashed line represents the large-pool limit η. The percentage of the pooling gain at 50 VBSs with respect to the maximum $1 - \eta$ is shown by vertical arrows in the figure. First, we can see that on the one hand, a medium-sized VBS pool can readily obtain considerable statistical multiplexing gain and the marginal gain diminishes fast. On the other hand, a huge number of VBSs is needed for the pooling gain to approach the large-pool limit. These observations imply that a C-RAN formed with multiple medium-sized VBS pools can obtain almost the same pooling gain as one formed by a single huge pool, which brings good news for the deployment cost, especially of the fronthaul. In other words, if

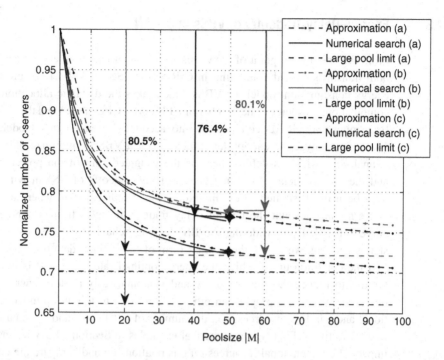

Figure 14.6 Knee-point position (in terms of the normalized number of c-servers) as a function of pool size. Parameters: (a) $a = 20$, $K = 30$, $P^{\text{bth}} = 0.01$; (b) $a = 32$, $K = 44$, $P^{\text{bth}} = 0.01$; (c) $a = 20$, $K = 30$, $P^{\text{bth}} = 0.02$.

we take the expenditure of the fronthaul network into consideration, the former choice may be far more economical than the latter one.

By contrasting the (a) and (c) curves in Fig. 14.6, we can see that stricter QoS requirements can increase the pooling gain. This is so because, on the one hand, one needs to increase the number of r-servers $|K|$ in order to reduce the blocking probability, which will in turn increase $M^T K$; on the other hand, the average number of c-servers occupied is always around $|M|\mu$, as indicated by Theorem 14.2. Therefore, the stricter the QoS requirement, the more idle c-servers there will be in the VBS pool and consequently the greater the pooling gain. Also, we can see that the increase in traffic load will reduce the pooling gain by making the "knee point" occur at a larger value. This observation indicates that we may need to dynamically adjust the size of the VBS pools in order to get a satisfactory pooling gain under a varying traffic load, which requires a software-defined fronthaul network with switching capabilities [13], so that the fronthaul can deliver the traffic of the VBSs to a data center of the desired size.

In short, since the pooling gain reaches a significant level even with a medium pool size (more than 75% of the pooling gain is reached with around 50 VBSs) and the marginal gain of a larger pool size tends to be negligible in the practical deployment of C-RANs; a medium size of VBS pool with 50 to 100 VBSs can save a notable amount of fronthaul cost while still enjoying most benefits of BBU consolidation.

14.3   Energy–Delay Tradeoffs of VBSs in C-RAN

From the operational point of view, this section focuses on the energy consumption of both transmission and computing in C-RANs. In Section 14.3.1 we propose a novel energy consumption model for VBSs which takes the dynamic allocation of both the computational resources (the number of CPU cores, CPU speed) and the radio resources (transmit power, RF circuit power) into account. Based on this model, in Section 14.3.2 we analyze the energy-delay tradeoff for a VBS to minimize the system operational cost, which is defined here as the weighted sum of the power consumption and the average delay. Note that energy–delay tradeoffs of BSs in wireless systems have been much studied in the literature, which indicates that when taking practical concerns into account the energy–delay tradeoff deviates from a simple monotonic curve [21–23]. By considering the BS sleeping behavior and the cost of BS on/off-state switching, we derive the explicit form of the optimal data transmission rate and find the condition under which the energy optimal rate exists and is unique. Opportunities to reduce the average delay and simultaneously achieve energy savings are observed. We further propose an efficient algorithm to jointly optimize the computational and wireless resources, i.e., the number of CPU cores and the data transmission rate respectively. Finally, via numerical examples in Section 14.3.3 we investigate the impact of computational resources on this relationship and compare the energy-saving performance of VBSs in cloud-based architectures with BSs in conventional cellular networks.

14.3.1   Energy Consumption Model of VBS

EARTH Model

The EARTH energy consumption model of BSs [14] has been widely used in the literature to analyze the energy efficiency of cellular systems. It has the form:

$$P_{in} = \begin{cases} N_{TRX}P_0 + \Delta_p P_{out}, & 0 < P_{out} \leq P_{max}, \\ N_{TRX}P_{sleep}, & P_{out} = 0, \end{cases} \tag{14.47}$$

where P_{in} is the total power supply of the BS and P_{out} is the power radiated at the antenna elements; N_{TRX} is the number of antennas at the BS, P_0 is the power consumption at the minimum non-zero load, Δ_p is the slope of load versus power consumption, and P_{sleep} is the energy consumption in sleep mode.

This model cannot be directly used for VBSs, for two reasons. One is that multiple BBUs reside in the cloud infrastructure, so the energy consumption of the BBU per BS should be reduced owing to the multiplexing of computational resources. The other is that in the BBU pool, the baseband computational resources can be dynamically allocated and the BBU application can be constructed on demand, thanks to virtualization technology. However, the EARTH model cannot reflect the variations in computational resources. As a result, a new model for the VBSs in C-RANs is required.

Computational-Resource-Aware Model

On the basis of the existing energy consumption model, we propose a computational-resource-aware energy consumption model for the VBSs in a C-RAN. Following the component-based methodology of the EARTH model, we calculate the power consumption of the BBU and the RRH in the VBS separately, and then take the sum as the total power consumption:

$$P = P_R + P_B, \tag{14.48}$$

where P_R and P_B are the power consumptions of the RRH and the BBU respectively.

Regarding the RRH term, we leverage the intermediate result from the EARTH model [14],

$$P_R = \frac{P_{out}}{\eta} + P_{RF}, \tag{14.49}$$

where η denotes the power amplifier (PA) efficiency and P_{RF} denotes the power consumption of the radio frequency (RF) circuits. Note that here we assume the single-antenna case, but the above equation can be easily applied to the multiple-antenna case.

As for the BBU term, we calculate its energy consumption as follows:

$$P_B = N_c(P_{Bm} + \Delta_{P_B}\rho_c s^\beta), \tag{14.50}$$

where

$$\Delta_{P_B} = \frac{P_{BM} - P_{Bm}}{s_0^\beta}, \tag{14.51}$$

and N_c denotes the number of *active* CPU cores used in the VBS of interest, P_{Bm} and P_{BM} are the minimum and maximum power consumptions of each CPU core, ρ_c denotes the CPU load from the BBU process for the N_c cores, which is usually expressed as a percentage; s is the CPU speed and s_0 is the reference CPU speed. In this model the power consumption of the BBU is linear with the number of CPU cores N_c as well as with the CPU load ρ_c [24, 25]. With speed-scaling, the dynamic part of the power consumption is polynomial with the CPU speed, with exponent β [26].

More importantly, we are able to describe the relationship between the utilized computational resources and the software tasks which process baseband signal samples for wireless communications. The CPU load can be expressed by the following equation:

$$\rho_c = \frac{f(r)}{sN_c} = \frac{c_0 + \kappa r}{sN_c}, \tag{14.52}$$

where $f(r)$ is the number of instructions per unit time and sN_c represents the maximum instructions available per unit time. We assume that $f(r)$ is linear in the data transmission rate r of the air interface, where c_0 and κ are the relevant coefficients. This assumption is supported by the profiling results in CloudIQ [27], which show a linear relationship between the processing time of an LTE subframe and the modulation and coding scheme (MCS) used as well as the available physical resource blocks (PRB).

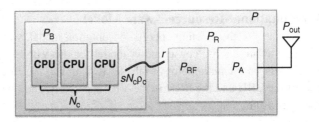

Figure 14.7 Block diagram of the VBS power model.

Substituting (14.52) into (14.50), we get

$$P_B = N_c P_{Bm} + \Delta_{P_B} c_0 s^{\beta-1} + \Delta_{P_B} \kappa r s^{\beta-1}, \tag{14.53}$$

which means that the computational power consumption is linear with the data rate.

In summary, the power consumption of a VBS is:

$$P = \begin{cases} P_B + P_R, & 0 < P_{out} \le P_{max}, \\ P_{sleep}, & P_{out} = 0. \end{cases} \tag{14.54}$$

Here P_{sleep} includes the power consumption of all components that are active in the sleep mode, either on the RRU side or in the BBU pool. Note that, in the BBU pool, even if computing resources are released when the VBS is in sleep mode, there may still be some computing needed to allow quick computing-resource allocation when the VBS is required to be turned on. In addition, in this model we have not counted the power consumption from the fronthaul, which is highly dependent on how the fronthaul network is designed. A block diagram of the proposed VBS power model is shown in Fig. 14.7.

Processor-Sharing-Based Service Model

Consider one VBS on a server with N_c active CPU cores with speed s. We model the system as an M/G/1 processor sharing (PS) queue. Traffic flows arrive at the BS with average rate λ, and each flow has an average file size L. The data transmission rate is r bps when the queue has customers; otherwise it is zero. Therefore the traffic load of the queue is $\rho = \lambda L/r$. According to queueing theory [28], the average queue length is

$$\mathbb{E}\{n\} = \frac{\rho}{1-\rho} = \frac{\lambda L}{r - \lambda L}. \tag{14.55}$$

By applying Little's law [28], the average of the delay D is $\mathbb{E}\{D\} = \mathbb{E}\{n\}/\lambda$.

In our model the VBS will go to sleep when there is no customer in the queue (the "off" state), and turn back to work when a new customer arrives (the "on" state). In one on–off cycle we denote by T_a the time duration in the busy period, by T_s the time in the consecutive idle period, and by $T_c = T_a + T_s$ the total time. We assume some power cost E_{sw} is incurred during each on–off-state switching, so the average power consumption in a cycle is as follows:

$$\mathbb{E}\{P\} = p_{active}(P_B + P_R) + p_{sleep}P_{sleep} + \frac{2E_{sw}}{\mathbb{E}\{T_c\}}, \tag{14.56}$$

where $p_{active} = \rho$ and $p_{sleep} = 1 - \rho$ are the time fractions of the busy and idle periods during one cycle, respectively.

As for the channel model, we consider the large-scale path loss while ignoring shadowing fading. The downlink signal-to-interference-plus-noise ratio (SINR) is given by

$$\text{SINR}(d) = gP_{out} = \frac{P_{out}}{L(d)FN_0W}, \tag{14.57}$$

where we denote by $L(d)$ the path loss, by F the noise factor of the user equipment (UE), by N_0 the noise spectral density, and by W the system bandwidth. For explicit analysis, we assume that the distance from every user to the RRH of the VBS is the same, so the overall channel gain g is the same for each user. Therefore the sum data rate at the BS can be expressed by

$$r = W \log_2(1 + gP_{out}). \tag{14.58}$$

Finally, to observe the tradeoff between the average power consumption and the average delay of the VBS system, the optimization problem is formulated as:

$$\underset{r, N_c}{\text{minimize}} \quad z = \mathbb{E}\{P\} + \alpha\mathbb{E}\{n\}$$

$$\text{s.t.} \qquad r \geq 0, \quad N_c \in \mathbb{Z}^+, \tag{14.59}$$

where we want to minimize the system cost z, which is a weighted sum of the average power consumption and average queue length, and the weighting factor is α. The decision variables are the data rate r and the number of CPU cores N_c.

14.3.2 Energy–Delay Tradeoff Analysis

It is easy to validate the convexity of the objective function of (14.59), and thus to get the optimal solution to problem (14.59). First, let $\partial z/\partial r = 0$ and thus find that the optimal rate r^* satisfies

$$\Omega\left(\frac{\alpha g \eta}{e}\left(\frac{r^*}{r^* - \lambda L}\right)^2 + \frac{g\eta P_s - 1}{e}\right) = \frac{r^* \ln 2}{W} - 1, \tag{14.60}$$

where

$$P_s = P_0 - P_{sleep} - 2\lambda E_{sw}, \tag{14.61}$$

$$P_0 = N_c P_{Bm} + \Delta P_B \cos^{\beta-1} + P_{RF}, \tag{14.62}$$

and $\Omega(\cdot)$ is the principal branch of the Lambert W-function. As r increases, the left-hand side of Eq. (14.60) decreases and the right-hand side increases, which implies that the optimal rate r^* is unique, if it exists.

We have the following proposition on the energy–delay tradeoff relationship of VBSs with varying data rate r. The derivation is straightforward and is not presented here.

PROPOSITION 14.4 For the relationship between the average power consumption and average delay, the following claims hold.

Algorithm 12 Algorithm to find the optimum (r^*, N_c^*)

1: Set N_{cM}, $N_c \leftarrow 1$, $S \leftarrow \Phi$
2: **while** $N_c \leq N_{cM}$ **do**
3:     $\hat{r}(N_c) \leftarrow \text{argmin}_r z(r, N_c)$
4:     **if** $\hat{r}(N_c) \leq r_M(N_c)$ **then**
5:         $S \leftarrow S \cup \{(\hat{r}(N_c), N_c)\}$
6:         Break out of the loop
7:     **else**
8:         $S \leftarrow S \cup \{(r_M(N_c), N_c)\}$
9:         $N_c \leftarrow N_c + 1$
    return $(r^*, N_c^*) = \text{argmin}_{(r,N_c) \in S} z(r, N_c)$

---

1. There exists a unique energy-optimal rate r_e^* when the following conditions are satisfied:

$$\lambda < \frac{P_o - P_{\text{sleep}}}{2E_{\text{sw}}},\tag{14.63}$$

$$L < \frac{W}{\lambda \ln 2}\left[\Omega\left(\frac{g\eta P_s - 1}{e}\right) + 1\right].\tag{14.64}$$

This energy-optimal rate is given by

$$r_e^* = \frac{W}{\ln 2}\left[\Omega\left(\frac{g\eta P_s - 1}{e}\right) + 1\right].\tag{14.65}$$

2. When the conditions (14.64), (14.4) are not satisfied, the average power consumption is monotonically decreasing with the average delay.
3. In both cases, when the average delay goes to infinity, the average power consumption approaches $P_o + \kappa \Delta_{P_B} s^{\beta-1}\lambda L + (2^{\lambda L/W} - 1)/(g\eta)$.

When the energy-optimal point exists, the average delay can be traded off for energy savings only when $r > r_e^*$; however, when $r < r_e^*$ we can reduce the average delay and save energy simultaneously. Interestingly, the proposition has the same mathematical structure as that in our previous work [23], and for VBSs the energy-optimal rate is not affected by the part of the computational power consumption which is a linear function the of data rate. The reason is that the effect of the computational power consumption is neutralized by the time fraction factor, which influenced by the traffic load, which in turn is inversely proportional to the data rate as expressed in (14.55).

Further, we investigated the impact of the computational resources, in particular the impact of the number of CPU cores N_c. On one hand, given N_c, the maximum possible rate is

$$r_M(N_c) = \frac{N_c s - c_0}{\kappa}.\tag{14.66}$$

On the other hand, we have $\partial z/\partial N_c = P_{Bm} > 0$, $\partial r^*/\partial N_c > 0$, which means that increasing the number of CPU cores will increase the system cost and the optimal rate.

Table 14.1 Simulation parameters

CPU speed, s	2 GHz
Reference CPU speed, s_0	2 GHz
Maximum power per CPU core, P_{BM}	20 W
Minimum power per CPU core, P_{Bm}	5 W
Exponential coefficient of CPU speed, β	2
Constant coefficient of instruction speed, c_0	7×10^8
Rate-varying coefficient of instruction speed, κ	35
Carrier frequency, f	2 GHz
Cell radius, R	0.5 km
UE noise figure, F	9 dB
Noise spectral density, N_0	−174 dBm/Hz
System bandwidth, W	20 MHz
RF circuit power, P_{RF}	12.9 W
PA efficiency, η	31.1%
Switch cost, E_{sw}	5 J
VBS or BS sleeping power, P_{sleep}	6.45 W
BS basic power at non-zero load, P_0	84 W
Slope of the load-dependent BS power, Δ_p	2.8
Number of antennas at a BS, N_{TRX}	1

On the basis of the above analyses, we propose an efficient algorithm, Algorithm 12, to find the optimal data rate and number of CPU cores (r^*, N_c^*) jointly. In the algorithm, N_{cM} is the maximum number of CPU cores in a practical system, and S is the set that stores the candidates for the optimal point. We denote the empty set by Φ. At first only one CPU cores is considered. When the current number of CPU core cannot support the local optimal rate, we mark the maximum supportable rate as one candidate for the global optimal solution and consider one more CPU core. When we find the number of CPU cores under which the local optimal rate can be achieved, we are able to mark the local optimal rate as the final candidate and exit the search, since the total cost always increases afterwards. Finally the global optimal rate and number of CPU cores can be obtained by comparing all the candidates. In this way it is unnecessary to search exhaustively all the possible numbers of CPU cores and make comparisons; thus the complexity, is reduced.

14.3.3 Numerical Results

In this section we present numerical results to show the energy-delay tradeoffs of VBSs. The simulation parameters for the VBSs are listed in Table 14.1. Among them the baseband parameters are based on commodity servers, and the cellular parameters are from the 3GPP document [29]. The parameters for conventional BSs correspond to the last four items.

Figure 14.8 shows the energy saving performance of VBSs and that of conventional BSs with the EARTH energy consumption model. We can find a more than 60% energy

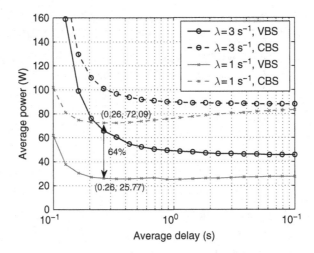

Figure 14.8 Power consumption of a VBS and of a conventional BS (CBS); $L = 2$ MB.

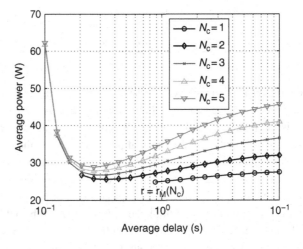

Figure 14.9 Energy–delay tradeoff with different numbers of CPU cores N_c; $\lambda = 1\,s^{-1}$ and $L = 2$ MB.

saving with C-RAN VBSs. For example, when $\lambda = 1\,s^{-1}$, about 64% savings can be achieved with the same average delay $\mathbb{E}\{D\} = 0.26\,s$ as that which optimizes the power consumption of conventional BSs. The savings come from traffic-aware computational resource adaptation: the computational power is matched to the temporal traffic variations. In other words, in C-RAN the BBU power consumption scales with the actual data rate rather than remaining static as in conventional BSs.

Figure 14.9 shows the relationship between the average power consumption and the average delay, given different numbers of CPU cores. For example, when $N_c = 5$ there exists a unique optimal point to minimize energy consumption. To the right of the optimal point there is an opportunity to reduce the average delay and achieve energy savings

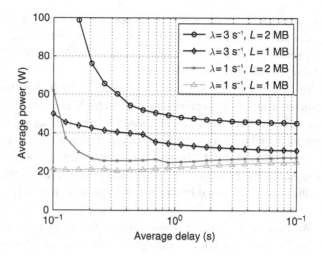

Figure 14.10 Energy–delay tradeoff with different arrival rates.

simultaneously. In addition, the impact of the number of CPU cores on the energy–delay tradeoff is also illustrated. The left end-points of the curves for smaller numbers of CPU cores N_c mark the maximum supportable data rate. Given the average delay, increasing the number of CPU cores will increase the average power consumption as well as the energy-optimal rate.

When both the data rate and the number of CPU cores are adjustable, there is a tradeoff relationship between the average power consumption and the average delay. This is shown for different traffic loads in Fig. 14.10. Note that each curve is divided into several zones, owing to the impact of the number of CPU cores. The algorithm to find the optimum (r^*, N_c^*) is illustrated by the figure. Take the third curve in the key ($\lambda = 1$ s^{-1}, $L = 2$ MB) as an example. We need to compare the rightmost turning point with $\mathbb{E}\{D\} = 0.89$ s and the local optimal point with $\mathbb{E}\{D\} = 0.34$ s to determine the global optimal solution. Furthermore, Fig. 14.10 depicts the impact of traffic load on the energy–delay tradeoff. Either a larger arrival rate or a larger average file size will increase the average power consumption for a given average delay. The power consumption when the average delay approaches infinity is monotonically increasing with λL. Also, the file size has a more notable impact when the delay requirement is stringent.

14.4 Conclusions and Outlook

In this chapter, we have explored green designs for deploying and operating C-RAN. From the viewpoint of saving the deployment cost of the fronthaul, we studied the relationship between the performance gain led by BS-function centralization and the VBS pool size. Using a multi-dimensional Markov model for the session-level behavior of VBS pools, we proved that the statistical multiplexing gain increases slowly as the pool size grows, with the residual gain diminishing at a speed between $|M|^{-3/4}$ and $|M|^{-1/2}$, where $|M|$ denotes the pool size. The result provides a clear picture of fronthauling cost versus the network performance in the C-RAN architecture and can be exploited to deploy C-RAN with cost constraints.

To save energy consumption in running C-RAN, we proposed a computational-resource-aware energy consumption model for VBSs, and investigated the energy–delay tradeoff of a VBS, considering a sleeping BS. Further we looked into the impact of computational resources and proposed an efficient algorithm to jointly optimize the data rate and the number of CPU cores that run the VBS functions. The analytical framework is validated by the numerical results, and it was revealed that more than 60% energy savings can be achieved by the use of VBSs in C-RANs compared with conventional BSs.

We now look at some possible research directions toward the green design of C-RANs. Green design and deployment has been an intensively studied problem for the Internet Cloud. However, it is essential to realize that the design space of C-RAN is fundamentally different from Internet Cloud: owing to the large aggregation cost on fronthaul, designs and deployments for a C-RAN can only be green and efficient when confined to a spatial locality. For this reason, it is valuable to have a fine-grained cost profile of the fronthaul and data center, so that promising tradeoff spots of the C-RAN can be identified in the design space.

In addition, the design of a C-RAN over the space domain and the corresponding deployment are dependent on the traffic and workload it serves. But both may change as networks evolve into the 5G era: it is reasonable to foresee that the traffic and workload may shift away from the current modes and features with the introduction of mobile cloud computing. When that happens, further modeling and analysis will be needed to maintain and improve the greenness and efficiency of C-RANs.

As for the operation of C-RANs, we expect some future work to extend BS sleeping mechanisms to more advanced ones, reflecting more practical factors, e.g., the CPU speed, the switching on/off delay, the use of multiple VBSs, etc. Besides, in the future the fronthaul might face innovative re-design, and fronthaul can play a significant role in terms of energy consumption. As a result, characterizing the energy footprint of the fronthaul needs to be addressed.

References

[1] C. V. N. Index, "Global mobile data traffic forecast update, 2014–2019." White Paper, February, 2011.

[2] G. Americas, "Meeting the 1000 × challenge: the need for spectrum, technology and policy innovations." White Paper, October 2013.

[3] C. M. R. Institute, "C-RAN: the road towards green RAN (version 4.0)." White Paper, June 2014.

[4] Y. Lin, L. Shao, Q. Wang, and R. Sabhikhi, "Wireless network cloud: architecture and system requirements," *IBM J. Res. Dev.*, vol. 54, no. 1, pp. 1–12, 2010.

[5] Z. Zhu, P. Gupta, Q. Wang, S. Kalyanaraman, Y. Lin, H. Franke, and S. Sarangi, "Virtual base station pool: towards a wireless network cloud for radio access networks," in *Pro. 8th ACM Int. Conf. on Computing Frontiers*, 2011, pp. 1–10.

[6] S. Bhaumik, S. P. Chandrabose, M. K. Jataprolu, G. Kumar, A. Muralidhar, P. Polakos *et al.*, "CloudIQ: a framework for processing base stations in a data center," in *Proc. 18th ACM Int. Conf. on Mobile Computing and Networking*, 2012, pp. 125–136.

[7] Q. Yang, X. Li, H. Yao, J. Fang, K. Tan, W. Hu *et al.*, "BigStation: enabling scalable real-time signal processing in large mu-mimo systems," in *Proc. ACM SIGCOMM 2013 Conf.*, 2013, pp. 399–410.

[8] NGMN Alliance, "Suggestions on potential solutions to C-RAN." White Paper, 2013.

[9] S. Namba, T. Matsunaka, T. Warabino, S. Kaneko, and Y. Kishi, "Colony-RAN architecture for future cellular network," in *Proc. of IEEE Future Network & Mobile Summit*, 2012, pp. 1–8.

[10] K. Sundaresan, M. Y. Arslan, S. Singh, S. Rangarajan, and S. V. Krishnamurthy, "Fluidnet: a flexible cloud-based radio access network for small cells," in *Proc. 19th ACM Int. Conf. on Mobile Computing & Networking*, 2013, pp. 99–110.

[11] J. Liu, T. Zhao, S. Zhou, Y. Cheng, and Z. Niu, "Concert: a cloud-based architecture for next-generation cellular systems," *IEEE Wireless Commun.*, vol. 21, no. 6, pp. 14–22, December 2014.

[12] "CPRI specification v6.0: interface specification." 2013.

[13] J. Liu, S. Xu, S. Zhou, and Z. Niu, "Redesigning fronthaul for next-generation networks: beyond I/Q samples and point-to-point links," *IEEE Wireless Commun.*, 2015, to be published.

[14] G. Auer, V. Giannini, C. Desset, I. Godor, P. Skillermark, M. Olsson *et al.*, "How much energy is needed to run a wireless network?" *IEEE Wireless Commun.*, vol. 18, no. 5, pp. 40–49, October 2011.

[15] R. Gupta, E. Calvanese Strinati, and D. Ktenas, "Energy efficient joint DTX and MIMO in cloud radio access networks," in *Proc. of 2012 IEEE 1st International Conf. on Cloud Networking*, November 2012, pp. 191–196.

[16] I. Gomez-Miguelez, V. Marojevic, and A. Gelonch, "Deployment and management of SDR cloud computing resources: problem definition and fundamental limits," *EURASIP J. Wireless Commun. and Netw.*, vol. 2013, no. 1, pp. 1–11, 2013.

[17] K. Ross and D. H. K. Tsang, "The stochastic knapsack problem," *IEEE Trans. Commun.*, vol. 37, no. 7, pp. 740–747, July 1989.

[18] J. M. Aein and O. S. Kosovych, "Satellite capacity allocation," *Proc. IEEE*, vol. 65, no. 3, pp. 332–342, 1977.

[19] J. Kaufman, "Blocking in a shared resource environment," *IEEE Trans. Commun.*, vol. 29, no. 10, pp. 1474–1481, Octtober 1981.

[20] R. W. Wolff, "Poisson arrivals see time averages," *Oper. Res.*, vol. 30, no. 2, pp. 223–231, 1982.

[21] Y. Chen, S. Zhang, S. Xu, and G. Li, "Fundamental trade-offs on green wireless networks," *IEEE Commun. Mag.*, vol. 49, no. 6, pp. 30–37, June 2011.

[22] J. Wu, Y. Wu, S. Zhou, and Z. Niu, "Traffic-aware power adaptation and base station sleep control for energy-delay tradeoffs in green cellular networks," in *Porc. 2012 IEEE Global Communications Conf.*, December 2012, pp. 3171–3176.

[23] J. Wu, S. Zhou, and Z. Niu, "Traffic-aware base station sleeping control and power matching for energy-delay tradeoffs in green cellular networks," *IEEE Trans. Wireless Commun.*, vol. 12, no. 8, pp. 4196–4209, August 2013.

[24] M. Blackburn, "Five ways to reduce data center server power consumption." White Paper, 2008. Online. Available at www.thegreengrid.org/Global/Content/white-papers/Five-Ways -to-Save-Power.

[25] A. Vasan, A. Sivasubramaniam, V. Shimpi, T. Sivabalan, and R. Subbiah, "Worth their watts? – an empirical study of datacenter servers," in *Proc. 2010 IEEE 16th Int. Symposium on High Performance Computer Architecture*, January 2010, pp. 1–10.

[26] K. Son and B. Krishnamachari, "Speedbalance: Speed-scaling-aware optimal load balancing for green cellular networks," in *Proc. of IEEE INFOCOM*, March 2012, pp. 2816–2820.

[27] S. Bhaumik, S. P. Chandrabose, M. K. Jataprolu, G. Kumar, A. Muralidhar, P. Polakos *et al.*, "Cloudiq: a framework for processing base stations in a data center," in *Proc. 18th ACM Int. Conf. Mobile Computing and Networking*, 2012, pp. 125–136.

[28] L. Kleinrock, *Queueing Systems, Volume I: Theory*. Wiley Interscience, 1975.

[29] "Further advancements for E-UTRA physical layer aspects." 3GPP TR 36.814, Technical Report, 2010.

15 Optimal Repeated Spectrum Sharing by Delay-Sensitive Users

Yuanzhang Xiao and Mihaela van der Schaar

15.1 Introduction

The spectrum is becoming an increasingly scarce resource, owing to the emergence of a plethora of bandwidth-intensive and delay-critical applications (e.g. multimedia streaming, video conferencing, and gaming). To achieve the gigabit data rates required by next-generation wireless systems, we need to manage efficiently the interference among a multitude of wireless devices, most of which have limited computational capability. Central to interference management are *spectrum-sharing* policies, which specify when and at which power level each device should access the spectrum. Given the heterogeneity and the huge number of distributed wireless devices, it is computationally hard to design efficient spectrum sharing policies.

Cloud-RANs present a promising network architecture for designing spectrum-sharing policies. They consist of two components, a pool of baseband processing units (BBUs) and remote radio heads (RRHs), and allocate most demanding computations to the BBU pool (i.e., the "cloud") [1–7]. In this way, C-RANs open up opportunities for designing efficient (even optimal) spectrum-sharing protocols. However, these opportunities come with the following challenges in C-RANs [1–7]:

1. How to allocate the computations between the BBU pool and the RRHs and minimize message exchange between them?
2. How to cope with dynamic entry and exit in large networks?
3. How to support the delay-sensitive applications that constitute a majority of the traffic in C-RANs?

This chapter presents advances made in the past years on a systematic design methodology for spectrum-sharing protocols that are particularly suitable for C-RANs. The spectrum-sharing protocols designed by the presented methodology can be implemented naturally in two phases:

- the first phase, of determining the optimal network operating point, which requires most computation and can be done in the BBU pool; and
- the second, phase of distributed implementation by RRHs with very limited computational capability.

Requiring limited message exchange between the BBU pool and the RRHs, the presented methodology results in provably optimal spectrum-sharing policies for C-RANs in interference-limited scenarios. More importantly, the presented methodology is general and can flexibly reconfigure the BBU pool to compute different optimal operating points in a variety of different C-RAN deployment scenarios.

In this chapter we start with a description of a general model for a C-RAN, with the focus on spectrum sharing, and formulate the spectrum-sharing design problem. We present our model for delay-sensitive applications and illustrate the implication of delay sensitivity on the structure of the optimal spectrum-sharing policy. We give a high-level overview of our design methodology and discuss the instantiation of the methodology in various deployment scenarios. Finally, we demonstrate the performance improvement achieved by our design methodology in comparison with state-of-the-art spectrum-sharing policies.

15.2   A General Model of Spectrum Sharing in C-RANs

15.2.1   Basic Setup

Consider a C-RAN with a number of cells, which can be either macro cells or small cells (see Fig. 15.1). At each time slot and at each frequency channel, there is one wireless device actively served by the base station. Depending on whether the downlink or the uplink is our consideration, an RRH can be the base station or the active device of a cell. To be general, we will refer to the pair consisting of the base station and the wireless device as the RRH's transmitter and its receiver. Hence, in the downlink (resp. uplink), RRH's transmitter is the base station (resp. the wireless device) and its receiver is the device (resp. the base station).

The channel gain from RRH i's transmitter to RRH j's receiver is g_{ij}. Each RRH i chooses its transmit power level p_i from the set $\mathcal{P}_i \triangleq \left[0, P_i^{\max}\right]$. Note that the set $\mathcal{P}_i$

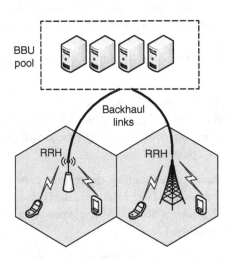

Figure 15.1 Illustration of C-RANs with a BBU pool and distributed RRHs.

contains 0, namely RRH i can choose not to transmit. Denote the joint power profile of all the RRHs by $p = (p_1, \ldots, p_N)$.

Given the power profile, each RRH i obtains a *reward* $r_i(p)$. This reward can be any general function that is decreasing in the power levels of the other RRHs. Two representative examples of the reward function are as follows.

Example 15.1   One example of a reward function is the Shannon throughput. Since the RRHs are distributed and cannot decode each other's messages, each RRH treats interference from the others as noise. Therefore, RRH i's throughput is

$$r_i(p) = \log_2 \left(1 + \frac{p_i g_{ii}}{\sum_{j \neq i} p_j g_{ji} + \sigma_i} \right), \tag{15.1}$$

where σ_i is the noise power at RRH i's receiver.

Example 15.2   Another example of the reward function is the ratio of throughput and power level:

$$r_i(p) = \frac{1}{p_i} \log_2 \left(1 + \frac{p_i g_{ii}}{\sum_{j \neq i} p_j g_{ji} + \sigma_i} \right). \tag{15.2}$$

This reward function captures the energy efficiency.

Note that we do not require the RRH's reward to be an increasing function of its own power level, as in Example 15.2. In other words, we allow very general reward functions. For illustration, we adopt the throughput as the reward in the rest of this chapter.

We define RRH i's local interference temperature I_i as the aggregate interference and noise power level at its receiver, namely

$$I_i = \sum_{j \neq i} p_j g_{ji} + \sigma_i. \tag{15.3}$$

Each RRH i measures the interference temperature with errors. The erroneous estimate is $I_i + \varepsilon_i$, where ε_i is the additive estimation error. Each RRH i quantizes the estimate and feeds the quantized estimate back to the transmitter. We require each RRH simply to use an unbiased estimator and a two-level quantizer. This results in the following one-bit feedback signal:

$$y_i = \begin{cases} 1, & \text{if } I_i + \varepsilon > \text{threshold,} \\ 0, & \text{otherwise.} \end{cases} \tag{15.4}$$

15.2.2   Spectrum-Sharing Policy

The system is time slotted at $t = 0, 1, 2, \ldots$ At the beginning of time slot t, each RRH i chooses its transmit power p_i^t and achieves the throughput $r_i(p^t)$. At the end of time

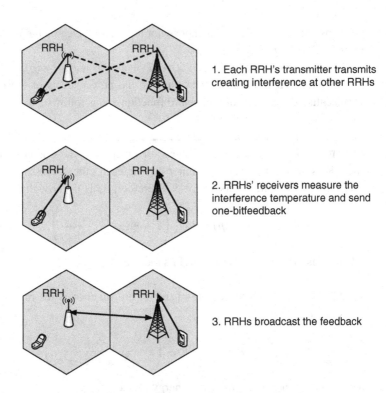

Figure 15.2 Illustration of the spectrum-sharing protocol in each time slot.

slot t, each RRH i broadcasts its feedback signal y_i^t.[1] We define a system distress signal to indicate whether there exists an RRH whose local interference temperature is above the threshold. We denote the (system) distress signal by $y^t \in Y \triangleq \{0, 1\}$, where $y^t = 1$ if there exists an RRH i such that $y_i^t = 1$. See Fig. 15.2 for an illustration of the above procedure at one time slot.

The spectrum-sharing policy π is the collection of all the RRHs' policies, namely $\pi = (\pi_1, \ldots, \pi_N)$. Each RRH i's strategy is a mapping from the history of past distress signals to its set of power levels, which is formally defined as[2]

$$\pi_i : \cup_{t=0}^\infty (Y)^t \to P_i,$$

$$h^t \triangleq \left(y^1, \ldots, y^{t-1}\right) \mapsto p_i^t, \qquad (15.5)$$

where h^t is the collection of past distress signals at time t.

Note that the above definition of spectrum-sharing policies is general and can represent all the existing policies. For instance, it can represent spectrum-sharing policies that require RRHs to transmit at fixed power levels (i.e., power control policies as in [20, 21]). We call such policies "constant policies" and formally define them as follows.

---

[1] We could reduce the amount of broadcasting by asking an RRH to broadcast only when its feedback signal is $y_i^t = 1$.

[2] Throughout this chapter, we use a^t to denote a at time t, and $(a)^t$ (resp. $(A)^t$) as the tth power (resp. power set) of real number a (resp. set A).

DEFINITION 15.1 (Constant policies)   In a constant spectrum-sharing policy, each RRH transmits at a fixed power level all the time, namely

$$\pi_i(h^t) = p_i^{\text{const}}, \quad \forall i, \forall t, \forall h^t. \tag{15.6}$$

Another class of existing policies of great interest are round-robin time-division multiple access (TDMA) policies [22]. We can use (15.5) to represent these policies as well.

Example 15.3   A simple round-robin policy with N RRHs can be written as

$$\pi_i(h^t) = \begin{cases} p_i^{\text{const}}, & \text{if } (t \bmod N) = i, \\ 0, & \text{otherwise.} \end{cases} \tag{15.7}$$

In this case, the RRHs transmit in cycles of N, and each RRH i transmits in the ith time slot in each cycle. Note that the policy $\pi_i(h^t)$ depends only on the time slot t, not on the history of distress signals h^t.

DEFINITION 15.2 (Round-robin policies)   In a round-robin spectrum-sharing policy, the RRHs transmit in cycles and choose the same power levels in each cycle, independently of the history. For a round-robin policy with cycle length L, we have

$$\pi_i(h^t) = \pi_i(h^{t+L}), \quad \forall i, \forall t, \forall h^t, \forall h^{t+L}. \tag{15.8}$$

Remark 15.1   Note that our definition of round-robin policies is more general than usual. It includes Example 15.3 as a special case. It also extends the simple round-robin policy in Example 15.3 by allowing the cycle length to be different from the number of RRHs (i.e., $L \neq N$) or allowing each RRH to have different numbers of transmission time slots in each cycle, and so on.

Remark 15.2   Despite the generality of our definition of round-robin policies, we will show later that the optimal spectrum-sharing policy is not round-robin.

DEFINITION 15.3 (Non-stationary policies)   Any policy defined by (15.5) that is neither constant nor round-robin is *non-stationary*.

Remark 15.3   Non-stationary policies are a class of general policies. In particular, some are time-division multiple access (TDMA) policies without cyclic (or periodic) structures. These non-stationary TDMA policies are optimal for delay-sensitive applications, as we will illustrate in Section 15.3.

15.2.3   Delay Sensitivity

We model the delay sensitivity of the applications run by the RRHs by a discount factor $\delta \in [0, 1)$ [8–11, 15, 16]. Assuming such a discount factor $\delta \in [0, 1)$, RRH i's expected discounted average throughput is

$$R_i(\pi) = \mathbb{E}\left\{(1-\delta)\sum_{t=0}^{\infty}(\delta)^t r_i\left[\pi\left(h^t\right)\right]\right\},\tag{15.9}$$

where the expectation is taken over the history h^t with respect to its distribution induced by the policy and the (random) distress signal. Similarly, RRH i's expected discounted average energy consumption is

$$P_i(\pi) = \mathbb{E}\left\{(1-\delta)\sum_{t=0}^{\infty}(\delta)^t \pi\left(h^t\right)\right\}.\tag{15.10}$$

The discount factor models the delay sensitivity by discounting future rewards. A more delay-sensitive RRH discounts the future throughput more (i.e. it has a smaller discount factor), because it has more urgency to transmit.

15.2.4   Problem Formulation

Each RRH i aims to maximize its own average throughput $R_i(\pi)$ or minimize its own average energy consumption $P_i(\pi)$, while fulfilling a minimum throughput requirement $R_i^{\min}$. Our goal is to design a general methodology for constructing optimal spectrum-sharing policies for C-RANs. The design problem can be formulated in a variety of forms, depending on the requirements of specific deployment scenarios. One essential feature is that the policy should guarantee some minimum throughput requirements for all the RRHs; these will be introduced as constraints in all the formulations. The objective function can be some energy efficiency criterion $E(P_1(\pi),\ldots,P_N(\pi))$, e.g., the weighted average of all the RRHs' energy consumptions, with each RRH's weight reflecting its importance, or some spectrum efficiency criterion $W(R_1(\pi),\ldots,R_N(\pi))$, e.g., the weighted average throughput. Next, we define two policy-design problems.

$$\begin{aligned}\text{MaxPayoff}: \quad &\text{maximize}_\pi \ W(R_1(\pi),\ldots,R_N(\pi)) \\ &\text{s.t.} \qquad\quad R_i(\pi) \geq R_i^{\min}, \quad \forall i = 1,2,\ldots,N.\end{aligned}\tag{15.11}$$

$$\begin{aligned}\text{MinCost}: \quad &\text{minimize}_\pi \ E(R_1(\pi),\ldots,R_N(\pi)) \\ &\text{s.t.} \qquad\quad R_i(\pi) \geq R_i^{\min}, \quad \forall i = 1,2,\ldots,N.\end{aligned}\tag{15.12}$$

15.3   The Optimal Spectrum-Sharing Policy is Non-Stationary

15.3.1   Intuitions

To better illustrate the structure of the optimal spectrum-sharing policies, we focus on the case in which the RRHs have strong multi-user interference. We suppose that the interference is so strong that it is optimal to let only one RRH to be active at each time slot, as in 802.11e MAC wireless networks [17]. Therefore, we will focus on TDMA policies.

All the existing TDMA policies are round-robin policies (and include weighted round-robin policies) [17–19]. In round-robin policies, the time slots are divided into cycles of a fixed predetermined length, and each RRH transmits in fixed predetermined positions within each cycle. The cyclic nature of round-robin policies simplifies the implementation, but imposes restrictions that render round-robin policies inefficient for delay-sensitive applications. The reasons are as follows.

For delay-sensitive applications, not all the transmission opportunities, i.e., positions in a cycle are created as equal: earlier TXOPs are more desirable because they result in higher chances to deliver packets before their delay deadlines [17–19]. To ensure that the RRH's throughput and delay constraints are met, round-robin policies need a long cycle and a careful sharing of TXOPs in a cycle.

First, a long cycle is necessary. Suppose that the cycle length is the shortest possible, namely it is equal to the number of RRHs, as in standard round-robin policies. Then the RRH allocated to the last TXOP suffers severely from delay. We can compensate this RRH for its delay by having a longer cycle and allocating some extra TXOPs to it. However, a long cycle results in an exponentially increasing (in the cycle length) number of possible policies from which to choose.

Second, a careful sharing of TXOPs is necessary; see Fig. 15.3 for an illustration of the following discussion. Suppose that the cycle length is twice the number of RRHs and that each RRH gets two positions in a cycle. For fairness, no RRH should get two advantageous (i.e., earlier) TXOPs. A possible fair sharing would ensure that the RRH gets both an earlier and a later TXOP. However, such a schedule is inefficient in the worst-case delay: the RRH who gets the first and the last TXOPs in a cycle will experience high delay between consecutive transmissions. As we will illustrate in our motivating example, in the next subsection, and by simulations in Section 15.6, round-robin policies cannot simultaneously achieve high system performance, e.g., max-min fairness and also fulfill the guarantees in terms of transmission delays required by the delay-sensitive RRHs.

In conclusion, the optimal spectrum-sharing policy for delay-sensitive RRHs is in general non-stationary (i.e., it is not cyclic).

15.3.2 An Illustrative Example

To illustrate the performance improvement of optimal non-stationary policies over stationary policies, consider a spectrum-sharing scenario in which the RRHs need to

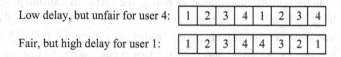

Low delay, but unfair for user 4: | 1 | 2 | 3 | 4 | 1 | 2 | 3 | 4 |

Fair, but high delay for user 1: | 1 | 2 | 3 | 4 | 4 | 3 | 2 | 1 |

Figure 15.3 Two simple round-robin schedules with cycle length 8 for four RRHs. The first has the same low delay of 4 for all four RRHs, but unfair sharing of transmission opportunities (TXOPs) (i.e., RRH 4 has later TXOPs). The second has a fair sharing of TXOPs but incurs a high maximum delay of 7 for RRH 1.

Table 15.1 Non-stationary and stationary policies and their performance

Policies	Transmit power (mW)	Scheduling	Average energy (mW)
Optimal constant [12, 13]	(186, 186, 186)	All the time	186
Optimal round-robin cycle length 3	(33, 144, 1432)	123 123 123 . . .	108
Optimal round-robin cycle length 4	(43, 212, 249)	1234 1234 1234 . . .	48
Optimal non-stationary (proposed)	(108, 108, 108)	123323213231 . . .	36

determine the transmission schedule and their own transmit power levels. Each RRH seeks to minimize its average energy consumption subject to its minimum throughput requirement. We have proved in [9] that the optimal policy, in the sense of minimizing the average energy, has the property that only one RRH transmits at a given time (i.e. the policy is TDMA) and at a fixed power level whenever it transmits. We caution the reader that although the optimal policy is TDMA, it is not round-robin. For a specific numerical example, suppose that there are three RRHs all having a direct channel gain of 1, a cross-channel gain of 0.25, a noise power of 5 mW, and a discount factor of 0.6 representing delay sensitivity. The RRHs have the same minimum throughput requirement of 1.5 bits/s per hertz.

We illustrate the policies and their performances in Table 15.1. The power levels are the transmit power levels of the three RRHs whenever they transmit. In the optimal constant policy, all three RRHs transmit all the time, at the same power level 186 mW.

In the round-robin and the proposed TDMA policies, the RRHs do not all transmit all the time. For the round-robin policies, we compute the optimal policy given the cycle length by determining the optimal (in terms of the average long-term energy consumption across RRHs) order of transmission in a cycle and the corresponding power levels. In the optimal round-robin policy with cycles of length 3, RRH 1 transmits first at a low power level (33 mW), RRH 2 transmits after RRH 1 at a higher power level (144 mW), to compensate for having to wait for transmission, and RRH 3 transmits last at an even higher power level (1432 mW), to compensate for having to wait even longer. In the cycle of length 4, again RRH 1 transmits at the lowest power level, RRH 2 transmits at a middle power level, and RRH 3 transmits at the highest power level, but the last two power levels are closer together than in the cycle of length 3 because RRH 3 transmits more often.

In the optimal non-stationary policy, the RRHs all transmit at the same constant power level (108 mW) whenever they transmit; this works because the order in which they transmit is constantly changing. In the last column of Table 15.1, the discounted average energy per RRH and per time slot is calculated. Notice that the cycle of length 3 is slightly more efficient than the constant policy, the cycle of length 4 is much more efficient, but the optimal non-stationary policy is more efficient still. Indeed, the optimal policy achieves 80%, 67%, and 25% energy savings compared with the optimal constant policy and the optimal round-robin policy with cycle is of 3 and 4, respectively.

Importantly, the energy savings are even more significant when the number of RRHs or the minimum throughput requirement is large (see Section 15.6).

Note that the optimal policy shown in the last row of Table 15.1 is obtained by our proposed distributed online algorithm. An RRH will determine its schedule online (i.e., whether it should be active at the beginning of each time slot), with low complexity.

15.4 New Design Methodology for Spectrum-Sharing Policies

Our general methodology can take a variety of forms in different deployment scenarios. Hence, the solutions are dependent on the scenarios considered. However, as illustrated in Fig. 15.4, the general methodology has two common key components under all scenarios: the optimal operating point selection (OOPS) algorithm, which is run by the BBU pool to determine the optimal operating point, and the longest distance first (LDF) scheduling algorithm, which is run by distributed RRHs to construct the policy. In this section we introduce the design framework in a baseline scenario [8, 9] where the presented methodology is provably optimal and is simple enough for a good understanding of its essence. In this way, we can appreciate intuition behind the design framework, which can be applied to a variety of other scenarios.

We illustrate the new design methodology for spectrum-sharing protocols in C-RANs in Fig. 15.4. The design toolbox takes as input the performance criterion $E(P_1, \ldots, P_N)$ or $W(R_1, \ldots, R_N)$ selected by the C-RAN. For example, when the criterion is the weighted sum of the energy consumption, the input will be the weights assigned by the designer to each RRH on the basis of on its importance. A C-RAN protocol designer can input any desirable performance criterion to the design toolbox, which will

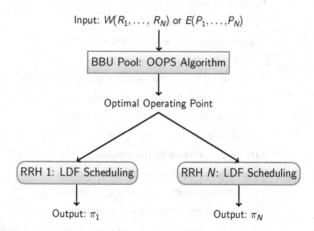

Figure 15.4 The design toolbox for spectrum-sharing protocols in C-RANs. Given the performance criterion as the input, the BBU pool runs the optimal operating point selection (OOPS) algorithm and sends the output to the distributed longest distance first (LDF) scheduling modules at each RRH.

then output the optimal spectrum-sharing protocol. The design toolbox provides two modules:

- the OOPS (optimal operating point selection) algorithm run by the BBU pool to determine the instantaneous throughput when an RRH accesses the spectrum, as well as the optimal operating point (i.e., the average throughput of each RRH); and
- the LDF (longest distance first) scheduling run by each RRH to determine whether it should access the spectrum at each time slot.

We give a brief description of the two modules in Fig. 15.5.

New briefly discuss the intuition behind the LDF scheduling algorithm. According to our definition of non-stationary policies, the scheduling decision should be based on the history of distress signals. The central key to, and most difficult part, of our construction is to prove that it is enough to summarize the history up to each time slot by a particular metric (see [8, 9] for an analytical expression for the metric). This metric can be easily computed by each RRH in a completely distributed way, and it has a nice interpretation of the "distance from target throughput". The scheduling decision is then based simply on the metric: the RRH "farthest away" from the target throughput transmits. The way in which the RRHs update the metric makes sure that the resulting scheduling can indeed achieve the target throughput.

When equipped with these two modules, the RRH can reach the optimal spectrum-sharing policy in a distributed manner. As proved in [8, 9], the operations performed by both modules converge in logarithmic time to the desired operating point. Importantly, we can prove theoretically that the dynamic entry and exit of devices will not affect the convergence speed of the spectrum-sharing policy [9]. Hence, the design toolbox is very suitable for C-RANs with frequent switch-on and switch-off of RRHs.

Distributed protocol implementation modules at each user

Optimal operating point selection algorithm:

    Input: design criterion, initial "price" (dual variable)

    Repeat until desired accuracy reached

        From the "price", find optimal operating point by Newton's method

        Broadcast the optimal operating point

        Update the price using the bisection method

    Output: the optimal operating point

LDF (longest distance first) scheduling:

    Input: optimal operating point

        Calculates "distances from targets" of each user

        The user with the largest distance is active

        Updates the distances analytically on the basis of the feedback signal

    Output: the optimal deviation-proof scheduling

Figure 15.5 The two modules deployed at the BBU pool and at each RRH to determine the optimal spectrum-sharing policy.

In the next two sections we will present some instantiations of our design methodology in several realistic C-RAN deployment scenarios.

15.5 Applications to Realistic C-RAN Deployment Scenarios

15.5.1 Large-Scale Heterogeneous Small-Cell Networks

We consider the first representative deployment scenario, that of very-large-scale heterogeneous small-cell networks [10, 11]. The unique features of large-scale heterogeneous small-cell networks impose the following requirements for efficient spectrum sharing:

- *Deployment of heterogeneous small-cell networks* Existing deployments of small-cell networks exhibit significant heterogeneity, such as different types of small cells (picocells and femtocells), different cell sizes, different throughput requirements for the RRHs, etc.
- *Interference avoidance and spatial reuse* Effective interference management policies should take into account the strong interference between neighboring RRHs as well as the weak interference between non-neighboring RRHs. Hence, the policies should effectively avoid interference among neighboring RRHs and employ spatial reuse to take advantage of the weak interference among non-neighboring RRHs.
- *Scalability to large networks* Small cells are often deployed over a large scale (e.g., in a city). Effective interference management policies should scale upto large networks, namely they should achieve efficient network performance while maintaining low computational complexity.

In large-scale heterogeneous small-cell networks, the design methodology achieves the following:

- A *spectrum-sharing policy* that schedules maximal independent sets (MISs)[3] of the interference graph to transmit in each time slot. In this way, we can avoid strong interference among neighboring RRHs (since neighboring RRHs cannot be in the same MIS), and efficiently exploit the weak interference among RRHs in a MIS by letting them to transmit at the same time.
- A *distributed algorithm* for the RRHs to determine a subset of MISs. The subset of MISs generated ensures that each RRH belongs to at least one MIS in this subset. Moreover, the subset of MISs can be generated in a distributed manner in logarithmic time (logarithmic in the number of RRHs in the network) for bounded-degree interference graphs.[4] The logarithmic convergence time is significantly faster than the time (linear or quadratic in the number of RRHs) required by existing distributed algorithms for generating subsets of MISs.

---

[3] Consider the interference graph of the network, where each vertex is a pair consisting of an RRH and its user, and each edge indicates strong interference between the two vertices. An independent set (IS) is a set of vertices in which no pair is connected by an edge. An IS is an MIS if it is not a proper subset of another IS.

[4] Bounded-degree graphs are graphs whose maximum degree can be bounded by a constant that is independent of the size of the graph.

- *A distributed algorithm* for each RRH to determine the optimal fractions of time occupied by the MISs with only local message exchange. Messages are exchanged only between neighboring RRHs. The distributed algorithm will output the optimal fractions of time for each MIS such that the given network performance criterion is maximized subject to the minimum throughput requirements.

More importantly, under a wide range of conditions, we can characterize analytically the competitive ratio of the proposed distributed policy with respect to the optimal network performance. We proved that the competitive ratio is independent of the network size, which demonstrates the scalability of our proposed policy in large networks. Remarkably, the constant competitive ratio is achieved even though our proposed policy requires only local information, is distributed, and can be computed fast, while the optimal network performance can be obtained only in a centralized manner with global information and NP (non-deterministic polynomial time) complexity.

Through simulations, we demonstrated significant (from 160% to 700%) performance gains over state-of-the-art policies.

15.5.2   C-RANs with Multimedia Applications

We now consider a second representative deployment scenario of C-RANs with delay-sensitive multimedia applications [12–14]. In this deployment scenario, it is important to provide hard delay guarantees.

On the basis of the new design methodology, we define a novel quality-of-service (QoS) metric, called "continuing QoS (CQoS) guarantees", as follows [14]

$$\text{CQoS: } R_i^t(\pi) \geq \gamma_i^{\text{cont}} r_i^{\text{max}}, \quad \forall t = 0, 1, \ldots, \tag{15.13}$$

where $r_i^{\text{max}} = \log_2\left(1 + P_i^{\text{max}}/\sigma_i\right)$ is the maximum achievable throughput.

Continuing QoS guarantees require an RRH's average throughput, starting from *every point in time*, to be higher than a threshold. Such guarantees are stricter requirements than the conventional QoS guarantees, which guarantee only the average throughput, starting from the beginning.

A byproduct of the CQoS guarantees is that, once they are satisfied, we can also provide upper bounds on the transmission delays of each RRH. First, we define RRH i's transmission delay at any time t as

$$\textbf{Transmission delay: } d_i^t(\pi) \triangleq \min_{\tau > t} \{\tau - t : \pi_i(\tau) > 0\}.$$

In words, the transmission delay $d_i^t(\pi)$ is the minimum wait time until the next transmission. An upper bound on the transmission delays is critical for delay-sensitive applications.

We proved in [14] that each RRH's CQoS guarantee leads to an upper bound on its maximum delay, $\sup_t d_i^t(\pi)$. Figure 15.6 illustrates the relationship between the delay and the CQoS guarantees.

We propose a systematic design methodology which constructs the optimal TDMA policy that maximizes the system performance (e.g. fairness) subject to the RRHs' CQoS guarantees.

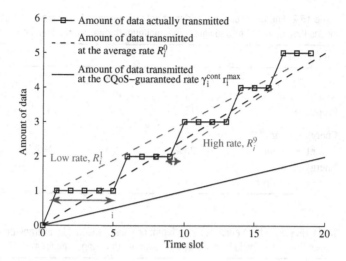

Figure 15.6 Relationship of delay and CQoS guarantees of RRH i. The solid curve with square data points is the amount of data transmitted; each jump in the curve corresponds to a transmission. The two straight lines through the origin are the amount of data transmitted as if the throughput was R_i^0 and $\gamma_i^{cont} r_i^{max}$, respectively. At each time t, if the continuation throughput R_i^t is higher then the RRH needs to transmit more after time t. Hence, the corresponding delay $d_i(t)$ is lower. The arrow shows a large delay, $d_i(1) = 4$; the short arrow shows a small delay, $d_i(9) = 1$.

Again, the key feature of the proposed policy is that it is not cyclic, as in round-robin policies. Instead, it adaptively determines which RRH should transmit according to the RRHs' remaining amounts of TXOPs needed to achieve the target throughput. We propose a low-complexity distributed algorithm to construct the optimal policy. Simulation results show that our proposed policy significantly outperforms the optimal constant policy and round-robin policies in peak signal-to-noise ratio (PSNR) for video streaming, by up to 6 dB and 4 dB, respectively.

15.6 Performance Gains

To illustrate the performance gain over existing policies, we considered a network of small cells (e.g. femtocells, picocells). Each small cell serves one device at each time. We randomly place small-cell base stations in a two-dimensional space with an average inter-site distance of 20 m, and randomly place devices with an average distance to their base stations of 5 m. The path loss exponent is 2. The maximum received SNR P_i/σ_i is 20 dB. The energy efficiency criterion is the average energy consumption, and the spectrum efficiency criterion is the average throughput. The discount factor is 0.95. The minimum throughput is $R_i^{min} = 1$ bit/s/Hz for all i. All data are averages over the results obtained from 10 000 random placements of small cells and devices.

Table 15.2 Energy efficiency of different policies measured by average energy expenditure (mW); N/A means that the policy fails to satisfy the minimum throughput requirements

Number of cells	7	9	11	13	15
Stationary	20	N/A	N/A	N/A	N/A
Round-robin	19	40	81	13	1320
Proposed	15	28	37	31	106
Energy saving w.r.t. stationary	25%	N/A	N/A	N/A	N/A
Energy saving w.r.t. round-robin	21%	30%	54%	77%	92%

Table 15.3 Spectrum efficiency of different policies measured by average throughput (bits/s/Hz); N/A means that the policy fails to satisfy the minimum throughput requirements

Number of cells	7	9	11	13	15
Stationary	0.9	N/A	N/A	N/A	N/A
Round-robin	2.0	1.8	1.3	1.1	N/A
Proposed	2.7	2.4	2.0	1.7	1.4
Improvement w.r.t. stationary	200%	∞	∞	∞	∞
Improvement w.r.t. round-robin	35%	33%	54%	55%	∞

From Tables 15.2 and 15.3, we can compare the proposed policies with a stationary policy (in which RRHs transmit at fixed powers simultaneously) [12, 13] and round-robin TDMA policies [14] in terms of energy efficiency and spectrum efficiency. The minimum throughput requirements and the weights are the same for all the RRHs. The proposed policy significantly improves the spectrum and energy efficiency of existing policies in most scenarios. In particular, when the number of RRHs is large, stationary policies quickly become infeasible (i.e. fail to achieve the minimum throughput requirements), while the proposed policy is feasible. Compared with existing policies, the proposed policy can achieve up to 92% energy saving and up to 200% spectrum efficiency improvement.

15.7 Related Work

The methodology presented in this chapter is based on two key insights [8–15]:

- it is more efficient to access the spectrum in a time-division multiple access (TDMA) fashion rather than to access the spectrum at the same time, owing to interference among RRHs, and

Table 15.4 The proposed methodology with network utility maximization (NUM) in comparison with the single-user Markov decision process (SU-MDP) and multi-user Markov decision process (MU-MDP), and the implications for spectrum-sharing scenarios

	Coupling	Interference regime	Resulting policy
NUM	weak	weak or no interference	stationary
SU-MDP	N/A	N/A	stationary
MU-MDP	weak	no interference	stationary
Proposed	strong	strong interference	non-stationary

- the optimal way to access the spectrum is not (weighted) round-robin but rather follows a carefully designed non-stationary schedule in which each RRH's transmit power level depends not only on its current state but also on the history of previous states and power levels.

15.7.1   Related Work in Spectrum Sharing

In contrast with the protocols developed on the basis of the new general methodology presented in this chapter, state-of-the-art spectrum-sharing protocols [18–22] use stationary policies, in which an RRH's transmit power level depends only on its current state. For example, some works [20, 21] propose physical-layer power control policies that require the RRHs to transmit simultaneously at fixed power levels over the time horizon in which they interact. Owing to strong multi-user interference, stationary power control policies can achieve only low spectrum efficiency and low energy efficiency.

Some works, e.g. [22], have proposed stationary medium-access-control(MAC)-layer-centric solutions, by modeling the physical layer with collision models and neglecting power control. These solutions adopt contention-free round-robin TDMA [22] protocols, which are suboptimal. The performance loss as compared with that of optimal policies is even larger when the RRHs are heterogeneous and experience different channel conditions, have different throughput requirements and demands, etc.

15.7.2   Related Theoretical Frameworks

Note that existing theoretical frameworks, such as network utility maximization (NUM) and standard Markov decision process (MDP), are not suitable for designing spectrum-sharing policies. The reason is that they focus on scenarios with "weakly coupled" RRHs, namely one RRH's action e.g., its transmit power, does not affect the others' payoffs, e.g., their throughput. In spectrum sharing in C-RANs the RRHs are strongly coupled, since the transmission of one RRH may cause strong interference to other RRHs. In addition, standard MDP is mostly used for single-user decision problems and is often suboptimal when applied to multi-user decision problems. We summarize the key differences from NUM and the MDP theory in Table 15.4. In addition, since both our methodology and MDP result in dynamic spectrum-sharing policies, we give a detailed comparison with MDP in Table 15.5.

Table 15.5 Detailed comparison with Markov decision process (MDP).

	Agents	Action	Value function	Policy
Single-user MDP	single	single action	single-valued	stationary
Multi-user MDP	multiple	action profile	single-valued	stationary
Non-stationary policy design	multiple	action profile	set-valued	non-stationary

15.8   Conclusion

In this chapter, we have introduced a novel general methodology for designing provably optimal spectrum-sharing policies, which are particularly suitable for C-RANs. In the protocols designed using this methodology, computationally demanding operations (i.e., determining the optimal operating points) are implemented by the BBU pool and are separate from the simple operations implemented by distributed RRHs with limited computational capability. Moreover, the protocols require limited message exchange between the BBU pool and the RRHs, reducing the burden of the backhaul. The protocols achieve high overall spectrum and energy efficiency and provide performance guarantees for each individual RRH. The presented methodology is general and can flexibly configure the cloud to optimize the system performance in different C-RAN deployment scenarios, such as large-scale heterogeneous small-cell networks, delay-sensitive multimedia communications, and the Internet of things. Initial experiments in specific deployment scenarios shows that, compared with existing protocols, the proposed protocols can significantly improve the spectrum and energy efficiency.

The design methodology can also be extended to domains other than spectrum sharing. Interested readers are referred to [16] for a treatment of general resource-sharing games and to [23] for an application to demand side management in smart grids.

References

[1] M. Peng, Y. Li, J. Jiang, J. Li, and C. Wang, "Heterogeneous cloud radio access networks: a new perspective for enhancing spectral and energy efficiencies," *IEEE Wireless Commun. Mag.*, vol. 21, no. 6, pp. 126–135, December 2014.

[2] Z. Ding and H. V. Poor, "The use of spatially random base stations in cloud radio access networks," *IEEE Signal Process. Lett.*, vol. 20, no. 11, pp. 1138–1141, November 2013.

[3] China Mobile. (2011). "C-RAN: the road towards green RAN." White Paper. Available: http://bit.ly/Ya1zuW.

[4] H. Harada, "Cognitive wireless cloud: a network concept to handle heterogeneous and spectrum sharing type radio access networks," in *Proc. conf. on IEEE Personal, Indoor and Mobile Radio Commun.*, 2009, pp. 1–5.

[5] H. T. Dinh, C. Lee, D. Niyato, and P. Wang, "A survey of mobile cloud computing: architecture, applications, and approaches," *Wireless Commun. Mobile Comput.*, vol. 13, no. 18, pp. 1587–1611, December 2013.

[6] J. Zander and P. Mhnen, "Riding the data tsunami in the cloud: myths and challenges in future wireless access," *IEEE Commun. Mag.*, vol. 51, no. 3, pp. 145–151, March 2013.

[7] K. Sundaresan, M. Y. Arslan, S. Singh, S. Rangarajan, and S. V. Krishnamurthy, "FluidNet: a flexible cloud-based radio access network for small cells," in *Proc. 19th Int. Conf. on Mobile Computing & Networking*, 2013, pp. 99–110.

[8] Y. Xiao and M. van der Schaar, "Dynamic spectrum sharing among repeatedly interacting selfish devices with imperfect monitoring," *IEEE J. Sel. Areas Commun.*, vol. 30, no. 10, pp. 1890–1899, March 2012.

[9] Y. Xiao and M. van der Schaar, "Energy-efficient nonstationary spectrum sharing," *IEEE Trans. Commun.*, vol. 62, no. 3, pp. 810–821, March 2014.

[10] K. Ahuja, Y. Xiao, and M. van der Schaar, "Efficient interference management policies for femtocell networks," *IEEE Trans. Wireless Commun.*, vol. 14, no. 9, pp. 4879–4893, September 2015.

[11] K. Ahuja, Y. Xiao, and M. van der Schaar, "Distributed interference management policies for heterogeneous small cell networks," *IEEE J. Sel. Areas Commun.*, vol. 33, no. 6, pp. 1112–1126, June 2015.

[12] J. Xu, Y. Andreopoulos, Y. Xiao, and M. van der Schaar, "Non-stationary resource allocation policies for delay-constrained video streaming: application to video over Internet-of-Things-enabled networks," *IEEE J. Sel. Areas Commun.*, vol. 32, no. 4, pp. 782–794, April 2014.

[13] Y. Xiao, K. Ahuja, and M. van der Schaar, "Spectrum sharing for delay-sensitive applications with continuing QoS guarantees," in *Proc. IEEE Global Communications Conf.*, 2014, pp. 1265–1270.

[14] Y. Xiao and M. van der Schaar, "Spectrum sharing policies for heterogeneous delay-sensitive users: a novel design framework," in *Proc. Allerton Conf. Communications Control, and Computation*, 2013, pp. 85–92.

[15] Y. Xiao and M. van der Schaar, "Optimal foresighted multi-user wireless video," *IEEE J. Sel. Topics Signal Process.*, vol. 9, no. 1, pp. 89–101, February 2015.

[16] M. van der Schaar, Y. Xiao, and W. Zame, "Efficient outcomes in repeated games with limited monitoring and impatient players," *Economic Theory*, vol. 60, no. 1, pp. 1–34, September 2015.

[17] M. van der Schaar, Y. Andreopoulos, and Z. Hu, "Optimized scalable video streaming over IEEE 802.11 a/e HCCA wireless networks under delay constraints," *IEEE Trans. Mobile Comput.*, vol. 5, no. 6, pp. 755–768, June 2006.

[18] D. Pradas and M. A. Vazquez-Castro, "NUM-based fair rate-delay balancing for layered video multicasting over adaptive satellite networks," *IEEE J. Sel. Areas Commun.*, vol. 29, no. 5, May 2011.

[19] P. Dutta, A. Seetharam, V. Arya, M. Chetlur, S. Kalyanaraman, and J. Kurose, "On managing quality of experience of multiple video streams in wireless networks," in *Proc. IEEE Infocom conf.*, 2012.

[20] R. D. Yates, "A framework for uplink power control in cellular radio systems," *IEEE J. Sel. Areas Commun.*, vol. 13, no. 7, pp. 1341–1347, September 1995.

[21] R. Etkin, A. Parekh, and D. Tse, "Spectrum sharing for unlicensed bands," *IEEE J. Sel. Areas Commun.*, vol. 25, no. 3, pp. 517–528, April 2007.

[22] M. Fang, D. Malone, K. R. Duffy and D. J. Keith, "Decentralised learning MACs for collision-free access in WLANS," *Wireless Networks*, vol. 19, no. 1, pp. 83–98, January 2013.

[23] L. Song, Y. Xiao, and M. van der Schaar, "Demand side management in smart grids using a repeated game framework," *IEEE J. Sel. Areas Commun.*, vol. 32, no. 7, pp. 1412–1424, June 2014.

Part IV

Networking in C-RANs

16 Mobility Management for C-RANs

Haijun Zhang, Julian Cheng, and Victor C. M. Leung

16.1 Introduction

Cloud-RAN is a promising wireless network architecture in 5G networks, and it was first proposed by the China Mobile Research Institute [1]. In C-RANs, baseband processing is centralized in a baseband unit (BBU) pool, while radio frequency (RF) processing is distributed in remote radio heads (RRHs). The C-RAN network architecture can reduce both the capital expenditure (CAPEX) and operating expenditure (OPEX) for mobile operators, because fewer BBUs are potentially required in the C-RAN architecture, and the consumed power is lowered [2].

Heterogeneous small-cell networks have attracted much attention owing to the explosion in demand of users' data requirements. In heterogeneous small-cell network, low-power small cells (such as pico-cells, relay cells and femto-cells), together with macro cells, can improve the coverage and capacity of cell-edge users and hotspots by exploiting the spatial reuse of spectrum [3]. Small cells can also offload the explosive growth of wireless data traffic from macro cells. For example, in an indoor environment WiFi and femtocells can offload most data traffic from macro cells [4]. For mobile operators, small cells such as femtocells can reduce the CAPEX and OPEX because of the self-installing and self-operating features of femto base stations.

The combination of a heterogeneous small-cell network and a C-RAN, which is called a heterogeneous cloud small-cell network (HCSNet), benefits from employing both C-RAN and a small-cell network [5]. First, C-RAN reduces the power and energy cost in HCSNet by lowering the number of BBUs in densely deployed heterogeneous small-cell networks. Second, BBUs can be added and upgraded without much effort in the BBU pool, and network maintenance and operation can also be performed easily. Third, many radio resource management functions can be facilitated in the BBU pool with little delay. In HCSNet, cloud-computing-enabled signal processing can be fully utilized to mitigate interference and to improve spectrum efficiency in 5G networks.

In the literature, HCSNet has been studied extensively. In [5], state-of-the-art research results and challenges were surveyed for heterogeneous C-RANs, and promising key techniques were investigated to improve both spectral and energy efficiencies. To mitigate the interference for cell-edge users, coordinated multi-point (CoMP) transmission and reception is also investigated in a C-RAN environment. The C-RAN network architecture is effective for implementing CoMP. Energy-efficient resource optimization was studied in [7] for C-RAN-enabled heterogeneous cellular networks. In [8] the authors investigated the joint transmission CoMP performance in a C-RAN implementation of an LTE-Advanced heterogeneous network with large CoMP cluster sizes; CoMP was also investigated in [5] with the aim of mitigating the interference in heterogeneous C-RANs.

Another important challenge for HCSNet is the seamless mobility handover of users. Since mobility management is a wide topic, we focus only on handover management in this chapter. Most traditional handover decision schemes are based on reference-signal receiving power and reference-signal receiving quality. The handover procedures and signaling flows in HCSNet will be different from those in traditional small cell networks [2]. In [9], the authors analyzed mobility of handover control in cloud-computing-enabled wireless networks and indicated that mobility is an inherent feature of today's wireless networks. The authors in [5] also discussed handover management in heterogeneous C-RANs; however, a quantitative analysis was missing. Generally, handover management for HCSNet has received little attention in the current literature.

Differently from the existing studies in HCSNet, we focus, in this chapter, on handover management in HCSNet. We first present the network architecture for a cloud-computing-based HCSNet, in Section 16.2. A handover management scheme, including handover procedures, is presented in Section 7.3 Finally, Section 7.4 concludes this chapter.

16.2   HCSNet Architecture

Inspired by [1, 5], we describe a network HCSNet architecture, as shown in Fig. 16.1 [6], which comprises both macro cells and small cells where CoMP is deployed. As shown in Fig. 16.1, macro base stations and small base stations are reduced to macro RRHs (MRRHs) and small RRHs (SRRHs) in HCSNet, respectively. The resource management and control capabilities of BBUs for macro cells and small cells are co-located and processed in the BBU pool. A BBU pool consists of general-purpose processors that perform baseband processing. Different BBU pools are connected by X2 interfaces. The RRHs are located in different sites to provide wireless signal coverage for the user equipment (UE). Millimeter-wave radio is used for the fronthaul links (between BBU pools to RRHs) in HCSNet.

The RRHs can be deployed on each floor of a building or office to provide enhanced coverage and capacity. They can also be deployed in a hotspot scenario, e.g., a stadium. Though the interference can be high, cooperative interference management is efficiently implemented in the HCSNet architecture. The BBU pool usually supports 100 MRRHs for a medium-sized urban network (coverage 5×5 km), and 1000 MRRHs for 15×15 km [2]. The number of SRRHs will be much greater than the number of MRRHs for the size mentioned above and it depends on the specific scenario.

In an HCSNet scenario with high mobility users, handovers may frequently happen between small cells because of the small cell size. Handover management is therefore essential and it is processed in the BBU pools.

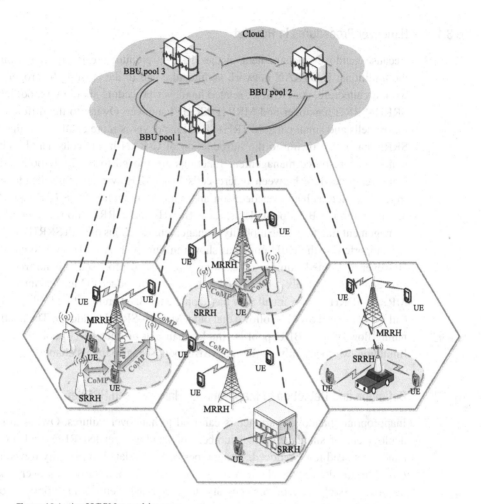

Figure 16.1 An HCSNet architecture.

16.3   Handover Management in HCSNet

Handover management is one of the key techniques to satisfy users' quality of service (QoS) requirements in mobile communications. However, handover management in HCSNet has not received enough attention in the existing literature. In [5], the authors conducted a survey for HCSNet and discussed how high-mobility UEs should be served by macro cells with reliable connections and low-mobility UEs should be served by SRRHs. In a densely deployed HCSNet, handovers occur frequently, causing a heavy burden to the fronthaul and core networks. Besides, mobility-handover-related radio link failure (RLF) and unnecessary handover (e.g., ping-pong handover) may happen because of the small size of an SRRH's coverage and severe co-channel interference. In C-RAN-enabled HCSNet, the interrupt time and delay of handover can be reduced because handover can be accomplished within the BBU pool.

16.3.1   Handover Procedures in HCSNet

Because handover management in the HCSNet architecture is different from that in the traditional E-UTRAN network architecture and in the existing heterogeneous network architecture, the HCSNet-related handover procedure should be modified in both SRRH–SRRH handover and MRRH–SRRH handover. Owing to the different sizes of macro cells and small cells, MRRH–SRRH handover is more challenging than SRRH–SRRH handover. owing to the introduction of C-RAN in the considered architecture, many radio resource management functions are moved to the BBU pool. Unlike the handover procedure between macro cells in an LTE system, many handover-related functions, such as handover decision and admission control for SRRHs and MRRHs, are moved to the BBU pool. Thus, these SRRHs and MRRHs do not support mobility management functions; the mobility management functions of both SRRHs and MRRHs are supported by the BBU pools, as shown in Fig. 16.1. A BBU pool contains layer 3. However, a limited number of existing works have proposed a handover procedure between the BBU pools in HCSNet. Therefore, we introduce in Fig. 2 an inter-BBU-pool MRRH–SRRH handover call flow based on the HCSNet architecture [10]. The handover call flow between SRRHs follows that of an MRRH–SRRH handover. The handover signaling flow of intra-BBU-pool is simpler than that of inter-BBU-pool. owing to space limitation, we focus only on the inter-BBU-pool scenario in this chapter.

16.3.2   Definition and Detection of Inappropriate Handovers in HCSNet

Inappropriate handover parameters can lead to handover failures. Owing to the dense deployment of small cells, the numbers of handover related RLFs and unnecessary handovers need to be reduced. Three types of RLF related to mobility robustness optimization are defined in HCSNet: too late handover, too early handover, and wrong handover. Two types of unnecessary handover are defined in mobility robustness optimization: ping-pong handover and continuous handover. The characteristics of mobility-related successful and unsuccessful handovers and unnecessary handovers are given in Fig. 16.3 [11]:

(a) *Regular handover*
(b) *Too late handover*: An RLF occurs under a serving RRH before handover or during the handover procedure, and then the UE reconnects to the target RRH (different from the serving RRH).
(c) *Too early handover*: An RLF occurs shortly after a successful handover to the target RRH, and then the UE reconnects to the serving RRH.
(d) *Wrong handover*: An RLF occurs shortly after a successful handover to the target RRH, and then the UE reconnects to another RRH (neither the serving RRH nor the target RRH).
(e) *Ping-pong handover*: Handover back to the serving cell from the target RRH occurs shortly after a successful handover to the target RRH.
(f) *Continuous handover*: Handover to another RRH (neither the original serving RRH nor the target RRH) occurs shortly after a successful handover to the target RRH.

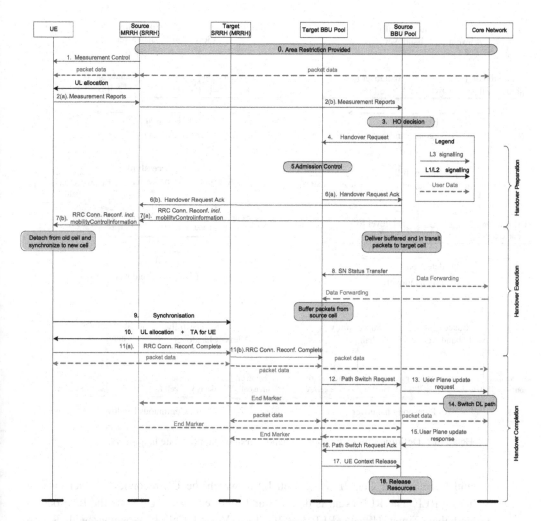

Figure 16.2 Handover procedure between SRRH and MRRH (Inter-BBU-pool) in HCSNet.

In order for the cloud to detect one of those scenarios, the following procedure is applied.

The RRH starts the timer for each UE at the moment of receiving the handover completion from the UE. During the connecting time period, if the RRH receives the RLF report from other RRHs then the RRH stops the timer. On the basis of the performance metric definitions, according to the UE's status after the RLF, the RRH categorizes the RLF as a call drop, a too late handover, a too early handover or a wrong handover.

The procedure of RLF detection related to handover and unnecessary handover is shown in Algorithm 13.

In Algorithm 13, Timer_UE_ID is the timer for the UE, RLF_UE_ID is the cell ID of the UE undergoing RLF, Last_Visited_Cell_ID is the cell ID the UE last visited,

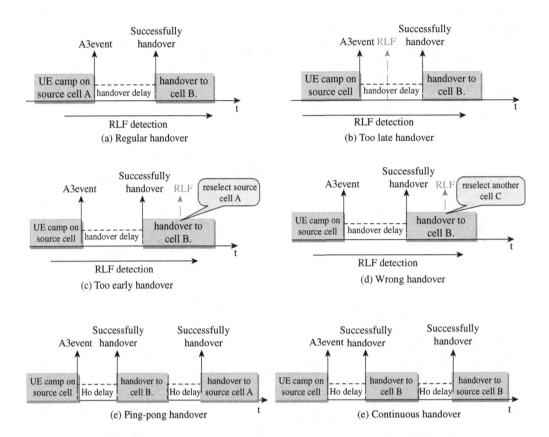

Figure 16.3 Definition of an appropriate handover and inappropriate handovers.

and Reconnected_Cell_ID is the cell ID to which the UE reconnected after encountering RLF. The RLF event is detected and reported by the UE once the RLF occurs, and then Timer_UE_ID, RLF_UE_ID, Last_Visited_Cell_ID, together with Reconnected_Cell_ID, are reported to the SRRH or MRRH and are finally collected by BBU pool.

16.3.3   A Low-Complexity Handover Optimization Scheme in HCSNet

High-speed macro cell UEs usually do not need to hand over to a small cell, while low-speed UEs may wish to hand over to a small cell, [12]. The traditional handover scheme lets the high-speed macro cell users hand over to a small cell, which may twice introduce unnecessary handovers for the user. We present a simple and effective low-complexity handover scheme to optimize the system performance. The main idea of this low-complexity scheme is as follows: Suppose that a handover is about to occur. A high-speed user does not hand over to an SRRH. For medium-speed users, those users with real-time service hand over to small cell and those users with non-real-time service do not hand over to a small cell. Low-speed users hand over to a small cell [13].

Algorithm 13 Detection of RLF and unnecessary handover.

```
 1:  Handover from cell A to cell B, start Timer_UE_ID;
 2:  if RLF Detected then
 3:      Get the RLF_UE_ID and the Reconnected_Cell_ID;
 4:      Stop Timer_UE_ID;
 5:      if Reconnected_Cell_ID==Last_Visited_Cell_ID then
 6:          if Timer_UE_ID<Time_Threshold then
 7:              Too_Early_Handover_Counter++;
 8:      else if Timer_UE_ID<Time_Threshold then
 9:          Wrong_Handover_Counter++;
10:      else
11:          Too_Late_Handover_Counter++;
12:  else if Handover from cell B to cell C then
13:      Stop Timer_UE_ID;
14:      if Timer_UE_ID<Timer_Threshold then
15:          if Target_Cell_ID==Last_Visited_Cell_ID then
16:              Ping-Pong_handover_Counter++;
17:          else
18:              Continuous_Handover_Counter++;
```

Note that the UE's speed can be estimated from the Doppler-spread frequency or the autocorrelation function [14].

In order to verify the performance of the optimized schemes presented in this subsection, we compare an optimized scheme with a traditional handover scheme in terms of the system signaling overhead in different scenarios.

Figure 16.4 shows the signaling overhead (which is unitless and proportional to the delay required to send or process a signaling message) versus the mean of the session holding time in HCSNet, with the proportion of high mobility state users α set to 0.1 [6]. As seen in Fig. 16.4, the total signaling overhead increases as the mean of the session holding time increases. This is so because the larger is the session holding time, the larger is the probability of cell-crossing, implying a high probability of handover.

Figure 16.5 plots the signaling overhead versus the proportion of high-mobility-state users [6]. As shown in Fig. 16.5, the number of handovers and the signaling overhead in the traditional scheme increases with higher α, while the signaling overhead decreases in the optimized handover scheme. In the scheme optimized in this section, we do not allow the high-speed users to hand over from an MRRH to an SRRH, while the low-speed users are allowed to do so. Therefore, when α increases to close to 1, the signaling cost in the optimized algorithm decreases to zero. From Figs. 16.4 and Fig. 16.5, it can be seen that, as the average session-holding time increases, the signaling costs in both the traditional algorithm and the optimized algorithm increase, since more handovers are expected with an increase in the session-holding time. In a traditional handover algorithm for SRRHs, high-speed users and low-speed users are treated in the same way.

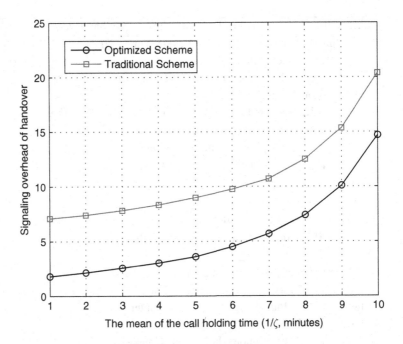

Figure 16.4 Signaling overhead versus the mean of the holding time $1/\zeta$.

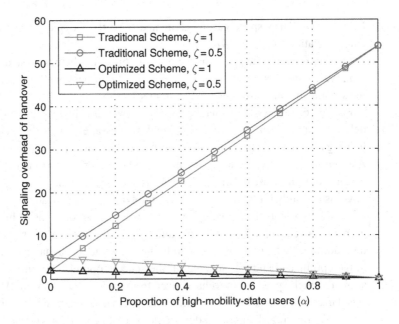

Figure 16.5 Signaling overhead versus high-speed user proportion α.

Two unnecessary handovers happen as the UE moves from MRRH to SRRH. As the optimized scheme does not allow high-speed users' to have two unnecessary handovers, the total cost of the handover is reduced. As a result, there is a big decrease in the optimized scheme's signalling overhead.

16.4   Conclusion

In this chapter, we examined handover detection and handover management in a HCSNet, where a C-RAN is combined with small cells. A handover management scheme was presented, and handover signaling procedures were analyzed for HCSNet. Numerical results demonstrate that the proposed network architecture and handover management scheme can significantly maintain users' quality of service. Moreover, self-organized HCSNet can also be considered as a future direction, where interference mitigation and handover management are controlled in a self-organizing way [15]. Handover management can be enhanced by automatic neighbor relations and physical-cell ID self-configuration in self-organized HCSNet. Handover management in HCSNet can also be optimized by a swarm intelligence algorithm [16].

References

[1] M. Peng, Y. Li, Z. Zhao, and C. Wang, "System architecture and key technologies for 5G heterogeneous cloud radio access networks," *IEEE Netw.*, vol. 29, no. 2, pp. 6–14, March 2015.

[2] A. Checko, H. L. Christiansen, Y. Yan, L. Scolari, G. Kardaras, M. S. Berger, and L. Dittmann, "Cloud RAN for mobile networks – a technology overview," *IEEE Commun. Surveys & Tutorials*, to appear.

[3] T. Q. S. Quek, G. de la Roche, I. Guvenc, and M. Kountouris, *Small Cell Networks: Deployment, PHY Techniques, and Resource Allocation*, Cambridge University Press, May 2013.

[4] M. Peng, X. Xie, Q. Hu, J. Zhang, and H. V. Poor, "Contract-based interference coordination in heterogeneous cloud radio access networks," *IEEE J. Sel. Areas Commun.*, vol. 33, no. 6, pp. 1140–1153, June 2015.

[5] M. Peng, Y. Li, J. Jiang, J. Li, and C. Wang "Heterogeneous cloud radio access networks: a new perspective for enhancing spectral and energy efficiencies," *IEEE Wireless Commun.*, vol. 21, no. 6, pp. 126–135, December 2014.

[6] H. Zhang, C. Jiang, J. Cheng, and V. C. M. Leung, "Cooperative interference mitigation and handover management for heterogeneous cloud small cell networks," *IEEE Wireless Commun.*, vol. 22, no. 3, pp. 92–99, June 2015.

[7] M. Peng, K. Zhang, J. Jiang, J. Wang, and W. Wang, "Energy-efficient resource assignment and power allocation in heterogeneous cloud radio access networks," *IEEE Trans. Veh. Technol.*, vol. 64, no. 11, pp. 5275–5287, November 2015.

[8] A. Davydov, G. Morozov, I. Bolotin, and A. Papathanassiou, "Evaluation of joint transmission CoMP in C-RAN based LTE-A HetNets with large coordination areas," in *Proc. 2013 IEEE Globecom Workshops*, pp. 801–806, December 2013.

[9] S. Chen, Y. Shi, B. Hu, and M. Ai, "Mobility-driven networks (MDN): from evolution to visions of mobility management," *IEEE Netw.*, vol. 28, no. 4, pp. 66–73, July 2014.

[10] H. Zhang, W. Zheng, X. Wen and C. Jiang, "Signalling overhead evaluation of HeNB mobility enhanced schemes in 3GPP LTE-Advanced," in *Proc. 2011 IEEE 73rd Vehicular Technology Conf.*, pp. 1–5, May 2011.

[11] W. Zheng, H. Zhang, X. Chu, and X. Wen, "Mobility robustness optimization in self-organizing LTE femtocell networks," *EURASIP J. Wireless Commun*, vol. 27, February 2013.

[12] H. Zhang, X. Wen, B. Wang, W. Zheng and Y. Sun, "A novel handover mechanism between femtocell and macro cell for LTE based networks," in *Proc. IEEE ICCSN, Conf.* pp. 228–232, Singapore, February 2010.

[13] H. Zhang, W. Ma, W. Li, W. Zheng, X. Wen, and C. Jiang, "Signalling cost evaluation of handover management schemes in LTE-Advanced femtocell," in *Proc. 2011 IEEE 73rd Vehicular Technology Conf.*, pp. 1–5, May 2011.

[14] L. Song, M. Peng, B. Lv, M. Wang, and H. Jiang, "Speed estimation in uplink frequency domain for mobile OFDM systems," in *Proc. 2014 IEEE/CIC Int Conf. Communications in China*, pp. 458–462, October 2014.

[15] H. Zhang, C. Jiang, Q. Hu, and Y. Qian, "Self-organization in disaster resilient heterogeneous small cell networks," *IEEE Network*, vol. 30, no. 2, pp. 116–121, March–April 2016.

[16] H. Zhang, H. Liu, W. Ma, W. Zheng, X. Wen and C. Jiang, "Mobility robustness optimization in femtocell networks based on ant colony algorithm," *IEICE Trans. Commun.*, vol. E95-B, no. 4, pp. 1455–1458, April 2012.

17 Caching in C-RAN

Kenza Hamidouche, Walid Saad, and Mérouane Debbah

17.1 Introduction

Mobile traffic has shown an exponential growth during the past decade and is expected to increase by five times in the upcoming five years [1]. This growth is mainly driven by the proliferation of smart devices such as smartphones, mobile tablets, and cameras, coupled with the rising popularity of social networks. Most of the generated traffic is for mobile videos; this is envisioned to represent 80% to 90% of the total generated traffic by 2019 [1, 2]. To support this rapidly growing traffic and meet the strict quality of serivce (QoS) requirements of video streaming users, wireless networks have evolved into a distributed heterogeneous architecture [3]. This structure is met by a dense deployment of low-power and low-coverage small base stations (SBSs), which offload the traffic from the conventional macro base stations [4]. The use of dense small base stations allows boosting of the network capacity by reducing the distance between the SBSs and the end-user. Moreover, it gives the possibility of sharing the spectrum more efficiently by an improved spectrum-reuse ratio across multiple cells, ensuring higher transmission rates for the users. However, as the network becomes denser, inter-cell and intra-cell interference become more important and cause the overall system performance to deteriorate [5]. Different approaches have been proposed to alleviate this mutual interference, especially leveraging the cooperation between base stations (BSs) through coordinated multi-point (CoMP) transmission techniques [6, 7]. In CoMP, the base stations exchange information about channel state and cooperate to form a group that serve a given user. Even though this approach can significantly improve the spectral efficiency and increase the system throughput, its performance depends on the number of base stations in the network as well as the capacity of the backhaul links. In fact, cooperation between the distributed base stations requires an important amount of signaling and exchange of control information between the BSs, which limits the performance of CoMP in capacity-limited backhaul networks [6].

Recently, a centralized (or cloud) network architecture known as C-RAN [8] was introduced as an energy-efficient and low-cost paradigm with high-processing capabilities. In C-RAN the cooperation between the BSs can be implemented in a centralized and efficient way. The idea of C-RAN consists in virtualizing the BSs by moving the processing units of the BSs to a central, virtual, baseband (BBU) pool in which all the operations and computations are performed. Remote radio heads (RRH), which are composed only of antennas, can then be deployed densely at a very low cost. The virtual baseband pool is connected to the remote radio heads via high-capacity fiber-optical fronthaul links to support the significant amount of data exchange between the BBU pool and the RRHs. However, with the increasing amount of video traffic with high QoS requirements, the fronthaul capacity becomes the network bottleneck owing to the difficulty of satisfying all the users. One promising way to overcome the fronthaul capacity limitation is through distributed caching. The idea consists of equipping the RRHs with storage units and storing, ahead of time, the most popular content, closer to the users in the RRHs. Thus, most users' requests can be served from the network edge without using the fronthaul. Distributed caching in small-cell networks has gained recently the interest of both academia and industry. An extensive overview can be found in [9]. However, only a few works have focused on the application of caching to C-RANs. In this chapter we provide an overview of recent works that have considered caching in C-RAN in order to create more CoMP opportunities in the network and thus simultaneously reduce, interference as well as the load on the fronthaul links. Moreover, two game-theoretic tools, namely mean-field games and matching games, are used to develop new efficient caching strategies that take into account the high density of the RRHs and exploit social networks to predict user requests.

This chapter is organized as follows. In Section 17.2 we introduce the architecture of C-RAN with its main advantages and challenges. Section 17.3 presents the general idea of caching in wireless networks and its impact on the different functionalities in C-RAN. Section 17.4 provides an overview of recent work that has studied caching in C-RAN networks. In Section 17.5 we address the caching problem in ultra-dense C-RAN from a game-theoretic point of view. Finally, we offer a summary in Section 17.6.

17.2   Generalities on C-RANs

Cloud-RANs provides a new centralized architecture of broadband wireless access that refers to clean, centralized, processing, collaborative radio, and a real-time cloud radio access network. This structure relies on the idea of shifting the baseband units that are located at the BSs to a centralized baseband pool in order to ensure a high degree of cooperation and communication between the BSs. By incorporating cloud computing capabilities into wireless networks, the BBUs can be fully exploited using appropriate dynamic task-scheduling approaches for large-scale cooperative signal processing and network functionalities. Moreover, both operating expenses and energy consumption are decreased by reducing the number of BS sites as well as the distance between the RRHs and the end-users. In the following, we present the architecture of a typical C-RAN network and its advantages compared with the currently deployed wireless networks.

17.2.1   Architecture of a C-RAN

A centralized radio access network is composed of a set of remote radio head sites equipped with antennas that are connected to a centralized BBU pool of high processing capabilities. The connection between the RRH sites and the BBU pool is ensured via

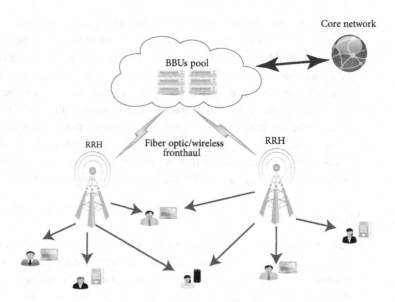

Figure 17.1 Architecture of a C-RAN.

a fronthaul network that requires a high bandwidth capacity to support all the traffic from the densely deployed RRHs. The general architecture of a C-RAN is provided in Fig. 17.1, and a detailed description of each component is given next.

- **The remote radio head** The RRH is the component of the base station that ensures the tranmission and reception of radio frequency signals from and to the users. This entity includes radio frequency amplification, up/down conversion, filtering, A/D and D/A conversion, and interface adaptation [10]. In traditional wireless networks the signals are transmitted to the baseband component of the base station for further processing. By removing the BBU from the base station, the RRH structure becomes simpler and RRHs can be deployed densely in the network to achieve high data rates with a wide area coverage in a cost-effective way.
- **The baseband unit (BBU) pool** The BBU pool is the part of the network that is in charge of signal processing, optimizing the network operations, and resource allocation. The BBU pool is a virtual BS that aggregates all the BBUs that are located at the BSs in traditional wireless networks. Depending on the complexity of the tasks, the network load, and the expected performance, the BBUs can be assigned to the RRHs in different ways. A distributed implementation in which one RRH is assigned to a corresponding BBU is the simplest approach, as it requires less coordination and synchronization between the different BBUs in the pool. However, the processing capabilities of the BBU pool are not fully exploited as some complex operations can be performed by one BBU while other BBUs are not operating. Moreover, centralized processing can be used for the efficient joint scheduling of tasks, mitigating inter- and intra-cell interference, through coordinated multi-point processing [7], and reducing

energy consumption by turning off inactive BBUs. Hence, centralizing the processing offers a high flexibility that allows the enhancing of the global network efficiency.

- **The fronthaul link**   The fronthaul link connects the RRHs to the BBU pool and can be a combination of different technologies and wired and wireless links. Since all the processing is done at the BBU pool, the fronthaul link need to be of high capacity to support all the traffic and signal exchanges between the BBU pool and the RRHs. Optical fiber is the ideal solution as it offers a high data rate but, owing to its expensive deployment, depending on the geographical areas in which the RRHs are deployed it is not always practical to use it. Thus, the deployment of wireless fronthaul including millimeter-wave and sub-6-GHz bands is also considered to support the traffic [11].

17.2.2   Advantages of C-RAN

- **Energy efficiency**   The virtualization of base stations in C-RAN and the creation of a centralized processing entity allows a reduction in the number of BS sites in the network. This results in a considerable reduction in the required energy for cooling the distributed sites and operating the network equipment. Moreover, by simplifying the BS operations and reducing the deployment cost, the RRHs can be densely deployed in order to shorten the distances between the RRHs and the users. Thus, different RRHs can cooperatively serve users with low signal-transmission power while allowing the users to extend their batteries' stand-by time. A further saving of power is achieved by sharing the processing resources among a large number of RRHs. In addition, the BBUs can be turned off selectively when they are inactive without affecting the service availability.

- **Cost saving on CAPEX and COPEX**   The centralized BBU pool architecture helps in achieving a considerable reduction in the operating and maintenance costs compared with those for the traditional distributed architecture of a RAN network. In fact, with the simplification of operations at the RRHs, which become smaller with a lower energy consumption, the RRHs require less maintenance and management after installation of the antennas.

- **Capacity improvement**   By reducing the CAPEX and COPEX costs, network operators can deploy the RRHs at a very low cost and achieve high coverage. Moreover, the joint-resource-allocation and interference-cancelation approaches can be implemented efficiently in a centralized way through the BBU pool to improve spectral efficiency. In fact, by allowing cooperative processing in a central entity, it becomes easier for the RRHs to coordinate and share information such as channel state information and traffic load.

- **Load balancing**   In a traditional RAN, depending on the geographical position of the base station and the time period, the load can be different from one BS to another BS. For instance, while some overloaded BSs cannot serve all their users, other BSs are idle and cannot take in charge some users or processing operations. The C-RAN architecture can efficiently cope with this heterogeneous distribution of the BS traffic load by implementing load-balancing protocols to distribute uniformly the processing load over the BBUs in the BBU pool.

17.2.3   Challenges in C-RAN

- **Heterogeneous fronthaul**   In order to support the huge amount of data exchanged between the BBU pool and the RRHs, fiber optical links are considered the most appropriate solution for C-RANs. However, the deployment of fiber optical links to connect the densely deployed RRHs may be very costly for the operators and also difficult, depending on the geographical area of the base stations. Thus, considering a deployment of heterogeneous fronthaul, including sub-6-GHz bands and millimeter waves that can coexist together via carrier aggregation techniques, is an inevitable solution. However, these different technologies offer different QoS and thus, depending on the fronthaul load and the sensitivity of the content, new approaches need to be considered that allow the transmission of data via the appropriate fronthaul links.

- **Low-latency applications**   The recent increasing interest in the Internet of Things (IoT) makes a latency requirement critical for many applications in communication networks, such as high-performance online gaming, online trading, and M2M sensory data communications. In C-RANs, it becomes difficult to meet such stringent QoS requirements, which are in the order of milliseconds, owing to the centralized processing and congested fronthaul. A hybrid architecture in which some BSs remain equipped with BBUs is more appropriate to support latency-sensitive applications [12–14]. However, different challenges rise and new approaches need to be developed, in such an infrastructure, for user–BS association and resources allocation [15].

- **Joint design and optimization of C-RANs**   The traditional resource allocation, user scheduling, and interference cancelation methods, which were performed in a distributed manner at the BSs, are not appropriate to the new architecture of radio access networks. In order to benefit fully from the virtualization of the BSs, through the possible cooperation between the virtual BSs at a very low cost, new dynamic and flexible approaches should be developed for the optimization of the operations in a C-RAN while taking into account the specific characteristics of its architecture.

- **Fronthaul-limited capacity**   By moving all the processing capabilities to a centralized BBU pool, all the signaling and information exchange between the traditional BSs is supported by the fronthaul links. Owing to the high density of RRHs, even fiber optical links that offer a high data rate are not able to support the induced communication load. Different approaches have been considered recently, such as allowing the BSs to be in charge of some of the processing operations locally before transmitting the content to the BBU pool for more complex operations. Another approach to deal with capacity-limited fronthaul is through distributed caching, which consists in equipping the RHHs with storage units in order to serve users locally without using the fronthaul.

In the next section, we address the problem of the fronthaul capacity in C-RANs through the idea of distributed caching.

17.3   General Idea of Distributed Caching

The idea of caching relies on the premise of equipping the RHHs with storage units to reduce the load on the fronthaul links. This is made possible by predicting users' requests and downloading the most popular content ahead of time through the fronthaul. By downloading the content during off-peak hours and serving users from the cache during peak hours, the network can achieve a high-capacity gain, and users experience a considerable data rate improvement. The idea of distributed caching, to overcome the limited capacity of the backhaul links in small cell networks has received considerable interest during last few years. Most proposed algorithms have focused on developing distributed approaches that allow small base stations to take their decisions independently for a further reduction in the signaling overhead on the backhaul links. However, with the virtualization of base stations in C-RAN, applying the existing caching approaches would not take advantage of the offered capabilities in C-RAN such as possible cooperation between the virtualized base stations and the centralized processing at the BBU pool, which is aware of the global system state. The impact of caching on the different network layers in C-RAN can be summarized as follows.

- **Impact of caching on physical layer**   Depending on the cache placement strategy that is used, the transmission mode for users can be different. For instance, the signal quality can be enhanced and the spatial interference reduced by serving a user via a joint transmission technique between a number of RRHs. In order to be able to employ such an interference-mitigation technique, all the cooperating RRHs must cache the requested file by the served user. Thus, there is a mutual dependence between the transmission mode and the cache state. Dynamic selection of cooperating RHH schemes based on the cache state as well as on the cache placement strategies, which depend on the RRHs' geographical placement, can improve considerably the achievable multiplexing gain.
- **Impact of caching on MAC control**   The scheduling strategies should be different for users that have their data cached at the RRHs and those who do not. For instance, by associating a given user with a group of RRHs that can cache its requested content and serve him jointly without using the backhaul, the user can experience an improved QoS. Thus, dynamic resource allocation protocols that take into account the cache state and the advantages of C-RAN through centralized processing can help in achieving efficient use of the network resources.
- **Impact of caching on the fronthaul load**   Fronthaul capacity is the main limiting bottleneck of C-RAN, owing to the huge signaling load that is exchanged by the BBU pool and the large number of RRHs at the network edge. Hybrid radio access networks are of solution to this issue, since they allowing the RHHs units and the BBU pool to share the processing operations. Another approach, which is the subject of this chapter is through caching. By caching the most popular content and serving users from the network edge, more fronthaul capacity can be allocated for signaling and thus latency can be reduced and the achievable throughput by the users increased.

The most important parameters that should be considered when designing caching approaches for wireless networks can be summarized as follows:

- **Cache-size-limited capacity** Owing to the limited capacity of the storage units at the RRHs, it is not possible to store all the content. Thus, dynamic caching and replacement approaches are required to refresh the storage units and serve the maximum amount of requests.
- **Fronthaul-limited capacity** As for the storage space, the capacity of the heterogeneous fronthaul is also limited and the content cannot be downloaded from the core network at any time. Thus, caching approaches should take into account the fronthaul load using traffic prediction approaches such as machine learning.
- **File popularity** By caching the most popular files at the RRHs, a large number of requests can be served locally, saving the fronthaul capacity. In order to predict user requests, social networks can be exploited and machine learning approaches used for this purpose.
- **QoS requirement of the users** Depending on the type of the application or the requested content, users can have different QoS requirements. Moreover, the caching problem can be defined to optimize many parameters, such as the total transmission delay, the total number of served requests, or the total energy consumption.

17.4 Cooperative Caching in C-RAN

In this section, we review two recent works that have considered caching in hybrid RAN and C-RAN networks [16, 17]. The proposed idea of caching in these works consists in equipping the RRHs with storage units and caching the content at the RRHs to create more coordinated multi-point transmission opportunities while serving the users. In such a model, in addition to the capacity gain from caching the content closer to the users, CoMP opportunities allow a reduction the inter-cell interference. In these RAN architectures, the opportunity of serving users through a CoMP transmission depends heavily on the caching strategy used. For instance, RRHs that cache a given requested file can cooperate and serve the user without employing the backhaul, while this cannot be performed if the requested file is not available in the cache of the RRHs. In [16] a two-time-scale joint optimization of power and cache control to serve requests for real-time video on demand was formulated and analyzed in a hybrid RAN network. Dynamic power control was introduced to exploit CoMP opportunities in order to serve a maximum number of users while satisfying the required QoS of users. The proposed dynamic model is thus adaptive to the instantaneous channel state information (CSI), the cache state at the RRHs, and the queue state information (QSI) at the mobile users. In another work [17], the authors considered a joint-time-scale optimization of MIMO precoding and cache control for C-RAN networks. A joint cache control and MIMO precoding algorithm was proposed to ensure a given QoS for users by reducing interference in MIMO networks using cache-enabled opportunistic CoMP. In the following, we give more details on the proposed solutions.

17.4.1   Joint Power and Cache Control

System State and Caching Model

Consider a hybrid RAN network composed of $2M$ users that can be served cooperatively by a $2M$-antenna BS and an M-antenna RRH, if the requested content is available at the cache of the RRHs, or directly through the fronthaul by the BS if the content is not available. The requested file by user k is denoted π_k and the global user-request profile is $\pi = \{\pi_1, \ldots, \pi_{2M}\}$. The global system state is denoted $\mathcal{X} = (\mathbf{Q}, S, \mathbf{H})$ and represent respectively the queue state, the cache state, and the channel state; $\mathbf{H} = \{\mathbf{h}_k, \forall k\}$ represents the channel vector between the $2M$ users and the $2M$ antennas, where $\mathbf{h}_k \in \mathbb{C}^{2M}$ is the channel vector of user k and the $2M$ transmit antennas. The cache state is denoted by a binary variable $S \in \{0, 1\}$, where $S = 1$ corresponds to the case in which all the requested video files by the $2M$ users are cached at the RRH. In this case, all the users can be served cooperatively by the BS and the RRH. The cache state $S = 0$ corresponds to the case in which the requested contents are not cached and the users can be served only by the BS.

Finally, each user is assumed to maintain a queue for its video file. The state of the queue of user k at time slot t, which is of size τ, is denoted $Q_k(t) \in \mathcal{Q}$, where $\mathcal{Q} \in [0, \infty)$ is the QSI state space. The queue state is given by the number of bits that are in the user's queue. The global QSI of all the $2M$ mobile users in the network is denoted $\mathbf{Q}(t) = (Q_1(t), \ldots, Q_K(t)) \in \mathcal{Q}^{2M}$. The playback rate of a given user k depends on the amount of video content in its own queue $Q_k(t)$ and is given by:

$$\mu_k(Q_k(t)) = \begin{cases} \dfrac{Q_k(t)\mu_0}{W_L}, & Q_k(t) < W_L, \\ \mu_0, & Q_k(t) \geq W_L, \end{cases} \tag{17.1}$$

where $W_L \geq \mu_0 \tau$ is a parameter. If $Q_k(t) \geq W_L$, then a constant display-rate μ_0 is maintained. Otherwise, the departure rate is $Q_k(t)\mu_0/W_L$ to avoid buffer underflow. The queue dynamics at user k is given by

$$Q_k(t+1) = Q_k(t) + \left[r_k(g_k(t), p_k(t)) - \mu_k(Q_k(t)) \right] \tau. \tag{17.2}$$

The quality of the watched video by a user is degraded if the amount of video content at the queue is not enough to ensure the required minimum display rate, i.e., $Q_k < W_L$. The QoS requirement of each user is expressed in terms of an interruption probability,

$$I_k = \Pr[Q_k < W_L]. \tag{17.3}$$

QoS Metric

In order to minimize the global interruption cost in the network, the values of the cache control variables, denoted $\mathbf{q} = [q_1, \ldots, q_L]$, where $q_i \in [0, 1]$ is the fraction of file i that is cached at the RRH, as well as the power control policy must be determined while taking into account the global system state. The power control policiy is defined as follows.

DEFINITION 17.1 A stationary power-control policy for the kth user ω_k is a mapping from the global system state $\mathcal{X}$ of the power-control actions of the kth user. Specifically, $p_k = \omega(\mathcal{X})$. Furthermore, let $\omega = \{\omega_k, \forall k\}$.

Smooth approximations to the interruption probability and queue-overflow probability are given respectively by

$$I_k^{(q,\pi,\omega)} = \lim_{T\to\infty} \sup \frac{1}{T} \sum_{t=1}^{T} \mathbb{E}^{(q,\pi,\omega)} \left\{ e^{-\alpha(Q_k(t)-W_L)^+} \right\}, \tag{17.4}$$

$$B_k^{(q,\pi,\omega)} = \lim_{T\to\infty} \sup \frac{1}{T} \sum_{t=1}^{T} \mathbb{E}^{(q,\pi,\omega)} \left\{ e^{-\alpha(W_H-Q_k(t))^+} \right\}, \tag{17.5}$$

where $W_H > W_L$ is the target maximum display-queue size at the user. The average transmit power at the BS and the RRH is given by

$$P^{(q,\pi,\omega)} = \lim_{T\to\infty} \sup \frac{1}{T} \sum_{t=1}^{T} \mathbb{E}^{(q,\pi,\omega)} \left\{ \sum_{k=1}^{2M} p_k(t) \right\}, \tag{17.6}$$

which is the summation of the average transmit power at the BS and the average transmit power at the RRH. The final system cost is given by

$$C(q,\pi,\omega) = \sum_{k=1}^{2M} \left(\beta_k I_k^{(q,\pi,\omega)} + \gamma_k B_k^{(q,\pi,\omega)} \right) + P^{(q,\pi,\omega)} \tag{17.7}$$

where $\beta_k > 0$ and $\gamma_k > 0$ are respectively the interruption price and the queue price of user k. The joint cache and power-control problem is formulated as the following two-time-scale optimization problem:

$$\mathcal{P} : \underset{q}{\text{minimize}} \quad \left(\mathbb{E}\left[\min_{\omega} C(q,\pi,\omega) \right] + \eta \sum_{l=1}^{L} F_l q_l \right)$$

$$\text{s.t.} \quad q_l \in [0,1], \quad \forall, l \tag{17.8}$$

where the expectation is taken with respect to the distribution of users' requests π. The term $\sum_{l=1}^{L} F_l q_l$ corresponds to the cache size constraint at the RRH and $\eta > 0$ is the cache price.

In order to solve this optimization problem, two tractable subproblems are derived; they are power-control and cache-control optimization problems and are solved as follows.

Power-Control Policy

This involves the optimization of ω for a given q and π:

$$\mathcal{P}_I(q,\pi) : C^*(q,\pi) = \min_{\omega} C(q,\pi,\omega). \tag{17.9}$$

THEOREM 17.2 *The optimal power control is given by*

$$p_k = \left(-\frac{\tilde{f}_k(Q_k)}{\ln 2} - \frac{1}{g_k} \right) \tag{17.10}$$

where g_k is the effective channel distrubution and $\tilde{f}_k$ is the unique solution of the approximate Bellman equation that ensures the optimality of $\mathcal{P}_I(q, \pi)$.

Cache Control Policy

This involves the optimization of q:

$$\mathcal{P}_O : \underset{q}{\text{minimize}} \quad \left(\mathbb{E}\{C^*(q, \pi)\} + \eta \sum_{l=1}^{L} F_l q_l \right)$$

$$\text{s.t.} \quad q_l \in [0, 1], \quad \forall l. \tag{17.11}$$

To solve this subproblem, an asymptotically accurate closed-form approximation $\hat{C}(q, \pi)$ for $C^*(q, \pi)$ is derived; $\hat{C}(q, \pi)$ is shown to be convex with respect to q and consequently the problem is convex. A stochastic subgradient algorithm that converges to the optimal solution of the approximate problem is then proposed without knowing the distribution of users' requests π. On the basis of this simplification, a stochastic subgradient algorithm is proposed to asymptotically determine the solution of $\mathcal{P}_O$.

17.4.2   Joint MIMO Precoding and Cache Control

In this model, distributed caching in an C-RAN network is considered. The network is composed of K users and each user is associated with one RRH. Each RRH and user are equipped with N_T and N_R antennas, respectively. The channel matrix between user k and RRH n on subcarrier m is denoted $H_{m,k,n} \in \mathbb{C}^{N_R \times N_T}$. The channel gain between RRH n and user k is denoted $g_{k,n}$. The BBU is assumed to be aware of the global CSI H. Depending on the cache state, the users are served in different manners. If the requested file by a given user is not cached at the RRHs, i.e $S = 0$, then the user is served through the fronthaul by its corresponding RRH. Linear precoding and MMSE receiving for inter-cell interference cancelation are considered. Thus, the received signal by a user k on subcarrier m in this case can be given by

$$y_{m,k} = H_{m,k,k} V_{m,k} s_{m,k} + \sum_{n \neq k} H_{m,k,n} V_{m,n} s_{m,n} + z_{m,k}, \tag{17.12}$$

where $s_{m,k} \in \mathbb{C}^{d_{m,k}} \mathcal{CN}(0, I)$ and $d_{m,k}$ are respectively the data vector and the number of data streams for user k on subcarrier m; $V_{m,k} \in \mathbb{C}^{d_{m,k}}$ is the precoding matrix for user k on subcarrier m; and $z_{m,k} \in \mathcal{CN}(0, I)$ is the AWGN noise vector. The MMSE receiver for user k on subcarrier m is given by

$$U_{m,k} = \left(\omega_{m,k} + H_{m,k,k} V_{m,k} V_{m,k,k}^\dagger H_{m,k}^\dagger \right)^{-1} H_{m,k,k} V_{m,k}, \tag{17.13}$$

where $\omega_{m,k} = I + \sum_{n \neq k} H_{m,k,n} V_{m,n} V_{m,n}^\dagger H_{m,k,n}^\dagger$ is the interference-plus-noise covariance matrix of user k on subcarrier m. For a given CSI H, cache state $S = 0$, and precoding matrices $V = \{V_{m,k} : \forall m, k\}$, the capacity achieved by user k is

$$R_k(H, V) = \frac{B_W}{M \ln 2} \sum_{m=1}^{M} r_{m,k}, \tag{17.14}$$

where B_W is the bandwidth capacity. The total transmit power of all the RRHs for cache states $S = 0$ and $S = 1$ is given by

$$P(V) = \sum_{k=1}^{K} \sum_{m=1}^{M} \mathrm{Tr}(V_{m,k} V_{m,k}^{\dagger}), \tag{17.15}$$

$$\tilde{P}(\tilde{V}) = \sum_{k=1}^{K} \sum_{m=1}^{M} \mathrm{Tr}(\tilde{V}_{m,k} \tilde{V}_{m,k}^{\dagger}), \tag{17.16}$$

where $\tilde{V} = \{\tilde{V}_{m,k} : \forall m, k\}$ are the precoding matrices when the cache state is $S = 1$. For a given set of control variables (q, V) and user-request profile π, the average total transmit power is given by

$$\bar{P}(q, V) = (1 - \min_{k}\{q_{\pi_k}\}) \mathbb{E}\{P(V(\pi, H))|\pi\} + \min_{k}\{q_{\pi_k}\} \mathbb{E}\{\tilde{P}(\tilde{V}(\pi, H))|\pi\}. \tag{17.17}$$

The feasible sets for cache control q, coordinated precoding $V(\pi, H)$, and CoMP precoding $\tilde{V}(\pi, H)$ are denoted respectively by $\mathcal{D}_q$, $\mathcal{D}_v(\pi, H)$, and $\mathcal{D}_{\tilde{v}}(\pi, H)$. The quality experienced by each user is expressed by the interruption probability, which corresponds to the probability that the display queue is empty. The display rate is assumed to be constant and is denoted μ_{π_k} for user k; it is constrained by the achievable rate $R_k(H, V) \geq \mu_{\pi_k}$, when $S = 0$, and $\tilde{R}_k(H, \tilde{V}) \geq \mu_{\pi_k} \geq \mu_{\pi_k}$ when $S = 1$, which guarantees that the display process does not encounter any interruption. Then the joint cache and precoding problem is formulated as

$$\mathcal{P} : \underset{q \in \mathcal{D}_v, V}{\text{minimize}} \quad \mathbb{E}\{\bar{P}_{\pi}(\mathcal{D}_v, V)\} \tag{17.18}$$

$$\text{s.t.} \quad V(\pi, V) \in \mathcal{D}_v(\pi, H), \quad \tilde{V}(\pi, H) \in \mathcal{D}_{\tilde{v}}(\pi, H), \tag{17.19}$$

where the expectation is taken with respect to the distribution of π. Due to the nonconvexity of the formulated problem $\mathcal{P}$, it first needs to be decomposed into two simpler optimization subproblems; then each problem is solved separately. The precoding control problem is solved for both cases of the cache state, $S = 0$ and $S = 1$, and then, using he optimal solution, the cache control problem is analyzed and solved.

Subproblem 1a Coordinated MIMO precoding for given π, H and $S = 0$:

$$\mathcal{P}_S(\pi, H) : \quad \underset{V}{\text{minimize}} \qquad \qquad P(V)$$

$$\text{s.t} \quad V \in \mathcal{D}_v(\pi, H). \tag{17.20}$$

Subproblem 1b CoMP precoding for given π, H and $S = 1$:

$$\tilde{\mathcal{P}}_S(\pi, H) : \quad \underset{\tilde{V}}{\text{minimize}} \quad \tilde{P}(\tilde{V})$$

$$\text{s.t} \qquad \qquad \tilde{V} \in \mathcal{D}_{\tilde{v}}(\pi, H). \tag{17.21}$$

The formulated problem involves the minimization of the total power under individual rate constraints in parallel interference networks. To solve the problem, a generalized weighted sum-rate maximization problem is formulated for both $\mathcal{P}_S$ and $\tilde{\mathcal{P}}_S$.

Subproblem 2: Cache control for given $\mathcal{V}$:

$$\mathcal{P}_L(\mathcal{V}) : \min_{q \in \mathcal{D}_q} \psi(q, \mathcal{V}) = \mathbb{E}\{\bar{P}_\pi(q, \mathcal{V})\}. \tag{17.22}$$

This problem is shown to be a convex stochastic optimization problem when $\mathcal{V}^*$ is a stationary point of $\mathcal{P}_S(\pi, H)$ and $\tilde{\mathcal{P}}_S(\pi, H)$. A stochastic subgradient algorithm is then proposed and is proved to converge to the optimal solution of $\mathcal{P}_L(\mathcal{V}^*)$ knowledge of the distribution of users requests π.

17.5   Game Theory for Distributed Caching in C-RAN

In this section we show how game theory can be used to model the caching problem in centralized and hybrid RANs. In particular, we use matching-games to design an adaptive caching approach that can be applied either at a centralized entity such as the BBU pool or in a distributed manner when BSs are integrated into a C-RAN. We explore mean-field games to deal with the expected high density of RRHs and to design an effective dynamic caching strategy in such networks. Moreover, we present another caching strategy in which the caching problem is formulated as a matching game while the virtual connection between users in a social network is exploited to predict users' requests.

17.5.1   Mean Field Games for Distributed Caching in Ultra-Dense Networks

System Model

Consider a C-RAN network composed of a set $\mathcal{U}$ of U user equipments (UEs) served by a set $\mathcal{N}$ of N RRHs. The RRHs are equipped with storage units that allow them to serve users' requests via radio links. The network model is illustrated in Fig. 17.2. Users can request videos from a set $\mathcal{V}$ of V videos. When the requested files are not available in the storage units, users are served from the core network via capacity-limited fronthaul links. In order to benefit from caching, users' requests must be predicted and cached at the network edge to serve users locally via the neighboring RRHs, without using fronthaul links. The main goal of each RRH i is to define the fraction of each file k that should be stored while optimizing a given cost. In order to avoid caching the same segments at all the RRHs, an RRH has to take into consideration the cache state of all the other RRHs as well as their caching strategy. Owing to the large number of RRHs in the network, we substitute all the individual interactions in this game by the distribution of the RRHs over the state space, which is given for file k at time t by the mean-field process $m_{k,t} = N^{-1} \sum_{i=1}^{N} \delta_{y_{k,t}^{(i)}}$, where $y_{k,t}^{(i)}$ is the state of RRH i at time t. The goal of a generic RRH is, then, to define the values of the cache control variables $\mathbf{n}_t = [n_{1,t}, \ldots, n_{k,t}, \ldots, n_{V,t}]$ while considering the occupancy measure of the $N-1$ other RRHs and the dynamics of their internal state $y_{k,t}$, where $n_{k,t} \in \mathbb{R}^+, \forall k \in \mathcal{V}$, is the rate at which a given RRH downloads bits of file k at time t. The instantaneous dynamic of an RRH's state vector $y_{k,t}$ is defined next.

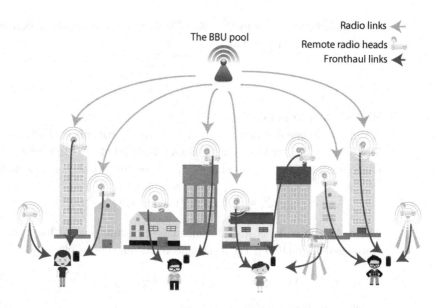

Figure 17.2 System model.

System Dynamics

The state $y_{k,t} = (h_t, s_{k,t})$ of a generic RRH at time t with respect to a given file k is defined by the dynamics of the channel h_t and the storage space $s_{k,t}$, given by:

$$\begin{cases} dh_t = \dfrac{\alpha}{2}(\mu - h_t)dt + \sigma_h d\mathcal{B}_t, \\ ds_{k,t} = [n_{k,t}q_k - \beta(1 - p_{k,t})\bar{\zeta}_t]dt + \sigma_s d\mathcal{B}_t. \end{cases} \tag{17.23}$$

The dynamics of the channel are modeled by a mean-reverting Ornstein–Uhlenbeck process, where $\mu > 0$, $\sigma_h > 0$, and $\mathcal{B}_{i,t}$ is a standard Brownian motion. The quantity $s_{k,t}$ represents the fraction of the bits from file k that is cached at the RRH at time t, where $n_{k,t}$ is the download rate of video file k by the RRH and q_k is the size of the file. The second expression in (17.23) gives the removal rate of file k at the RRH, where $\beta > 0$ is a parameter, $p_{k,t}$ is the popularity of file k at time t, and $\bar{\zeta}_t$ is the mean number of bits downloaded by all the users up to time t. In fact, this term models the tradeoff between the popularity of the file and the amount of bits that have been downloaded by the users in the time duration $[0, t]$. The instantaneous amount of bits downloaded by a given UE i corresponds to the minimum between the achievable transmission rate $\kappa_{i,t}$ from the RRH to this user and the available bits from file k at the RRH which have not been downloaded yet by the served UE. This download rate is given by

$$\zeta_{i,t} = \min\left\{\kappa_{i,t}, s_{k,t_l} - \int_{t_l}^{t} \zeta_{i,z}dz + \int_{t_l}^{t} n_{k,z}q_k \, dz\right\}, \tag{17.24}$$

where $\kappa_{i,t}$ is given by $w_i \log(1 + \gamma_{i,t})$, with w_i the wireless bandwidth capacity between the SBS and UE i; $\gamma_{i,t}$ is the signal-to-interference-plus-noise ratio (SINR) experienced by UE i due to the other SBSs during the downlink transmission; t_l is the time at which user i starts being served by the RRH. Thus the average number of downloaded bits up to time t can be formulated as

$$\bar{\zeta}_t = \frac{1}{U} \sum_{i=1}^{U} \zeta_{i,t}.$$

RRHs' Cost Function

A user can be served from the cache of many RRHs, but only the RRH having the highest SINR value is selected to serve the non-cached part of file k through the fronthaul. The goal of each RRH is to determine the cache control variables $n_{k,t}$ that minimize

$$\mathcal{J}_k = \mathbb{E}\left\{\int_0^T J_{k,t}(n_{k,t}, m_{k,t}) \, dt\right\} + \psi(\lambda_{k,T}), \tag{17.25}$$

where $\psi(\lambda_{k,T})$ is the cost of downloading a fraction $\lambda_{k,T}$ of file k at the end of the considered period $[0, T]$. The instantaneous caching cost is given by

$$J_{k,t}(n_{k,t}, m_{k,t}) = c_{k,t}(m_{k,t}) + g_t(n_{k,t}) + v\left(s_{k,t} - q_k\right) + \left(\sum_{k=1}^{V} s_{k,t} - o\right)^2, \tag{17.26}$$

which contains by the following terms:

- $c_{k,t}(m_{k,t})$ represents the cost of caching the same content as the other RRHs. This dependence between the states of the RRHs appears through the mean-field process.
- $g_t(n_{k,t})$ is the cost of downloading a fraction $n_{k,t}$ of file k through the allocated backhaul $B_{k,t}$ for this file.
- $v\left(s_{k,t} - q_k\right)$ is the cost of caching the same number of bits from the same file k at a given RRH. Thus, we limit the maximum cached bits from file k at the RRH to the size of the file, q_k.
- $(\sum_{k=1}^{V} s_{k,t} - o)^2$ models the limited storage capacity of the RRH, which should not exceed o.

Now, we introduce the value function for a generic RRH:

$$v_{k,t}(y_k) = \inf_{n_{k,t}}\{\mathcal{J}_k\}. \tag{17.27}$$

Finding the optimal control of a given RRH and file k amounts to solving the following coupled system of Hamiltonian–Jacobi–Belman (HJB) and Fokker–Planck–Kolmogorov (FPK) equations in v_k and m_k, respectively:

$$\partial_t v_{k,t}(y_k) + \left[n_{k,t}q_k - \beta(1 - p_{k,t})\bar{\zeta}_t\right]\partial_s v_{k,t}(y_k)$$

$$+ \frac{\alpha}{2}(\mu_h - h_t)\partial_h v_{k,t}(y_k) + \frac{\sigma_s^2}{2}\partial_{ss}^2 v_{k,t}(y_k) + \frac{\sigma_h^2}{2}\partial_{hh}^2 v_{k,t}(y_k)$$

$$+ J_{k,t}(n_{k,t}, m_{k,t}) = 0,$$

$$m_{k,0}(y_k) = \rho_0(y_k), \quad \forall y_k,$$

$$\partial_t m_{k,t}(y_k) + \left[n_{k,t}q_k - \beta(1 - p_{k,t})\bar{\zeta}_t\right]\partial_s m_{k,t}(y_k)$$

$$+ \frac{\alpha}{2}(\mu - h_t)\partial_h m_{k,t}(y_k) - \frac{\sigma_s^2}{2}\partial_{ss}^2 m_t(y) - \frac{\sigma_h^2}{2}\partial_{hh}^2 m_{k,t}(y_k) = 0,$$

where $m_{k,0}(y_k)$ is a given initial density distribution.

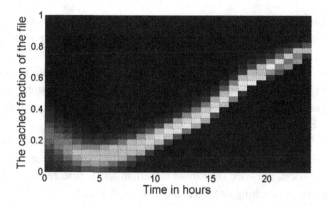

Figure 17.3 Density solution $m^*(s_k, t)$ as a function of time t and of the cached fraction of file k.

Numerical Result

To solve the HJB-FPK system of equations, we proceed by iteration, – using a simple fixed-point algorithm, until convergence. We present preliminary results for the case of a static channel model. The file size is normalized to 1 and the storage capacity of the RRHs is set to $o = 2/5$. The initial distribution of the RHHs follows the normal distribution $\mathcal{N}(0.2, 0.1)$.

In Fig. 17.3, we depict the evolution of m^* for one file whose popularity increases over 24 hours. We observe that the cached fraction of the file at the RRHs decreases when the file's popularity is low, making available the storage space for other, more popular, files. When more requests are expected for that file, all the RRHs cache a higher fraction of the file, at the limit of their storage capacity. This allows all the users to be served by any random subset of RRHs in their proximity, improving the quality experienced in terms of download time.

17.5.2 Matching Games for Distributed Caching in C-RAN

In this subsection we leverage the social interactions between users to predict their requests and then define the best caching policy accordingly.

17.5.2.1 System Model

We assume two networks, a virtual network and a real network. The virtual network represents an online social network through which N UEs from the set $\mathcal{N}$ are connected to one another. In such networks, users can communicate and share information with their connected friends. The users can request, share, and send videos selected from a set $\mathcal{V}$ of V video files that are proposed by the service providers. The service providers cache these videos in a set of servers (SPSs) denoted $\mathcal{C}$. In order to improve the perceived QoS by the end-users, service providers aim to cache a maximum number of videos at the RRHs to avoid serving the users through the capacity-limited fronhaul links. The physical network, however, is composed of the N UEs, served by K SPSs from the set $\mathcal{C}$

and M SBSs in the set $\mathcal{M}$. Moreover, the SPSs are connected to the RRHs via capacity-limited fronthaul links. Each RRH j downloads the files to be cached from a given SPS i via a capacity-limited link b_{ij}.

The storage capacities of the RRHs are given by the vector $\mathbf{Q} = [q_1, q_2, \ldots, q_M]$, expressed as the number of videos that each RRH can store. Thus, the service providers can cache their videos in the RRHs in such a way that each RRH s_i can locally serve a UE u_j via a radio link of capacity r_{ij}.

The aim in this model is to cache the video content at the RRHs while capturing users' requests for the shared videos on the basis of users' interests and interactions in the social network. The most important social features that are used to design efficient proactive caching strategies are discussed next.

- *Social interactions* The videos that a user watches typically depend strongly on the friends who share them. In fact, a user is more likely to request a video, shared by one of his friends, if this user is accustomed to watching videos shared by that friend [18]. This induced popularity can be given by $I_{social} = \alpha_{ln}/\sum_{j=1}^{F_l} \alpha_{jl}$, where α_{ln} is the number of videos previously shared by user u_n and viewed by user u_l; F_l is the number of u_l's friends.

- *Sharing impact:* Given that each video file can belong to a distinct category (e.g., news, music, games, etc.), we let S_{gl} be the number of videos of category g shared by a user u_l. Whenever a user u_l's request for a specific video is predicted and cached in its serving SBS s_m, sharing this video with this user's friends can have an important impact on the traffic load. This sharing impact depends on the number of user u_l's friends that are connected to the same SBS and the probability that user u_l shares the video. More formally, the sharing impact is given by $I_{sharing} = F_l^m S_{gl}/\sum_{i=1}^{G} S_{il}$, where F_l^m is the number of u_l's friends who are connected to the SBS s_m and G is the total number of considered video categories.

- *Users' interests* Whenever a user is interested in a certain topic, they can request a video that belongs to their preferred categories irrespective of the friend who shared it [18, 19]. On the basis of the categories of the videos previously watched by a user u_l, an SBS can predict user u_l's interests. The impact of this parameter is computed using $I_{interests} = V_{gn}/\sum_{i=1}^{H} V_{il}$, with V_{gn} the number of videos of category g viewed by a user u_n and H the total number of videos in the history of user u_l.

On the basis of these factors, the goal is to predict users' requests and define a caching strategy that assigns a set of videos to the RRHs. This assignment problem is formulated as a *many-to-many matching game*, in which the SPSs aim to cache their videos in the RRHs that offer the smallest download time for the requesting users, while the RRHs prefer to cache the videos that can reduce the fronthaul load.

Caching Problem Formulation

To model the system as a many-to-many matching game [20], we consider the two sets $\mathcal{C}$ of SPSs and $\mathcal{M}$ of RRHs as two teams of players. The *matching* is defined as an assignment of SPSs in $\mathcal{C}$ to RRHs in $\mathcal{M}$. The SPSs acts on behalf of the video files and each of them decides on its own videos. However, the number of videos stored by the RRHs depends on their storage capacity. In a matching game, the storage capacity of an RRH s and the number of RRHs in which an SPS c would like to cache a given file v are known as quotas, q_s and $q_{(c,v)}$, respectively [21]. Since an SPS decides in which RRH it caches a video v independently of the other owned files, for ease of notation we use v instead of the pair (c, v).

DEFINITION 17.3 A *many-to-many matching* μ is a mapping from the set $\mathcal{M} \cup \mathcal{V}$ into the set of all subsets of $\mathcal{M} \cup \mathcal{V}$ such that, for every $v \in \mathcal{V}$ and $s \in \mathcal{M}$ [22]:

1. $\mu(v)$ is contained in $\mathcal{S}$ and $\mu(s)$ is contained in $\mathcal{V}$;
2. $|\mu(v)| \leq q_v$ for all v in $\mathcal{V}$;
3. $|\mu(s)| \leq q_s$ for all s in $\mathcal{S}$;
4. s is in $\mu(v)$ if and only if v is in $\mu(s)$,

with $\mu(v)$ the set of player v's partners under the matching μ.

The definition states that a matching is a many-to-many relation in the sense that each stored video in an SPS is matched to a set of RRHs, and vice versa. In other words, an SPS can decide to cache a video in a number of RRHs and an RRH can decide to cache videos originating from different SPSs. Before setting an assignment of videos to RRHs, each player needs to specify its *preferences* over subsets of the opposite set based on its goal in the network. We use the notation $S \succ_m T$ to imply that RRH m prefers to store the videos in the set $S \subseteq \mathcal{V}$ rather than to store them in $T \subseteq \mathcal{V}$. A similar notation is used for the SPSs to set a preference list for each video. Faced with a set S of possible partners, a player k can determine to which subset of S it wishes to match. We denote this choice set by $C_k(S)$. Let $A(i, \mu)$ be the set of $j \in \mathcal{V} \cup \mathcal{M}$ such that $i \in \mu(j)$ and $j \in \mu(i)$. The preferences of the two sets of players are determined by their utility functions, which we give as follows.

- *Preferences of the remote radio heads:* On the basis of the social features discussed previously, we define the local popularity of video v_i at the mth RRH as follows:

$$P_{v_i} = \sum_{l=1}^{F_n^m} I_{\text{sharing}}[\gamma \cdot I_{\text{social}} + (1 - \gamma) I_{\text{Interests}}], \qquad (17.28)$$

where $\gamma \in [0, 1]$ is a weight that balances the impact of social interactions and users' interests on the local popularity of a video.
- *Preferences of the service provider servers* The goal of service providers is to enhance the quality experienced by the end-users. In fact, an SPS c_i would prefer to cache a video v_k at the RRH s_j that offers the smallest download time for the expected requesting UEs. The download time depends on the capacity of the fronthaul link b_{ij} and the radio links r_{jn} that connect RRH s_j to UE u_n. The video file is first downloaded by s_j which then serves the UEs. Thus, in the worst case, downloading a video stored in c_i takes the required time to pass by the link with the poorest capacity. When many

UEs are expected to request the same file from s_j, the download time is given by:

$$T_D = \left(\min \left\{ b_{ij}, \frac{\sum_{n=1}^{N} r_{jn}}{N} \right\} \right)^{-1}. \tag{17.29}$$

Since each video might be requested by different UEs, an SPS defines its preferences over the RRHs for each owned video file.

To solve the matching game, we look at a stable solution in which there are no players that are not matched to one another; they all prefer to be partners. In many-to-many models, various stability concepts can be considered depending on the number of players that can improve their utility by forming new partners among one another. However, the large number of RRHs, which is expected to exceed the number of UEs [23], makes it more difficult to identify and organize large coalitions than to consider pairs of players and individuals. Hence, we will restrict ourselves to the notion of pairwise stability, defined as follows [24]:

DEFINITION 17.4   A matching μ is *pairwise stable* if there does not exist a pair (v_i, s_j) with $v_i \notin \mu(s_j)$ and $s_j \notin \mu(v_i)$ such that if $T \in C_{v_i}(A(v_i, \mu) \cup \{s_j\})$ and $S \in C_{s_j}(A(s_j, \mu) \cup \{v_i\})$ then $T \succ_{v_i} A(v_i, \mu)$ and $S \succ_{s_j} A(s_j, \mu)$.

In the system under study, the RRHs and SPSs are always interested in the gain they can get from individuals of the opposite set. For instance, an RRH would always like to cache first the most popular file as long as that file is proposed to it. Thus, even though the set of stable outcomes may be empty [20], SPSs and RRHs have *substitutable* preferences, defined as follows [25]:

DEFINITION 17.5   Let T be the set of player i's potential partners and $S \subseteq T$. Player i's preferences are called *substitutable* if, for any players $k, k' \in C_i(S)$, $k \in C_i(S \setminus \{k'\})$.

In fact, *substitutability* is the weakest condition needed for the existence of pairwise stable matching in a many-to-many matching game [24].

In [26], a many-to-many matching game is formulated to model a resource allocation problem in wireless networks. The proposed algorithm deals with *responsive* preferences, which constitute a stronger condition than *substitutability*. Thus, the algorithm could not be applied to the formulated caching problem. Under substitutable preferences, a stable matching algorithm has been proposed in [21] for many-to-one games. Pairwise stable matching in the many-to-many problem has been proved to exist between firms and workers when salaries (money) are explicitly incorporated in the model [24]. Here, we extend and adapt these works to our model in order to propose a new matching algorithm and prove its convergence to pairwise stable matching.

Caching Algorithm
After formulating the caching problem as a many-to-many game, we propose an extension of the *deferred acceptance* algorithm [27] to the current model with SPS-proposing. The algorithm consists of three phases. During the first phase, SPSs and RRHs discover

Table 17.1 Proposed proactive caching algorithm

Phase 1: Network discovery
Each SPS discovers its neighboring RRHs and collects the required network parameters.

Phase 2: Specification of the preferences
Each SPS and RRH sets its preference list(s).

Phase 3: Matching algorithm
The SPSs propose each owned video to preferred RRHs in $C_{v_i}(\mathcal{M})$.
Each RRH rejects all but the q_s most preferred videos.

Repeat
The SPSs propose each file to the related most preferred set of RRHs that includes all those RRHs to whom it previously proposed the file and who have not yet rejected it (*Substitutability*).
Each RRH rejects all but the q_s most preferred videos.

Until convergence to a stable matching.

---

their neighbors and collect the required parameters to define preferences such as the fronthaul and radio link capacities. This can be done, for instance, by exchanging *hello* messages periodically. In the second phase SPSs define a preference list for each owned file over the set of RRHs, while the RRHs define their preferences over the set of videos that would be proposed by the SPSs. The last phase consists of two steps. In the first step, every SPS proposes an owned video to the most preferred set of RRHs, which offer the shortest download time for that video. Afterwards, each RRH s_j rejects all but the q_j most popular videos from the set of alternatives proposed to it. In the second step of the last phase, the SPSs propose an owned video to the most preferred set of RRHs, which includes the RRHs to which it previously proposed that video and which have not yet rejected it (*substitutability*). Each RRH rejects all but its choice set from the proposed videos. This second step is repeated until no rejections are issued. The algorithm is summarized in Table 17.1.

17.5.2.2 Pairwise Stability

Let $P_{s_j}(k)$ be the set of proposals received by an RRH s_j at step k and $C_{v_i}(\mathcal{M}, k) \subseteq \mathcal{M}$ be the choice set of RRHs to which a video v_i has been proposed at step k. By analyzing the algorithm, we can state the following propositions.

PROPOSITION 17.6 (Offers remain open) For every video v_i, if an RRH s_j is contained in $C_{v_i}(\mathcal{M}, (k-1))$ at step $k-1$ and did not reject v_i at this step then s_j is contained in $C_{v_i}(\mathcal{M}, k)$.

Proof Note that $C_{v_i}(\mathcal{M}, k-1) = C_{v_i}(C_{v_i}(\mathcal{M}, k-1) \cup C_{v_i}(\mathcal{M}, k))$, since $C_{v_i}(\mathcal{M}, k-1)$ is video v_i's choice set from those sets whose elements have not been rejected prior to step $k-1$ while $C_{v_i}(\mathcal{M}, k)$ is the choice set from the smaller class of sets whose elements have not been rejected prior to step k. Here, substitutability implies that if $s_j \in C_{v_i}(\mathcal{M}, k-1)$ then $s_j \in C_{v_j}(C_{v_i}(\mathcal{M}, k) \cup \{s_j\})$. So, if $s_j \in C_{v_i}(\mathcal{M}, k-1)$ is not rejected at step $k-1$ then it must be contained in $C_{v_i}(\mathcal{M}, k)$ since otherwise $C_{v_i}(C_{v_i}(\mathcal{M}, k) \cup \{s_j\}) \succ_{v_i} C_{v_i}(\mathcal{M}, k)$, violating the requirement that $C_{v_i}(\mathcal{M}, k)$ is the most preferred set whose elements have not been rejected. □

PROPOSITION 17.7 (Rejections are final) *If a video v_i is rejected by an RRH s_j at step k then, at any step $p \geq k$, $v_i \notin C_{s_j}(\mathcal{P}_{s_j}(p) \cup \{v_i\})$.*

Proof Assume that the proposition is false, and let $p \geq k$ be the first step at which $v_i \in C_{s_j}(\mathcal{P}_{s_j}(p) \cup \{v_i\})$. Since $C_{s_j}(\mathcal{P}_{s_j}(p))$ contains $C_{s_j}(\mathcal{P}_{s_j}(p-1))$, by Proposition 17.6, substitutability implies that $v_i \in C_{s_j}(C_{s_j}(\mathcal{P}_{s_j}(p-1)) \cup \{v_i\})$ and thus $v_i \in C_{s_j}(\mathcal{P}_{s_j}(p-1) \cup \{v_i\})$, which contradicts the definition of p and completes the proof of the proposition. □

THEOREM 17.8 *The proposed matching algorithm between SPSs and RRHs is guaranteed to converge to a pairwise stable matching.*

Proof We make the proof by contradiction. Suppose that there exists a video v_i and an RRH s_j with $v_i \notin \mu(s_j)$ and $s_j \notin \mu(v_i)$ such that for $T \in C_{v_i}(A(v_i, \mu) \cup \{s_j\})$, $S \in C_{s_j}(A(s_j, \mu) \cup \{v_i\})$ we have $T \succ_{v_i} A(v_i, \mu)$ and $S \succ_{s_j} A(s_j, \mu)$. Then, the video v_i has been proposed to the SBS s_j, which rejected it at some step k. So $v_i \notin C_{s_j}(A(s_j, \mu) \cup \{v_i\})$ and thus μ cannot be unstable. □

Numerical Results

For our simulations we considered a network in which we set the number of SPSs, RRHs, UEs, and videos to $K = 80$, $M = 150$, $N = 400$, and $V = 100$, respectively. We assumed that all the SBSs have the same storage capacity, while the backhaul links capacities are lower than the radio link capacities, which captures the real network characteristics [28]. We set the total backhaul link capacities and radio link capacities to $B = 80$ Mbit/time slot and $R = 180$ Mbit/time slot, respectively. Owing to the unavailability of a social dataset that includes the parameters addressed in this chapter (e.g., the video categories accessed and videos shared by users), because of privacy reasons, we generated file popularity and user requests pseudo-randomly. The popularity of files is generated at each RRH with a Zipf distribution, which is commonly used to model content popularity in networks [29]. Users' requests are generated using a uniform distribution.

To evaluate the performance of the proactive caching algorithm, we implemented the proposed matching algorithm (MA) as well as a random caching policy (RA), in which the RRHs are filled randomly with videos from the SPSs to which they are connected, until their storage capacity limit is reached. We ran the two algorithms for different values of a storage ratio β, which represents the number of files that each RRH has the capacity to store. More formally, $\beta = q_i/V, \forall i \in V$. In the numerical results we show the evolution of the satisfaction ratio, which corresponds to the total number of requests served by the RRHs over the total number of requests, as the total number of requests increases. Moreover, we compared the mean times required for downloading all the videos by the requesting users.

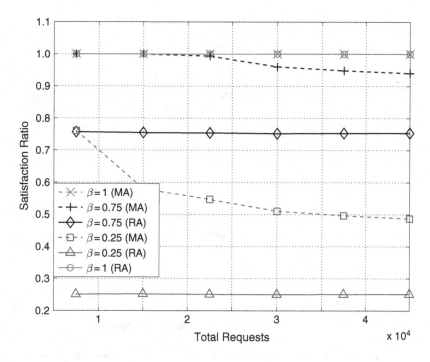

Figure 17.4 Satisfaction evolution for MA and RA.

In Fig. 17.4 we show the proportion of served users for $\beta \in \{0.25, 0.75, 1\}$. The satisfaction ratio decreases when the storage capacity of the SBSs decreases. This is evident, as fewer files are then cached in the RRHs. Figure 17.4 shows that, when each SBS has the capacity to store all the videos proposed by the SPSs ($\beta = 1$), the *satisfaction ratio* remains equal to 1 irrespective of the caching policy used. This result stems from the fact that all the requests are served locally by the RRHs. When the RRHs have the capacity to store 25% and 75% of the videos proposed by the SPSs, i.e., $\beta = 0.25$ and $\beta = 0.75$ respectively, the satisfaction of the MA is up to three times higher than the that of RA ($\beta = 0.25$). Under the MA, the number of users served by the RRHs decreases with an increasing number of requests. This is due to the fact that the RRHs choose to cache first the files with a higher local popularity. Figure 17.4 shows that, as the number of requests increases, the satisfaction ratio of the RA changes only slightly, owing to the uniform selection of videos under the same distribution of file popularity.

Figure 17.5 shows the expected download time of all the files in the network for $\beta \in \{0.25, 1\}$. The download time is lower when the requests are served by the RRHs ($\beta = 1$) than when some requests are served by the SPSs ($\beta = 0.25$). This is due to the fact that the latter videos require more time to pass through the fronthaul links, which are of lower capacity than the radio links. In Fig. 17.5 we can see that, when $\beta = 1$, the expected download time is identical for both RA and MA. When $\beta = 0.25$, although the download time increases as the number of requests in the network increases, owing to

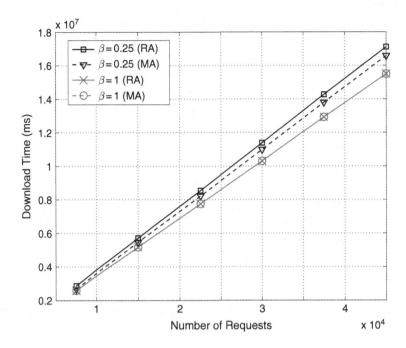

Figure 17.5 Change in download time as the number of requests increases.

network congestion, the MA outperforms the RA. In fact, the higher satisfaction ratio of the MA compared to that of the RA leads to a smaller expected download time because most of requests are served via the radio links.

As seen from Figs. 17.4 and 17.5, clearly the proposed MA allows the fronthaul bottleneck to be efficiently overcome and so improves the performance of video download in small-cell networks.

17.6   Conclusion

In this chapter we have studied the impact of caching on C-RAN. We first presented C-RANs, their architecture, advantages, and challenges. The main bottleneck of C-RANs is their capacity-limited fronthaul links, which can degrade significantly the network performance at peak times. To deal with this problem, distributed caching at the RRHs was introduced to reduce the load on the fronthaul links. Only a few works have studied the impact of caching on C-RANs. We have provided an overview of recent works that have addressed this problem and have proposed new caching strategies to induce more CoMP opportunities in C-RAN networks. Moreover, we have presented two game-theoretic frameworks to help develop new caching strategies in which we take into account the high density of the RRHs and also leverage the virtual connections between users in social networks to predict user requests.

References

[1] Cisco, "Cisco visual networking index: global mobile data traffic forecast update, 2013–2018," White Paper. Online. Available at http://goo.gl/l77HAJ, 2014.

[2] Ericsson, "5G radio access – research and vision." White Paper. Online. Available at goo.gl/Huf0b6, 2012.

[3] J. G. Andrews, "Seven ways that HetNets are a cellular paradigm shift," *IEEE Commun. Mag.*, vol. 51, no. 3, pp. 136–144, March 2013.

[4] J. G. Andrews, H. Claussen, M. Dohler, S. Rangan, and M. C. Reed, "Femtocells: past, present, and future," *IEEE J. Sel. Areas Commun.*, vol. 30, no. 3, pp. 497–508, 2012.

[5] T. Q. S. Quek, G. de la Roche, I. Guvenc, and M. Kountouris, *Small Cell Networks: deployment, phy Techniques, and Resource Allocation*. Cambridge University Press, 2013.

[6] R. Irmer, H. Droste, P. Marsch, M. Grieger, G. Fettweis, S. Brueck *et al.*, "Coordinated multipoint: concepts, performance, and field trial results," *IEEE Commun. Mag.*, vol. 49, no. 2, pp. 102–111, February 2011.

[7] D. Lee, H. Seo, B. Clerckx, E. Hardouin, D. Mazzarese, S. Nagata *et al.*, "Coordinated multipoint transmission and reception in LTE-Advanced: deployment scenarios and operational challenges," *IEEE Commun. Mag.*, vol. 50, no. 2, pp. 148–155, April 2012.

[8] C. Mobile, "C-RAN: the road towards green RAN." White Paper, vol. 2, 2011.

[9] E. Bastug, M. Bennis, and M. Debbah, "Proactive caching in 5G small cell networks," 2015.

[10] N. Alliance, "Suggestions on potential solutions to C-RAN." White Paper, January, 2013.

[11] J. Bartelt, G. Fettweis, D. Wubben, M. Boldi, and B. Melis, "Heterogeneous backhaul for cloud-based mobile networks," in *Proc. 2013 IEEE 78th Vehicular Technology Conf.*, 2013, pp. 1–5.

[12] M. Peng, Y. Li, J. Jiang, J. Li, and C. Wang, "Heterogeneous cloud radio access networks: a new perspective for enhancing spectral and energy efficiencies," *IEEE Wireless Commun.*, vol. 21, no. 6, pp. 126–135, June 2014.

[13] M. Peng, Y. Li, Z. Zhao, and C. Wang, "System architecture and key technologies for 5g heterogeneous cloud radio access networks," *IEEE Netw.*, vol. 29, no. 2, pp. 6–14, 2015.

[14] M. Peng, X. Xie, Q. Hu, J. Zhang, and H. V. Poor, "Contract-based interference coordination in heterogeneous cloud radio access networks," *IEEE J. Sel. Areas Commun.*, vol. 33, no. 6, pp. 1140–1153, June 2015.

[15] M. Gerasimenko, D. Moltchanov, R. Florea, S. Andreev, Y. Koucheryavy, N. Himayat *et al.* "Cooperative radio resource management in heterogeneous cloud radio access networks," *IEEE Access*, vol. 3, no. 6, pp. 397–406, 2015.

[16] A. Liu and V. K. Lau, "Cache-enabled opportunistic cooperative MIMO for video streaming in wireless systems," *IEEE Trans. Signal Process.*, vol. 62, no. 2, pp. 390–402, January 2014.

[17] A. Liu and V. K. Lau, "Mixed-timescale precoding and cache control in cached MIMO interference network," *IEEE Trans. Signal Process.*, vol. 61, no. 24, pp. 6320–6332, 2013.

[18] H. Kwak, C. Lee, H. Park, and S. Moon, "Beyond social graphs: user interactions in online social networks and their implications," *ACM Trans. Web (TWEB)*, vol. 6, no. 4, November 2012.

[19] H. Li, H. T. Wang, J. Liu, and K. Xu, "Video requests from online social networks: characterization, analysis and generation," in *Proc. of IEEE Int. Conf. on Computer Communications*, Turin, April 2013, pp. 50–54.

[20] M. Sotomayor, "Three remarks on the many-to-many stable matching problem," *Math. Social Sci.*, vol. 38, pp. 55–70, 1999.

[21] A. E. Roth and M. Sotomayor, *Two-Sided Matching: A Study in Game-Theoretic Modeling and Analysis*, Cambridge University Press, 1990.

[22] A. E. Roth, "A natural experiment in the organization of entry level labor markets: regional markets for new physicians and surgeons in the UK" *Amer. Econ. Rev.*, vol. 81, pp. 415–425, 1991.

[23] D. P. Malladi, "Heterogeneous networks in 3G and 4G," in *Proc. IEEE Communication Theory Workshop*, Hawaii, 2012.

[24] A. Roth, "Stability and polarization of interests in job matching," *Econometrica*, vol. 52, pp. 47–57, 1984.

[25] A. S. Kelso and V. Crawford, "Job matching, coalition formation and gross substitutes," *Econometrica*, vol. 50, no. 6, pp. 1483–1504, 1982.

[26] H. Xu and B. Li, "Seen as stable marriages," in *Proc. IEEE Int. Conf. on Computer Communications*, Shanghai, 2011, pp. 586–590.

[27] D. Gale and L. Shapley, "College admissions and the stability of marriage," *Amer. Math. Monthly*, pp. 9–15, 1969.

[28] JDSU, "Small cells and the evolution of backhaul assurance." White Paper, May 2013.

[29] L. Breslau, C. Pei, F. Li, and G. Phillips, "Web caching and Zipf-like distributions: evidence and implications," in *Proc. IEEE Conf. on Computer and Communications Societies*, March 1999, pp. 126–134.

18 A Cloud Service Model and Architecture for Small-Cell RANs

Sau-Hsuan Wu, Hsi-Lu Chao, Hsin-Li Chiu, Chun-Hsien Ko, Yun-Ting Li, Ting-Wei Chang, Tong-Lun Tsai, and Che Chen

The exponentially growing wireless data traffic has pushed wireless network vendors and researchers to think of a revolutionizing architecture for fifth-generation (5G) wireless networks is that is able to support 100 times the network capacity of the current 3G and 4G systems and yet is capable of reducing the net energy consumption of the systems by up to 90 percent. Numerous concepts and technologies have been proposed to approach this goal from various system-design aspects. A consensus reached so far is that 5G is going to be a collection of technology evolutions that include soft core networks, small cellular networks, pervasive but invisible base stations, and heterogeneous yet self-organizing radio access networks. Given the need for advanced radio access technologies, the success of such an ambitious goal relies on a flexible and scalable network infrastructure to accommodate and harmonize these technologies. In this chapter, we introduce a cloud service model and system architecture that has the potential to serve the numerous system requirements for future 5G networks.

18.1 A Cloud Service Model for Radio Access Networks

According to statistics released by [1], global mobile broadband subscriptions will reach 4.4 billion by 2016. The global mobile data traffic has almost doubled every year in the past four years, and is expected to reach 10.8 exabytes per month by 2016. This wireless tsunami has hit the global wireless infrastructure, and is causing wireless service providers to deploy many more base stations (BSs) and their access networks to satisfy the ever increasing service demands. This, however, gives rise to other threats such as increased energy consumption and radio pollution to our environment. According to statistics released by China Mobile, the number of its BSs exceeded one million by 2012 and continued to increase by more than 200 thousand a year. The overall power consumption to support such a huge wireless network reached 14 billion kilowatt hours (kW h) a year by 2012, and increased to 1.5 billion kW h a year on average over the period 2004–2012 [2]. Mobile wireless services have grown to a point such that even the network of one service provider needs to accommodate tens of millions of BSs and yet must be able to regulate the tremendous amount of traffic running through it. Under the mobile traffic boom of such a wireless era, not only are mobile users (MUs) the customers of cloud services, but BSs are also becoming the subject of cloud services. In this section, we will present a cloud service model to support the radio-access and core-network functions of mobile communications. To begin, we briefly review the issues that face the current system. Then we introduce a cooperative service model for a system to improve its throughput by joint channel and power allocation (CPA) among the BSs. Using the model we will develop a cloud architecture to support the CPA service in a densely deployed RAN.

18.1.1   Cloud-Based Data Processing for Densely Deployed Small-Cell RANs

To improve system throughput and power coverage under the RAN architecture of the current 4G system, the first thought that comes to mind is perhaps to increase the density of BSs. This would not only help bring BSs closer to their users but would also reduce their power consumption since, after all, radiation power decreases exponentially with its traveling distance. In another statistic released by China Mobile [3], 65 percent of a BS's power consumption is on average used for radio transmissions with another 17 percent used for BS cooling. Only 10 percent of the power is used in backend data transmission and signal processing. If the computation and radio transmission power of a BS can be reduced to that of a WiFi access point (AP) then there could be no extra power needed for BS cooling. Following this argument, a dense small-cell RAN (DSN) can be a win–win solution for both users and network operators. The downside of DSN, however, is the aggravated mutual interference inside the network and the much heavier workloads for its system's core network (CN). It may require a design overhaul for a system really to benefit from the deployment of DSN.

First, operators will need to deploy much larger amounts of small- or femto- cell BSs and APs to support the radio accesses of MUs. The number of such BSs and APs can easily reach a few hundred thousand or more for a wireless metropolitan area network (WMAN) with millions of mobile subscribers. This makes it unlikely that a well-planned RAN model designed for macro-cell BSs can still work for such kinds of densely and nearly randomly deployed BSs and APs. Second, under the technology development trends of universal frequency reuse and carrier aggregation for 4G-and-beyond systems, it requires an effective system to manage the strong interference within the DSN. This implies that not only do the massive numbers of MUs and BSs and APs need to be managed by a cloud center but their mutual interference also has to be controlled with a powerful cloud computing center. A smart and adaptive management system is also required to distribute the available spectrum resources to the mutually overlapping small-cell self-organizing networks (SONs) inside a WMAN, since the traffic loading is often unbalanced and varies in different areas and at different times of the day. In view of this technology evolution trend and communication paradigm shift, cloud-based RAN (C-RAN) has recently been proposed and prototyped by NTT DoCoMo and China Mobile [3]. The functions of the BS for the 3G or 4G system are partitioned into two parts in C-RAN: one for the baseband processing unit (BBU) in the cloud and the other for remote radio heads (RRHs) to provide radio access. The intensive signal processing on a typical BS can thus be largely relegated to a custom-designed cloud computing center for a C-RAN.

Under the framework of C-RANs [3], a cloud center not only needs to process the channel access, power control and user scheduling functions of a typical RAN; it also needs to execute the CN functions for mobility management and data routing. All data traffic in and out of the network are, thus, exchanged through the cloud. The advantage of this architecture is that cross-layer data processing and network security can be more effectively implemented, in a centralized manner with custom-designed hardware and software stacks, in the cloud. The downside is, however, that this centralized architecture relies on dense high-speed network links between the C-RAN and RRHs, and the cloud may become a bottleneck for mobile networks in face of the exponentially increasing data traffic. More importantly, this architecture forms an even more closed RAN and CN and, hence, is less flexible and agile to the rapidly changing wireless networking technologies. Inspired by the open architecture of software defined networks (SDNs) [4], we present herein a cognitive and open-cloud architecture for the RAN (CoC-RAN) and CN functions of dense small-cell networks, making use of the open architectures of SDN and the OpenStack cloud operating system [5].

A significant feature difference between our CoC-RAN and a typical C-RAN is that only control traffic is delivered to and processed in the CoC-RAN; it does not process both the control and data traffic like a typical C-RAN in a custom-designed cloud computing center. Data traffic in the CoC-RAN is directly scheduled on small or femto-cell APs and routed through the Internet. The service functions for spectrum resource sharing (SRS) and CPA of the RAN are implemented in the cloud, with remote APs providing the medium access control (MAC) and physical (PHY) transceiving functions for radio access. The mobility management (MM) function of CN, however, is implemented with OpenFlow controllers (OFCs) in the cloud and OpenFlow switches on, or *en route* to, the APs. This computation and network architecture of CoC-RAN allows us to more efficiently partition the functional modules that require different processing speeds and management tasks into the cloud and the AP sides, hence, largely relaxing the strict real-time computing requirements on the cloud. This opens up the possibility for us to realize the CoC-RAN with open-source cloud service platforms and software on commodity hardware. The entire services of CoC-RAN can, thus, be viewed as those of a tenant in a cloud, a tenant which employs several stacks of virtual machines (VMs) in the cloud via a cloud operating system such as OpenStack to execute the different functions of RAN and CN required for CoC-RAN. As such, more diversified functions for SRS, CPA, MM, network management, smart charging, and many others can be easily virtualized in the cloud. A functional block diagram for our concept and prototype is illustrated in Fig. 18.1. In the next section we will introduce the CPA function that underlies the mechanisms we propose for cooperative interference control, channel allocation, and user scheduling in DSN.

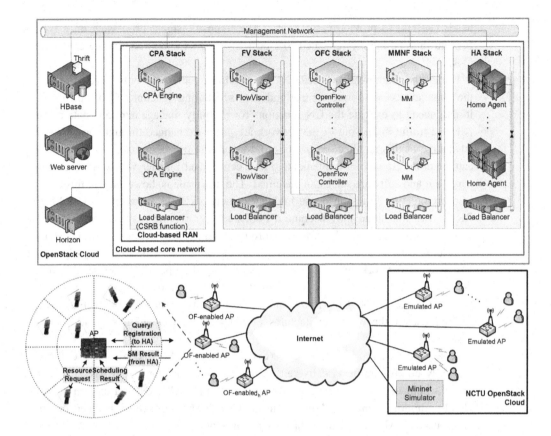

Figure 18.1 The functional block diagram of CoC-RAN.

18.2   Joint Channel and Power Allocation in Dense Small-Cell RANs

We suppose that APs and MUs are randomly distributed inside the signal coverage range of a DSN and are partitioned into several clusters. Each cluster of APs and their associated MUs is considered a self-organized network (SON) inside the DSN. The locations of the MUs and APs are assumed to be provided to the RAN and the CN of the DSN. Given that the numbers of MUs and APs inside the DSN are likely to be large, it follows that providing CPA for each single channel access request of an MU inside the signal coverage range of every AP will make the complexity of a CPA algorithm too high to be implemented in real time. To resolve this difficulty, the signal coverage range of an AP is partitioned into 12 sectors, as shown in Fig. 18.1, and the channel access requests inside a sector of an AP are lumped and transformed into a number of SU requests, say N. Each sector's SU request is associated with an SINR requirement $\overline{\gamma}_i$, $i \in \mathcal{I}_N$[1], and will occupy one physical channel (PCH), if granted, over a frame duration of the MAC layer [6]. As a result, each SU may support a number of channel access requests issued by the same or different MUs in the same sector.

---

[1] For brevity, an index set $\{1, \ldots, N\}$ is denoted by $\mathcal{I}_N$.

The location of an SU is thus considered as the geographical center of its serving MUs. More details about the translations of channel access requests into SUs are introduced in Section 18.3.

Assume there are L APs inside a cluster. Since the number of channels for APs inside or on the boundary of a cluster may vary, for succinctness, the available channel set for an AP j is denoted by $\mathcal{M}_j, j \in \mathcal{I}_L$. The number of channels, namely the cardinality of $\mathcal{M}_j$, denoted by $|\mathcal{M}_j|$, is less than or equal to the maximum number of available channels, M.

Supposing that each SU will be allocated to an AP only if its SINR requirement is satisfied and that it may carry channel access requests from a number of users, we therefore consider a CPA algorithm that will maximize the number of granted SUs subject to (s.t.) a transmit power limitation on each PCH. In addition, to reuse the spectrum resources of a cluster most efficiently, the SUs in the overlapping coverage ranges of APs will be reallocated to their best APs, taking with consideration the relative interference among SUs and their individual SINR requirements even if they are issued from different APs. Therefore, to help formulate such a CPA problem we define for each SU_i, $i \in \mathcal{I}_N$, an allocation variable $X_{i,j,k}$. The value of $X_{i,j,k}$ is set to 1 when SU_i is allocated to AP_j on the kth channel, with $k \in \mathcal{M}_j$. Otherwise, we have $X_{i,j,k} = 0$.

In addition to the above system requirements, we consider a simple multiple access scheme in our system where each channel of an AP can be assigned to one SU only, namely, $\sum_{i=1}^{N} X_{i,j,k} \leq 1$. As a result, we have $\sum_{k \in \mathcal{M}_j} \sum_{i=1}^{N} X_{i,j,k} \leq |\mathcal{M}_j|$. In the same spirit, each SU can be allocated to at most one channel of an AP, namely, $\sum_{k \in \mathcal{M}_j} X_{i,j,k} \leq 1$ and $\sum_{k \in \mathcal{M}_j} \sum_{j=1}^{L} X_{i,j,k} \leq 1, \forall i \in \mathcal{I}_N$.

In order to incorporate a quality-of-service (QoS) function in to our system design and performance evaluations, we consider a typical mapping between the QoS and the channel capacity $C = \log(1 + SINR)$, which leads to a capacity constraint and, hence, an average SINR constraint on each channel, given by

$$\gamma_{i,j,k} \triangleq \frac{P_{j,k} G_{i,j,k}}{N_0 + \sum_{i'=1, i' \neq i}^{N} \sum_{j'=1, j' \neq j}^{L} X_{i',j',k} P_{j',k} G_{i,j',k}} \geq \overline{\gamma}_i,$$

where $\gamma_{i,j,k}$ stands for the received SINR at SU_i on the kth channel of AP_j, and N_0 is the noise power. The transmit power of AP_j on its kth channel is denoted by $P_{j,k}$, whose propagation pathloss $G_{i,j,k}$ along the path from AP_j to SU_i is defined in decibels (dB) as

$$10 \log_{10}(G_{i,j,k}) = -\overline{PL}_k(d_0) - 10n \log_{10}(d_{i,j}/d_0), \tag{18.1}$$

where $\overline{PL}_k(d_0)$ is the reference free-space pathloss in dB measured at a close-in reference distance d_0 and n is the pathloss exponent, which indicates the rate at which the pathloss increases with distance. The separation distance between SU_i and AP_j is denoted by $d_{i,j}$. Finally, the transmit power on each channel is limited to P_k^{max}, leading to a power constraint $0 \leq P_{j,k} \leq P_k^{max}$.

On the basis of the above system requirements and operating constraints, we formulate our downlink CPA criterion to distribute the channel resources as follows:

$$\text{Find } \underset{X_{i,j,k},\, P_{j,k}}{\arg\max} \sum_{i=1}^{N} \sum_{j=1}^{L} \sum_{k \in \mathcal{M}_j} X_{i,j,k}$$

$$\text{s.t. } \sum_{i=1}^{N} X_{i,j,k} \le 1 \quad \text{and} \quad \sum_{j=1}^{L} \sum_{k \in \mathcal{M}_j} X_{i,j,k} \le 1,$$

$$0 \le P_{j,k} \le P_k^{\max}, \quad \gamma_{i,j,k} \ge \overline{\gamma}_i, \quad \forall i \in \mathcal{I}_N, j \in \mathcal{I}_L, k \in \mathcal{M}_j. \tag{18.2}$$

From this CPA formulation, it follows immediately that a valid solution set of $X_{i,j,k}$ satisfies $\sum_{i=1}^{N} \sum_{j=1}^{L} \sum_{k \in \mathcal{M}_j} X_{i,j,k} \le \min\{N, \sum_{j=1}^{L} \mathcal{M}_j\}$. Besides, for a set of valid SU–AP link pairs $(i,j)_k$ in channel k, defined as $\mathcal{L}_k \triangleq \{(i,j)_k | X_{i,j,k} = 1, \forall i \in \mathcal{I}_N, j \in \mathcal{I}_L\}$, it follows that

$$P_{j,k} - \sum_{j_t \ne j,\, (i_t,j_t)_k \in \mathcal{L}_k} P_{j_t,k} \frac{\overline{\gamma}_i G_{i,j_t,k}}{G_{i,j,k}} \ge \frac{N_0 \overline{\gamma}_i}{G_{i,j,k}}, \quad \forall (i,j)_k \in \mathcal{L}_k.$$

Renumbering the SU–AP pairs of $\mathcal{L}_k$ as $(i_1,j_1)_k, \ldots, (i_m,j_m)_k$, where $m \triangleq |\mathcal{L}_k|$, we may define an $m \times 1$ vector

$$\mathbf{u}_k \triangleq \left[\frac{N_0 \overline{\gamma}_{i_1}}{G_{i_1,i_1,k}}, \ldots, \frac{N_0 \overline{\gamma}_{i_m}}{G_{i_m,i_m,k}} \right]^T,$$

whose sth entry corresponds to the $(i_s,j_s)_k$ pair of $\mathcal{L}_k$. Accordingly, we can also define $\mathbf{p}_k \triangleq [P_{j_1,k} \cdots P_{j_m,k}]^T$ and an $m \times m$ matrix $\mathbf{A}_k$ as

$$\mathbf{A}_k(s,t) = \begin{cases} 0 & , \quad s = t \\ \dfrac{\overline{\gamma}_{i_s} G_{i_s,j_t,k}}{G_{i_s,j_s,k}} & , \quad s \ne t \end{cases}, \quad s,t \in \mathcal{I}_m. \tag{18.3}$$

Let $\mathbf{I}_m$ be an $m \times m$ identity matrix. The SINR constraint can thus be rewritten as $(\mathbf{I}_m - \mathbf{A}_k)\mathbf{p}_k \ge \mathbf{u}_k$. Given that the equality holds for the optimum solution set $\mathcal{L}_k$, the corresponding transmit power of APs equals $\mathbf{p}_k = (\mathbf{I}_m - \mathbf{A}_k)^{-1}\mathbf{u}_k$. The computational complexity of solving this matrix equality with the Gaussian elimination method is $\frac{1}{3}m^3 + m^2 - \frac{1}{3}m$. Since the maximum value of m is equal to the number L, of APs, the complexity is bounded from above by $\frac{1}{3}L^3 + L^2 - \frac{1}{3}L$.

The entire algorithm for solving (18.2) is, however, quite involved. Readers who are interested in it may refer to [7] where the optimal algorithm and an heuristic algorithm for solving it are discussed in detail.

18.3   A QoS-Based User Scheduling in Dense Small-Cell RANs

Following the proposed CPA method, in this section we present how channel access requests from MUs are lumped and translated into SUs and how the granted SU requests are employed for user-level resource scheduling (US) in the DSN.

Considering the characteristics of delay tolerance, traffic volume, and the interactivity of different Internet services, we assume that Q service types are provided in the DSN and that each service type requires a data rate r_i, $i = 1, \ldots, Q$, with $r_1 > r_2 > \cdots > r_Q$, to fulfill its QoS demand and furthermore that it has a corresponding utility function to define its QoS satisfaction degree. The utility functions studied in [8] are used herein for both real-time and non-real-time services.

Let the numbers of sessions of the Q service classes in a specific sector be $n_1, \ldots, n_Q$, respectively. To provide channel access for these service requests, the transmission and reception times are framed for multiple access control (MAC), and each frame duration is further partitioned into several time slots. The frame duration is defined as T seconds, and the time duration of a slot is t seconds. To simply our discussions in this chapter, we consider downlink transmissions only, and define each frame to have z time slots for downlink deliveries per data channel.

According to the above system settings, we also assume that there are Q levels of channel capacits provided by the physical layer (PHY), which are $C_1, \ldots, C_Q$ with $C_1 > C_2 > \cdots > C_Q$. Considering the issues of queue stability and the QoS requirements for a session, the total capacity of the PCHs mapped to the sessions of a specific service class must be greater than or equal to their incoming data sum rate. For simplicity, we assume a one-to-one mapping between session data rates and physical channel capacities, i.e., $r_i \rightarrow C_i$, $i = 1, \ldots, Q$. Nevertheless, the translation can easily be extended to one-to-many or many-to-one mappings.

Using on this mapping, the service sessions are translated into SU, namely PCH, requests in a decreasing order of data rate. In other words, we first process the r_1-to-C_1 mapping, followed by the r_2-to-C_2 mapping, $\ldots$, and the r_Q-to-C_Q mapping. To improve channel utilization, the traffic volume of a class which utilizes less than half its mapped physical channel bandwidth will be counted in the translation of SU requests of the next class. We use δ_i to represent the excess traffic volume of service class i which will be merged to service class $i + 1$. It is clear, then, $\delta_0 = 0$.

Following the aforementioned translation rule, the number of SU requests of capacity C_i in a sector, denoted by m_i, is given by

$$m_i = \text{Round} \left(\frac{n_i r_i T + \delta_{i-1}}{C_i tz} \right), \quad i \in \mathcal{I}_Q, \tag{18.4}$$

$$\delta_i = \max \left\{ (n_i R_i T + \delta_{i-1} - m_i C_i tz), 0 \right\}, \quad i \in \mathcal{I}_Q, \tag{18.5}$$

where Round(x) is defined as the function that returns the integer nearest to x. The translated SU requests of each sector of an AP are then provided to the CPA engine in the cloud. We have one CPA engine to processes the CPA task (18.2) for a cluster of APs. The granted SU requests and their associated PCHs for every sector of every AP are informed of their corresponding APs for user scheduling.

18.3.1   User-Level Resource Scheduling

Given the granted SUs and their associated PCHs and transmission powers for every sector of every AP, the objective of the US algorithm is to maximize the summed utility of all sessions and, in the meantime, to provide service and occupancy guarantees. As described earlier, the basic unit for resource allocation is a slot; thus the major dissimilarity from traditional utility-based resource scheduling problems is the discrete feasible allocation rates in our system. In the following formulation, we assume that the number of allocated PCHs of capacity C_i in a sector is $\hat{m}_i$, $i = 1, \ldots, Q$, and that US is performed on a per sector basis.

Let $X_{i,j}^{k,q}(l)$ be an indicator of slot allocation, where i is the service class index and j is the flow index of the specific service class; k is the PCH category index, and q is the channel index of that specific PCH category. Furthermore, l represents the time slot index. From our system settings we have $i \in \mathcal{I}_Q$, $k \in \mathcal{I}_Q$, and $l \in \mathcal{I}_z$; for a specific service class i and PCH category k, $j \in \mathcal{I}_{n_i}$ and $q \in \mathcal{I}_{\hat{m}_k}$. The indicator $X_{i,j}^{k,q}(l)$ is 1 if slot l of the qth channel with capacity C_k is allocated to the session j of service class i; otherwise it is 0. The user-level downlink scheduling problem is formulated on a per sector basis as follows:

$$
\text{Find} \quad \underset{X_{i,j}^{k,q}(l)}{\arg\max} \left\{ \sum_{i=1}^{Q} \sum_{j=1}^{n_i} U_i \left(\frac{t}{T} \sum_{k=1}^{Q} \sum_{q=1}^{\hat{m}_k} \sum_{l=1}^{z} X_{i,j}^{k,q}(l) C_k \right) \right\}
$$

$$
\text{s.t.} \quad
\begin{cases}
l \sum_{k=1}^{Q} \sum_{q=1}^{\hat{m}_k} X_{i,j}^{k,q}(l) \leq 1, \quad \forall (i,j), \\[2mm]
\sum_{i=1}^{Q} \sum_{j=1}^{n_i} X_{i,j}^{k,q}(l) \leq z(\sum_{k=1}^{Q} \hat{m}_k), \\[2mm]
\dfrac{\sum_{k=1}^{Q} \sum_{q=1}^{\hat{m}_k} \left(\sum_{l=1}^{z} X_{i,j}^{k,q}(l) C_k t \right)}{r_i T} \geq \alpha, \quad \forall (i,j) \\[4mm]
\dfrac{r_i T}{\sum_{k=1}^{Q} \sum_{q=1}^{\hat{m}_k} \left(\sum_{l=1}^{z} X_{i,j}^{k,q}(l) C_k t \right)} \geq \beta, \quad \forall (i,j), \quad 0 \leq \beta \leq 1.
\end{cases}
\tag{18.6}
$$

The first constraint ensures that a specific flow cannot be allocated more than two slots on different data channels at the same time. The second constraint prevents the slots allocated to the served flows per frame from exceeding the allocated resources, which are granted by the CPA engine. The third constraint provides each flow with a minimum service guarantee. Each flow shares the granted resources of at least a proportion α. The value of α should be less than or equal to the ratio of the allocated resources to the requested resources, that is,

$$
\alpha \leq \frac{\sum_{k=1}^{Q} C_k \hat{m}_k z t}{\sum_{i=1}^{Q} n_i r_i T}.
$$

The last constraint ensures that the per-slot utilization is greater than or equal to a predefined threshold β, with $0 \leq \beta \leq 1$. This avoids allocating high capacity channels to low rate flows, and thus yielding a high sum utility that seriously wastes resources.

Again, the entire algorithm for solving (18.6) is very involved. Readers who are interested in it may again refer to [7], where the optimal algorithm is given and an heuristic algorithm are provided for solving it.

18.4   The MAC Protocol for Joint Resource Sharing in the CoC-RAN

On the basis of the proposed channel and power resource sharing methods among APs and MUs, we design herein a frequency-division and time-division multiple access (FD-TDMA) MAC protocol for femto APs to coordinate channel access among MUs. The prototyped femto AP and MUs are illustrated in Fig. 18.1. In our design, the available spectra are channelized into one control and multiple data channels. The control channel allows femto APs to collect sensing reports and coordinate data channel accesses; the data channels are assigned to MUs for data deliveries. Multiple MUs can be assigned to different data channels or share the same channel in a TDMA manner.

In the MAC protocol, time is framed and several frames are grouped into a super-frame. Each frame on the control channel consists of five fields, which are Beacon (B), Join period (J), Quiet period (Q), Resource Request (RR), and Report Collection (RC), as shown in Fig. 18.2. A frame on the data channel consists of three fields, including Quiet period (Q), DownLink data transmissions (DL), and UpLink data transmissions (UL). For synchronization purposes, a 10-ms guard interval is inserted between the DL and UL fields.

To announce their existence, start new frames, and manage channel activities, femto APs periodically broadcast beacons on the control channel. By means of receiving beacons and initiating association requests in the Join period, newly join-ing MUs are able to become associated and synchronized with a femto AP. Only associated MUs can hop to data channels and perform spectrum sensing in the Quiet period. These MUs can then utilize the RR period to send their associated

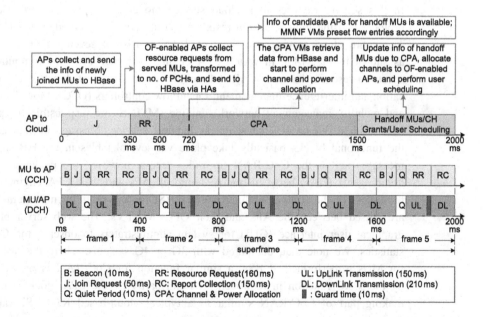

Figure 18.2 The frame structure of the designed MAC protocol and the task processing schedules at the femto APs and the cloud.

femto APs resource requests, while the femto APs utilize the RC period to collect the sensing results. Note that, to improve channel detection accuracy, all MUs are forbidden to transmit data in this time duration. Following the QP, the DL and UL periods are for a femto AP to respectively, transmit data to and receive data from its scheduled MUs. Moreover, the timing of communications between the femto APs and the CN, and the task processing times are also highlighted in Fig. 18.2. To alleviate the burden of backhaul traffic loading, message exchange between femto APs and the CN is on a per-superframe basis. Specifically, femto APs summarize and send the collected information, including the locations of the newly associated MUs and the resource requests of served MUs, to the HBase database in the cloud, and this process takes at most 350 ms. In the following 150 ms, for each sector the femto APs lump and translate the received channel access requests into the PCH (SU) requests, and further relay the requests to the HBase. The CPA engines then start to perform operations and finish the HBase updates (i.e., the PCH allocation results) at 1500 ms. Following this, the femto APs query the channel and power allocation results from the HBase via HAs in the cloud, and perform user scheduling. Both tasks take at most 500 ms, so that the femto APs can carry the DL together with the UL scheduling results in the beacons for coordinating MUs' channel access.[2]

18.5   A Cloud Service Model for the CNs and RANs of Dense Small-Cell Networks

In this section we present a cloud service model to support the proposed MAC protocol and the joint spectrum-resource sharing and US method in the CoC-RAN. The entire cloud services for the CoC-RAN consist of seven functional modules, as shown in Fig. 18.3, in which CPA engines, together with the computation and spectrum resource balance (CSRB) function, are in charge of dynamic channel allocation and interference management; OpenFlow controllers (OFC), FlowVisors (FVs), and mobility management network function (MMNF) are responsible for providing MUs with a 3G or 4G-like seamless handoff service. Message exchanges among the functional blocks basically take place via several tables in the HBase. Control messages across the CoC-RAN for radio access, however, are exchanged through home agents (HAs) between remote femtocell APs and their CPA engines. Control messages for the software-defined network (SDN) and mobility management (MM) are passed between the APs and their OFCs via FVs. The web server provides a graphical user interface (GUI) for system administraters to observe the CoC-RAN statistics. We note that, compared with 3G or 4G systems, where the task of CPA for MUs is performed at BSs, a similar function for CoC-RAN is provided partly by the MAC function in femtocell APs and partly by the CPA block in the cloud, making part of CoC-RAN similar to a typical cloud-based RAN. Similarly, HA,

---

[2] We note that the processing times for CPA and US depend on their algorithms' complexity and the CPU speed.

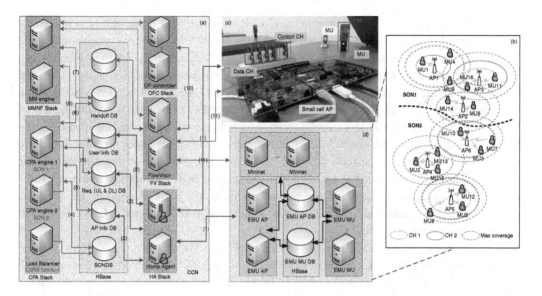

Figure 18.3 The operation model of the cloud-based core network and RAN for DSN.

OFC, FV, and MMNF jointly provide a function analogous to the mobile switching center (MSC), home location register (HLR), and visiting location register (VLR) of a 3G system, making another part of the cloud services a cloud-based core network (CCN).

To perform the real-time task of dynamic channel allocation and interference management, CoC-RAN adopts a two-tier scheme to handle MUs' radio access requests, which consists of the CSRB and CPA functions. Specifically, the CSRB first partitions the densely deployed femtocell APs into SONs, and allocates the available spectrum resources to the boundary femtocell APs in a mutual exclusion manners, to avoid inter-SON interference. A clustering algorithm for this purpose can be found in [7]. An illustration of the CSRB function is shown in Fig. 18.3(b). In this example, femto APs are partitioned into two SONs, and the coverage areas of AP2 and AP6, which belong to different SONs, overlap with each other. If both CPA engines 1 and 2, which process the tasks of CPA for SONs 1 and 2 independently, allocate the same channel to AP2 and AP6 then co-channel interference will occur at the boundary. To avoid this situation, the CSRB function not only performs APs clustering but also allocates channel 1 to AP2 and channel 2 to AP6 to guarantee no interference across the SON boundaries. Nonetheless, non-boundary APs of different SONs can still be allocated the same channels without interfering with each other.

Once the SONs are formed and the available channels for different SONs on the boundaries are determined, the CPA engines then take over to allocate the available channels to the femtocell APs, and in the mean time to determine adequate transmission power for them. Since CPA engines perform channel and power allocation independently for different SONs, inter-SON interference avoidance is achieved by the CSRB while intra-SON interference is controlled by the CPA engine of the SON.

One question that may arise in our CoC-RAN architecture is that of how many femtocell APs are appropriate for a SON. Unlike to the SONs of 3G or 4G systems with static network sizes, the SON size in our CoC-RAN is dynamically adjusted according to the traffic density in its serving area and the computation capability of a CPA engine. In general, to meet the delay requirements of real-time communications, as the traffic density of an area increases then the number of SONs formed and hence the number of CPA engines utilized for the area increase as well. According to statistics released by the National Taiwan University [9], the traffic density of an area does not vary frequently so the CSRB may perform SON formations only a few times each day, e.g., once in the day time and the other at the midnight. The scalable SON network size makes our CoC-RAN greener by shutting down unused femtocell APs and CPA engines. The concepts of SON formation and the rules of auto-scaling and load balancing for CPA engines will be introduced in Section 18.6.

Regarding the MM function, in order to provide MUs seamless handoff services the deployed femtocell APs are in fact OpenFlow-enabled (represented by OF-enabled AP in Fig. 18.1), and each OF-enabled AP is controlled and managed by an OFC in the CCN. In our CoC-RAN, handoff services are triggered by the MMNF block. Specifically, the CPA engines first determine, on the basis of the MUs' locations, which MUs are going to incur handoffs and then update this information to the Handoff database in the HBase. The entire process for this is finished in 750 ms in a superframe. When obtaining information of going-to-handoff MUs and their candidate target APs, an MMNF engine then retrieves from the HBase the IP addresses of the corresponding OFCs for these candidate APs and requests these OFCs to preset flow entries in the APs' flow tables. A flow table is, however, maintained on a soft-state basis. When the final target AP is determined, the redundant flow entries preset in the other APs will expire after a period of time and be deleted from the tables automatically without additional refreshing. More details of the MM function will be provided in Section 18.7.

18.6   Cloud Operating Systems for Core and Radio Access Networks

As stated in the previous section, a two-tier channel allocation and interference control mechanism, i.e., CSRB followed by CPA, is adopted in our CoC-RAN. The CSRB engine in the cloud partitions the DSN into disjoint clusters of SONs according to: (1) the computation capability of a CPA engine, (2) the number of femto APs, (3) the statistics of SU requests, and (4) the response time required for MAC [7]. Once the partitioning is done, a CPA engine is in charge of channel and transmission power allocation to the femto APs inside a specific SON. To meet real-time communication requirements, CPA engines work in parallel and independently to serve all the SONs together. The computational time for the CPA scheme (18.2) varies with the numbers of APs, available channels, MUs and the corresponding SU requests. When the traffic load increases, both the numbers of active APs and activated CPA engines increase to meet the service demands; in contrast, some APs and CPA engines can be turned off during off-peak hours. Consequently, adequate rules for auto-scaling and load balancing are essential to a greener CoC-RAN. Moreover, as the number of active APs varies, the traffic of control message exchanges between APs and the CCN also changes dynamically. Thus, appropriately scaling-out or scaling-in HAs affects the performance of the CoC-RAN.

As shown in Figs. 18.1 and 18.3, our CCN and cloud-based RAN were prototyped on the OpenStack cloud computing software platform. The OpenStack community has collaboratively identified nine key components as the core of OpenStack, including Nova, Neutron, Heat, Ceilometers, etc. These components together not only allow users to deploy VMs and instances to handle different computation tasks on the fly, but also make horizontal scaling easy through the setting of scaling parameters. Online resources for OpenStack can be found at [5]. Owing to space limitations, we focus here on the components related to auto-scaling and load balancing only, i.e., Nova, Heat engine, Ceilometer, and HAproxy. For other components, please refer to the OpenStack white paper [5].

Nova is the primary computing engine behind OpenStack for deploying and managing VMs and other instances to process computing tasks, while Neutron provides networking capability for OpenStack. Heat is the main orchestration program in OpenStack for launching multiple composite cloud applications based on templates and, when integrating with Ceilometer, Heat can also provide an auto-scaling service. HAproxy is a fast yet reliable solution offering high availability, load balancing, and proxying for TCP and HTTP-based applications. Before introducing the scaling rules designed for HAs and CPA engines, we first use Fig. 18.4 to illustrate the procedure of orchestrating these components to provide auto-scaling and load balancing on OpenStack.

For a provisioned stack of VMs, e.g. an HA or CPA Stack, thresholds to scale-out or scale-in VMs are determined on the basis of an adopted performance metric, such as CPU utilization and network flow. Next the Ceilometer is in charge of monitoring the status of the stack (step 1). Once the measured average exceeds (is below) the threshold, it informs Heat engine through the OpenStack application programming interface (API) to open up (shutt down) some VMs (step 2). Then, the Heat engine relays this request to the Nova server through the advanced message queueing protocol (AMQP) (step 3). Upon receiving this message, the Nova server correspondingly issues a command to the Nova agent and then the Nova agent opens VMs up (step 4). Once the Health Monitor, a program in the Load Balancer of a stack, observes a VM addition (or a VM removal), it requests HAproxy, through AMQP, to distribute loadings to all the activated VMs on the basis of the adopted balancing rule (step 5).

The scaling parameter utilized by the HA Stack is "network flow". When the Ceilometer observes that the incoming traffic loading exceeds or is below defined thresholds, a similar procedure to that highlighted in Fig. 18.4 is performed to add or remove VMs in the HA Stack as well as launching these VMs into the HAproxy group. With regard to distributing loads among HAs, HA Stack adopts the "least connections" policy. In other words, when a femto AP tries to establish a connection to one HA, the HAproxy will direct this request to the HA which serves the least number of APs. Note that AP may

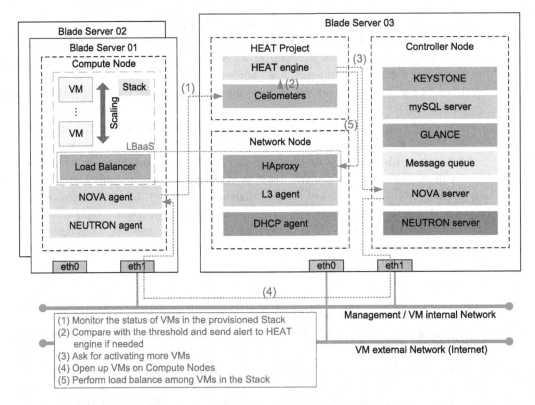

Figure 18.4 The software architecture for the OpenStack system and the procedure to perform auto-scaling and load balancing.

be served by another HA when it reconnects to the HA Stack after the timeout of the original connection.

However, the adopted scaling parameter for the CPA Stack is "mean CPU utilization". Once again, the threshold setting is tightly coupled to the computation complexity of the CPA algorithm. Indeed, several parameters, including the computational capability of the CPA engines, the requested channels, and the required response time, affect the threshold setting. How to set the threshold appropriately is beyond the scope of this chapter; the reader should refer to [7] for some details of the CPA algorithm. Regarding load balancing among CPA engines, to obtain balancing rule is exactly the CSRB function. The reason is that the mean CPU utilization of the provisioned CPA Stack exceeding the threshold is a sign that some CPA engines are handling too many channel requests and failing to meet the response time requirement. In such a situation, more CPA engines should be added into the CPA Stack, which triggers the CSRB function to re-form the SONs. Once this re-formation is done, not only the number of VMs to be further launched but also the members of the femto APs in each SON are known, and thus the load balancing is simultaneously achieved. Unlike to the HA Stack, where HAproxy is responsible for distributing connections to HAs, each CPA engine receives its served femto AP from the SON DB (see steps 3–5 in Fig. 18.3).

18.7   A Cloud Service Model for SDN-Based Mobility Management

In comparison with the typical macro or micro cellular networks, user roaming among femtocell APs occurs much more frequently because of the much smaller cell coverage area of each AP. In dense small-cell networks, the coverage radius of an AP is about 10–100 meters. Taking the average human walking speed as an example, an MU moving at about 3–4 km/hr may travel across the coverage areas of several APs in 5 minutes. Therefore, how to provide a handoff service and let MUs switch connections smoothly among APs becomes a crucial and challenging issue in our CoC-RAN. In particular, this mobility management (MM) service also has to be very flexible and able to handle the highly dynamic TCP traffic across femtocell boundaries. OpenFlow-enabled APs plus cloud-based OpenFlow controllers emerge as a promising technology to provide such an MM service. As a matter of fact, this MM service can be regarded as an SDN application. In addition, a cloud-based architecture is able to provide a scalable managing service for large amounts of femtocell APs. The marriage of these two features gives birth to a cloud service model and MM architecture for our CoC-RAN prototype.

18.7.1   Mobility Management as a Network Function of SDN

The architecture of the SDN-based MM function in CoC-RAN is illustrated in Fig. 18.5. When an MU is in a communicating session with a corresponding node (CN) while traveling, the CN will keep sending packets to the AP where this session originally started. To forward packets to the MU's new serving AP, the mobility management network functions (MMNF) will ask a certain OpenFlow controller, which is a Floodlight controller (FLC) in our prototype, to push MM flow entries to the original and the new serving APs. The command of the MMNF is sent to the FLC via a northbound interface, which is REST API in our prototype, between the MMNF and the FLC. The complete procedure to provide such a handoff service in our CoC-RAN prototype is described as follows.

Step 1: When a CPA engine allocates channel resources for a new MAC superframe interval, the CPA also stores the IDs of the MUs that need handoff services (called handoff MUs in the sequel) in an HBase table.

Step 2: The MMNF's health monitor (HM) retrieves the IDs of the handoff MUs from HBase, and distributes the MM tasks to the VMs in the MMNF stack.

Step 3: The MMNF VMs retrieve from HBase the IDs of the APs that serve the handoff MUs.

Step 4: The MMNF VMs retrieve the IP addresses of the FLCs that control these APs from HBase.

Step 5: The MMNF VMs design flow entries and command FLCs to insert the entries to their APs.

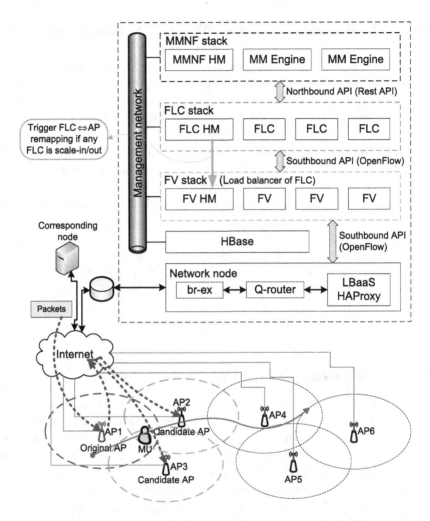

Figure 18.5 The functional blocks for SDN-based mobility management in CoC-RAN.

18.7.2   FlowVisors: Serving as FLC's Load-Balancer

In addition to an effective procedure for MM, the CoC-RAN also needs to make the MM service scalable to the wireless traffic in the system, given that the numbers of MUs and APs may vary significantly with time and local conditions. This implies that the number of VMs for FLCs needs to scale with the wireless traffic too. To offer such a scalable service with OpenFlow controllers, we use FlowVisor (FV), which is also a special-purpose OpenFlow controller, to define the mapping between FLCs and APs. For simplicity, we have one FLC to control the flows of certain number of APs on the basis of its processing speed.

FV Auto-Scaling and Load-Balancing

When an FV receives an OpenFlow control message, it determines to which FLC or AP to forward the message according to a preexisting mapping table. Therefore, the loading of the FV comes from (1) the number of mappings between FLCs and APs, and (2) the frequency of receiving control messages from the FLCs and APs, which is usually proportional to the number of connected APs. To avoid a processing bottleneck occurring at an FV, we propose in Fig. 18.5 a parallel architecture for FV load balancing and use the default load balancer, HAproxy, for TCP traffic in OpenStack to distribute the connections to multiple FVs.

Under this parallel load-balancing architecture the scaling rule for OpenStack to determine the number of VMs in an FV stack becomes easy, and it set to be proportional to the average incoming network traffic to all the FVs. Nevertheless, a synchronization issue will soon occur in this architecture when multiple FVs work together to pass control messages to the same set of FLCs and APs. We resolve this problem by storing all the mappings in an HBase table. Each FV then retrieves the mapping it needs from the table on the basis of its connected APs. As such, not only the AP connections but also the mappings can be shared and distributed to all the FVs. Using this method, the scaling and load-balancing functions for FVs can be directly implemented in OpenStack, using its Heat and Load Balancing as a Service (LBaaS) modules, respectively.

FLC Auto-Scaling and Load-Balancing

The loading of FLCs, however, depends on the number of connected APs, which is in turn proportional to the network traffic. Therefore, we also define the average incoming network traffic as the metering data for an FLC stack. However, the incoming traffic of each FLC must come from the APs which are defined by the latest FLC-to-AP mappings. In other words, the FLC-to-AP mapping actually defines the load-balancing policy for the FLCs. As shown Fig. 18.5, when the FLC stack is scaled-in/out, the FLC's health monitor will trigger FV's health monitor to redefine the FLC-to-AP mapping, and then the FVs will update the latest mappings into their local databases. Therefore, the whole FV stack can be regarded as the load-balancer of the FLC stack and the load-balancing policy can be easily defined and updated in FV's health monitor, instead of developing a new project in OpenStack.

MMNF Auto-Scaling and Load-Balancing

Intuitively, the number of handoff MUs determines the loading of the MMNF, and, thus, it is defined as the scaling metering data for the MMNF stack. We let the MMNF's health monitor keep monitoring the number of handoff MUs maintained in HBase, trigger scaling-in/out, and assign the handoff MUs to each MMNF. It is worth mentioning that the flow entries are designed with a timeout mechanism. To the original AP, the flow entries for its handoff MUs will be automatically removed after they have not been used for a while. This can save the AP's memory, and no extra mechanism is required for the cloud to remove the entries.

18.7.3   Remaining Issues for Mobility Management in Dense Small-Cell Networks

Actually, there are still many issues needing to be discussed and resolved for mobility management in dense small cell networks, such as Internal protocol (IP) address management. Since the 5G network is expected to be an all IP-based system, and TCP/IP is an end-to-end protocol, the usual type of network service has to start over again if its corresponding MU's IP address is changed. In other words, redirecting packets to candidate APs with the SDN functions is not enough to support handoff if an MU's IP address is changed after re-association. To avoid the above issues, in this work we provide mobility management in one subnet with one DHCP server only. User roaming among APs in different subnets will be an important issue for future SDN-based wireless networks.

18.8   CoC-RAN Prototype and Emulation Results

Experimental results on the CoC-RAN prototype are presented in this section. The CCN and cloud-based RAN are implemented on an OpenStack cloud software platform that runs on an IBM blade center that consists of two HS22 servers and one HS23 server. The database is built with HBase [10]. However the femto APs and MUs are implemented with TI's OMAP and transceiver modules CC1111 and CC1110. The frequency channels used in the experiments are set in the ISM bands right above the UHF TV band, including the 866, 868, 870, and 872 MHz bands. In this CoC-RAN prototype the frame time is 400 ms; a superframe consists of five frames, so its duration is 2 seconds. The time settings for each field in a control frame and a data frame are shown in Fig. 18.2.

There are four femto APs and 20 MUs physically implemented with OMAP development boards and TI CC1111/CC1110 transceiver modules. In order to test the various auto-scaling rules and load balancing policies designed for our CoC-RAN, we also developed a network emulator on the cloud platform of National Chiao Tung University (NCTU), Taiwan, to generate the network traffic for the CoC-RAN. A photograph of an AP and its associated MUs is shown in Fig. 18.3(c), and the network emulator's block modules are illustrated in Fig. 18.3(d). In the experiments, two sets of traffic are tested by the emulator, one with 24 APs and 100 MUs and the other with 48 APs and 200 MUs, both simulated in the NCTU cloud. In spite of the simulated traffic, the radio access requests to and the control messages required from the CoC-RAN for them are exactly the same as those for the real four femto APs and 20 MUs. This is achieved by periodically sending the newly simulated traffic into the HBase at the NCTU cloud. Following that, the simulated channel requests and the location information of the APs and MUs are periodically delivered via HAs to the CN and RAN functional stacks in the cloud.

In addition to the emulated radio access requests, we also use Mininet to emulate the OpenFlow control messages for the APs since they are all OpenFlow-enabled APs. To this end, the mobility of the MUs and the network topology of the APs are also simulated and stored in the HBase at the NCTU cloud. The simulated OpenFlow-enabled APs in Mininet retrieve the network topology from the Hbase and establish connections with the FVs in the cloud of CoC-RAN. Since there is no wireless link supported in Mininet

Figure 18.6 The statistics of the system resources used for CoC-RAN in OpenStack.

version 2.0, we add new *x*-axis and *y*-axis parameters for each MU in Mininet, and arrange for Mininet periodically to get access to the HBase to retrieve information on the MUs' locations and AP associations and update the topology to emulate the MUs' mobility and associations in Mininet.

However the flow entries issued by the OFCs through the MMNF engines are pushed into the flow tables of the simulated APs in the Mininet. The experimental results on the utilized VMs in the cloud are shown in Fig. 18.6. On the left-hand side are results for the set of 24 + 4 APs and 100 + 20 MUs, and on the right-hand side are results for the set of 48 + 4 APs and 200 + 20 MUs. When doubling the numbers of APs and MUs, Heat stacks with different purposes automatically scale out one or two more VMs in their individual stacks. Thus the total utilized VMs as well as the CPU cores increase from 21 and 26, respectively, to 28 to 33.

18.9 Conclusions

We have presented a cloud service model and system prototype for SDN-based dense small-cell RANs. Integrating the various functions of the core network and RAN on OpenStack, we demonstrated the feasibility of deploying a WMAN based on dense and cheap femtocell APs. The spectrum-resource sharing and mobility management requests among the dense APs can all be adequately provided in real time under the proposed infrastructure. Although more performance metrics of the CoC-RAN, such as the processing times of the CPA engines, HA servers, FLCs, FVs, and MMNF, need to be provided to verify the effectiveness of the prototyped CoC-RAN, the emulation results do suggest that the conflicting goals of a greener yet faster 5G network could possibly be realized with an SDN-based femtocell RAN running over a regular cloud platform. Notwithstanding, the true performance of a CoC-RAN depends on the system architecture, protocol design, and CPU speed as well, a discussion of which is beyond the scope and space limitations of the chapter. Interested readers are encouraged to explore these issues further, using the reference list.

References

[1] Cisco, "Visual networking index," in *Global Mobile Data Traffic Forecast Update, 2011–2016*, Cisco Systems, 2011, available online at www.cisco.com.
[2] C.-L. I, "Green evolution of mobile communications," in *Proc. Telecommunications Industry Association Conf.*, (Dallas, TX U.S.A), 2012.
[3] K. Chen and R. Duan, "C-RAN: the road towards green RAN' White Paper Version 2.5, China Mobile Research Institute, 2011, available online at labs.chinamobile.com/cran/.
[4] "Software-defined networking: the new norm for networks" ONF White Paper. Open Networking Foundation, 2013, available online at: www.opennetworking.org.
[5] "OpenStack architecture design guide." OpenStack Document, 2014, available online at docs.openstack.org/arch-design/arch-design.pdf.
[6] S.-H. Wu, H.-L. Chao, C.-H. Ko, S.-R. Mo, C.-F. Liang, and C.-C. Cheng, "Green spectrum sharing in a cloud-based cognitive radio access network," in *Proc. IEEE Int. Conf. on Green Computing and Communications*, (Beijing, China), 2013.
[7] H.-L. Chao, S.-H. Wu, Y.-H. Huang, and S.-C. Li, "Cooperative spectrum sharing and scheduling in self-organizing femtocell networks," in *Proc. IEEE Int. Conf. on Communications*, (Sydney), 2014.
[8] S. Shenker, "Fundamental design issues for the future internet," *IEEE J.Sel. Areas Commun.*, vol. 13, no. 7, pp. 1176–1187, September 1995.
[9] "NTU wireless network statistics." Computer and Information Networking Center, National Taiwan University, available online at http://ccnet.ntu.edu.tw/wireless2/flow_rate.html.
[10] "Apache HBase official website," available online at: http://hbase.apache.org.

19 Field Trials and Testbed Design for C-RAN

Chih-Lin I, Jinri Huang, Min Yan, and Xiaogen Jiang

19.1 Introduction

Since the proposal of C-RAN [1–3] in 2009, China Mobile (CMCC) has been committed to developing various kinds of proof-of-concept (PoC), testbeds, and field trials to demonstrate C-RAN's benefits and verify the key enabling technologies. This chapter gives a comprehensive introduction to these activities. In particular, we will demonstrate not only the feasibility and reliability of wavelength-division-multiplexing (WDM)-based fronthaul (FH) solutions but also how a noticeable coordinated multiple-points (CoMP) gain can be achieved with the C-RAN architecture. In addition, a virtualized C-RAN system is elaborated, including the design principles, the architecture, and the field trial results.

19.2 Field-Trial Verification of FH Solutions

19.2.1 Centralization Field Trials in 2G and 3G Networks

The first step toward C-RAN was baseband unit (BBU) centralization which is relatively easy to implement and can be tested with the existing 2G, 3G, and 4G systems. In the past few years, extensive field trials have been carried out in more than 10 cities in China using commercial 2G, 3G, and pre-commercial TD-LTE networks with different centralization scales. The main objective of C-RAN deployment in 2G and 3G is to demonstrate the deployment benefits of centralization, including accelerated site construction and reduced power consumption. For example, one trial took place in the city of Changchun where 506 2G BSs in five counties were upgraded to a C-RAN-type architecture centralized in several sites. In the largest of these, 21 BSs were aggregated to support 101 RRUs with a total of 312 carriers. It was observed that power consumption was reduced by 41% owing to shared air-conditioning. In addition, system performance in terms of the call-drop rate as well as the downlink data rate was enhanced using multiple RRU-co-cell technologies. For the results and benefits from using centralization in 2G and 3G trials, the reader is referred to [4].

When it comes to TD-LTE, centralization becomes more challenging owing to the high data rate in the FH connection. For example, the data rate of the most widely used FH interface in the industry, the common public radio interface (CPRI), could be as high as 9.8 Gb/s for an TD-LTE carrier with a 20 MHz bandwidth and eight antennas. As a result, a considerable number of fibers is required for centralization deployment, which unfortunately is unaffordable for most operators. To address the fiber consumption issue several solutions have been proposed, including compression techniques, single fiber bi-direction (SFBD), wavelength-division multiplexing (WDM), and microwave solutions. In [4] the authors demonstrated that compression solutions with 2 : 1 compression ratios are adequate and yet do not have impact on the system performance. With SFBD, which allows simultaneous uplink and downlink transmission on the same fiber core, there is a fourfold saving in fiber usage. The reader is again referred to [4] for more information.

In the next section we will present our field trial results on the verification of a WDM-based FH solution.

19.2.2   Verification of WDM FH Solutions

Wavelength-division multiplexing (WDM) is a mature technology widely adopted in transportation. The basic idea of WDM is to use different, orthogonal, wavelengths to transport separate data. The capacity of a WDM system depends on the number of wavelengths that the system has. Making use of the maturity of WDM, vendors can develop WDM equipment tailored to fronthaul transmission within a short period of time. Currently a few operators have adopted this solution to realize large-scale C-RAN deployment. Some commercial products can support as many as 60 2.5 Gbps CPRI links in one fiber pair, which significantly reduces fiber consumption. Also, 1+1 or 1:1 ring protection is supported and several low data rate links can be multiplexed into one high data rate link [1].

In this section, a field trial is introduced to demonstrate the feasibility and performance of WDM FH solutions. We set up a C-RAN in China Mobile's TD-LTE commercial networks in a dense urban area. The system configuration is shown in Table 19.1. The RRUs in this trial were of the two-antenna type; therefore, the CPRI data rate was 2.5 Gbps. Seven sites were centralized in total. To save fiber resources, WDM equipment was introduced. The seven sites (21 carriers) were connected in two WDM rings, as shown in Fig. 19.1. It should be pointed out that, from a capacity perspective, the WDM equipment in the trial could support up to twelve 10 Gbps or 2.5 Gbps wavelengths. In other words, one such WDM ring could support twelve LTE carriers. The total length of the two WDM rings was 20 kilometers. Figure 19.2 shows the centralized baseband and WDM equipment in the C-RAN central office. The benefits and potential impact on system performance of WDM were carefully examined, and the following results were obtained.

With the WDM solution in this trial, one pair of fiber cores was needed for each ring, meaning that four fiber cores were used in total. Compared with a dark fiber solution, which requires 42 fiber cores (21 carriers, UL and DL), WDM reduces fiber consumption significantly. The processing delay of the WDM nodes is an important metric of WDM performance. If it is too large, it may impact CPRI transmission.

Table 19.1 System configuration of TD-LTE C-RAN field trial with WDM FH

Frequency	2.85 GHz
Bandwidth	20 MHz
Frame structure	UL/DL configuration type 1 Normal CP Special subframe configuration type 7 DwPTS for data transmission
CPRI	2 : 1 compression
Optic module	Single-fiber bidirection
UL	SIMO
DL	Adaptive MIMO
QCI	9
Scheduler	PF

Figure 19.1 C-RAN WDM field trial areas.

Figure 19.2 The centralized baseband and WDM equipment.

In this trial, the processing latency was found to be less than 1 μs, which is small enough to have no impact on CPRI transmission. Another key feature of the WDM solution is the protection switch capability. The operators require this to be less than 50 ms. In the trial the fiber was pulled out to simulate link failure and to trigger the automatic link switch. It was found that the switch time was less than 30 ms, meeting the requirement.

In addition to the performance of the WDM network itself, we further tested the whole wireless system performance in terms of throughput, coverage, end-to-end latency, and handover success rate. The key finding was that all the performance metrics were almost the same with or without WDM. The trial results show no impact on radio performance with WDM FH. For example, Fig. 19.3 gives downlink and

Table 19.2 User-plane ping delay with and without WDM FH

32-byte ping delay		Good point	Middle point	Weak point
	Max.	24	29	33
Without WDM FH	Min.	16	16	16
	Ave.	18	19	20
	Max.	22	27	37
With WDM FH	Min.	15	15	17
	Ave.	18	19	20

Table 19.3 Handover signaling latency with and without WDM FH (unit: seconds)

	Min.	Max.	Ave.
Without WDM FH	0.017	0.02	0.018
With WDM FH	0.014	0.022	0.018

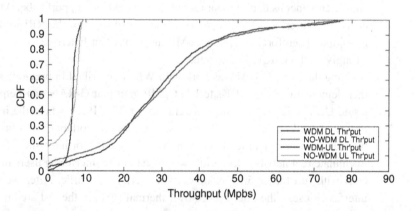

Figure 19.3 Throughput with and without WDM FH.

uplink throughput CDF with and without WDM FH. The results show very similar throughput performances. Table 19.2 shows the user-plane ping delay with 32-byte packages with and without WDM FH. The results also indicate similar user-plane delay performance.

We also tested the handover success rate and the handover latency. The results are shown in Tables 19.3 and 19.4. It can be seen that the handover signaling latency is almost the same with and without WDM. In addition, the handover rate is also the same.

Finally, it is worth pointing out that, commercially, the network has been functioning properly for more than a year without failure.

Table 19.4 Handover success rate with and without WDM FH

	No. of HOs	No. of successes	No. of failures	HO success rate (%)
With WDM	24	24	0	100
Without WDM	60	60	0	100

19.3   CoMP Demonstration in C-RANs

19.3.1   Uplink CoMP Verification

LTE networks suffer from severe interference issues. Uplink CoMP has the potential to improve uplink capacity and reduce inter-cell interference by jointly processing the signals received from more antennas of adjacent nodes.

Uplink CoMP needs real-time operation with a great deal of data exchange between cells, and it also requires centralized data processing. Intra-site CoMP is the easiest to deploy in contemporary networks and there is no need to update radio network architecture that has sufficient coordination between cells of the same eNodeB. Inter-site CoMP and heterogeneous deployment require ideal backhaul support (fibers) to guarantee low delay, low jitter, and high capacity CPRI transport from the RRUs to the centralized baseband. Therefore, inter-site CoMP and CoMP in heterogeneous networks depend strongly on the C-RAN architecture.

Using the same C-RAN network with WDM described in previous sections, we further demonstrated the inter-site UL CoMP gain that C-RAN can help to achieve. We tested UL CoMP in the Ring 1 coverage area (Fig. 19.1), where the inter-site distance is around 1 kilometer. Six cells were selected as one collaborative cluster out of which two cells can be dynamically chosen to perform intra- or inter-site UL CoMP. In order to emulate real network scenarios, we allocated five interfering users in the surrounding cells with full buffer 50% resource block (RB) UL traffic. After we activated all the interfering users, the interference over thermal (IoT) in the test area increased from 13 to 16 dB.

During the trial, one user walked around the test area with full buffer uplink traffic. Uplink throughput, serving cell RSRP, and supporting cell RSRP were captured.

Figures 19.4 and 19.5 show the uplink CoMP trial results. Figure 19.4 gives the absolute uplink throughput and Fig. 19.5 gives the uplink CoMP throughput gain. The results indicate that uplink CoMP has a strong potential to strengthen the uplink signal and suppress interference. It shows that the uplink CoMP gain can reach ∼40%–100% in the weak coverage area (that is, the area where the RSRP is lower than −95 dBm), and in some areas the gain can exceed 100%. Because of the interference, the uplink throughput cannot reach its peak in high coverage areas (with RSRP higher than −90dBm) when there is no uplink CoMP. Uplink CoMP can bring in a ∼12 Mbps throughput boost and results in a ∼20%–50% throughput gain in these areas.

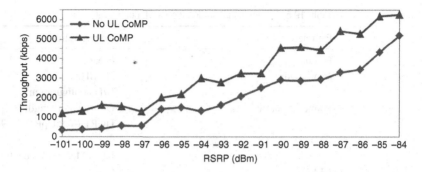

Figure 19.4 Uplink throughput.

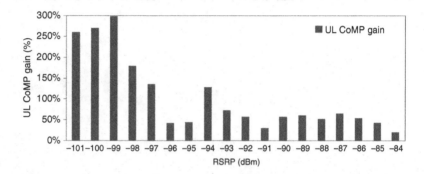

Figure 19.5 Uplink CoMP gain.

19.3.2   Downlink CoMP Verification

Downlink (DL) CoMP has been viewed as one of the main 5G technology candidates to improve system performance. It can be divided into two classes: MAC-layer coordination and physical-layer coordination. For example, collaborative scheduling and beamforming (CS/CB) is one of the MAC-layer-coordinated mechanisms. Joint transmission (JT) is a physical-layer-coordinated technology. This test aimed at DL CoMP function and performance verification in the scenario of C-RAN centralized deployment. It was expected that with the use of CoMP in dense urban areas, the edge-user experience could be improved without significantly impacting the overall cell throughput.

In the test, two CoMP technologies including CS and CB were verified. The test area is shown in Fig. 19.6. It consisted of five TD-LTE sites (15 cells) with an operating frequency in the F-band and RRUs with eight antennas. The BBUs in the test area were centralized in the same equipment room with switches to connect them together and allow direct signaling and data interaction among the BBUs. We defined the overlapping coverage as the regions in which the difference between the reference signal received power (RSRP) in the servicing cell and that in the neighboring cell is consistently less

Table 19.5 System configuration for the DL CoMP trial

Parameter	Configuration
Frequency	F-band
Bandwidth	20 MHz
Frame Structure	U/D configuration 2 CP: Regular length DwPTS:GP:UpPTS = 9 : 3 : 2
CFI	3
Antenna pattern	DL?TM3/7, self-adaption
ULPC	ON
HARQ	ON
AMC	ON
Base station transmission power	8 × 5 W

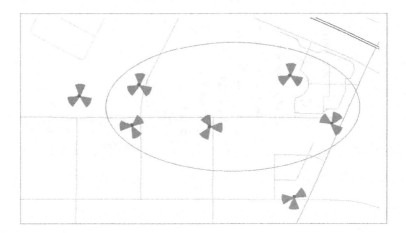

Figure 19.6 Test area for C-RAN DL CoMP verification.

than or equal to 6 dB. The overlapping coverage degree is defined as the number of neighboring cells whose coverage overlaps that of the serving cell.

The configuration of the trial network is shown in Table 19.5.

CoMP CS Test

CoMP CS performance was first evaluated under different network load conditions. For each cell in the test area, 3UEs which acted as interference sources were put in good, middle, and edge points respectively. Three other terminals were moved around within the test area. The movement paths are shown in Fig. 19.7. Through the configuration at the network side, different traffic loads such as 20%, 50%, 70%, and 100% could be set for the CS tests to be performed.

The test results for the average user throughput gain as well as the average cell throughput under different load conditions are given in Table 19.6. As shown in the table, when the load was 20% the average gain for the edge users could be as high as

Table 19.6 Edge user and cell throughput gain under different traffic loads in CS test

	20% traffic load	50% traffic load	70% traffic load	100% traffic load
Edge user gain	131.40%	87.70%	75.10%	55.90%
Cell throughput gain	19.50%	9.00%	6.20%	1.30%

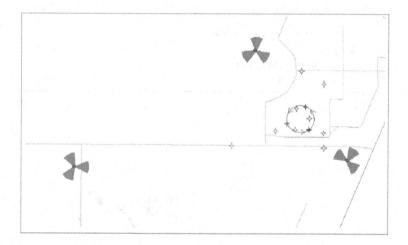

Figure 19.7 Movement paths for CS test.

131.4%, which proved that the CS could improve the edge user's experience noticeably. However, the overall cell throughput gain was relatively small, only about 19.5%. The reason is that in CS mode, although the scheduling information is decided by the multiple coordinated cells, the UE DL data is transmitted only by the serving cell. In the serving cell, the scheduled frequency and time resources are used by only a single user, instead of multiple users. Therefore, the spectrum efficiency of the serving cell is almost constant. The overall throughput gain is therefore bottlenecked by the throughput gain of the single CS edge user, which is much smaller. Moreover, other coordinated cells should try not to schedule the same frequency and time resources for the users in the overlapping area. With increasing traffic load for the coordinated cells, it is harder and harder to schedule different frequency and time resources for different edge users. It is also seen in the table that the performance gain of the edge user and the overall cell throughput decreases with an increase in traffic load. In conclusion, although the performance of the edge user can be improved significantly by collaborative scheduling, the overall cell throughput gain is considerably smaller.

Single-User CB Performance Test

The CB test for single users was then performed. In the test area, a test UE, UE1 was placed at the edge of Cell 1. Another UE, UE2, was in Cell 2 and acted as an interference source for UE1. There was an overlap area between Cell 1 and Cell 2. With CB on and

Table 19.7 Performance gain of CB

Position	P1	P2	P3	P4	P5
Edge user's gain (%)	43.97	52.73	56.48	30.81	36.01
Cell throughput gain (%)	8.65	14.72	13.62	3.60	6.37
Position	P6	P7	P8	P9	P10
Edge user's gain (%)	49.74	33.49	56.97	52.81	99.39
Cell throughput gain (%)	23.83	1.60	11.68	17.85	16.83
Average edge user's gain over the test points (%)			49.30		
Average cell throughput gain over the test points (%)			11.60		

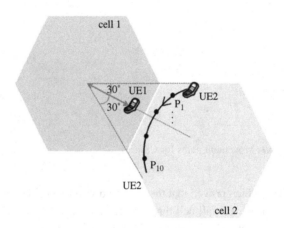

Figure 19.8 Movement path for CB test.

UE1 static, UE2 moved within an angular range of 30°, which extended between the radial direction of UE2 relative to Cell 1 and the radial direction of UE1 relative to Cell 1, as shown in Fig. 19.8. During the movement, UE2 was receiving full-buffer DL service with 100% traffic load.

During the movement of UE2, the throughput of UE1 and UE2 was recorded in ten different positions, P1, P2 to P10, uniformly distributed on the test route. After that, the CB was switched off and the test was repeated.

The test results are shown in Table 19.7. It can be seen that the highest gain for edge user throughput was 99.39% while the lowest gain was 30.18%. The average gain over the 10 positions was 49.3%. The peak gain in cell throughput was 24.83% and the lowest was 11.6%. In CB the beamforming (BF) direction of the edge users is decided by the coordinated cells in conjunction. In this way interference among the edge users could be reduced and the throughput could be improved. However, users near the midpoint may need to make adjustments in order to guarantee a performance gain for edge users. Their throughput decreases as a result. It could be concluded that, as with CS, CB brings a sharp rise in edge user throughput while providing a modest improvement in cell average performance.

19.4   COTS and Accelerator-Based Virtualized C-RAN System

The core feature of C-RAN, as mentioned in Chapter 1, is to realize resource cloudification. It has been pointed out in [1] that the realization of this feature could rely on the virtualization technology widely adopted in the IT industry. This philosophy is indeed in line with the concept of network function virtualization (NFV), which is to "consolidate many network equipment types onto industry-standard high volume servers, switches, and storage, which could be located in data centers, network nodes, and in the end-user premises" [5].

With the development of telecom and IT cloud convergence the commercial-off-the-shelf (COTS) open platform has achieved huge improvements in the past few years in terms of processing capability, which enables NFV's introduction in the telecom field; NFV-compatible platforms can greatly improve hardware efficiency with resource virtualization and sharing. They can fully support multiple radio access technologies (RATs) as 3G, 4G and the future 5G converge, improve network reliability, optimize network efficiency and, reduce the network capital expense (CAPEX) and operating expense (OPEX). In the meantime, it is relatively straightforward for an IT server-based baseband platform to provide open application programming interfaces (APIs) to accommodate mobile internet applications and provide better experiences for the end-users.

In 2012, supported by the National High-Tech R&D Program of China, CMCC and several partners set up a dedicated C-RAN project which aimed to study the key technologies for realizing virtualization in real-time radio processing. The key technologies include:

- real-time (RT) signal processing;
- operating system (OS) enhancement on RT performance, including RT interrupt response and I/O performance enhancement;
- resource virtualization and I/O virtualization;
- system architecture design for baseband pools;
- intelligent load monitoring and dynamic resource management, e.g. load balancing, seamless live migration, automatic resource scale-in/scale-out, etc.

With virtualization technologies, the baseband (BB) pool or cloud is expected to offer the benefits of low cost, low latency, high bandwidth, high reliability, flexibility and scalability. Resources can be managed in an effective and efficient way. Multi-RAT could be easily supported in the same platform and ultimately the evolution of wireless communication systems could be achieved in a faster, more cost-effective, and smoother way.

19.4.1 System Architecture

Figure 19.9 shows the system architecture of a virtualized C-RAN testbed developed in the project. The testbed realized two RATs in TD-LTE and GSM. In the trial network there were six TD-LTE sites of 18 cells deployed outdoors while a TD-LTE site with three sectors was deployed for indoor coverage. For the baseband pool, an accelerator called the baseband acceleration element (BAE) was developed to deal with the LTE PHY processing. The upper layers were processed in the COTS platform, which consisted of standard servers connected with each other through standard switch networks.

The key components of the system include the following.

- *Computing process unit*   The platform mainly consists of high-performance commercial servers. The servers are based on Intel Xeon 2670 dual QPI and afford computing resources for diverse processing functions, including TD-LTE L2 or L3 protocols, networking, S1–X2 interfaces, virtualization, and cloud management.
- *Baseband accelerator element (BAE)*   The BAE works as a radio engine, supplying CPRI ports for the BBU pool to connect to the remote radio head and for 10GE–PCIe interfaces to connect to the COTS platform. It realizes complete LTE PHY functions and provides multiple interfaces including 10GE, CPRI Rate 7 and PCIe Gen2.
- *Shared storage*   Currently, storage area networks (SANs) and network attached storage (NAS) are mainstream solutions for shared storage. The virtualized base station resource pool takes the benefits from both solutions: it was built on SAN while adopting NAS to provide file system services.
- *Networking*   The logic units, such as the virtual BBU, in the system communicate with each other via the internal switching network, which is composed of 10GE Ethernet switches. The network chose two-level fat trees [1] to support flexible configuration and smooth expansion of the BAE and the processing unit.

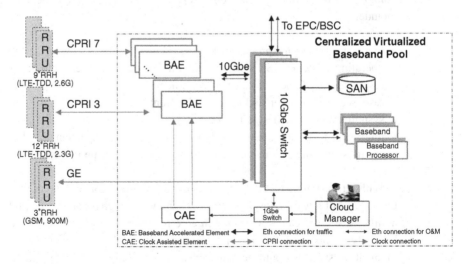

Figure 19.9 System architecture of a virtualized C-RAN testbed.

Table 19.8 Latency performance in traditional BBU platforms

Latency (μs)	Min.	Ave.	Max.
Legacy BBU	3	4	16

- *Clock-assisted element (CAE)* The CAE acts as the centralized clock source and provides the timing information for the platform to meet the frequency and phase synchronization requirements of the radio air interface.

19.4.2 System Design and Optimization

Optimization on Real-Time Performance

Traditional BBUs are developed on the basis of vendors' customized platforms with dedicated design on the embedded real-time (RT) OS for LTE baseband processing. According to internal measurements, the OS scheduling latency in the legacy BBU platform is shown the Table 19.8.

With open platforms which are based on general-purpose processors and OSs, it is a major challenge to achieve an RT performance comparable with traditional platforms. The latency performance on an open platform with a general Linux OS is shown in Table 19.9.

With the introduction of virtualization technology the latency becomes worse, as shown in Table 19.10.

This project designed the following solutions to optimize the system real-time performance to meet the BBU latency requirement:

- *Hardware parameter optimization* The hardware configuration parameters were optimized for improved RT performance. For example, different servers may have different features designed by the vendors and such features could be realized by parameter configuration. Some features such as intelligent temperature control are unnecessary from a signal processing perspective. Switching off such features could improve RT performance to some extent.
- *OS optimization including both host and guest OS optimization* First, an RT OS kernel was adopted. Next, customized optimizations were performed including the insertion of advanced programmable interrupt controller (APIC) timer interrupts for virtual machines (VMs), CPU core isolation, interrupt affinity configuration and software simulator replacement, etc.
- *Hypervisor optimization* to support CPU core affinity configuration, virtualized processor priority ranking, interrupt affinity configuration in VMs, host interrupt allocation, and virtualization memory locking.

Virtualization of Processing Resources

The virtualization of processing resources is an important way to enable the migration to a COTS platform from traditional customized platforms. The processing resources

Table 19.9 Latency performance on general platforms

Latency (μs)	Min.	Ave.	Max.
General platform	1	2	3555

Table 19.10 Latency performance on general platforms with virtualization

Latency (μs)	Min.	Ave.	Max.
General platform with virtualization	5	47	4072

in a COTS platform include CPUs, memory storage and I/O networks. In the testbed, the hardware platform was based on the $x86$ architecture while the virtualization architecture used kernel-based virtual machines (KVMs) with a hypervisor virtual machine (HVM) instruction set, which could use the $x86$ instruction set directly. In this way, the virtualization overhead is reduced.

Traditionally KVM processing requires coordination between the kernel mode, user mode, and guest mode which cannot meet the RT requirement of baseband processing. Therefore in our design some customization had to be made to improve the RT performance, as described below.

1. Customization of CPU virtualization The customization and optimization measurements include the following

- isolating the CPU cores from the host OS and assigning the RT application tasks to the isolated CPU cores: the host OS does not schedule the isolated CPU cores and interruptions are redirected to other, non-isolated, cores.
- Binding the processing cores of the virtual machines with the physical cores in the host to prevent the virtual CPU cores from frequently switching between the physical CPU cores.
- Priority adjustment on the virtual CPU cores to achieve the desired RT performance.

2. KVM memory virtualization Kernel-based virtual machine memory virtualization is achieved with shadow page tables (SPTs) in combination with translation lookaside buffers (TLBs). To satisfy the RT requirement of BBU applications, optimization and enhancement mechanisms on virtual machine memory have been made, including the following.

- Turning off the kernel samepage merge (KSM) function on host OSs.
- Memory locking: one-time allocation in VM memory and mapping to the physical memory at boot stage. In addition, the host OS is prohibited from swapping out the locked memory to buffer.
- Using huge pages to enhance the translation lookaside buffer (TLB) cache hit rate.
- Optimizing the Nonuniform memory access (NUMA). It is preferable to allocate VM memory and CPU from the same NUMA zone. For VMs with extremely high RT

requirements, the allocation of VM memory and CPU from the same NUMA zone is mandatory.

3. KVM networking resource virtualization Kernel-based virtual machine network resources include virtual network interface cards (NICs) and virtual switches. The dominant virtual switch solutions currently used in IT industry such as Linux Bridge or OpenvSwitch cannot meet the LTE BBU virtualization requirements from the performance point of view. In the testbed we developed a data plane development kit (DPDK)-based accelerated virtual switching technology to achieve high-performance virtual networking. In our solution, the accelerated virtual switch is a simplified switch based on DPDK. The packets are received and sent through physical NICs and accelerated virtual NICs in polling mode, and distributed to the destination port according to rules based on the MAC address, IP address, or customized rules, or any combination of these. The upper-layer applications receive the packets from their own receiving queues and send packets to their own sending queue after processing. Communication between VMs within a host is accomplished via the accelerated virtual switches and need not involve the physical NIC. As the accelerated virtual switches isolate the hardware resources of the VMs and the host, the BBU VM can implement seamless live migration.

On-Demand Dynamic Resource Scheduling
In a virtualized environment, multiple VMs share the same physical resources. Because of this resource sharing, resource contention may occur among different VMs. A resource conflict will introduce jitter and significant latency increase and will badly impact RT system performance. To reduce the possibility of resource conflicts, the resource status needs to be monitored and conflicts avoided through dynamic coordinated scheduling. The testbed adopted a dynamic-processing workload-monitoring mechanism which used CPU usage as a key metric. By setting CPU usage rate check intervals, thresholds, scale-up/down thresholds, scale-up durations, scale-down durations, and so on, the VM processing resource load threshold is dynamically determined and the resources are allocated accordingly. The mechanism can apply not only to resource sharing within a single physical machine for multiple VMs but also to resource sharing and load balancing for different VMs on multiple physical machines.

19.4.3   Test Results

19.4.3.1   Verification of Virtualization Technologies for C-RAN
Benchmark test
(a) To evaluate the capability of COTS in terms of the processing of wireless stacks, we then evaluated the processing benchmark for TD-LTE L1 functions. The results are presented in Tables 19.11 and 19.12. The figures for the two-antenna case were tested based on the platform while those for the eight-antenna case were estimated for the same configuration.

From Tables 19.11 and 19.12, it can be seen that the processing time for some module functions such as modulation and HARQ Merge are irrelevant to the number of antennas

Table 19.11 Benchmark for UL physical functions processing (for two and eight antennas)

		20 MHz	
	Uplink process	2 ant. (μs)	8 ant. (μs)
PUSCH(single UE with UL 100PRB)	7.5K shift+FFT	105.9	189.9
	Channel evaluation (IRC)	147.7	524
	PUSCH frequency domain processing	52	218
	MIMO/EQU	179.8	384
	IDFT	34.6	34.6
	Demodulation	71.9	71.9
	HARQ Merge	59.7	59.7
	Turbo Decode	113	113
	CRC	9.6	9.6
PUSCH	PUCCH format 1 per UE	2.3	4.1
	PUCCH format 2 per UE	9.7	17.2

Table 19.12 Benchmark for DL physical functions processing for two and eight antennas

		20 MHz	
	Downlink Process	2 ant. (μs)	8 ant. (μs)
PDSCH (Single UE, 200PRB, TM3/TM8, peak throughput)	CRC	13.8	13.8
	Turbo Encode	70	70
	Scrambling	14.9	14.9
	Modulation	36.4	36.4
	Power control, precode, beamforming	21	190
	iFFT	32	116

while some are not; IRC and MIMO are the two components which consume the most CPU resources with the highest processing times. With the configuration and the evaluation results in the table, it can be further calculated that on average around 2.5 CPU cores are required for one eight-antenna cell to meet the commercial capacity requirement.

Resource virtualization

(b) Resource virtualization is a key feature and a major implementation target for C-RAN. Resource virtualization means that the resources are virtualized and separated from the upper-layer applications and can be dynamically managed. To verify the feature we tested two scenarios: VM live migration and dynamic load balance between hosts.

In the first scenario, we first initiated an LTE FTP download call in one COST host (VM tdd003c01 in h012), as shown in Fig. 19.10, and then triggered a VM live migration to another host (VM tdd003c01 in h014) by manual. During the migration, the FTP download service kept active without any service outage (Fig. 19.11).

In the second scenario, we first created two eNB VM in one COST host (VM BBU003 and BBU004 in h015), as shown in Fig. 19.12, then initiated the LTE FTP download

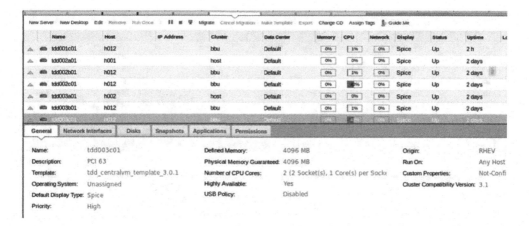

Figure 19.10 Resource view before migration.

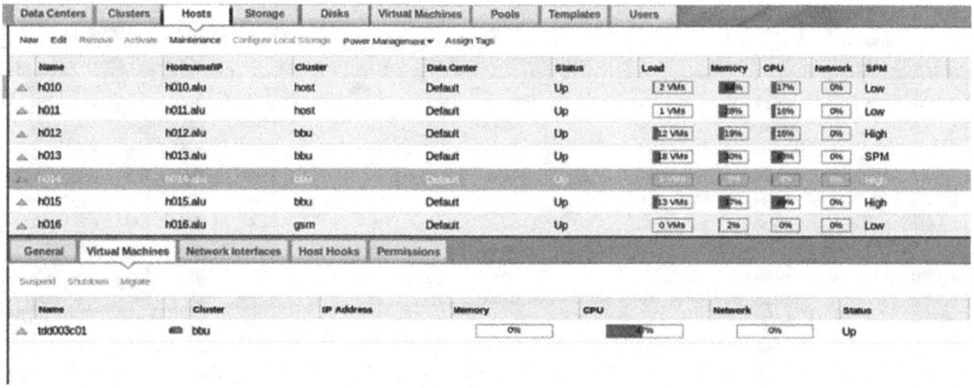

Figure 19.11 Resource view after migration.

calls respectively in two BBUs, and set CPU traffic topline as 75% in the management system. Using the script to overload the host CPU, it was found that the BBU VMs migrated to another host (h012) automatically when the CPU load reached 75% (Fig. 19.13).

19.4.3.2 Functional and Performance Test

To verify the performance of the virtualized C-RAN system, we first tested the air interface KPI by using an Agilent 89600 VSA analyzer to catch the air interface signal. The framework can be seen in Fig. 19.14, which gives a 64 QAM constellation diagram for TD-LTE; each physical channel is decoded successfully, which demonstrates that the testbed is fully 3GPP R8 compliant.

Peak user throughput was also tested, with the results shown in Fig. 19.15. For TD-LTE with virtualization implemented on the basis of COTS, the peak throughput was as high as 112.5 Mbps on average. It is worth pointing out that even with commercial TD-LTE products the peak throughput is around 120 Mbps. The difference is very small,

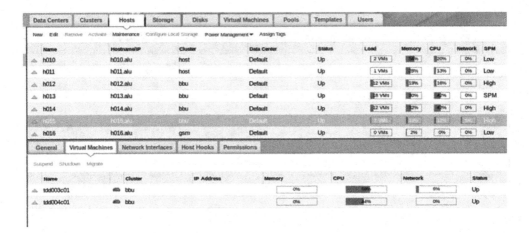

Figure 19.12 Resource view before load balance.

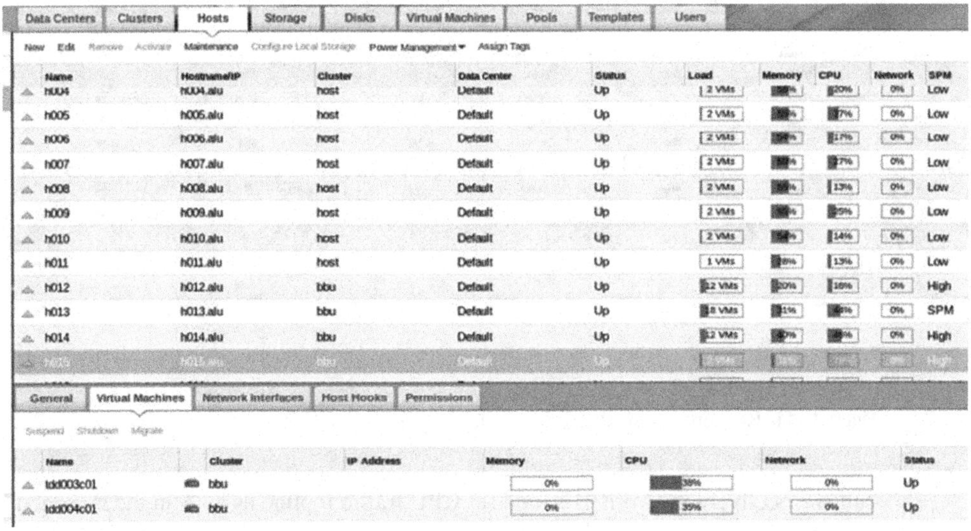

Figure 19.13 Resource view after load balance.

which is mainly due to the considerable optimization of every key component including the operating system, hypervisor, and so on.

For a multi-user multi-cell scenario, as described at the beginning of the section, a field trial network was set up with support of 18 TD-LTE cells and three GSM carriers.[1] Fifteen UEs were activated, which then connected to the network and ran file-download services. The 15 UEs were all commercial in type. The data rate is shown in Table 19.13. Simultaneously, two commercial GSM terminals were calling each other. The test ran for more than two hours and it was found that the voice quality maintained a high level.

[1] Note: The capacity of the system in the trial was limited by the number of RRUs instead of by BBU pool. The BBU pool consists of standard COTS platform and it is easy for it to obtain capacity expansion.

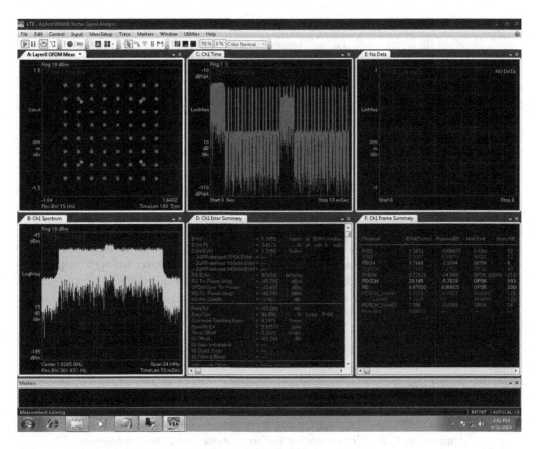

Figure 19.14 LTE TDD 64 QAM constellation diagram.

Figure 19.15 Peak throughput for TD-LTE with virtualization implemented.

Other functional tests, including handover, load balancing, and live migration, were all carried out in the trial to demonstrate the feasibility, reliability, and flexibility of the COTS-based RT virtualized platform. In the handover test, two users connected to two cells and received file download services. As UE1 moved toward to the target cell, the handover took place. It was observed that there was a slight pause in the download

Table 19.13 User data rate in the virtualized C-RAN field trial

UE no.	DL rate (Mbps)	UE no.	DL rate (Mbps)
1	71.2	9	88.1
2	68.4	10	65.1
3	96.5	11	56.8
4	95.1	12	96.2
5	96.2	13	57.7
6	65.9	14	68.0
7	95.4	15	92.4
8	93.4	Total	1.21 Gbps

of the data file for UE1 but the radio link did not break. In the meantime, the UE2 in the destination cell maintained normal operation. The handover was well supported by the RT virtualized baseband platform. Another key observation is that after the handover finished the source server automatically released the processing resources to the resource pool.

19.5   Conclusions

In this chapter we presented a comprehensive C-RAN implementation not only to demonstrate C-RAN's benefits but also to verify key enabling technologies.

We first presented the field trial of C-RAN centralization. The WDM-based FH technology was verified with the capability of transmitting twelve 9.8 Gb/s CPRI links in one fiber pair, which greatly reduced the required fiber consumption. In addition, the CPRI-over-WDM solution showed ideal performance with no impact on radio performance; WDM could therefore be one of the dominant solutions for future large-scale C-RAN deployment.

The performance of CoMP under a C-RAN architecture was well demonstrated through field trials, in which a significant CoMP gain was observed. The results indicate that CoMP can effectively suppress the interference of neighboring cells, efficiently enhance signal strength, and improve the overall user experience. For example, it was observed that with the JR scheme the uplink CoMP gain is 20%–50% at good coverage areas and can reach 50%–100% at cell edge areas. As for the downlink CS and CB, it was found that both the CS and CB could greatly improve the edge users' throughput, e.g. giving around 100% increase under a 20% network load condition. There was also a slight improvement in the average cell throughput.

One of the key features of C-RAN is resource cloudification. To verify the feasibility of the use of a COTS platform, a testbed was developed. Several mechanisms from hardware parameter optimization to hypervisor optimization were designed in order to improve the RT performance in support of radio signal processing. A field trial network was further built and extensive tests were performed. Various functions, including multi-RAT support, load balancing, and live migration were all verified. In addition the testbed demonstrated a similar level of performance to traditional DSP/FPGA-based systems, which gives a positive indication for BBU evolution towards standard IT platforms.

19.6 Acknowledgments

We would like to express our sincere gratitude to our partners, including Alcatel-Lucent for development of the COTS-based C-RAN testbed, and Huawei and Ericsson for the joint C-RAN field trials. We also owe a special thanks to all C-RAN team members in China Mobile for helpful discussion and valuable comments. This work was partly supported by the National High-Tech R&D Program of China (Grants No. 2014AA01A703 & No. 2014AA01A704).

References

[1] C. M. R. Institute, "C-RAN: the road towards green ran." Online. 2014.
[2] C. I, C. Rowell, S. Han, Z. Xu, G. Li, and Z. Pan, "Toward green and soft: a 5G perspective." *IEEE Commun. Mag.*, vol. 52, no. 2, pp. 66–73, February 2014.
[3] S. R. Jinsong Wu and H. Zhang, *Green Communication*. CRC Press, 2013.
[4] C. I, J. Huang, R. Duan, C. Cui, J. Jiang, and L. Li, "Recent progress on C-RAN centralization and cloudification," *IEEE Access*, vol. 2, pp. 1030–1039, 2014.
[5] ETSI NFV ISG, "Network functions virtualisation." Online. Available at http://portal.etsi.org/portal/server.pt/community/NFV/367, 2012.

Index

Printed in the United States
by Baker & Taylor Publisher Services

Printed in the United States
by Baker & Taylor Publisher Services